;er Series in Chemical Physics

'italii I. Goldanskii Fritz P. Schäfer J. Peter Toennies

ditor: H. K. V. Lotsch

olution Spectroscopy of
t Molecules
ota

olution Spectral Atlas of
Dioxide 559–597 nm
ıara and H. Sasada

and Reaction Centers of
thetic Bacteria
Interactions, and Dynamics
E. Michel-Beyerle

-Atom Potential Method
ns to Organic Molecular

rtsin and A. I. Kitaigorodsky

y Ion Mass Spectrometry

. Benninghoven, R. J. Colton,
ns, and H. W. Werner

opic Liquid Crystals,
tals
ogen and W. H. de Jeu

'henomena V
R. Fleming and
ıan

'hemical Reaction Systems
cal Modelling
tion
Warnatz and W. Jäger

'henomena VI
Yajima, K. Yoshihara,
s, and S. Shionoya

nteractions in Molecules
ls
suker and V. Z. Polinger

50 **Molecular and Laser Spectroscopy**
By Zu-Geng Wang and Hui-Rong Xia

51 **Space-Time Organization in Macromolecular Fluids**
Editors: F. Tanaka, M. Doi, and T. Ohta

52 **Multiple-Photon Laser Chemistry**
By. R. V. Ambartzumian, C. D. Cantrell, and A. Puretzky

53 **Ultrafast Phenomena VII**
Editors: C. B. Harris, E. P. Ippen, G. A. Mourou, and A. H. Zewail

54 **Physics of Ion Impact Phenomena**
Editor: D. Mathur

55 **Ultrafast Phenomena VIII**
Editors: J.-L. Martin, A. Migus, G. A. Mourou, and A. H. Zewail

56 **Clusters of Atoms and Molecules**
Salvation and Chemistry of Free Clusters, and Embedded, Supported and Compressed Clusters
Editor: H. Haberland

57 **Radiationless Transitions in Polyatomic Molecules**
By E. S. Medvedev and V. I. Osherov

58 **Positron Annihilation in Chemistry**
By O. E. Mogensen

59 **Soot Formation in Combustion**
Mechanisms and Models
Editor: H. Bockhorn

60 **Ultrafast Phenomena IX**
Editors: P. F. Barbara, W. H. Knox, G. A. Mourou, and A. H. Zewail

are listed at the end of the book

60 Springer Series in Cher

Edited by Fritz P. Schäfer

Sprin

Editors:

Managing

40 **High-R**
Transie
By E. H

41 **High Re**
Nitroge
By K. U

42 **Antenna**
Photosy
Structure
Editor: N

43 **The Ato**
Applicati
Solids
By A. J. I

44 **Seconda**
SIMS V
Editors: A
D. S. Sim

45 **Thermot**
Fundame
By G. Ve

46 **Ultrafast**
Editors: C
A. E. Sieg

47 **Complex**
Mathemat
and Simul
Editors: J.

48 **Ultrafast**
Editors: T.
C. B. Harr

49 **Vibronic**
and Cryst
By I. B. B

Volumes 1–3

P. F. Barbara · W. H. Knox
G. A. Mourou · A. H. Zewail (Eds.)

Ultrafast Phenomena IX

Proceedings
of the 9th International Conference,
Dana Point, CA, May 2-6, 1994

With 384 Figures

Springer

Professor Paul F. Barbara
Department of Chemistry, University of Minnesota
Minneapolis, MN 55455-0431, USA

Wayne H. Knox, Distinguished Member of Technical Staff
AT&T Bell Laboratories, Crawford Corner Road
Holmdel, NJ 07733, USA

Professor Gerald A. Mourou
Electrical Engineering and Computer Science Department
University of Michigan, Ann Arbor, MI 48109, USA

Professor Ahmed H. Zewail
Department of Chemistry, California Institute of Technology
Pasadena, CA 91125, USA

ISBN 3-540-58455-2 Springer-Verlag Berlin Heidelberg New York

Printed in Germany

Typesetting: Camera ready by authors/editors
SPIN: 10130530 54/3140 - 5 4 3 2 1 0 - Printed on acid-free paper

Preface

This volume contains papers presented at the Ninth International Conference on Ultrafast Phenomena held at Dana Point, CA, USA, from May 2 to 6, 1994. The biannual Ultrafast Phenomena Conferences provide a forum for discussion of the latest advances in ultrafast optics and applications in science and engineering. The vitality and excitement of the various disciplines sharing common interest in ultrafast phenomena were well represented at the meeting by the 438 participants from 18 countries, including 98 students. Cross-fertilization of ultrafast concepts and techniques among the various scientific and engineering disciplines continues to be the primary driving force behind this successful meeting.

Progress was reported in the technology of generating ultrafast pulses, including extensions in pulse width, output power, wavelength range, and intensity. Ultrafast spectroscopy continues to impact and expand the knowledge base of fundamental processes in physics, chemistry, biology, and engineering. A new series of lively, late-night panel discussions were introduced at this meeting, reflecting the maturing of the field into applications, while at the same time keeping a strong interest in fundamentals.

Acknowledgements. Many people and organizations contributed to the success of this meeting. The members of the program committee deserve special thanks for reviewing all the papers and organizing the final program. The staff of the Optical Society of America very expertly took care of the conference arrangements. We would like to acknowledge the Optical Society of America for sponsorship and AFOSR, ARO and NSF for providing funding. Additional grants from the following companies are appreciated and contributed greatly to the success of the conference: Clark MXR, Cleveland Crystals, Excel/Quantronix, IMRA America, Lightning Optical, Newport, Spectra-Physics and Thorlabs.

Ann Arbor, MI — *Paul F. Barbara*
Pasadena, CA — *Wayne H. Knox*
Minneapolis, MN — *Gérard A. Mourou*
Holmdel, NJ — *Ahmed H. Zewail*
October 1994

Contents

Part I Overview and General Topics

Part III Spectroscopy and Advances in Measurements

Part IV Sources

Part VI **Metals, Surfaces, and Materials**

Part VIII **Biology: Primary Dynamics, Electron and Energy Transfer**

Part I

Overview and General Topics

Ultrafast Solid-State Lasers: Status and Prospects

F. Krausz[1], *Ch. Spielmann*[1], *P.F. Curley*[1], *T. Brabec*[1], *S.M.J. Kelly*[1], *A. Stingl*[1], *R. Szipöcs*[2], *E. Wintner*[1], *and A.J. Schmidt*[1]

[1]Quantenelektronik und Lasertechnik, Technische Universität Wien, Gusshausstrasse 27-29, A-1040 Wien, Austria

[2]Research Institute for Solid State Physics, P.O.Box 49, H-1525 Budapest, Hungary

Abstract An overview of the basic principles, major theoretical aspects and technological issues of self-mode-locked solid state lasers is presented, with particular emphasis on recent advances in femtosecond Ti:sapphire lasers.

Introduction The last few years have brought about significant advances in ultrashort-pulse laser physics. The development of novel all-optical modulation techniques along with the appearance of ultrabroad-band solid state gain media opened up a new era in femtosecond pulse technology. We have recently witnessed the emergence and evolution of a new generation of ultrashort-pulse lasers based exclusively on solid state components. This paper provides a brief synopsis of the basic operation principles and the major performance limitations of femtosecond solid state lasers (for a deeper insight and extensive reference lists the reader is referred to [1–3]), and summarizes recent technological advances allowing the generation of nearly bandwidth-limited sub-10-fs optical pulses directly from laser oscillators.

Although the first picosecond optical pulses were generated in a solid state (Nd:glass) laser in the Mid-60es [4], progress in ultrashort-pulse solid-state laser technology came to a standstill shortly after the first pioneering experiments. It was the picosecond relaxation time of the saturable absorbers used for passive mode locking that prevented researchers from pushing the pulse duration below the picosecond limit. Hence interest shifted to organic dye lasers [5], which, as a result of their nanosecond upper-state lifetimes, were capable of actively participating in short pulse formation [6], allowing intracavity pulse shortening down to the femtosecond regime by using "slow" (picosecond)-relaxation-time absorbers [7]. Operation of femtosecond dye lasers, however, continued to be a highly sophisticated art up to the present day, mainly because of the large number of inaccessible system parameters. One of the demands of highest priority for a variety of (potential) application fields: high reproducibility of femtosecond laser performance could not be met until the recent development of femtosecond solid state systems.

Major techniques and physical mechanisms The emergence of a femtosecond solid state laser technology has been possible by utilizing the intensity-induced change in the refractive index. This ultrafast Kerr-effect can be transformed into an almost instantaneous-response saturable absorber effect by introducing appropriate linear optical components into the cavity. The two most successful embodiments of this general concept have been *additive-pulse mode locking* [8], and *self (or Kerr-lens) mode locking* [9]. The effect of these passive Kerr modulators on the envelope of the electric field $v(t)$ can, to a first approximation, be written as [1,2],

$$\Delta v(t) = (\kappa/2 + i\phi)|v(t)|^2, \tag{1}$$

Springer Series in Chemical Physics, Vol. 60 **Ultrafast Phenomena IX**
Editors: P. F. Barbara · W. H. Knox · G. A. Mourou · A. H. Zewail

where κ and ϕ measure the strengths of self-amplitude and self-phase modulation (SAM & SPM), respectively. Other than SAM and SPM, intracavity group delay dispersion (GDD) is a major pulse shaping effect, which is characterized by the parameter

$$D = \left(\frac{\partial T_r(\omega)}{\partial \omega} \right)_{\omega=\omega_0}, \tag{2}$$

i.e. by the first derivative of the cavity round-trip time T_r with respect to the optical frequency at ω_0, the center of the laser spectrum. In order to avoid a strong pulse broadening due to the interaction of a pulse carrying a positive chirp (as a consequence of $\phi > 0$) with normal dispersion ($D > 0$) the intracavity GDD must be negative.

Build-up of mode locking Efficient SAM with $\kappa > 0$ is an indispensable requirement for the formation of a single mode-locked pulse out of the free running laser oscillation. The SAM coefficient κ must exceed a threshold value in order that the saturable absorber effect can overcome dynamic gain saturation [10] just as the product of κ and the cw circulating intracavity power need to exceed a critical value in order that the most intense initial fluctuation can evolve into a mode-locked pulse before it is disrupted due to the finite mode coherence time in the free-running laser [11]. The former condition is easily satisfied in most solid state lasers, while meeting the second criterion is often difficult and thus additional mechanisms (e.g. active modulation, synch-pumping) may be required for starting the mode-locking process. Unfortunately the use of these starting mechanisms adds complexity and often impairs the steady-state mode locking performance. Hence, in laboratory systems one often resorts to a most simple method: with optimized SAM uninterrupted self mode locking operation over many hours can be achieved after starting the passive pulse formation by tapping one of the resonator mirrors.

Steady-state operation Both additive-pulse and Kerr-lens modulators exhibit SPM coefficients that are much larger than the corresponding SAM parameters. $\phi >> \kappa$ implies that steady-state pulse formation is dominated by a soliton-like interplay between SPM and negative GDD. Hence, assuming i) a linear variation of $T_r(\omega)$ as a function of ω, i.e. $D(\omega) = D(\omega_0)$, over the mode-locked spectrum, ii) evenly distributed SPM and GDD in the cavity, and iii) a sufficiently broadband gain medium, the pulse duration τ is expected to approximately obey the well-know soliton formula

$$\tau \approx \tau_s = \frac{3.53|D|}{\phi W}, \tag{3}$$

where W stands for the intracavity pulse energy. The soliton-like behavior of the steady-state pulse has been verified in a self-mode-locked fused-silica-prism-controlled Ti:sapphire laser down to the 10 fs regime, as is shown in Fig. 1. Although κ does not significantly affect τ, optimized SAM is essential for stabilizing the pulse against noise and perturbations as well as preventing the emergence of a narrow-band cw background, which tends to

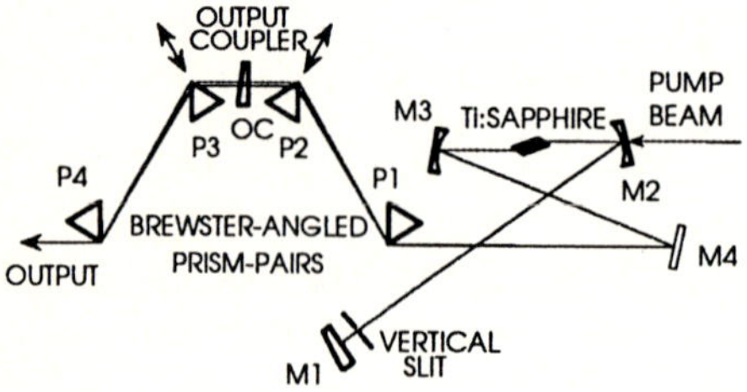

FIG. 1. Pulse duration and time-bandwidth product as a function of glass insertion in a fused-silica-prism-controlled Ti:S laser (Ref [1]).

co-exist with the mode-locked pulse in soliton-like (solitary) systems under non-optimized conditions [3,12].

Performance limitations In femtosecond solid state lasers the pulse duration can be reduced by decreasing the magnitude of negative intracavity GDD until one of the assumptions or approximations i)-iii) leading to Eq. (3) fails. The most severe limitation in practical broadband solid-state lasers originates from the increasing deviation of $T_r(\omega)$ from a linear function as the oscillation spectrum broadens. The lowest-order contribution to this deviation is referred to as third-order dispersion (TOD). As the pulse duration is decreased and/or the pulse energy is increased the separated action of GDD and SPM increasingly modulates the pulse parameters (duration, bandwidth) as the pulse circulates in the cavity [1]. This modulation can be regarded as a periodic perturbation to the ideal soliton-like pulse and manifests itself in an additional term $\Delta\tau = \alpha(z)\phi W$, where α depends on the position in the cavity [1]. Other potential limitations to pulse shortening are the gain and resonator bandwidths, which, however, have turned out to be insignificant in practical systems so far. For a more detailed and quantitative analysis of the major effects limiting the performance of practical systems the reader is referred to [3]. The considerations presented above generally apply to both additive-pulse and self mode-locking. The former is considered the most powerful technique for ultrashort pulse generation in *fiber* lasers, whereas self mode-locking is far superior to any other techniques if femtosecond pulses from *bulk* lasers are to be generated. In what follows we concentrate on self-mode-locked systems and refer to two recent reviews of APM fiber lasers [13,14].

Prism-dispersion-controlled systems Until recently broadband negative GDD has been introduced in short-pulse laser oscillators almost exclusively by a pair of Brewster-angled prisms. The layout of e.g. a prism-controlled self-mode-locked Ti:sapphire laser is shown in Fig. 2. To obtain the shortest output pulse duration, the circulating pulse is coupled out of the cavity after traversing the dispersive delay line [1,15,16], and the extracavity prism pair allows GDD control outside the cavity. Since the prism pair introduces also high-order dispersion, a careful selection of the prism material is needed if the pulse duration is to be minimized [1]. For Ti:sapphire the optimum choice turned out to be fused silica, which, by minimizing TOD in the cavity allowed the generation of nearly bandwidth-limited pulses of 11-12fs in duration around 0.8 μm [18,19]. The spectrum of a 12.3fs pulse [18] (trace (a) in Fig. 3) clearly exhibit a strong asymmetry as a result of residual negative TOD in the cavity, which sets a limit to further pulse shortening. Recent investigations [3,17] predicted a vanishing TOD in the Ti:sapphire/fused silica laser around 0.85 μm. As a matter of fact, tuning the laser to this wavelength extremely broad and yet symmetric mode-locked spectra (Fig. 4) can be generated. A careful evaluation [3] of the spectrum and the corresponding fringe-resolved autocorrelation (Fig. 5) yields, however, only a mi-

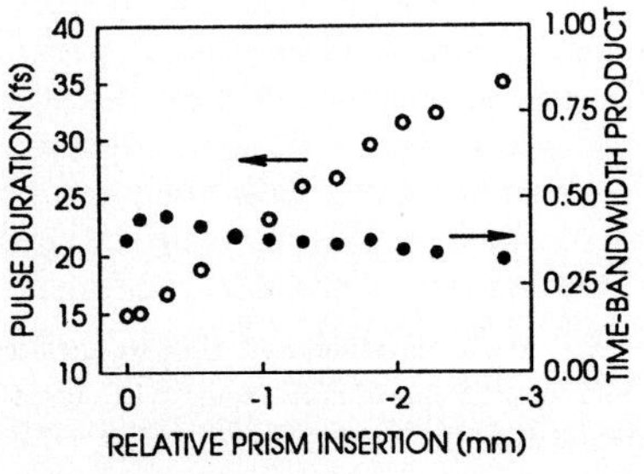

FIG. 2. Schematic of a prism-controlled self-mode-locked (hard aperture) Ti:S laser. For more details see Ref. [3].

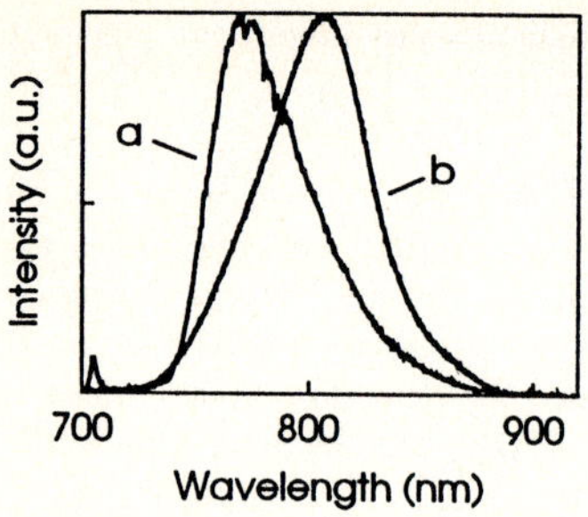

FIG. 3

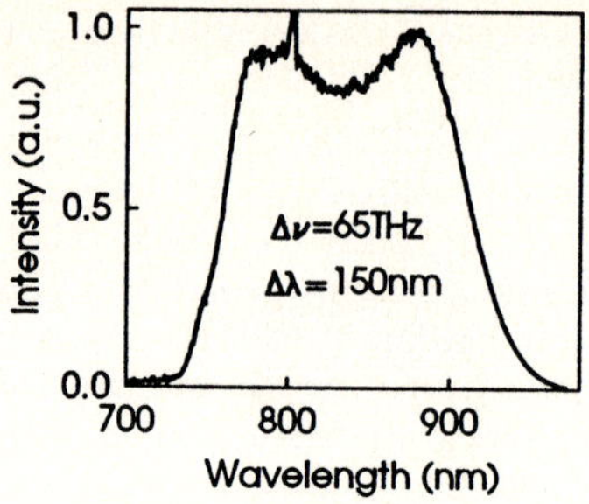

FIG. 4

FIG. 3. Spectra obtained with (a) a prism-pair-compensated, and (b) with a MDC Ti:S laser.

FIG. 4. Mode-locked spectrum of the FS-prism-controlled Ti:S laser around 850nm.

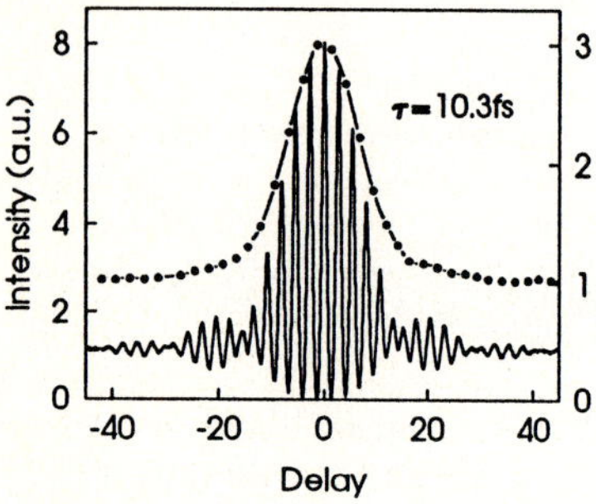

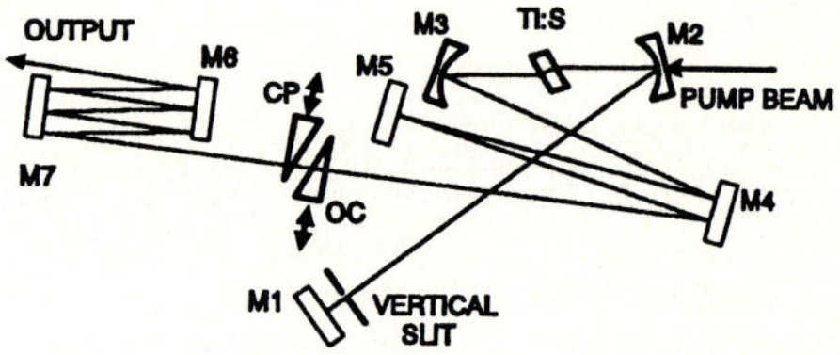

FIG. 5. Measured interferometric and evaluated intensity autocorrelation. Note the disappearance of the structure in the wings of the intensity autocorrelation.

FIG. 6. Schematic of the MDC Ti:sapphire laser using chirped dispersive mirrors (Ref. [20]) for intracavity (M4-M5) and extracavity (M6-M7) GDD control. For more details see Ref. [21].

nor improvement in pulse duration because of the strong deviation of the evaluated pulse shape from a $sech^2$ profile. In this context it is important to notice that the intensity autocorrelation (upper trace in Fig. 5) shows little sensitivity to this deviation and hence inappropriate for a reliable and accurate determination of pulse parameters.

Mirror-dispersion-controlled systems Recently a novel technique has been proposed and demonstrated for intracavity dispersion control. Broadband, high-reflectivity multilayer dielectric mirrors have been developed with their multilayer period modulated during the evaporation process. This modulation not only broadens the high-reflectivity bandwidth but, more importantly, offers the possibility of engineering the dispersion properties of the mirrors [20]. As a first embodiment of this general concept, chirped dielectric mirrors exhibiting a nearly constant, i.e. high-order-dispersion-free, GDD over the wavelength range 720-890 nm have been fabricated [20], where the wavelength range can be easily shifted to match the emmision spectrum of other lasers by simply rescaling the layer thicknesses. Using these mirrors, compact, mirror-dispersion-controlled (MDC) self-mode-locked oscillators can be constructed, which are exemplified by the MDC Ti:sapphire laser shown in Fig. 6 [21]. The significantly reduced TOD in this system as compared to its prism-controlled counterpart is clearly demonstrated by a symmetric $sech^2$-shaped spec-

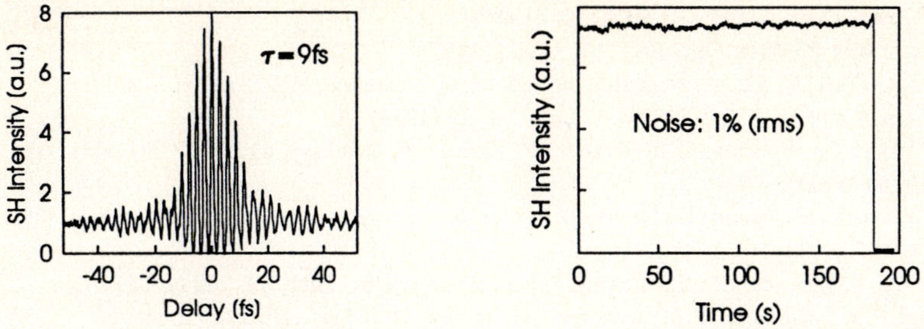

FIG. 7. Autocorrelation trace (left side) and second-harmonic stability (right side) of a 100 MHz, 9fs pulse train generated from an MDC Ti:S laser. (The amplitude noise of ≈1% rms is evaluated for the frequency band 0.01-20Hz.)

trum in the 10fs domain (trace (b) in Fig. 3). As revealed by Fig. 7, the MDC Ti:sapphire laser is capable of producing highly stable, nearly bandwidth-limited sub-10 fs pulses.

Reliability & reproducibility In strong contrast to dye lasers, ultrashort pulse formation in solid-state lasers using Kerr-modulators can be described and controlled in terms of a few uniquely defined and experimentally easily accessible parameters (κ, ϕ, D, and W). As a result, the reliability of femtosecond solid-state lasers and the reproducibility of their performance is far superior to those of any previous femtosecond source. Whilst prism-controlled systems offer the advantage of continuously tunable intracavity dispersion, MDC oscillators exhibit a cavity dispersion that is completely independent of resonator alignment, resulting in an unprecedented long-term stability and day-to-day reproducibility of femtosecond pulse parameters. These advances are expected to give rise to further expansion of ultrafast techniques into new areas of science and technology.
This research was supported by the FFWF Austria and the Jubiläumsfonds der ÖNB under grants P9710, P8566, and 4910.

[1] F. Krausz, M. E. Fermann, T. Brabec, P. F. Curley, M. Hofer, M. H. Ober, Ch. Spielmann, E. Wintner, A. J. Schmidt, IEEE J. Quantum Electron. , **QE-28**, 2097 (1992)
[2] E. P. Ippen, Appl. Phys. B, **B58**, 159 (1994)
[3] Ch. Spielmann, P. F. Curley, T. Brabec, F. Krausz, "Ultrabroad-band femtosecond lasers," IEEE J. Quantum Electron. , **QE-30**, April 1994 (in press)
[4] A. J. DeMaria, D. A. Stetser, and A. Heynau, Appl. Phys. Lett. **8**, 174 (1966)
[5] E. P. Ippen, C. V. Shank, and A. Dienes, Appl. Phys. Lett. **21**, 348 (1972)
[6] G. H. C. New, Opt. Commun. **6**, 188 (1972)
[7] R. L. Fork, B. I. Greene, and C. V. Shank, Appl. Phys. Lett. **38**, 671 (1981)
[8] J. Mark, L. Y. Liu, K. L. Hall, H. A. Haus, and E. P. Ippen, Opt. Lett. **14**, 48 (1989)
[9] D. E. Spence, P. N. Kean, and W. Sibbett, Opt. Lett. **16**, 42 (1991)
[10] E. P. Ippen, L. Y. Liu, and H. A. Haus, Opt. Lett. **15**, 183 (1990)
[11] F. Krausz, T. Brabec, and Ch. Spielmann, Opt. Lett. **16**, 235 (1991)
[12] T. Brabec, S. M. J. Kelly, F. Krausz, "Passive mode locking in solid-state lasers," in *Compact Sources of Ultrashort Pulses*, Ed. I. N. Duling, Cambridge Press, 1994
[13] H. A. Haus, E. P. Ippen, and K. Tamura, IEEE J. Quantum Electron. **QE-30**, 200 (1994)

[14] M. E. Fermann, Appl. Phys. B **B58**, 197 (1994)
[15] B. Proctor and F. Wise, Opt. Lett. **17**, 1295 (1992)
[16] J. Zhou, G. Taft, C. Shi, H. C. Kapteyn, and M. M. Murnane, ICUP, paper MA2-1
[17] B. E. Lemoff and C. P. J. Barty, Opt. Lett. **18**, 57 (1993)
[18] P. F. Curley, Ch. Spielmann, T. Brabec, F. Krausz, E. Wintner, and A. J. Schmidt, Opt. Lett. **18**, 54 (1993)
[19] M. T. Asaki, C. P. Huang, D. Garvey, J. Zhou, H. C. Kapteyn, and M. M. Murnane, Opt. Lett. **18**, 977 (1993)
[20] R. Szipöcs, K. Ferencz, Ch. Spielmann, and F. Krausz, Opt. Lett. **19**, 202 (1994)
[21] A. Stingl, Ch. Spielmann, F. Krausz, and R. Szipöcs, Opt. Lett. **19**, 204 (1994)

Applications of the Chronocyclic Representation of Ultrashort Light Pulses

J. Paye

Commisariat à l'Energie Atomique, Centre d'Etudes de Limeil-Valenton, F-94195 Villeneuve-Saint-Georges Cedex, France

Abstract. The chronocyclic representation can be applied to the response function of linear optical devices. It provides an intuitive understanding of their action on ultrashort pulses.

1. Introduction

The chronocylic representation [1] is a way to display an ultrashort light pulse in a time-frequency space. It uses the Wigner distribution function applied to the complex electric field of the pulse. It thus contains both the amplitude and phase information but is a two-dimensional real function, although it can take negative values. Its mathematical properties make it as close as possible to a representation of the energy distribution of the pulse as a function of both time and frequency.

So far, the chronocyclic representation was used to describe ultrashort pulses in a more intuitive way than using the complex electric field. In particular, chirp properties and effects of the uncertainty relation appear clearly in this representation. It was also proven that the chronocylic representation of any pulse (or equivalently its amplitude and phase) could be measured by spectrally-resolved cross-correlation with a perfectly known reference pulse [1]. Later, the measurement of the chronocyclic representation of a pulse was obtained (without reference pulse) by spectrally-resolved autocorrelation [2].

In this paper, the chronocyclic representation will be used to define a spectrally-resolved temporal response function for the analysis of linear optical devices commonly used in femtosecond optics, such as dispersive elements, Fabry-Perot étalons, Gires-Tournois interferometers. This approach provides an intuitive understanding of the action of these linear devices.

2. Chronocyclic Response Function

If the complex electric of an ultrashort light pulse is written as $\tilde{E}(\omega)$ in the spectral domain, its chronocyclic intensity is defined as the following function of time and frequency:

$$W(t,\omega)=\frac{1}{2\pi}\int_{-\infty}^{+\infty}\tilde{E}\left(\omega+\frac{\omega'}{2}\right)\tilde{E}^{*}\left(\omega-\frac{\omega'}{2}\right)\exp(-i\omega' t)d\omega' \qquad (1)$$

Springer Series in Chemical Physics, Vol. 60 **Ultrafast Phenomena IX**
Editors: P. F. Barbara · W. H. Knox · G. A. Mourou · A. H. Zewail

All linear devices can be described by a response function defined in the temporal domain. If the input complex electric field is $E_{in}(t)$ in the temporal domain, the output electric field $E_{out}(t)$ can be written as:

$$E_{out}(t) = \int_{-\infty}^{+\infty} \chi(\tau) E(t-\tau) d\tau \tag{2}$$

$\chi(\tau)$ is the temporal impulse response function. Causality imposes that:

$$\chi(\tau) = 0 \quad \text{if} \quad \tau < 0 \tag{3}$$

The Fourier transform of Eq.(2) reads:

$$\tilde{E}_{out}(\omega) = \tilde{\chi}(\omega) \tilde{E}_{in}(\omega) \tag{4}$$

where $\tilde{\chi}(\omega)$ is the spectral response function.

Using a well-known property of the Wigner function, the chronocylic intensity of the output pulse can be written as:

$$W_{out}(t,\omega) = \int_{-\infty}^{+\infty} X(\tau,\omega) W_{in}(t-\tau,\omega) d\tau \tag{5}$$

where $X(\tau,\omega)$ is the Wigner function of the spectral response function $\tilde{\chi}(\omega)$ [3]. Equation (5) shows that the linear response of a system can be expressed in chronocyclic representation. The chronocyclic intensity of the output pulse is the temporal convolution of the chronocyclic intensity of the input pulse with the chronocyclic representation of the response function. In other words, each spectral slice of the chronocyclic representation of the input pulse is transformed by a linear system with a response function defined as the corresponding spectral slice of the chronocyclic representation of the complex response function. In chronocyclic representation, a separate temporal response function can be defined for each frequency. The chronocyclic representation of the response function is indeed a *spectrally-resolved temporal response function*.

3. Examples

Let us apply these concepts to a few simple examples. The first one is the propagation in a dispersive medium. The response function of this linear system is simply:

$$\tilde{\chi}_{GVD}(\omega) = \exp[ik(\omega)z] \tag{6}$$

where z is the propagation distance in the dispersive medium and $k(\omega)$ is the wave

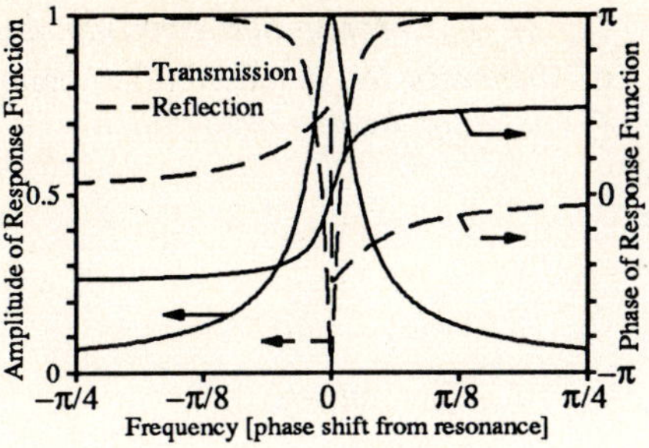

Figure 1. Response function of a Fabry-Perot étalon both in transmission (solid line) and in reflection (dotted line). The amplitude (left-hand scale) and the phase (right-hand scale) are represented.

vector. If the wave vector can be expanded in the vicinity of ω_0 to second order according to:

$$k(\omega) = k_0 + k_0' \cdot (\omega - \omega_0) + \frac{1}{2} k_0'' \cdot (\omega - \omega_0)^2 \tag{7}$$

then the chronocyclic response function is:

$$X_{GVD}(\tau, \omega) = \delta\left(t - k_0' z - k_0'' \cdot (\omega - \omega_0) z\right) \tag{8}$$

where δ is the Dirac function. This is interpreted obviously in the following way: each frequency component of the input pulse is delayed by the dispersive medium, but this delay depends linearly on the frequency itself.

A more interesting example is the action of a Fabry-Perot étalon on an ultrashort pulse. The étalon is made of two identical absorptionless mirrors with a reflectivity coefficient R=95%, the optical path length between the two mirrors is a. It is a well-known result that the response function of this device used in transmission is:

$$\tilde{\chi}_{trans}(\omega) = \frac{(1-R)\exp(i\omega\, a/c)}{1 - R\exp(2i\omega\, a/c)} \tag{9}$$

where $2\omega\, a/c$ is the round-trip phase shift. This complex response function is plotted in Fig.1 (solid line), both in amplitude and phase.

From the complex response function in Fig.1, it is difficult to have an intuitive feeling of the action of such a device on an ultrashort light pulse, apart from amplitude filtering. The chronocyclic representation of this response function is plotted in Fig.2. As expected, it is zero at negative times, according to causality requirements. It also shows immediately that the response of the étalon is much longer at resonance than off-resonance. A pulse with a chronocyclic intensity concentrated near resonance is lengthened by such a device much more than when its frequency is moved away from resonance. This is clear, since at resonance the spectrum of the pulse is narrowed by the étalon much more than off-resonance,

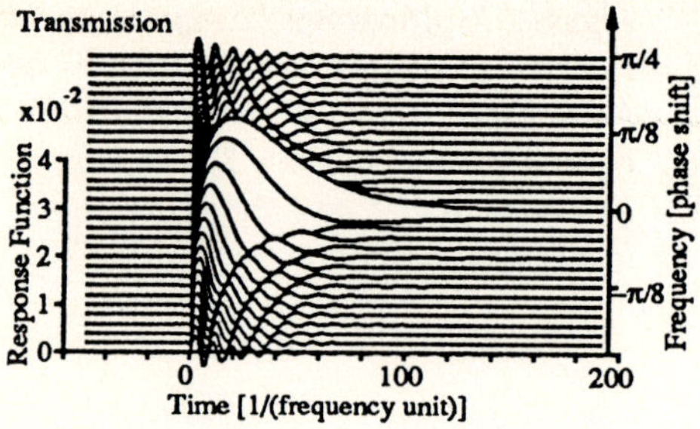

Figure 2. Chronocyclic representation of the response function of a Fabry-Perot étalon used in tranmission.

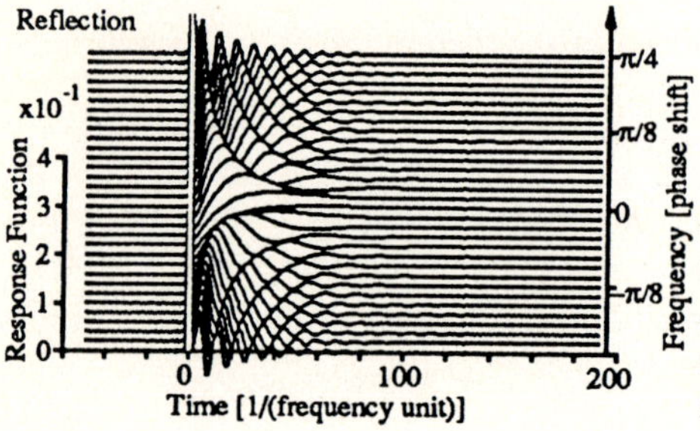

Figure 3. Chronocyclic representation of the response function of a Fabry-Perot étalon used in reflection. Each curve exhibits a Dirac δ-function at zero time.

therefore its duration must be increased. This example shows how the effects of the time-frequency uncertainty relation appear naturally in chronocyclic representation.

If the étalon is used in reflection, its spectral response function is:

$$\tilde{\chi}_{refl}(\omega) = \sqrt{R}\,\frac{1-\exp(2i\omega\, a/c)}{1-R\exp(2i\omega\, a/c)} \tag{10}$$

which is also plotted in Fig.1 (dotted line). The chronocyclic representation of this response function is plotted in Fig.3. Here again, the action of the device appears very clearly. For all frequencies, the response function exhibits a delta function at $t=0$. It is the signature of the direct reflection on the input mirror of the étalon. Around resonance, the reflection response function shows more features, mainly negative, which correspond to the partial cancellation of the direct reflection by waves trapped in the device exiting through the input mirror at later times.

As a last example, a Gires-Tournois interferometer (GTI) [4] consists of two reflective surfaces separated by a distance *a*, the first one with a low reflectivity *R* and the second one with a maximum reflectivity. It is used as a chirp compensating element. Its spectral response function is :

$$\tilde{\chi}_{GTI}(\omega) = \frac{\sqrt{R}+\exp(2i\omega\, a/c)}{1+\sqrt{R}\exp(2i\omega\, a/c)}. \tag{11}$$

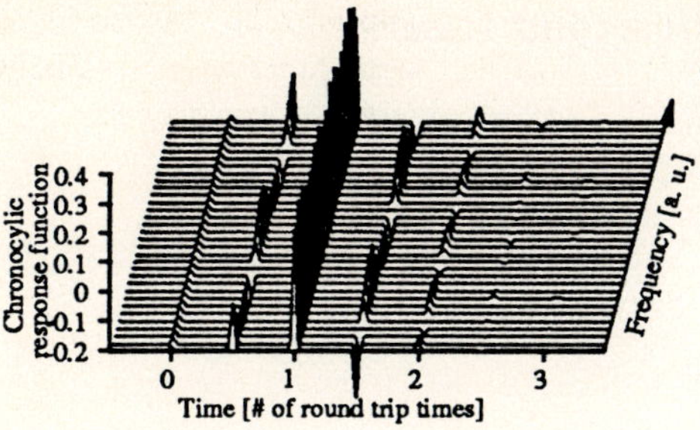

Figure 4. Chronocyclic representation of the response function of a Gires-Tournois interferometer. Each peak is a Dirac δ-function. R=4 %

Figure 4 shows the chronocyclic response function for this device. Its is made of Dirac δ-functions located in time at every half-round-trip delay. From this picture one can immediatetly see that for some wavelengths most of the positive values of the chronocyclic response function occur during the first round-trip, whereas for other wavelenghts, the response is delayed beyond the first round-trip. Thus it is clear that this device acts adds a frequency dependent delay on an ultrashort light pules. It can therefore correct the chirp of a pulse provided that its spectrum is narrower the spectral period of the response function of the GTI.

4. Conclusion

In conclusion, it is possible to define a spectrally-resolved response function for any linear device by applying the chronocyclic representation to the complex response function. In this way it is possible to analyze, in a very simple way, the action on ultrashort pulses of commonly used linear devices such, as for example, a dispersive medium, a Fabry-Perot étalon and a Gires-Tournois interferometer.

References

[1] J. Paye, IEEE J. Quant. Electron. **28**, 2262 (1992).
[2] J. Paye, M. Ramaswamy, J. G. Fujimoto and E. P. Ippen, Opt. Lett. **18**, 1946 (1993).
[3] R. Gase, Optics. Comm. **79**, 397 (1990)
[4] F. Gires and P. Tournois, C. R. Acad. Sci. Paris, **258**, 6112 (1964).

Development of Practical Femtosecond Lasers

P.M.W. French

Femtosecond Optics Group, Physics Department, Imperial College, London SW7 2BZ, UK

Abstract. The development of low-cost and potentially compact femtosecond laser oscillators and amplifiers is discussed.

The development of the Kerr Lens Mode-locked (KLM) argon-ion pumped Titanium-doped sapphire lasers [1] has had a significant impact on many areas of femtosecond laser physics. The same techniques have been used to achieve sub-100 fs pulse generation from other vibronic solid-state lasers such as Cr:LiSAF, Cr:Forsterite and Cr:YAG, thereby extending the fundamental tuning range of this class of laser from ~ 650 nm to ~ 1600 nm. Harmonic generation and optical parametric oscillators can considerably extend this spectral coverage. Possible "real-world" applications for ultrafast lasers include time-gated imaging through turbid media such as living tissue, two photon microscopy, optical ranging, electro-optic sampling, instrumentation for time-resolved spectroscopy and all-optical signal processing. Time-gated imaging using ultrashort pulses tunable through 700 - 1000 nm is particularly attractive for imaging through tissue. Ti:sapphire lasers can provide such pulses but unfortunately Ti:sapphire is not suitable for direct pumping by semiconductor diodes or arc lamps and so the high cost and power/cooling requirements of the necessary large frame argon ion or frequency-doubled Nd:YAG pump laser preclude its widespread practical application.

Any real-world applications will require compact, reliable and low cost ultrafast lasers to be practicable. This means low-cost, all-solid-state diode-pumped lasers with the minimum degree of complexity and therefore the minimum number of components. Reliability implies that the lasers should be self-starting and require no active stabilisation. Ideally the lasers should be widely tunable and able to provide a broad range of output pulse energies. For many applications, it is desirable to have pulse energies at or above the microjoule level - in particular to have the capability to generate a femtosecond white light continuum for transient absorption measurements. This implies a further requirement for low-cost, compact, all-solid-state laser amplifiers. If such laser systems could be available for ~$10,000 - 20,000, it is likely that their proliferation in routine laboratory instrumentation would yield many scientific and technological advances - perhaps rather more than will be realised by simply continuing to further push the ultimate performance of laboratory systems.

Figure 1(a) is a simple schematic of a state-of-the-art femtosecond oscillator/amplifier laser system providing fs pulses in many ultrafast laboratories. The cost of such a system may easily exceed $200,000. For the emergence of widespread real-world applications of femtosecond lasers, it is imperative that this initial outlay be drastically reduced in order that non-femtosecond laser specialists may be able to afford to develop such applications. Figure 1 (b) shows an elegant

Springer Series in Chemical Physics, Vol. 60 **Ultrafast Phenomena IX**
Editors: P. F. Barbara · W. H. Knox · G. A. Mourou · A. H. Zewail

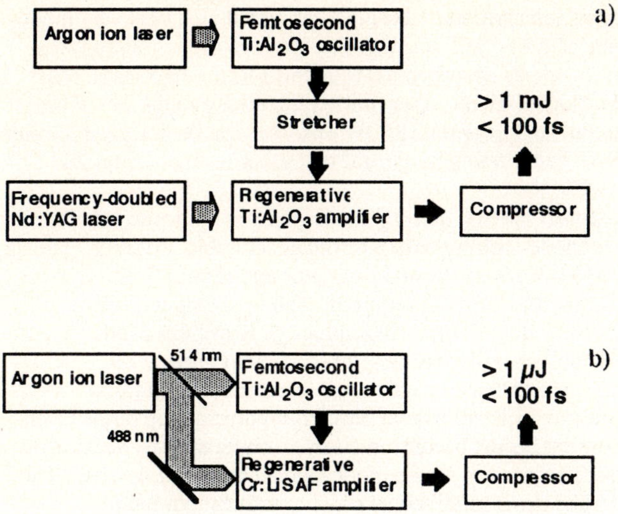

Figure 1. Schematics of (a) mJ femtosecond laser system and (b) modest μJ femtosecond laser system

way to use only one (argon ion) laser to pump both amplifier and oscillator. This approach was initially demonstrated by Norris [2] and developed by Coherent Radiation , Inc. The use of Ti:sapphire as the gain medium in oscillator and amplifier requires a relatively powerful (and expensive) argon ion laser. Replacing the amplifier gain medium with Cr:LiSAF reduces the pump power requirement to ~ 2 W at 488 nm, owing to the longer upper state lifetime (67 μs). In a hybrid oscillator/amplifier system with only a single argon ion pump laser operating at 11 W output power, we have obtained over 1 W average power femtosecond pulses from a Spectra-Physics Tsunami laser oscillator and over 4 μJ pulse energy at 10 kHz repetition rate from a Cr:LiSAF regenerative amplifier [3], which was based on the design by Balembois et al. [4]. For only ~ 8 W of all-lines argon ion laser pump power, it is possible to generate a few 100 mW of average output power from the Ti:sapphire laser while still achieving ~ 5 μJ from the 10 kHz Cr:LiSAF regenerative amplifier pumped by 488 nm radiation . This laser system represents a considerable step towards providing a lower cost laser system for time-resolved spectroscopy applications but is clearly not suitable for widespread deployment in the real world. Cr:LiSAF does, however, provide a means to develop such a (diode-pumped) laser system.

Cr^{3+}:$LiSrAlF_6$ (Cr:LiSAF) is an alternative ultrafast laser medium to Ti:sapphire exhibiting an absorption profile which permits it to be efficiently pumped by laser diodes at 670 nm. Accordingly we demonstrated the first mode-locked all-solid-state diode-pumped tunable (Cr:LiSAF) laser using broad stripe (160μm) 670 nm InGaAlP laser diodes [5]. The current status for this c.w. laser is a threshold of ~ 70 mW and the tuning from 786 to 884 nm with up to 150 mW of true c.w. output power. Active and passive mode-locking have been demonstrated to yield pulses of 100's ps duration [5-7] and now at this meeting we describe the first all-solid-state, diode-pumped femtosecond vibronic laser, based on a passively mode-locked Cr:LiSAF laser [see also 8]. Self-starting, c.w. passive mode-locking was obtained with the intracavity MQW absorber and transform-limited pulses as short as 220 fs were generated. The MQW absorber limited the tuning range to ~ 5 nm around 846 nm. We note that an appropriate MQW absorber can be designed for any wavelength in the Cr:LiSAF laser tuning range and that other resonant saturable absorbers may be employed.

Thus we have demonstrated the first *directly* diode-pumped femtosecond vibronic solid-state laser by passive mode-locking with a resonant saturable absorber. This is a potentially low-cost (< $10,000) femtosecond oscillator and should find application where only limited tunability is required. The next step will be to realise widely tunable femtosecond pulse generation though Kerr Lens Mode-locking. The principal difficulties will be to develop sufficient intracavity intensities to access the nonlinear refractive index and to achieve self-starting operation. The first difficulty will be alleviated once high power pump diode lasers with diffraction -limited output beams become available. These may be realised by injection-locked diode laser arrays or by single-mode tapered amplifier devices. The second issue may be addressed using relatively expensive active modulation techniques with regenerative mode-locking, moving mirror mode-locking or broadband resonant saturable absorbers. It is likely that products using various approaches will be developed. These issues can be side-stepped by using frequency-doubled, diode-pumped Nd lasers to synchronously pump vibronic solid-state lasers. This approach is certainly technically feasible but it always involves a higher degree of complexity than direct diode-pumping with corresponding penalties of cost, size and reliability.

In conclusion, we have demonstrated the first picosecond/femtosecond all-solid-state diode-pumped vibronic lasers using Cr:LiSAF pumped by InGaAlP laser diodes. We anticipate that the same technology will permit the demonstration of an all-solid-state diode-pumped regenerative amplifier delivering μJ femtosecond pulses in the near future. Low-cost femtosecond laser systems will thus be feasible if real-world applications generate the demand.

References

1) D. K. Negus, L. Spinelli, N. Goldblatt, G. Feugnet, in *Advanced Solid State Lasers* , G. Dube and L. L. Chase, eds., Vol. 10 of OSA Proceedings Series (Optical Society of America, Washington, D.C., 1991), pp. 120-124
2) T. B. Norris, Opt.Lett. **17**, 1009 (1992).
3) S. C. W. Hyde, N. P. Barry, R. Mellish, P. M. W. French and J. R. Taylor, *To be published*
4) F. Balembois, P. Georges and A. Brun, Opt.Lett. **18**, 1250 (1993)
5) P. M. W. French, R. Mellish, J. R. Taylor, P. J. Delfyett and L. T. Florez Electron. Lett. **29**, 1263 (1993)
6) F. Balembois, P. Georges and A. Brun, Opt.Lett. **18**, 1730 (1993)
7) R. Mellish, P. M. W. French, J. R. Taylor, P. J. Delfyett and L. T. Florez, Opt. Lett. **22**, 1934 (1993)
8) R. Mellish, P. M. W. French, J. R. Taylor, P. J. Delfyett and L. T. Florez, Electron. Lett. **30**, 223 (1993)

Brave New Nanoworld: Probing the Realm of Ultrafast, Ultrasmall Electronics

D.M. Bloom, A.S. Hou, F. Ho, and B.A. Nechay

Edward L. Ginzton Laboratory, Stanford University, Stanford, CA 94305, USA

ABSTRACT. Although scanning probe microscopy (SPM) is limited to slow mechanical responses, techniques utilizing nonlinear tip-to-sample interactions can be used to capture very fast temporal signals (voltages, in our case). Such SPM-based techniques may deliver unrivaled spatial and time resolution. We have developed a scanning force microscope probe for measuring ultrafast voltage signals with picosecond time resolution. Measurements of integrated devices are shown.

1. Introduction

To keep pace with device technology as it pushes toward 100-nm line widths and gigahertz operating speeds, new measurement tools are needed. These tools are needed to aid in the study of mesoscopic device physics as well as in the diagnosis and design of new, ultradense integrated circuits. In the past, high-speed characterization of nano-scale devices posed a formidable challenge because no single measurement technique simultaneously addressed the requirements of ultra-high speed and ultra-small scale. The spatial resolution of electro-optical testing is limited by the focused laser spot size, on the order of a few microns [1], whereas electron-beam testing encounters difficulties in probing signals with time resolution less than about 50 ps [2].

Recently, however, powerful new scanning probe microscope (SPM) techniques have been invented which combine the extraordinary spatial resolution of traditional SPM's with ultrafast time resolution [3-6]. It has been realized that ultrafast signals can be measured using SPM's by exploiting various nonlinear properties of tip-to-sample interactions. For instance, in a scanning tunneling microscope (STM), there is a nonlinear dependence of tunneling current on the voltage applied between the tip and the sample. In a scanning force microscope (SFM) that is designed to measure electrical forces, the force acting upon the tip varies with the square of the voltage difference between tip and sample. By applying an ultrafast electrical stimulus to the probe tip, both STM's and SFM's can be adapted to investigate ultrafast phenomena on the nanometer scale.

In addition to being non-invasive and having both high spatial and temporal resolution. other features of an ideal electrical probe should include ease-of-use and robustness. These factors favor SFM rather than STM, even though the ultimate spatial resolution of an SFM is not as great. Previous work on SFM-based potentiometry and capacitance force microscopy have achieved spatial resolutions down to 25 nm [7,8]. To achieve gigahertz measurement bandwidths, our probe system relies on mixing (heterodyning) due to square-law electrical force present between the SFM tip and sample. The mixing allows the measurement of frequencies far above the cantilever resonance.

Springer Series in Chemical Physics, Vol. 60 **Ultrafast Phenomena IX**
Editors: P. F. Barbara · W. H. Knox · G. A. Mourou · A. H. Zewail

2. Theory of Operation

To measure voltages, we operate an SFM in a non-contact mode and use a conductive tip. A diagram of the SFM sampling probe system is shown in Figure 1. A gap ranging from 10 to 100 nm is maintained between the tip and the sample. A voltage difference across the gap will cause an attractive electric force to act upon the tip, bending the cantilever. Suppose there is a periodic voltage waveform V_S on the sample that we wish to measure. We supply a strobe voltage V_P consisting of a train of identical voltage pulses, to a conductive tip. Both V_S and V_P are referenced to a common ground; thus the voltage across the gap is V_P - V_S The measured force is a function of the square of this voltage difference, given by

$$\begin{aligned} F &= \frac{\varepsilon_o A}{2z^2}(V_p - V_s)^2 \\ &= \frac{\varepsilon_o A}{2z^2}(V_p^2 - 2V_pV_s + V_s^2) \end{aligned} \tag{1}$$

where C is the capacitance between the tip and sample and z is the gap distance in the direction normal to the sample surface, and we have modeled the tip and sample as two parallel plates with an effective interaction area A.

The cross-product term in Equation (1) generates mixing between V_P and V_S, giving rise to sum and difference frequency force terms. The difference frequency, or intermediate frequency (IF), can be chosen to lie below the mechanical resonance of the cantilever, typically 10 to 100 kHz, so that it is slow enough for the cantilever to respond to, even if V_P and V_S are high frequency signals. All other terms produce either DC forces or extremely high frequency forces beyond the cantilever's range. In essence, the SFM functions as a high-speed electrical mixer with two high-frequency inputs and a relatively slow mechanical IF output.

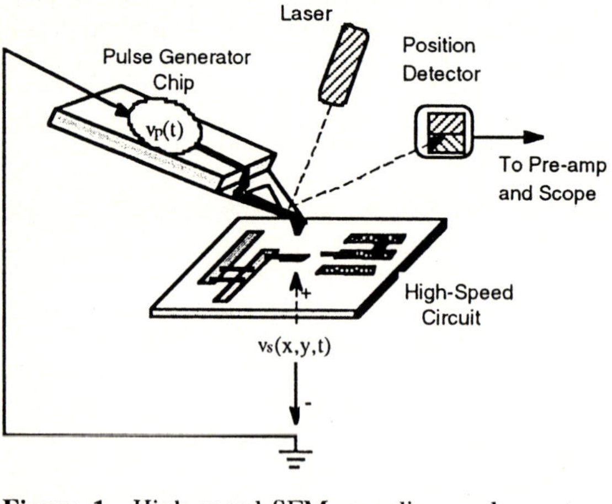

Figure 1. High-speed SFM sampling probe system.

3. Voltage Sensitivity

To assess and optimize the voltage sensitivity of an SFM probe, each candidate noise source in the system first must be identified. There exist laser shot noise and laser intensity noise for the optical detection scheme, electrical noise in the detection electronics, and 1/f noise everywhere. With good design, laser and electrical noise can be made negligibly small. However, we are still left with a fundamental noise source in the SFM which is the mechanical noise vibration of the cantilever due to its equilibrium with the finite temperature environment. Above about 1 kHz, the sum of the different noises is usually dominated by thermal mechanical noise, and we can use this fact to derive a noise-equivalent voltage, or minimum detectable voltage for an SFM probe. By equating RMS signal deflection due to electric forces with RMS noise deflection, we find the minimum detectable voltage V_{min} to be:

$$V_{min} \approx \frac{z^2}{\varepsilon_o A V_p (\tau/t)} \sqrt{\frac{32k(k_B TB)}{\omega_o Q}} , \qquad (2)$$

where k_B is Boltzmann's constant, T is temperature, B is the measurement bandwidth, ω_o is the cantilever resonant frequency, Q is the quality factor of the cantilever, k is the cantilever spring constant, A is the effective tip area, and z is the separation height. We have also included the duty cycle factor (τ/t) of the strobe pulses, where τ is the pulse width and t is the pulse train period. Specifications for typical microfabricated cantilevers are as follows: tip radius=50nm, A=0.016μm^2, z=10nm, k=0.05N/m, $f_0=\omega_0/2\pi$=33kHz, and Q=6. Assuming room temperature operation, a duty cycle factor of 0.01, B = 1 Hz, and V_p = 5 volts,

$$V_{min} \approx 1 \text{ mV}. \qquad (3)$$

This indicates that better than millivolt sensitivities are expected for measurements with averaging times equal to or longer than 1 second. Significant improvements in voltage sensitivity can be obtained by using more flexible cantilevers and higher probe voltages or in more extreme circumstances, going to lower temperatures or vacuum operation, which increases Q by several orders of magnitude.

4. Experimental Results

We performed several experiments that demonstrated high-speed performance in the time domain and equivalent-time sampling. Some of the results are shown in Figure 2, which are measurements of internal voltages on an Intel 80486 microprocessor chip. The chip had been mounted in a standard pin-grid-array package and depassivated. Fig. 2 shows two 20 MHz clock signals of opposite phases. The sampling pulses used were 2 ns wide. Also shown is a sampled result of a 2 µs-long test vector that was produced by applying a reset signal to the chip every 20 clock cycles. The bit pattern of zeros and ones can be clearly discerned.
The preceding experiments were performed using gold-coated commercially available cantilevers, a scheme which will not work at frequencies greater than several

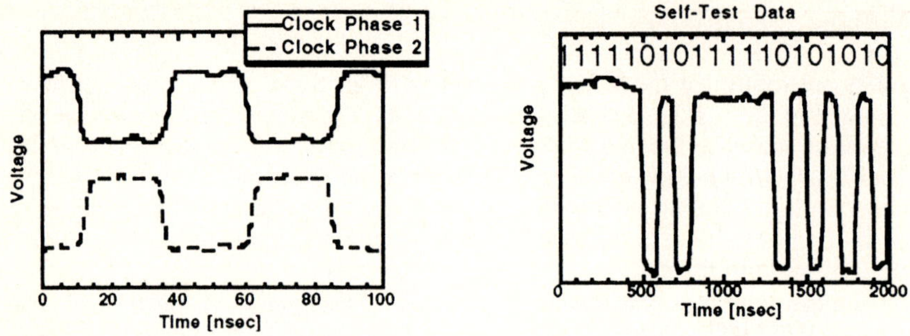

Figure 2. (Left) Two 20 MHz clock phases measured on the 80486 processor. (Right) A 20-bit vector is probed with the 80486 in self-test mode.

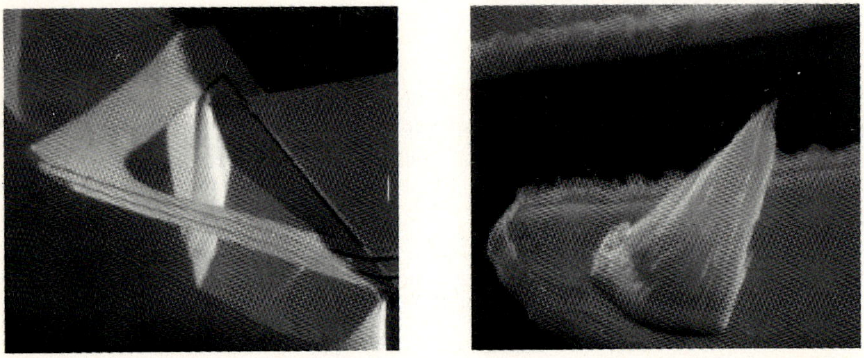

Figure 3. SEM images of microfabricated high-speed cantilever and tip.

gigahertz. In order to obtain better temporal resolution, we microfabricated cantilevers with integrated transmission lines. SEM images of the cantilever and the tip are shown in Figure 3. The cantilever was 0.3 μm thick and made of silicon nitride, and had been deposited with a gold film which formed a transmission line. At the end of the center conductor, a tip was fabricated by Spindt evaporation [9].

Using this high-speed cantilever, we were able to observe voltage steps as short as 5 picoseconds on circuits. This is shown in Figure 4. We used nonlinear transmission line (NLTL) circuits, which can generate subpicosecond voltage steps [10], to serve both as the source of the strobe pulses and as the circuit under test. Although equivalent-time sampling normally requires voltage pulses for the strobe signal, we were able to use voltage steps as the strobe signal by post-processing (numerical differentiation) the sampled signal. In this experiment, the NLTL circuit and the cantilever chip were wire-bonded together, and the two NLTL's were driven with 9 GHz sine waves. We anticipate that by integrating the NLTL circuit and the cantilever on the same substrate, we will obtain sub-picosecond resolutions.

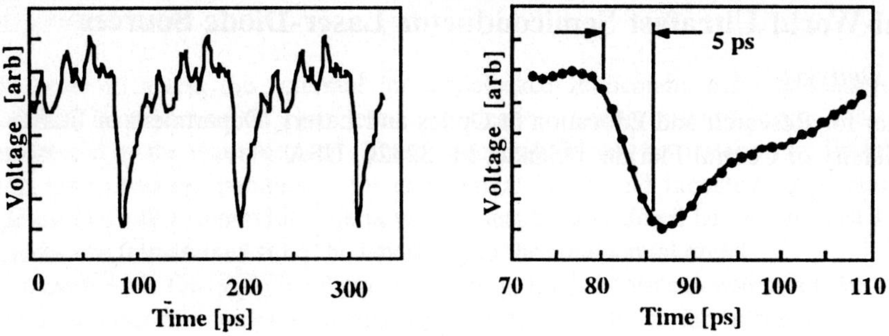

Figure 4. (Left) Measurement of nonlinear transmission line output signal. (Right) Magnified view of 5 picosecond falling edge

5. Conclusion

Scanning probe microscopes are being developed which combine very high spatial resolution with the temporal resolution of the fastest electronic pulse generator technology. We have measured 5 ps voltage steps using equivalent-time sampling, and probed a 80486 microprocessor. In the future, we anticipate being able to probe with sub-picosecond, nanometer resolution by integrating high-speed circuitry with a microfabricated cantilever and tip.

This work is supported by ONR/ARPA under the Ultra program. We are grateful for the assistance of Bob Rao and Debbie Cook (both at Intel), and Pauline Prather.

References

[1] K.J. Weingarten, M.J.W. Rodwell, and D.M. Bloom, IEEE J. Quantum Electron. **QE-24**, 198 (1988).

[2] P. Girard, Scanning Microscopy **2**, 151 (1988).

[3] A.S. Hou, F. Ho, and D.M. Bloom, Electronics Letters **28,** 2302 (1992).

[4] G. Nunes Jr. and M.R. Freeman, Science **262,** 1029 (1993).

[5] S. Weiss, D.F. Ogletree, D. Botkin, M. Salmeron, and D.S. Chemla, Appl. Phys. Lett. **63,** 2567 (1993).

[6] K. Takeuchi and Y. Kasahara, Appl.Phys. Lett. **63,** 3548 (1993).

[7] Y. Martin, D.W. Abraham, and H.K. Wickramasinghe, Appl. Phys. Lett. **52**, 1103 (1988) .

[8] C.C. Williams, W.P. Hough, and S.A. Rishton, Appl. Phys. Lett. **55,** 203 (1989) .

[9] C.A. Spindt, I. Brodie, L. Humphrey, and E.R.Westerberg, J. Appl. Phys. **47**, 5248 (1976).

[10] M.S. Shakouri, A. Black, B.A. Auld, and D.M. Bloom, Electronics Letters **29,** 557 (1993).

ASE background: For high output power from modelocked laser diodes, typically a power amplifier is incorporated into the system. This normally leads to a significant amount of amplified spontaneous emission, depending on the pulse repetition frequency. The ASE levels can be several percent or more. unless steps are taken to minimize this effect. Spectral filtering and pulse biasing the power amplifier help to minimize ASE, with the drawback of adding complexity to the laser system.

Packaging: Packaging is by far the most critical area in commercializing modelocked semiconductor laser diodes. The device package and other components must be resistant to misalignment and isolated from environmental fluctuations, e.g., temperature, humidity, dust, vibrations, electrical transients, etc.

Cost: The cost of a modelocked laser diode system has the potential to be far cheaper than other modelocked laser systems, owing to the potentially small cost for the active components. In addition, no external active modelocking element is required owing to the direct electrical injection nature of diodes. This feature also implies that laser diode systems will always be more efficient than a diode pumped solid state laser system, since the all diode approach has only one electrical to optical conversion step, while all other systems have at least two conversion steps.

Reliability: Device lifetimes normally approach 5000 hours or more. This avoids costly maintenance and down time with associated with most other ultrafast laser systems.

Power requirements: An all diode ultrafast laser is capable of running from a standard wall plug outlet, i.e., 110 VAC. Cooling will be from Peltier elements, avoiding any water cooling systems.

4. Conclusions

In summary, the realization of a compact and efficient source of ultrashort optical pulses will most likely occur with the utilization of semiconductor laser diodes. While alternative technologies exist, these approaches are still at the laboratory/bench-top stage. In the future, these alternative approaches will find viable markets and commercial applications. The salient feature to recognize is that the application, which will define the laser system's specification, will suggest the appropriate solution, whether diode based, or diode pumped solid state based.

5. References

[1] A. Mar, et al., (to be published in IEEE Photonic Tech. Lett., 1994).
[2] P. J. Delfyett, et al., IEEE J. Quant. Electron., vol. 28, 2203, (1992).
[3] L. Goldberg, et al., CLEO Postdeadline CPD 17, (1994).

"Coherent Control?" Is not a Question

W.S. Warren

Department of Chemistry and the Princeton Center for Photonics and Opto-Electronic Materials, Princeton University, Princeton, NJ 08544, USA

Abstract. The uses of shaped laser pulses for characterization and control of molecular dynamics and reactivity are detailed.

Coherent control–the idea of using controlled radiation fields to alter molecular states or drive chemical reactions–was once extremely controversial, and even considered quite unlikely. After all, many hundreds of experiments in the 1960s and 1970s showed that, in essentially every large molecule, energy can only be deposited in a specific bond for a short time, after which the energy randomizes due to intramolecular energy redistribution (IVR). Thus the best chance for beating IVR seemed to be to work on a still faster timescale. This belief in fact contributed greatly to the historical quest in the ultrafast community for ever-shorter pulses[1].

Today, however, there is no longer any doubt, either experimentally or theoretically, that controlled radiation fields alter molecules and molecular energy levels in useful ways[2,3,4]. In the last four years there have been three dedicated conferences and a dozen conference sessions in this field (now more commonly called "quantum control"),[2] and the community of scientists doing theoretical or experimental work on this subject numbers well into the hundreds. Quantum control became possible because the technology for controlling optical fields and for detecting shaped pulses has improved dramatically. By the time of the last Ultrafast meeting, programmable pulse shaping and phase shifting capabilities had evolved to <100 fs time resolution with amplified pulses. Substantial new developments have been reported in several laboratories in the last year[5,6]. For example, we have demonstrated a new technique, acousto-optic modulation inside a zero-dispersion line, which can dramatically improve capabilities for generating complex optical waveforms (including phase and amplitude modulation).[6] In addition, frequency-resolved optical gating (FROG) has developed rapidly into a very useful method for measuring the shapes of ultrafast pulses.

These technological developments have also spurred theoretical interest in quantum control. Calculations have shown that appropriately shaped optical pulses can prepare selected rotational states, pump up anharmonic vibrational ladders, transfer energy into selected bonds, and steer reactions between different pathways–at least for simple model molecules[2,3]. Unfortunately, simplifications in the early models often made them useless for designing experiments. For example, it seemed reasonable a few years ago start by ignoring rotational states, but our experimental and theoretical work has shown that strong-response experiments are dramatically affected by these states (even for fs pulses). Calculations have recently become more realistic; they now often include multiple potential energy surfaces, rotational states; and pulse fluence limitations to prevent uncontrolled multiphoton processes.

Springer Series in Chemical Physics, Vol. 60 **Ultrafast Phenomena IX**
Editors: P. F. Barbara · W. H. Knox · G. A. Mourou · A. H. Zewail

In my opinion, the *real* questions in this field are:

–How will we determine experimentally useful waveforms?

We could, with existing modulators, essentially generate a 2000 point approximation to any waveform we want, limited only by the risetime of the initial femtosecond laser pulse (and, in fact, one could even use the shaper to *shorten* the pulse itself). Trial and error will not identify useful waveforms with that many degrees of freedom. However, computations on model systems provide little guidance for real experiments, and calculating the evolution of even a general triatomic molecule is quite difficult.

We are pursuing several different approaches to address these issues. One approach, which we call molecular pulse shaping[7], uses an absorption cell inside an interferometer to optimize waveforms for specific molecules. In the weak-response limit, the optimum pulse for exciting any molecule with discrete spectral lines would contain only those frequency components; such waveforms are useful as a starting point for population inversions in the strong-response limit as well. Currently we are using a modified Sagnac interferometer[Figure 1] which creates pulses with discrete spectral components in the destructive direction. We have verified that the shaped pulse is absorbed much more efficiently in a second I_2 cell than was the original ps pulse; for example, a 2 ps pulse centered at λ=605 nm is attenuated in a second iodine cell only one-sixth as rapidly as is the modulated pulse out of the Sagnac.

Another approach is to use waveforms with intrinsically robust excitation profiles. For example, we have used frequency swept pulses to simultaneously invert dozens of transitions with different dipole moments. The effects of such sweeps are often independent of the exact pulse amplitude or precise details of the molecular Hamiltonian.

A final approach is to feed molecular responses directly into an optimization algorithm. In effect, any molecule is a "special purpose computer" which accurately solves Schrödinger's equation for one particular Hamiltonian- its own. Thus, changes in a simple observable (such as fluorescence) as the pulse shape is varied can be used to optimize waveforms. Variations of this idea are being explored in several laboratories.

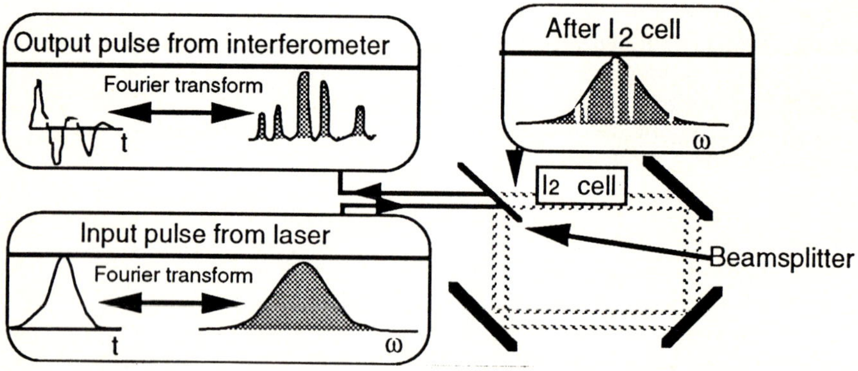

Figure 1. An absorption cell inside one arm of a modified Sagnac interferometer turns an ultrafast laser pulse into a highly modulated optical free induction decay.

–What are the ultimate limitations of coherent control?

The most important limitations of shaped pulse applications also apply to ultrafast laser spectroscopy in general. For instance, shaped laser pulses will not be used to produce cheaper gasoline in the foreseeable future. The cost of a mole of laser photons (an einstein) still makes coherent radiation a very expensive reagent. Moreover, we cannot solve Schrödinger's equation exactly for any reasonably sized molecule, so we do not know precisely how to predict useful waveforms or analyze their effects–but the same problem arises in all of laser spectroscopy, yet there are many papers in this book on ultrafast applications in large molecules.

In practice, the most important role of coherent control for the near future will be to give us more information on molecular dynamics, not necessarily to change those dynamics. Tailored pulses can be exquisitely sensitive probes of molecular potential energy surfaces. In the longer term, there are reasons to believe that full-fledged coherent control of large molecules is feasible, despite competition with relaxation and intramolecular dynamics. The engineering control literature shows that relaxing (damped) systems are far more controllable than undamped ones. Intelligent choices of target molecules will probably be crucial; for example, the low-frequency vibrational modes of enzymes provide a particularly intriguing target for "optimal control" with molecular feedback.

The quantum control community tends to distinguish between weak-response and strong-response experiments[2], and in fact the ultimate limitations may well be different in these two cases. For example, the density of states in even moderately sized molecules makes electronic transitions or overtones of vibrational transitions unresolvable. This suggests that they will be difficult to control by schemes which, for example, exploit interference between two well-defined "paths".[4] However, relaxation rates can be altered by strong-reponse schemes using with modest powers ($\approx 10^8$ W/cm^2 for strong transitions); in some cases such fields reduce IVR by turning off the couplings between bright and dark states.

In summary, the future development of ultrafast spectroscopy depends at least as much on quantum control (pulse shape and phase effects) as it does on the "traditional" technological thrusts of ever-shorter and ever-stronger pulses. Using quantum control as a tool for understanding dynamics is already feasible. Full-fledged alteration of molecular dynamics with shaped pulses is clearly possible in simple systems. In the longer term, it surely will lead to a better understanding of molecules and devices, and may ultimately change the way chemists approach molecular design.

Our work is supported by NSF and ONR.

References

[1]See A. H. Zewail, Physics Today, November 1980, p. 27 and references therein.
[2]W. S. Warren, H. Rabitz and M. Dahleh, Science 259, 1581 (1993); W. S. Warren, Science 262, 1008 (1993)
[3]S. Rice, Science 258, 412 (1992); D. J. Tannor, in Molecules in Laser Fields (Marcel Dekker, New York, 1994), p. 403

[4]P. Brumer and M. Shapiro, in Molecules in Laser Fields (Marcel Dekker, New York, 1994), p. 287
[5]See papers by Kent Wilson and coworkers; Wefers and Nelson; and Kane, DeLong and Trebino in this book
[6]C.W. Hillegas, J.X. Tull, D. Strickland, D. Goswami and W.S. Warren, Opt. Lett. 19, 10 (1994); see also the paper by Tull, Dugan, Hillegas and Warren in this book.
[7]D. Goswami, C. W. Hillegas, J. X. Tull and W. S. Warren, in Femtosecond Reaction Dynamics (D. A. Wiersma, editor; Royal Dutch Academy of Sciences, 1994); D. Goswami and W. S. Warren, J. Chem. Phys. 99, 4509 (1993).

High-Harmonic Generation: The Route to Attosecond Pulses

P.B. Corkum, N.H. Burnett, and M.Y. Ivanov

National Research Council of Canada, Ottawa, Ontario, Canada

Abstract: Experiment and theory both show that high-harmonic emission, double ionization and other strong field atomic effects are very sensitive to the polarization of a fundamental laser. This physics can be used to make "high harmonic" pulses with duration less than half of the period of the fundamental.

The discrete nature and the broad envelope spectrum of the high harmonic emission implies that a train of very short pulses separated by one half of the fundamental period can, in principle, be constructed [1]. However, for train of short pulses to occur, it is not only necessary to have uniformly spaced harmonics, but the harmonics must be appropriately phased. In analogy with lasers, it is not sufficient to have equally spaced modes of an oscillator, they must be locked in the appropriate relative phase.

Along with phase matching issues, the harmonic emission of a quantum system is determined by the squared modulus of the second time derivative of the dipole moment. In the intense field regime where the energy of the free electron oscillation in the laser field is greater than the ionization potential, the time dependent dipole moment of an atom can be understood [2,3,4] by considering the following three step: (1) In the process of ionizing, the atom releases an electron wavepacket into the continuum. (2) The electron follows a trajectory which, in the direction of the laser electric field, is first away from the ion and then comes back towards the ion. Quantum effects (spreading of the electron wave packet) dominate in the direction perpendicular to the field. (3) Those electrons that return to the vicinity of the ion emit harmonic radiation in a coherent interaction between the portion of the wavefunction remaining in the ground state and the returning portion (coherent spontaneous recombination to the ground state).

This semi-classical model indicates that, viewed through a filter that passes only the highest harmonics, a sequence of short pulses (Fig. 1a), would be seen; much like the output of a mode locked laser. The calculation was performed assuming a 25 fs pulse of 800 nm light interacting with neon with a peak laser intensity of 6×10^{14} W/cm^2 and the filter had the transmission characteristics of a 2000 Å thick silver film [5]. The filter insures that frequencies below 100 eV are filtered out. The dependence of the square of the electron wavefunction in the direction transverse to the electric field of the laser was taken as Gaussian $\exp(-(r/r_0(t))^2)$ with $r_0(t)=3.5t$ angstroms where t is the time difference between the moment of birth of the electron into the laser electric field and the moment of its re-encounter with the nucleus in femtoseconds. This value, $r_0(t)=3.5t$, is consistent with experiment and with quantum calculations

Springer Series in Chemical Physics, Vol. 60 **Ultrafast Phenomena IX**
Editors: P. F. Barbara · W. H. Knox · G. A. Mourou · A. H. Zewail

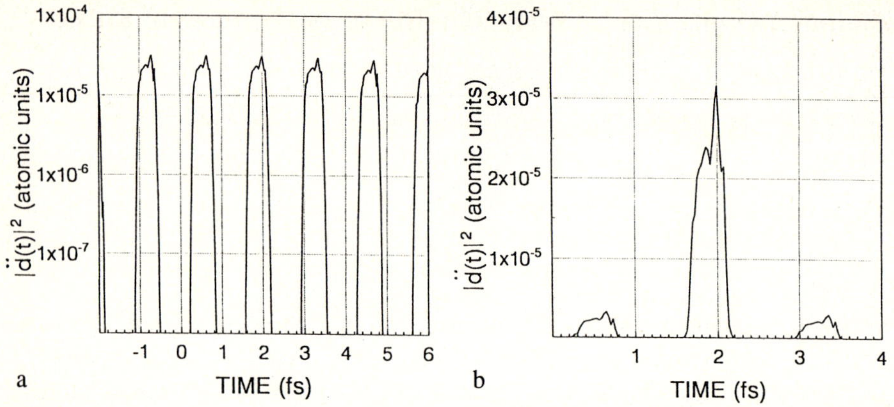

Fig. 1: Square modulus of the second derivative of the time dependent dipole moment plotted for linear polarized light 800 nm interacting with neon at an intensity of $6x10^{14}$ W/cm^2 viewed through a 2000 **Å** thick silver filter. The field of the light was given by $\mathbf{E}=E_0 f(t)[\cos(\omega_1 t)\mathbf{i}+\cos(\omega_2 t)\mathbf{j}]$ where $f(t) = \exp(-(t/T)^2)$ and T = 10 periods of the fundamental and $\omega_2/\omega_1 = 1.0$ (a) and $\omega_2/\omega_1 = 0.9$ (b).

[4,3]. Figure 1a shows $|\ddot{d}(t)|^2$ where d(t) is the time dependent dipole moment plotted as a function of time for a few periods near the peak of the pulse. Individual pulses within the train of pulses are not transform limited.

In order to obtain a single sub-femtosecond pulse from this train it is necessary to modulate the induced dipole moment on a time scale of the period of the fundamental wavelength. Surprisingly this is relatively simple thing to do at least conceptually. Returning to the analogy of a mode-locked laser, we introduce an atomic Pockels cell.

The semiclassical model outlined above predicts a strong ellipticity dependence for high harmonic emission. This is because, in an elliptically polarized field, an electron born at rest into the laser electric field when the field is near the maximum is displaced in the direction of minor axis of the polarization ellipse of the fundamental at the time of is first re-encounter with its parent ion. If this displacement exceeds an amount that can be compensated for by the initial transverse momentum distribution of the ionizing electron, harmonic production ceases. It is possible to modulate the ellipticity of the fundamental by combining two orthogonally polarized pulses of slightly different frequencies. Such a pulse is described by:

$$\mathbf{E}=E_0 f(t)(\cos(\omega_1 t)\mathbf{x}+\cos(\omega_2 t)\mathbf{y})/\sqrt{2} \qquad (1)$$

and can be constructed using femtosecond pulse shaping techniques [6]. In Eq. 1 $f(t)$ describes the pulse envelope.

In a reference frame x',y' with x' as the direction of the laser polarization at time t=0 we can write the laser electric field as:

$$\mathbf{E}=E_0 f(t)[\cos(\omega t)\cos(\Delta\omega t/2)\mathbf{x'}+\sin(\omega t)\sin(\Delta\omega t/2)\mathbf{y'}\} \qquad (2)$$

where $\omega=(\omega_1 + \omega_2)/2$ and $\Delta\omega=\omega_1 - \omega_2$.. At times $t=0$,$n\pi/\Delta\omega$ where n is an integer

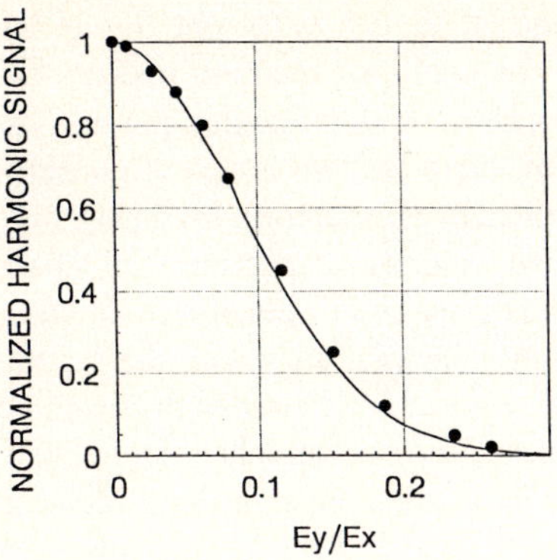

Fig. 2: Ellipticity dependence of the harmonic emission observed from helium using 800 nm light. The data points plot the experimental results while the solid curve plots the results of the semi-classical theory presented here.

number, the pulse is linearly polarized and the linear polarization alternates between perpendicular directions when n is even and odd. For other times the pulse is elliptically or circularly polarized. To obtain a single sub-femtosecond pulse we require that the fundamental pulse envelope $f(t)$ describes an ultrashort pulse. Then for $|t| = \pi/\Delta\omega$ the pulse generates harmonics less efficiently since the ionization rate has fallen and the maximum frequency of the harmonic emission will fall.

The approach to ultra-short pulse generation we have just outlined relies on a predicted strong ellipticity dependence for high harmonic generation. In our most recent experiments [4] we observe harmonic radiation extending out to $N > 61$ in neon. We observe an approximately universal ellipticity dependence for harmonics in the plateau region. In Fig. 2 we show the 37 th harmonic from neon. There are no nearby strong second order lines and so the 37th harmonic is relatively easy to isolate with a slit. Harmonic yield as a function of the ellipticity ε is shown as data points in Fig. 2. The curves plot theoretical results from of the quasi-static model that we discussed above and are completely consistent with the quantum model [3].

Based on the experimental results we can now state the necessary values of ω and $\Delta\omega$ in Eq. 2. If $\Delta\omega << \omega$, we have light with time dependent ellipticity, $\varepsilon \equiv E_{y'}/E_{x'} = \tan \Delta\omega t/2$. If we choose $\Delta\omega/\omega = 0.1$, then the ellipticity $\varepsilon = 0.15$ is reached after one half of a laser period. According to experimental results, for 800 nm light interacting with Ne (see Fig. 2) at ε=0.15 the emission will drop by a factor of 5 compared to $\varepsilon = 0$.

We now calculate the square of the time dependent dipole moment using the pulse described in Eq. 1. We use the semi-classical model described above with the expansion of the wave function transverse to the laser electric field that is determined by the quantum theory [3] and the above experiment. Figure 1b shows $|\ddot{d}(t)|^2$ vs t for the same conditions as Fig. 1a except that now the pulse has a time dependent ellipticity determined by $\Delta\omega/\omega$=0.1. To ensure that only high harmonic radiation is present, low harmonics were filtered by the same 2000 Å thick silver filter used in Fig 1a.

It is not sufficient to fabricate a sub-femtosecond pulse such as the one in Fig. 1b. We must ensure that the pulse will emerge from the nonlinear medium undistorted. That is, we must comment on dispersion of our ultrashort, high frequency pulse in the nonlinear medium and in the filter. In the medium, there are two issues of concern: group velocity and phase velocity dispersion. In typical experiments on high-harmonic generation, there are two main sources for phase velocity dispersion: the phase advance due to the focusing geometry of the incident laser and dispersion in the laser produced plasma. Whereas the first problem is, in principle, technical, the second problem is unavoidable: the high laser intensities necessary for efficient high-harmonic generation always lead to substantial ionization. Indeed, the two are inherently related. We now show that, if the phase matching condition is satisfied, no significant increase in the duration of the ultrashort pulse will occur.

In low-density plasmas the refractive index is $n(\Omega)=1-\omega^2_{pl}/2\Omega^2$ where Ω is the optical frequency in question. Therefore, the change in the phase velocity of $\delta v_{ph}(\omega_L)$ at the fundamental frequency ω_L is N^2 times larger than at $N\omega_L$. One can also show that, for low plasma densities the change in the group velocity in the plasma is: $\delta v_{gr}(\Omega)=-\delta v_{ph}(\Omega)$. Hence, (1) the main factor increasing the duration of the high frequency pulse is the difference between the group velocity of the generated signal and the phase velocity of the fundamental wave, and (2) this effect is of the same order as the phase dispersion between the incident and harmonic waves. The phase matching condition reads that the dispersion between incident and harmonic wave should not exceed one half of the *harmonic* radiation wavelength. Therefore, even in the event of poor phase matching the dispersion of the ultrashort pulse is much smaller than the wavelength of the *fundamental* radiation, ensuring that dispersive broadening can never approach an optical period of the fundamental. Another source of dispersive broadening is the dispersion in the filter, which is used to eliminate low harmonics. In our calculations we used a 2000 Å silver filter, which is transparent for the radiation with Ω exceeding 100 eV. All dispersion characteristics of this filter are known [5], and a 2000 Å silver film will increase the pulse duration by a maximum of 0.1 fs.

It is important to emphasize that the conditions assumed above are experimentally feasible. Specifically, phase related beams with frequencies ω_1 and ω_2 can be selected from a ultrashort laser pulse. 800 nm was chosen because, at this wavelength, pulses with duration of only 10 optical periods are now generated and amplified. Two frequency bands around ω_1 and ω_2 can be selected using standard techniques [6]. The bandwidth of the two components can be easily broad enough to allow the synthesised pulse to fall sufficiently in intensity before linear polarization condition can reappear thus ensuring that harmonic emission is very weak at all times except near t=0. Measurement of the time duration of sub-femtosecond pulses also seems feasable. Since it is as complex as the production process it will be addressed elsewhere [7].

The approach to attosecond pulse generation that we have described is a special case of a much more general phenomenon, namely coherent control of strong field atomic processes. Many multiphoton atomic phenomena can be controlled by the phase dependent restructuring of the laser field that results from the superposition of two or more laser beams. To fabricate attosecond pulses, we use the second pulse to control the electron trajectory in the continuum. In addition, it is possible to use a second pulse to control the formation of the initial electron wave-packet, or it is possible to control the electrons phase along the trajectory. Any physical process that depends on electrons can be manipulated. We discuss this more general way of understanding coherent control in another paper[8].

References:

(1) S. E. Harris, J.J. Macklin and T. W. Hansch, Opt Commun. **100**, 487 (1993).

(2) P. B. Corkum, Phys. Rev. Lett. **71**, 1994 (1993).

(3) M. Lewenstein, P. Balcou, M. Y. Ivanov, A. L'Huillier and P.B. Corkum, Phys. Rev. **A49**, 2117 (1994).

(4)P. Dietrich, N.H. Burnett, M. Yu. Ivanov and P.B. Corkum, Phys. Rev. **Axx**, (1994).

(5)H.-J. Hagemann. W. Gudat and C. Kunz, Deutsches Electronen-Synchrotron DESY SR-74/7, May 1974.

(6) C. Froehly, B. Colombeau and M. Vampouille in *Progress in Optics XX,* E. Wolf, ed (North Holland, Amsterdam, 1983), pp 65-153 ; A. Weiner, J.P. Heritage and E. M. Kirschner, J. Opt. Soc. Am. **B5**, 1563 (1988).

(7)P. B. Corkum, V. Taranukhin and E. Constant, presented at International Quantum Electronics Conference, May, 1994 (advanced program, pg 107).

(8) M. Y. Ivanov, P. B. Corkum, T. Zou and A. Bandrauk, submitted to Phys. Rev. Lett.

Time-Resolved Spectroscopic Techniques in Laser Medicine

S. Svanberg

Division of Atomic Physics, Lund Institute of Technology and
Lund University Medical Laser Center, P.O.Box 118, S-221 00 Lund, Sweden

Abstract. The observation of time-resolved laser-induced fluorescence allows a facilitated identification of chromophores that are useful for tissue diagnostics, e.g. in cancer and atherosclerotic-plaque identification. Optical photon propagation studies provide information on scattering as well as absorptive properties of tissue and alleviate image blurring due to multiple scattering. Optical mammography avoiding the use of ionizing radiation might become a reality. New possibilities in X-ray imaging arise using laser-produced plasmas. Ultra-fast and ultra-sharp tissue imaging has been demonstated.

1. Introduction

Laser spectroscopic techniques are increasingly being used for medical diagnostics (See, e.g. [1]). Generally, signals are observed time integrated. However, additional valuable information can be obtained by time-resolved observation. Different aspects of time-resolved spectroscopy in medicine will be discussed in the present paper. Tissue chromophores can be identified with higher accuracy if fluorescence decay curves are recorded in addition to spectra. Enhanced plaque identification and contrast enhanced tumor detection using time-resolved techniques will be discussed in Sect. 2. Time resolved photon propagation studies provide information on scattering as well as absorption properties of tissue and alleviate image blurring due to multiple scattering (Sect. 3). Optical mammography avoiding the use of ionizing radiation is an attractive possibility. Finally, laser-produced X-rays introduce new aspects in diagnostic radiology as demonstrated in Sect. 4.

2. Tissue Diagnostics using Time-Resolved Fluorescence Monitoring

Fluorescence spectra from biomolecules are frequently rather structureless and similar, which sometimes makes their diagnostic use difficult. Generally, fluorescence intensity ratios for suitably chosen wavelengths are utilized. Time-resolved spectroscopy observing fluorescence decay provides improved

Springer Series in Chemical Physics, Vol. 60 **Ultrafast Phenomena IX**
Editors: P. F. Barbara · W. H. Knox · G. A. Mourou · A. H. Zewail

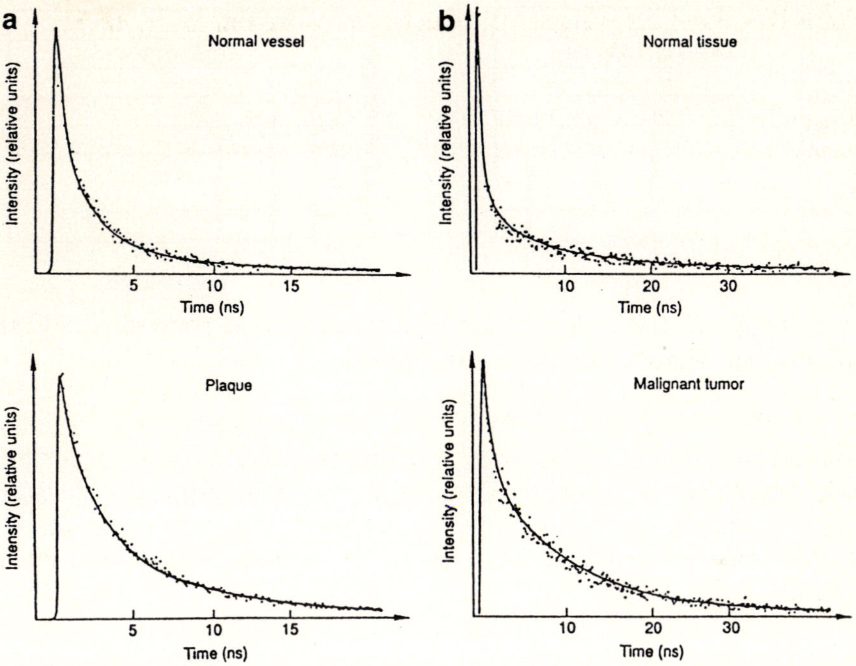

Fig.1. Fluorescence decay curves for normal human vessel wall and atherosclerotic plaque (left) and malignant tumor and normal surrounding tissue (right) in a rat that had been injected with haematoporphyrin derivative (HPD) (From [4]).

molecular identification and allows studies of the dynamic behaviour of biomolecules. Mode-locked and frequency-doubled picosecond pulses from a quasi-CW laser in conjunction with single-photon counting can be conveniently used for laboratory measurements. Such spectroscopy enables atherosclerotic plaque to be distinguished from normal vessel wall, which can allow spectroscopic guidance in transluminal laser ablation of atherosclerotic plaque [1-3]. It also allows enhanced demarcation of malignant tumors following injection of tumor-seeking agents such as haematoporphyrin derivative [4]. Examples of spectra are shown in Fig.1. Atherosclerotic plaque exhibits a longer effective decay time than normal tissue wall as seen in the left part of the figure. In order to understand this observation we have performed measurements on the pure major chromophores in such tissue. Decay curves can be fitted to up to three exponentials with different amplitudes A_i and lifetimes τ_i. Data for elastin and collagen are shown in Fig.2. The observations on the tissue specimen can be explained by a change in the relative concentrations of collagen and elastin with a higher abundance of more long-lived collagen in atherosclerotic plaque.

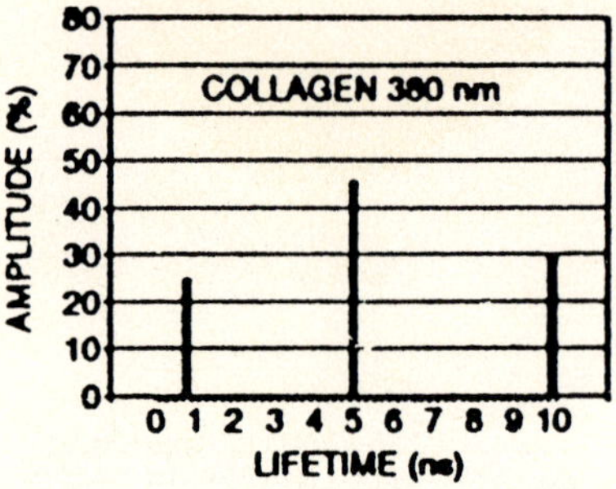

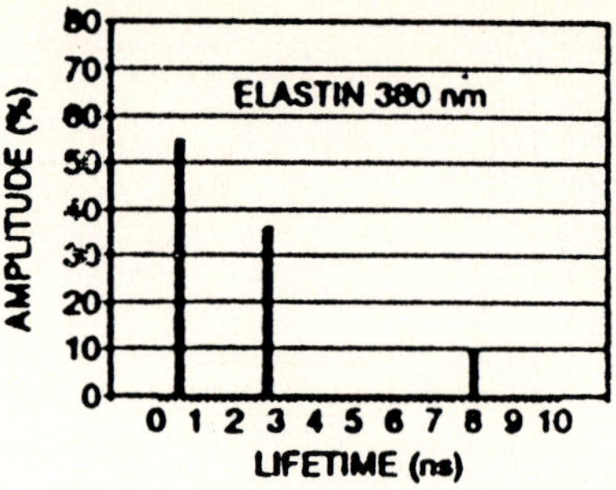

Fig.2. Amplitudes and decay constants for the time-resolved decay of collagen (left) and elastin (right). (From [3]).

In a similar way, the difference in appearance between the tumor and normal tissue decay curves in Fig. 1 (right) can be explained by a higher tumor concentration of HPD molecules with a longer lifetime than the short-lived normal tissue fluorescence, mainly due to NADH. The difference in lifetimes can be utilized In fluorescence imaging of tissue for early cancer detection. By time gating the image intensifier normally used in equipment for the purpose, observing only "late" fluorescence, the tumor signal can be enhanced [4].

3. Photon Propagation in Tissue with Application to Optical Mammography

Time-resolved studies of optical photon propagation in tissue is a rapidly evolving field with applications in brain oxygenation assessment and the development of optical mammography avoiding ionizing radiation [5]. Optical techniques have become even more attractive after the discovery of the *ataxia telangiectasia (AT)* gene in 1 per cent of the population, resulting in hypersensitization to X-rays [6]. The main problem in diagnostic tissue transillumination is image blurring due to heavy multiple scattering. The problems can be alleviated by restricting the light observation to "early" photons, gating out the multiply scattered light that spends a longer time on random walk in the tissue. We have performed extensive studies on tissue transillumination [7-11], employing lasers ranging from a terawatt unit [12] for white-light generation [10] to simple diode lasers [11]. A time dispersion curve through breast tissue and breast tumor imaging are shown in Fig. 3.

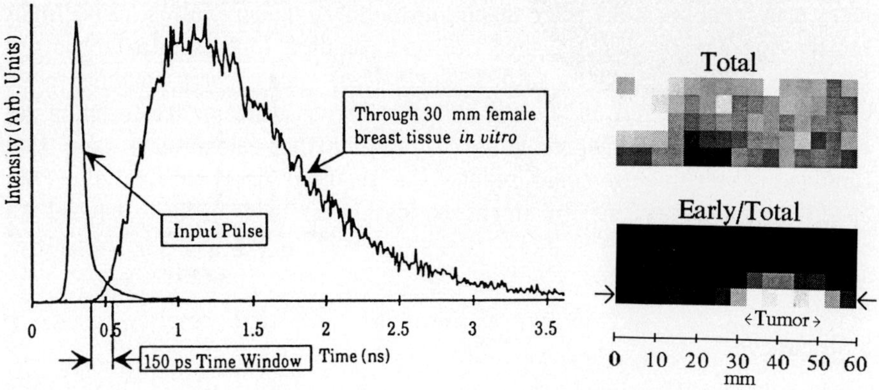

Fig.3. Temporal dispersion curve for a 30 mm thick female breast tissue *in vitro* using a pulsed diode laser as a transmitter for transillumination (left). Scan through a resected female breast specimen displaying the early penetrating light divided by the total detected light. (From [11]).

4. Ultra-Fast Imaging Using Laser-Produced X-rays

Laser-induced X-rays provide ultra-fast and ultra-sharp tissue imaging. We have used a terawatt laser based on chirped-pulse amplification in titanium-doped sapphire to generate picosecond pulses of X-rays by focussing on a solid target [13,12]. The experimental set-up is shown in Fig. 4. Since the

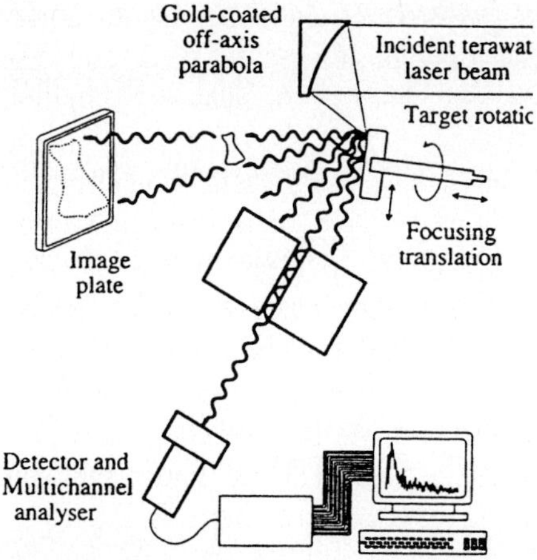

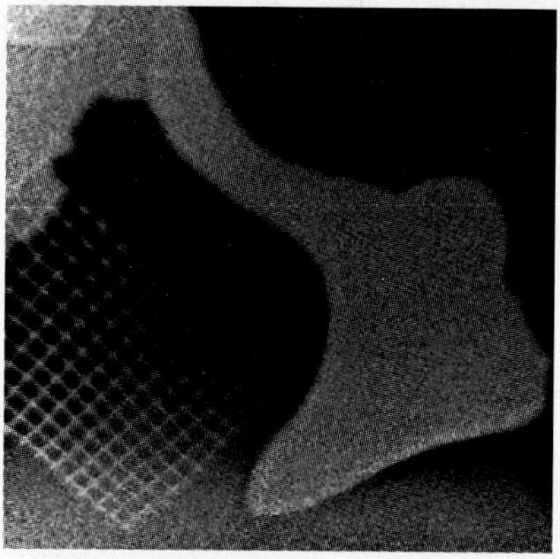

Fig.4 (left). Set-up for the recording of X-ray images using a laser-produced plasma (From [13]). Fig.5 (right). Single-shot X-ray magnification radiography recording of a human ear-bone (*Incus*) (From [15]).

focal spot can be made very small, ultra-sharp magnification radiography radiography can be accomplished utilizing standard image plate technology. An X-ray image recorded with a single laser pulse, corresponding to an exposure time of about 10^{-12} s is shown in Fig. 5. Although scattering is not a major problem in imaging at X-ray wavelengths, calculations show that time-gating techniques could reduce the required dose substantially [14]. Spectroscopic X-ray imaging might be developed using laser-produced X-rays.

5. Discussion

As illustrated by our and other groups, time-resolved laser-spectroscopic techniques have several potential applications in the field of medicine. The development of reliable, compact all solid-state short-pulse lasers will facilitate the practical implementation of the methods that could provide new diagnostic possibilities and thus help relieve human suffering.

1. S. Svanberg, in *Lasers in Medicine*, eds. G. Pettit and R. Waynant, (Plenum, New York), to appear.
2. S. Andersson-Engels, J. Johansson, U. Stenram K. Svanberg and S. Svanberg, J. Photochem. Photobiol. B4, 363 (1990).
3. S. Andersson-Engels, A. Gustafson, J. Johansson, U. Stenram, K. Svanberg and S. Svanberg, Lasers Life Sci. 5, 1 (1992)
4. S. Andersson-Engels, J. Johansson, K. Svanberg and S. Svanberg, Anal. Chem. 62, 19A (1990).
5. G. Müller *et al.* (eds), *Medical Optical Tomography: Functional Imaging and Monitoring* SPIE Institute Vol. IS11 (Bellingham, 1993).
6. M. Swift, D. Morrell, R.B. Massey, and C.L. Chase, New. Engl. J. Med. 352, 1831 (1991).
7. S. Andersson-Engels, R. Berg, S. Svanberg and O. Jarlman, Opt. Lett. 15, 1179 (1990).
8. S. Andersson-Engels, R. Berg and S. Svanberg, J. Photochem. Photobiol. 16, 155 (1992).
9. R. Berg, S. Andersson-Engels and S. Svanberg, in Ref. 5, p. 397.
10. R. Berg, S. Andersson-Engels and S. Svanberg, Opt. Lett. 15, 1697 (1993).
11. R. Berg, O. Jarlman and S. Svanberg, Appl. Opt. 32, 574 (1993).
12. S. Svanberg, J. Larsson, A. Persson and C.-G. Wahlström, Phys. Scr. 49, 187 (1994).
13. K. Herrlin, G. Svahn, G. Olsson, H. Petterson, C. Tillman, A. Persson, C.G. Wahlström and S. Svanberg, Radiology 189, 65 (1993).
14. C.L. Gordon III, C.P.J. Barty and S.E. Harris, in this volume (1994).
15. S. Svanberg *et al.*, in Proc. Euroconference, Salamanca, March 1994.

Sub-10 fs Pulse Generation in Ti:Sapphire: Capabilities and Ultimate Limits

J. Zhou, G. Taft, C.-P. Huang, I.P. Christov, H.C. Kapteyn, and M.M. Murnane*

Department of Physics, Washington State University, Pullman, WA 99164-2814, USA

*Permanent address: Department of Physics, Sofia University, 1126 Sofia, Bulgaria

We generate sub-10 fs pulses by operating near the zero second- and third-order dispersion point in a self-modelocked Ti:sapphire laser. Our numerical simulations show that the pulse duration is limited by fourth-order dispersion, and that shorter pulses will be possible if this can be optimized. In addition, we measured the intra-cavity pulsewidth at various points in the laser.

Recent advances in ultrashort-pulse Ti:sapphire laser technology[1] have made it possible to routinely generate optical pulses of ~10 fs duration using stable, reliable, and easy-to-use lasers. In previous work, we demonstrated the generation of 11 fs pulses at 800 nm, with pulse duration limited by third-order dispersion (TOD).[2] For this work, we used a 2 mm long Ti:sapphire crystal to reduce higher order dispersion, and operated at 850 nm, where second and TOD can be near zero.[3]

For our intra-cavity measurements, we placed a thin pellicle near Brewster at various positions in the cavity, and measured the pulse duration with a near-zero dispersion correlator. We determined that the pulse is negatively chirped, with a duration of 20±2 fs, as the pulse leaves the prism pair and propagates towards the crystal. On the end-mirror side of the 2 mm crystal, the pulse is positively chirped to a value of 21±2 fs. When the pulse exits the crystal on the return trip to the prisms, it is positively chirped to 47±5 fs. We found that the pulse was shortest at 8.5 fs somewhere within the crystal, and the exact position could be moved by translating one of the prisms.

To measure the shortest pulse duration from the laser, we couple the pulse out of the laser between the prism pair and the Ti:sapphire crystal, where it is negatively chirped. Then, by inserting ≈1 mm fused silica into the beam path, we can measure an autocorrelation(Figure 1) width of 13.2 fs, corresponding to a pulse duration of 8.5±0.5 fs assuming a $sech^2$ pulse shape. The corresponding spectrum had a bandwidth of 151 nm, centered at 856 nm. A Fourier transform of the spectrum yields a pulse FWHM of 8.2 fs, with satellite pulses of ~7% the peak amplitude, spaced 15 fs from the main pulse.

We use a model that self-consistently considers the interplay between the intra-ctive dispersion, gain lineshape, and self-phase modulation[3,4]. A comparison with data is shown in Figure 2. Fig. 2(a) shows two experimental spectra, for two different prism insertions. Figure 2(b) shows the corresponding simulation curves.

We conclude that our model includes the most significant pulse-shaping effects, and that more exact dispersion compensation should make possibly even shorter pulses (≈ 5 fs) than are demonstrated in this paper.

Springer Series in Chemical Physics, Vol. 60 **Ultrafast Phenomena IX**
Editors: P. F. Barbara · W. H. Knox · G. A. Mourou · A. H. Zewail

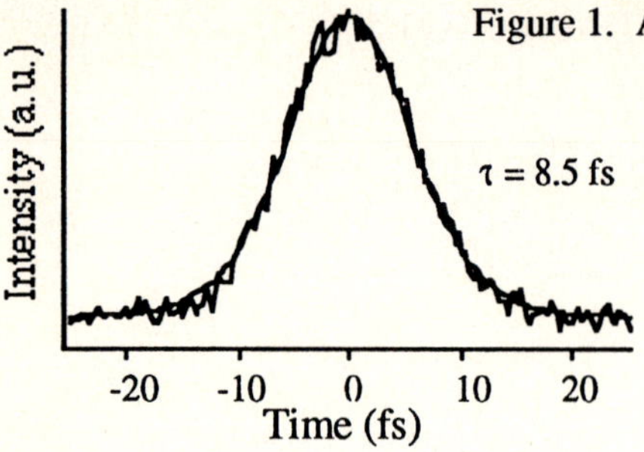

Figure 1. Autocorrelation trace and fit of 8.5 fs pulse.

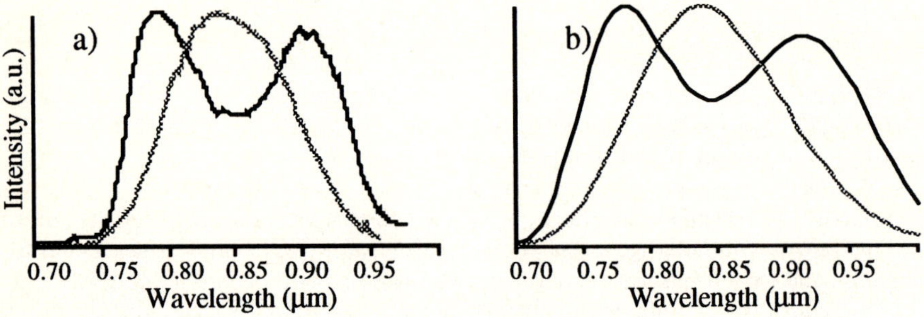

Figure 2: Experimental (a) and calculated (b) spectral measurements of the output of the laser for two different prism insertions, differing by ≈ 340 μm.

We gratefully acknowledge support from the National Science Foundation, the Washington Technology Centers, Union Carbide, and Coherent, Inc.. M. Murnane acknowledges support from a Sloan Foundation Fellowship.

References.

1. D. E. Spence, P. N. Kean, W. Sibbett, Opt. Lett. **16**, 42 (1991).
2. M. T. Asaki, C. P. Huang, D. Garvey, J. Zhou, H. C. Kapteyn, M. M. Murnane, Opt. Lett. **18**, 977 (1993).
3. J. Zhou, G. Taft, C.-P. Huang, M. M.. Murnane, H. C. Kapteyn, and I. P. Christov, to be published in Optics Letters (1994); H. C. Kapteyn, M. M. Murnane, Opt. Photon. News **5** (3), 20 (1994).
4. I. P. Christov, M. M. Murnane, H. C. Kapteyn, J. Zhou, C. P. Huang, to be published in Optics Letters (1994).

Pulse Shaping and Spectral Filtering for High-Speed Optical Networks

A.M. Weiner

School of Electrical Engineering, Purdue University,
West Lafayette, IN 47907-1285, USA

Abstract. Ultrashort light pulses have extremely large inherent bandwidths. Pulse shaping and spectral filtering techniques can help to access this bandwidth for high-speed information networks. We discuss the potential for applying these methods for three different types of optical networks.

Optical fibers offer an extremely wide channel for data transmission. The low loss window around 1.55 μm in silica fibers has a bandwidth of roughly 20 THz. The passband of erbium-doped fiber amplifiers (EDFAs) is narrower but still on the order of 4 THz. Either way, current lightwave systems exploit only a small fraction of the potential bandwidth. Mining this vast bandwidths important in order to support future advanced network services and information superhighways.

There are of course several roadblocks to mining the bandwidth. In the following we discuss how ultrafast pulse shaping methods based on spectral filtering can be used to overcome some of the roadblocks for ultrafast optical networks based on time-division multiplexing (TDM), wavelength-division multiplexing (WDM), and code-division multiplexing (CDM).

Several methods for shaping and filtering ultrashort pulses have been advanced. One successful method uses a simple bulk optics setup to separate an ultrashort pulse into its constituent spectral components [1]. A spatially patterned mask is then used to manipulate the phase and/or amplitude of the dispersed spectral components. In addition to fixed masks, programmable modulator arrays have also been utilized [2, 3]. Reassembling all the frequencies results in a shaped ultrafast waveform, as determined by the Fourier transform of the pattern transferred by the mask onto the spectrum. Within broad limits, nearly arbitrary control over the waveform is possible.

We first describe an ultrafast CDM network based on spectral phase filtering [4] (Fig. 1). Multiple stations are connected via a fiber star coupler. Each transmitter is equipped with an ultrashort pulse source which it encodes into a pseudonoise burst by phase filtering. Multiple-access is achieved by assigning different minimally interfering codes (or "keys") to the various transmitters. The transmitted data are combined and broadcast to the receiving stations. Each receiver is equipped with a spectral phase decoder (or "lock"). Data encoded with a key matched to a receiver's lock (access code) are restored to intense ultrashort pulses which can be detected using an optical thresholder. Data encoded using an incorrect key are rejected by the thresholder. Each transmitter-receiver pair operates at rates up to Gbits/sec compatible with current device technology. CDM allows tens to hundreds of simultaneous users for aggregate throughputs up to several hundred Gbit/sec. To date we have demonstrated only coding and decoding; we are currently setting up a simple testbed to test the ability to encode, transmit over moderate fiber lengths, decode,

Springer Series in Chemical Physics, Vol. 60 **Ultrafast Phenomena IX**
Editors: P. F. Barbara · W. H. Knox · G. A. Mourou · A. H. Zewail

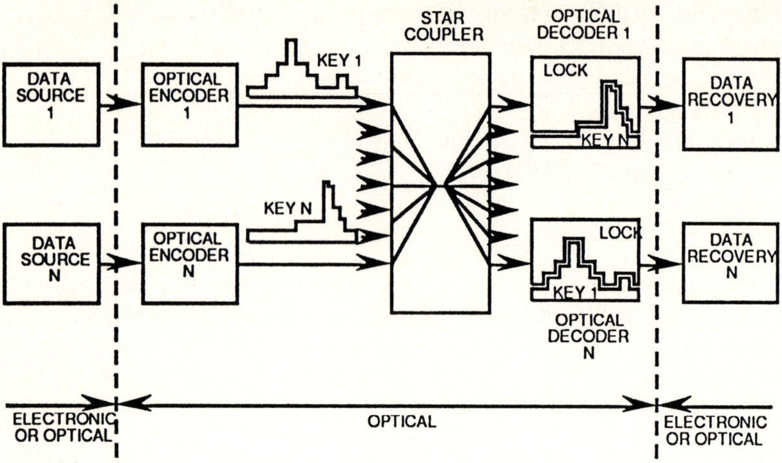

Fig.1 Schematic representation of an ultrafast code-division multiple-access network. Encoding and decoding are achieved by spectral phase filtering in a pulse shaping apparatus.

detect desired data, and reject interference. Key advantages of CDM are that device operating speeds and synchronization are required only at the individual bit rates, not at the much higher aggregate data rate. The CDM scheme may also simplify some important aspects of network control. Challenges include the need for dispersion compensation (since phase coding is used), which will likely limit ultrafast CDM to short-haul networks, and for practical ultrafast nonlinear threshold devices.

Spectral amplitude filters can be used to advantage for WDM networks in which the spectrum is divided into separate wavelength channels, each operating at bit rates up to Gbits/sec Although considerable effort is currently focused on wide-area WDM optical networks, the need to stabilize the wavelengths of multiple independent and geographically separated laser sources is a difficult challenge. One possible solution (Fig. 2) is to use a modelocked laser as a self-stabilized source of multiple optical wavelengths. An amplitude modulator array placed within a pulse shaping apparatus can be used to place independent data on the individual wavelength channels. The application of pulse shaping to dense wavelength-division multiplexing was mentioned briefly by Weiner *et al* [1]; recent exciting experiments by Nuss *et al* [5] have now demonstrated this approach's feasibility. Key advantages are that multiple lasers at different wavelengths are avoided and the phase of individual wavelengths does not enter (therefore dispersion is much less important than in CDM). The main challenge is the need for arrays of high-speed modulators with low crosstalk and optical bandwidths spanning the 30 nm EDFA passband or for integrated devices fulfilling the same function as the pulse shaper. Additionally, one should note the use of femtosecond pulses as a source may offer less flexibility than the use of multiple independent lasers, which can for example be modulated at different bit rates.

A third way to mine the fiber bandwidth is through TDM, telephony's traditional multiplexing approach. As the bit rate increases in TDM, one goes to shorter optical pulses packed more closely together. Challenges to implementing TDM for ultrafast

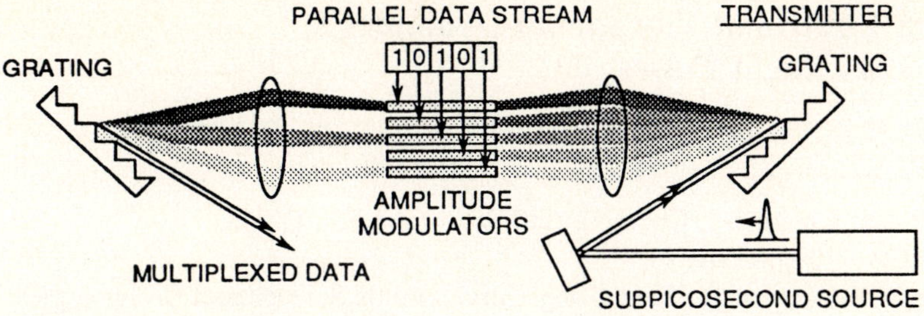

Fig.2 Dense wavelength-division multiplexing transmitter implemented by spectral amplitude filtering in a grating and lens pulse shaping apparatus.

optical networks include the need for ultrahigh-speed devices operating at extremely high rates and the need to maintain very precise synchronism in a geographically distributed network. One way that pulse shaping can contribute is to use parallel frequency domain modulation to form ultrafast sequences of bits in the time domain. Such optical bit sequences can serve as the address or perhaps even the data itself for ultrafast packet-switched TDM networks [6].

In addition, an extension of pulse shaping called *spectral holography* can be applied for important signal processing tasks in optical networks [7, 8]. In spectral holography the pulse shaping filter is replaced with a holographic material, which records a hologram of the incoming ultrashort pulse. The result is a self-aligning process in which the filter is automatically matched to the characteristics of the input pulse; with dynamic holographic materials the filter can even track time variations in the incoming signal. One can consequently compensate for dispersion and timing jitter, without having to know the amount of the dispersion or jitter. Applications to time-domain demultiplexing have also been described [7, 9]. The major challenges for making spectral holography practical are the need for synchronized reference pulses (similar to TDM's synchronism requirements) and for dynamic holographic materials sensitive in the 1.5-μm optical communications band.

References

1. A. M. Weiner, J. P. Heritage, E. M. Kirschner, *J. Opt. Soc. Amer. B* **5**, 1563-1572 (1988).
2. A. M. Weiner, D. E. Leaird, J. S. Patel, J. R. Wullert, *IEEE J. Quantum Electron.* **28**, 908-920 (1992).
3. M. C. Wefers, K. A. Nelson, *Opt. Lett.* **18**, 2032-2034 (1993).
4. A. M. Weiner, J. P. Heritage, J. A. Salehi, *Opt. Lett.* **13**, 300-302 (1988).
5. M. C. Nuss, W. H. Knox, D. A. B. Miller, *to be submitted to Electron. Lett.*
6. M. N. Islam, *Ultrafast Fiber Switching Devices and Systems.* (Cambridge University Press, Cambridge, 1992).
7. Y. T. Mazurenko, *Opt. Eng.* **31**, 739-749 (1992).
8. A. M. Weiner, D. E. Leaird, D. H. Reitze, E. G. Paek, *IEEE J. Quantum Electron.* **28**, 2251-2261 (1992).
9. M. C. Nuss, M. Li, T. H. Chiu, A. M. Weiner, A. Partovi, *submitted to Opt. Lett.*

Quantum Control of Vibrational Dynamics with Tailored Light Pulses

B. Kohler[1], *J.L. Krause*[1], *R.M. Whitnell*[1], *K.R. Wilson*[1], *V.V. Yakovlev*[1], and *Y.J. Yan*[2]

[1]Department of Chemistry, University of California, San Diego, La Jolla, CA 92093, USA

[2]Department of Chemistry, Hong Kong University of Science and Technology, Clear Water Bay, Kowloon, Hong Kong

Abstract: We have demonstrated that light pulses with essentially the same intensity versus time and versus frequency, but different time-dependent phase, can produce different quantum dynamics in matter. By adjusting the chirp of an ultrafast pump pulse we are able to control the vibrational dynamics of an iodine molecule.

The prospect of using tailored, ultrafast light pulses to go beyond observing to actually controlling the quantum dynamical evolution of atoms and molecules is of considerable current interest [1]. Quantum control of matter with light pulses requires the ability to tailor optical phase on the femtosecond time scale. Methods have recently been developed for computing the globally optimal light pulse to attain a given target in the weak response limit [2]. Significantly, it has been shown that control of vibrational dynamics in diatomics should be possible with pulses of moderate complexity [3]. Using the theory as a guide, we have synthesized the pulses needed to control wave packet evolution in iodine. Here we present measurements that indicate that vibrational quantum dynamics can indeed be altered by controlling the time-dependent phase of an ultrafast pump pulse.

In these experiments a pump or "control" pulse with adjustable chirp prepares a wave packet on the B excited state of iodine. To monitor the resulting dynamics we use a detection technique pioneered by Zewail and co-workers for studying the wave packet dynamics of iodine and other molecules [4-6]. After a variable delay, a second femtosecond pulse is used to probe the wave packet population within a window of internuclear distance and time on the B state via transitions to the higher lying e state. The laser-induced fluorescence (LIF) from this state is detected as a function of the delay time between the control and probe pulses.

The experimental setup is shown in Figure 1. A kHz Ti:sapphire chirped pulse regenerative amplifier (regen) yields after compression approximately 75-fs, 300 microjoule pulses centered at 790 nm. The control pulse is generated using a novel optical parametric amplifier we have recently developed. A small amount (< 2 microjoules) of 790 nm light is used to generate a white light pulse in sapphire. The infrared portion of this continuum is parametrically amplified in a BBO crystal pumped by a portion of the 790 nm light from the regen. The resulting 1.0 - 1.4 micron signal light is frequency-doubled in BBO to produce a visible wavelength pulse, tunable from 500-700 nm. Following compression of this light with an SF10 glass prism compressor, pulse durations below 45 fs are routinely measured. Finally, by translat-

Springer Series in Chemical Physics, Vol. 60 **Ultrafast Phenomena IX**
Editors: P. F. Barbara · W. H. Knox · G. A. Mourou · A. H. Zewail

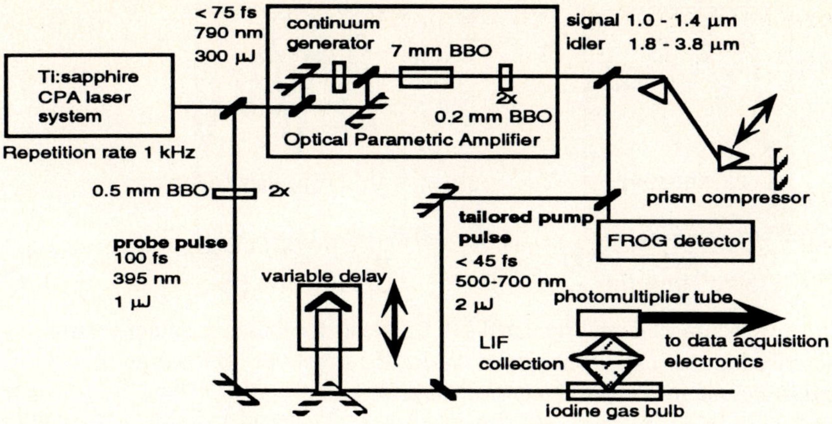

Figure 1. Apparatus for quantum control experiments.

ing one of the compressor prisms into or out of the optical beam, it is possible to adjust the chirp of the control pulse. The amount of chirp is measured by frequency-resolved optical gating [7]. This technique accurately measures the intensity and phase of an ultrashort pulse and is invaluable for comparison of experiment with theory.

A probe pulse centered at 395 nm is produced by directly doubling a smaller portion of the regen output with a 0.5 mm BBO crystal. Cross-correlation measurements between the control pulse and the probe pulse indicate negligible jitter. The iodine is contained in a sealed quartz tube which is heated to produce the appropriate vapor pressure. Fluorescence from the e state is detected through a 420-nm interference filter (10-nm bandpass) placed in front of a photomultiplier tube.

Figure 2 shows the experimental LIF signals versus delay time for control pulses with negative and positive chirps and a center wavelength of 565 nm. The curves are normalized to the amplitude of the first peak. The oscillating signal reflects the periodic motion of the wave packet through the region of internuclear separation selected by the probe pulse wavelength[4]. The period is a measure of the vibrational recurrence time in this region of the B state potential energy well. The open circles show the data when the control pulse is negatively chirped. The filled circles were measured by reversing the sign of the chirp of the control pulse by translating a compressor prism a suitable distance out of the beam. The second and fourth peaks increase in amplitude relative to peaks one and three in the case of the negatively chirped pulse. This is consistent with greater focusing of the wave packet (predicted elsewhere [3]) as it passes the probe pulse observation window on its way to shorter internuclear distances. The two control pulses used have virtually identical intensity distributions in both the time and frequency domains, and yet the wave packet dynamics are different, demonstrating that control of the time-frequency correlation of the pulse can be used to control quantum dynamics. More elaborate pulse schemes can be used to control chemical reactions.

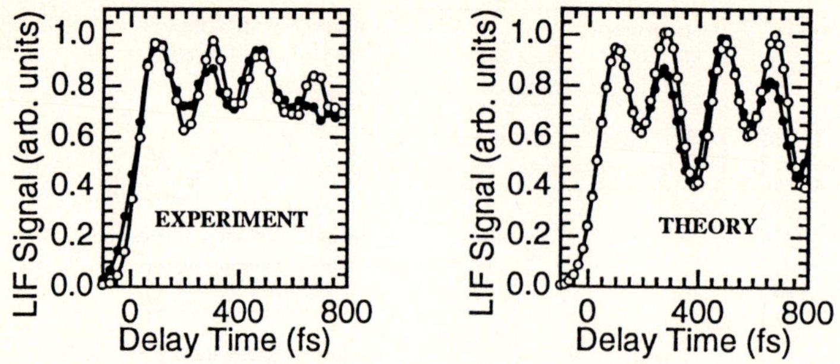

Figure 2. Experimental and simulated LIF transients for iodine quantum control as described in the text. Open circles are for the negatively chirped pulse and closed circles are for the positively chirped pulse.

The change in peak heights observed experimentally is also seen in numerical simulations of the LIF signal in Figure 2. These calculations use values of the control pulse bandwidth, chirp rate, and probe pulse duration appropriate for the experiment. The simulations indicate that the differences between the positively and negatively chirped cases become more pronounced as the bandwidth of the control pulse is increased. A more detailed analysis, which includes the effects of temperature and of higher-order chirp (cubic and higher-order phase terms) is in progress [8].

In conclusion, we have presented measurements which indicate that wave packet dynamics can be controlled by manipulating the time-dependent phase (i.e. chirp) of a pump pulse in a manner predicted by theory. This work is a promising first step toward an exciting new area of experimental inquiry in which ultrashort pulse shaping and measurement techniques will be used in close collaboration with theory to control and modify the dynamics of matter.

References

1. W. S. Warren, H. Rabitz, M. Dahleh, Science **259**, 1581 (1993).
2. Y.-J. Yan, R. E. Gillilan, R. M. Whitnell, K. R. Wilson, and S. Mukamel, J. Phys. Chem. **97**, 2320 (1993).
3. J. L. Krause, R. M. Whitnell, K. R. Wilson, Y.-J. Yan, S. Mukamel, J. Chem. Phys. **99**, 6562 (1993).
4. R. M. Bowman, M. Dantus, and A. H. Zewail, Chem. Phys. Lett. **161**, 297 (1989).
5. For a comprehensive review see, A. H. Zewail, J. Chem. Phys. **97**, 12427 (1993).
6. In their studies of ultrafast dynamics in iodine, A. H. Zewail, M. Gruebele and I. Sims observed qualitative changes in laser-induced fluorescence signals with changing pulse coherence. M. Gruebele, private communication.
7. R. Trebino and D. J. Kane, J. Opt. Soc. Am. A **10**, 1101 (1993).
8. B. Kohler, J. L. Krause, F. Raksi, C. Rose-Petruck, R. M. Whitnell, K. R. Wilson, V. V. Yakovlev, and Y.-J. Yan, J. Phys. Chem. **97**, 12602 (1993).

Coherent Control: Optics, Molecules, and Materials

K.A. Nelson

Department of Chemistry, Massachusetts Institute of Technology, Cambridge, MA 01239, USA

The development of versatile, automated femtosecond pulse-shaping capabilities has given experimental expression to rekindled interest in coherent optical control over the behavior of matter. In the chemical community, the focus of interest has been on coherent control over molecular behavior and reactivity, including "mode-selective chemistry". Less attention has been paid to the prospects for extensive optical control over *collective* behavior in materials. In this brief discussion some of the possibilities for coherent control over material behavior and structure, including "collective mode-selective chemistry", will be discussed.

We may begin by asking whether current femtosecond pulse-shaping capabilities are sufficient for a reasonable range of applications in control over molecular or material behavior. To this it seems that pulse shaping based on frequency filtering [1,2] offers an unqualified affirmative answer. It is now possible for a spectroscopic "user" to specify to a computer what are the desired amplitude and phase profiles of a shaped waveform, and for the waveform to be generated automatically [3]. Amplification of such waveforms to high light intensities poses no special problems [4]. Within some practical constraints the status of multiple-pulse NMR spectroscopy, in which a user may specify pulse sequences with little or no knowledge of the machinery involved in generating them, is approaching.

Application of programmable femtosecond pulse-shaping methods to control of electronic phase coherence in atoms and molecules has just begun [4,5]. Earlier efforts [6,7] involving beamsplitting and interferometric control over relative optical path lengths provided important demonstrations of concept, but the possibilities for more complex experiments were limited by the difficulty of generating more than two phase-related pulses through these methods.

Substantial theoretical work has been reported concerning molecular control with shaped ultrashort wavefors [8]. Examples range from prescriptions for controlled molecular reactivity using shaped far-IR or visible waveforms [9] to coherent wavepacket dynamics and spectroscopy of two-photon ("dark") states [10] to manipulation of molecular chirality [11]. In addition, considerable theoretical effort has been directed toward development of algorithms for optimization of the waveform to best elicit the desired molecular response [12].

Although much of the theory and speculation concerning optical control has emphasized molecules and their reactivity, the first spectroscopic applications of femtosecond pulse shaping were directed toward manipulation of collective vibrational motion in organic and inorganic crystal lattices [13,14]. Frequency filtering was used to generate sequences of more than ten evenly spaced femtosecond pulses whose timing matched the lattice vibrational oscillation periods. The selected vibrations were repetitively driven and amplified through impulsive stimulated Raman scattering (ISRS). In a similar vein, electron oscillations in coupled

Editors: P. F. Barbara · W. H. Knox · G. A. Mourou · A. H. Zewail

quantum wells, and Terahertz radiation emitted by them, were driven at the oscillation frequency by timed pulse sequences [15]. These multiple-pulse experiments indicate the broad spectroscopic applicability of the pulse-shaping method.

There have been only a few theoretical efforts directed at guiding optical control over material behavior. For example, prescriptions have been presented through which phase-related pulses can be used to generate large coherent vibrational amplitudes while minimizing electronic excitation which may cause damage to condensed materials [16,17]. Optical control over electronic coherences in coupled quantum wells [18] and in plasmas has been discussed recently [19]. Far more theoretical guidance is needed, especially considering our limited knowledge of the Hamiltonian for any of complex material. For example, theoretical treatment of nonlinear lattice dynamics and pulsed excitation of "local" lattice vibrations (analogous to local modes of anharmonic molecules) could guide experimental efforts toward control over crystalline chemical reactions and phase transitions.

To help motivate further experimental and theoretical work in this area, several possibilities for coherent control in condensed materials are discussed below: generation of large-amplitude coherent lattice vibrations in polar crystals; driving of structural phase transitions and domain switching in crystalline solids; and coherent far-IR generation and spectroscopy of polar liquids and crystals.

Phonon-polariton amplification and shaped vibrational waveforms. In noncentrosymmetric crystals, polar optic phonon modes which are both Raman and IR-active may be excited through ISRS [20]. These modes may couple to far-IR electromagnetic modes, forming mixed phonon-polariton modes which propagate in space. Just like a propagating acoustic wave, the phonon-polariton response induced by an initial pulse which is focussed to a linear or circular region of the crystal could be amplified by further pulses delayed in time and shifted to adjacent linear or smaller concentric circular regions of the crystal to match the phonon-polariton group velocity. In this manner the light intensity can be spread over a large sample region, reducing the risk of sample damage, while the vibrational energy remains localized in a much smaller volume to yield a large vibrational amplitude. Each of the light pulses, directed to its linear or circular region through an echelon or prism arrangement, could be a shaped pulse sequence so that the time-dependent form of the amplified vibrational response could be controlled. Thus shaped vibrational waveforms can be generated.

Structural phase transitions and domain switching. The possibility of inducing structural phase transitions in crystals through large-amplitude coherent phonons was discussed as early as 1985 [21]. Here we mention that the schemes discussed above could be particularly favorable for that possibility since larger amplitudes than usual should be reached. Vibrational waveform shaping also may play a useful role in driving long-lived collective structural change. Some ferroelectric crystals, whose propagating phonon-polariton modes have been studied through ISRS [14], offer especially attractive prospects. In the ferroelectric perovskites such as $PbTiO_3$, ferroelectric ordering occurs through ordered occupation of off-center sites by the Ti ion. It may be possible to induce transitions from a metastable (supercooled or superheated) state into a stable state, or from one domain orientation into another in which the Ti ions have been driven from their initial sites into other sites.

Coherent far-IR waveform generation and spectroscopy. As mentioned above, a shaped pulse sequence has been used to manipulate Terahertz (far-IR) emission [15,18]. In some cases, a single visible pulse yields a far-IR burst consisting of less than one cycle. Thus an arbitrary far-IR waveform could be crafted cycle by cycle. If very high far-IR intensity could be reached, chemical bond breaking might be induced [9]. Short of this, far-IR photon echo and other coherent spectroscopies could be conducted on orientational and intermolecular vibrational motions in polar liquids, glasses, and crystals. Far-IR waveforms could even be used for ESR spectroscopy of transient species or of very broad spin resonances.

In summary, coherent control over collective material behavior offers a rich field of possibilities whose exploration has just begun. Manipulation of electronic and vibrational coherences in a wide range of complex materials, in some cases yielding permanent structural change (i.e. "collective mode-selective chemistry"), is within reach with current femtosecond pulse-shaping capabilities.

This work was supported in part by NSF grant CHE-8901722.

1. A. M. Weiner, J. P. Heritage, E. M. Kirschner, J. Opt. Soc. Am. B 5, 1563 (1988)
2. A. M. Weiner, D. E. Leaird, J. S. Patel, J. R. Wullert, IEEE J. Quan. Elec. QE-28, 908 (1992)
3. M.M. Wefers and K.A. Nelson, Opt. Lett. 18, 2032; Science 262, 1381 (1993).
4. M.M. Wefers, H. Kawashima, and K.A. Nelson, elsewhere in this volume.
5. P. Bucksbaum, D. Schumacher, D. Pinkos, J. Hoogenraad, J. Krause, and K. Wilson, elsewhere in this volume.
6. J.T. Fourkas, W.L. Wilson, G. Wackerle, A.E. Frost, and M.D. Fayer, J. Opt. Soc. Am. B 6, 1905 (1989).
7. N.F. Scherer, A.J. Ruggiero, M. Du, and G.R. Fleming, J. Chem. Phys. 93, 856 (1990).
8. See W. S. Warren, H. Rabitz, and M. Dahleh, Science 259 1581 (1993).
9. S.H. Shi and H. Rabitz, J. Chem. Phys. 92, 2927 (1990).
10. H. Kawashima and K.A. Nelson, J. Chem. Phys. 100, 6160 (1994).
11. J.A. Cina and R.A. Harris, J. Chem. Phys. 100, 2531 (1994).
12. See e.g. R. Judson and H. Rabitz, Phys. Rev. Lett. 68, 1500 (1992).
13. A. M. Weiner, D. E. Leaird, G. P. Wiederrecht, and K. A. Nelson, Science 247, 1317 (1990)
14. See L. Dhar, J.A. Rogers, and K.A. Nelson, Chem. Rev. 94, 157 (1994).
15. I. Brener, P.C.M. Planken, M.C. Nuss, L. Pfeiffer, D.E. Leaird, and A.M. Weiner, Appl. Phys. Lett. 63, 2213 (1993).
16. J.A. Cina and T.J. Smith, J. Chem. Phys. 98, 9211 (1993).
17. R. Kosloff, A.D. Hammerich and D. Tannor, Phys. Rev. Lett. 69, 2172 (1992).
18. A.M. Weiner, elsewhere in this volume.
19. D. Umstadter, E. Easarey, and J. Kim, elsewhere in this volume.
20. T.P. Dougherty, G.P. Wiederrecht, and K.A. Nelson, J. Opt. Soc. Am. B 9, 2179 (1992).
21. Y.-X. Yan, E.B. Gamble, and K.A. Nelson, J. Chem. Phys. 83, 5391 (1985).

Part II

Elementary Dynamics: Chemistry, Biology, and Physics

Electronic-Relaxation Dynamics in Solution

P.J. Rossky, B.J. Schwartz, and Wen Shyan Sheu

Department of Chemistry and Biochemistry, The University of Texas at Austin, Austin, TX 78712-1167, USA

Abstract. Results obtained via quantum molecular dynamics simulation with electronic transitions are presented for the systems of a hydrated excited state halide ion and the hydrated electron. The ability to simultaneously describe the molecular processes and directly evaluate the ultrafast spectral observables is shown to provide a unique window on the underlying microscopic processes.

1. Introduction

Due to the significance of local electric fields and rapid solvent response for the ultrafast processes involved in chemical reactions, explicit molecular descriptions of the solvent and solute have an essential role in understanding the underlying microscopic dynamics. Recent progress in laser technology is enabling direct experimental observation of the fundamental events. At the same time, significant advances have been made in theoretical methods. The present contribution focusses on the direct connection now accessible between the two at the spectroscopic level, and the molecularly detailed pictures that are emerging. Here, we outline selected theoretical results for two experimentally investigated systems, namely aqueous halide ions and the hydrated electron.

The detailed computer simulations considered here employ a classical model for water [1] but explicitly treat the quantum electronic character of the solute, with a pseudopotential for the electron-solvent interaction [2]. From a simulated sequence of solvent configurations and the respective eigenstate manifolds, the vertical excitation energies and the corresponding dipole transition matrix elements provide direct access to steady state and transient spectra. Our algorithm for non-adiabatic dynamics [3] prescribes a semiclassical simulation of the electronic and solvent nuclear dynamics via a sequence of time steps composed of time propagation of the wavefunction, a statistical branching among adiabatic electronic eigenstates, and a time propagation of the nuclear coordinates subject to an appropriate total force.

2. Photoexcited Aqueous Halide Ions

We first consider studies of photoexcited aqueous halides [4-6]. These ions exhibit strong, broad UV absorption spectra in solution (charge-transfer-to-solvent (CTTS) states) that are absent in the gas phase, and solvated electrons can result from photoexcitation [7-9]. However, due to the limited tools available, the theoretical description of these states predated our studies by about thirty years [8,9]. Femtosecond spectroscopic data have also recently been reported [10,11]. Long et

Springer Series in Chemical Physics, Vol. 60 **Ultrafast Phenomena IX**
Editors: P. F. Barbara · W. H. Knox · G. A. Mourou · A. H. Zewail

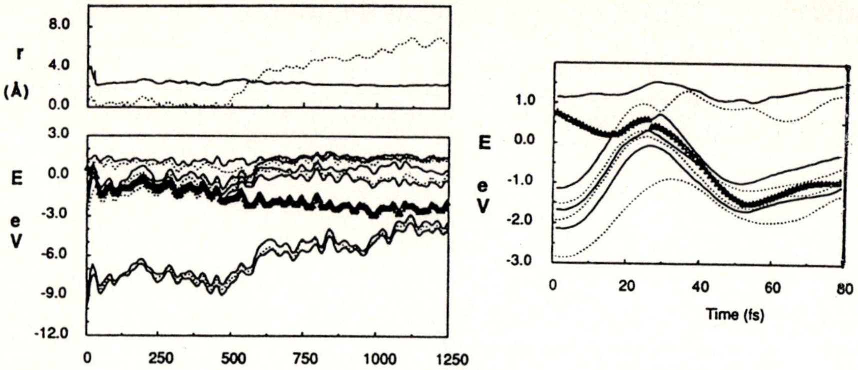

Figure 1. Dynamics of two photon excited aqueous halide. Top left: electronic radius of gyration (solid), separation of electronic center of mass from nucleus (dotted). Bottom left: adiabatic one electron eigenstates vs time (fs); occupied state indicated by bold symbols. Right: short time behavior of excited eigenstates.

al. [11] have carried out transient multiphoton absorption experiments on an aqueous iodide ion and concluded that the relaxation proceeded via an initial CTTS state.

Using a pseudopotential model for the halide, the properties of the solute ground and excited electronic states have been analyzed [4]. The CTTS spectra are comprised of several overlapping subbands: the principal excited state band consists of six states where a single electron is excited from the p-like valence orbital to an orbital of mixed s and d character. Above, this lies a band of three states with the single excitation of predominantly p to p type character. The former would be strongly one photon allowed and the latter would exhibit relatively strong two photon absorption.

Direct dynamical simulation of the halide following two photon excitation [5,6] (corresponding to experiment) shows that, although direct, nonadiabatic, electron transfer to a preexisting solvent cavity can occur, the predominant excited state relaxation channel observed exhibits direct electronic relaxation, as shown in Fig. 1 for a typical trajectory. Within 20 fs, the initially excited p-like excited electron crosses into the principal CTTS band and reaches the lowest excited state via a series of transitions after about 450 fs. Within another 50 fs, one observes the onset of a gradual electronic detachment, leading after about an additional 1 ps to a contact halogen atom-hydrated electron pair.

This sequence of events has an identifiable spectroscopic signature both in absorption and emission. As shown in Fig. 2, the (experimentally unresolved) red absorption spectra of the initially excited p-like electron (3 fs) decays very rapidly to that characteristic of the states of the principal CTTS band (103 fs) which lies somewhat to the blue of the hydrated electron and is narrower as well. With detachment, the spectrum red shifts and broadens so that the electron-atom pair is characterized by a spectrum that is rather similar to that of an isolated hydrated electron. The emission spectra, shown on the right in Fig. 2 (and not yet observed experimentally) are much more distinct in that the CTTS emission occurs in the UV and completely disappears with the separation into *ground* states of the halogen and electron.

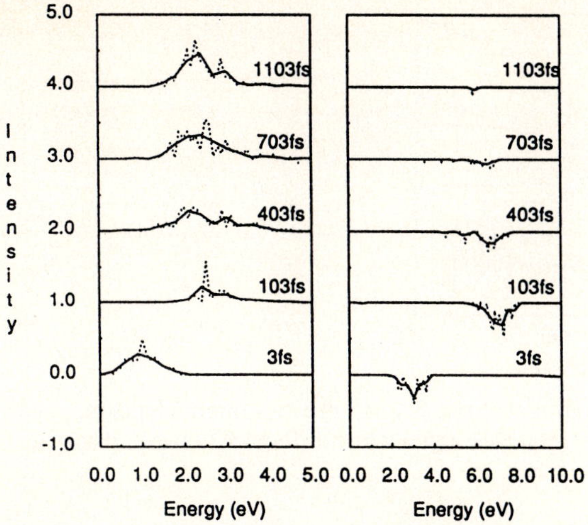

Figure 2. Simulated transient absorption (left) and emission (right) spectra for relaxation of two photon excited aqueous halide. (Calculated energies shown are expected to lie about 0.7 eV above observed values).

3. The Hydrated Electron

The hydrated electron is ubiquitous in aqueous photochemistry [12]. However, a deeper understanding of the process of energetic relaxation and electron localization has been the subject of much recent progress in the period since the original ultrafast experiments in water [12,13] demonstrated spectroscopically a stepwise solvation process. Simulation of the relaxation of an initially energetic excess electron in water [14] has led to results that are consistent with contributions from intermediate excited states. However, if the spectral properties were characterized simply by static excited and ground state spectra, a simple two-state kinetic model, as used to interpret these experiments [10,13,15], would explain the observed spectral dynamics. Whether this is indeed the case has been the subject of some controversy [13,15,16].

More recent hole-burning experiments have examined the transient spectroscopy of photoexcited equilibrium hydrated electrons [17]. These experiments employ a three pulse sequence: a UV pulse to generate electrons, a near infrared pump pulse to promote the electrons to their first excited state, and a third, probe, pulse. This experiment has a distinct advantage over earlier photoinjection studies [10,13,15] as the initial and intermediate states involved are clear. Although the results of these experiments [17] are generally consistent with the expected red shifted excited state spectrum [13-15], the experiments detect complicated spectral transients, in contrast to a simple two state model: the bluest probe wavelengths reveal a net transient bleach, at red probe wavelengths, a net transient absorption is seen, but, in between, just to the red of the pump laser, an initial bleach which overshoots into a longer lived transient absorption is observed.

Quantum dynamical simulations corresponding precisely to these experiments [18] yield calculated transient spectra in near quantitative agreement with experiment (see

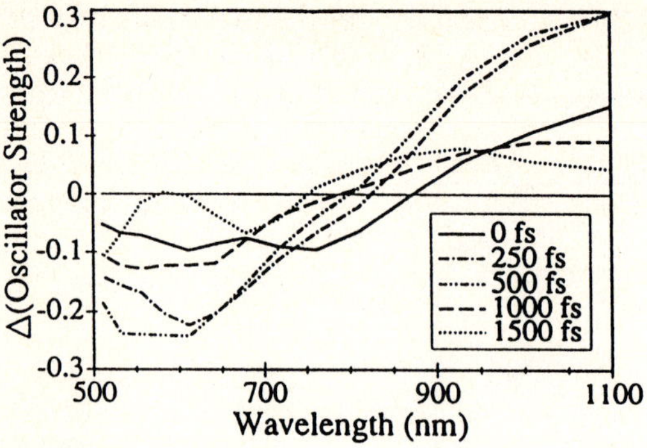

Figure 3. Calculated transient spectroscopy of the hydrated electron. The wavelength scale reflects an energetic shift equal to the difference between theory and experiment for ground state absorption.

Fig. 3). In particular, the computed spectra manifest each of the observed blue spectral bleach, red transient absorption, and biphasic temporal behavior in the intermediate spectral region. Further, the time scales for spectral evolution correspond very well to those observed, although, for times greater than about 1 ps, the simulated transients decay more rapidly than experiment. This comparison indicates that the simulation correctly captures the essential physics of the relaxation.

By evaluating the spectra that would result in the absence of various dynamical factors [18], we find that the complex spectral dynamics are the result of both excited state absorption and ground state bleach evolution, with the latter making a somewhat larger contribution. Slower spectral changes on the picosecond time scale are the result of electrons which still remain in the excited state. Ground state cooling after the transition is not found to play a critical role, in contrast to earlier proposals based on spectral modelling[17b], although we cannot rule out that cooling may contribute in the longer time regime where, as noted, the simulation and experiment decay at different rates.

4. Conclusions

As the work presented here demonstrates, via the use of a combination of measured ultrafast spectroscopy and simulated results, substantially greater microscopic insight into molecular dynamics can be obtained then is accessible by either technique alone. We anticipate that with further development of the theoretical techniques for molecular solutes, such close interaction of simulation and measurement will become an increasingly important tool in the area of ultrafast chemical dynamics.

Acknowledgements. Support for this work has been provided by grants from the NSF, the Office of Naval Research, and the R. A. Welch Foundation (F-0761) and is gratefully acknowledged, as is the computational support of the Center for High Performance Computing of The University of Texas System. B.J.S. gratefully acknowledges support by an NSF Postdoctoral Research Fellowship (CHE-9301479) and the allocation of computational resources from the San Diego Supercomputer Center.

References

1. K. Watanabe and M. L. Klein, Chem. Phys. **131**,157 (1989).
2. J. Schnitker and P.J. Rossky, J. Chem. Phys. **86**, 3462 (1987).
3. F. Webster, P. J. Rossky, and R. A. Friesner, Comput. Phys. Commun. **63**, 494 (1991).
4. W.-S. Sheu and P. J. Rossky, J. Am. Chem. Soc. **115**, 7729 (1993).
5. W.-S. Sheu and P. J. Rossky, Chem. Phys. Lett. **302**, 186 (1993).
6. W.-S. Sheu and P. J. Rossky, Chem. Phys. Lett. **213**, 233 (1993).
7. J. Jortner, M. Ottolenghi, and G. Stein, J. Phys. Chem. **68**, 247 (1964).
8. T. R. Griffiths and M. C. R. Symons, Trans. Faraday Soc. **56**, 1125 (1960).
9. J. Jortner, and A. Treinin, Trans. Faraday Soc. **56**, 1503 (1961).
10. Summaries of a number of recent measurements are provided in : H. Lu, F. H. Long, and K. B. Eisenthal, J. Opt. Soc. Am. B **7**, 1511 (1990); Y. Gauduel, S. Pommeret, A. Migus, N. Yamada, and A. Antonetti, *ibid.* **7**, 1528 (1990).
11. F. H. Long, H. Lu, X. Shi, and K. B. Eisenthal, Chem. Phys. Lett. **169**, 165 (1990); F. H. Long *et al.* (to be published).
12. see, for example, J. C. Mialocq, J. Chim. Phys. **85**, 31 (1988).
13. A. Migus, Y. Gauduel, J. L. Martin, and A. Antonetti, Phys. Rev. Lett. **58**, 1559 (1987).
14. F. J. Webster, J. Schnitker, M. S. Friedrichs, R. A. Friesner, and P. J. Rossky, Phys. Rev. Lett. **66**, 3172 (1991).
15. F. H. Long, H. Lu, and K. B. Eisenthal, Phys. Rev. Lett. **64**, 1469 (1990).
16. S. Pommeret, A. Antonetti, Y. Gauduel, J. Am. Chem. Soc. **113**, 9105 (1991); F.H. Long, H. Lu, H., X. Shi, K. B. Eisenthal, Chem. Phys. Lett. , **185**, 47 (1991).
17. (a) J. C. Alfano, P. K. Walhout, Y. Kimura, P. F. Barbara, J. Chem. Phys. **98,** 5996 (1993); (b) Y. Kimura, J. C. Alfano, P. K. Walhout, and P. F. Barbara, J. Phys. Chem. **98**, 3450 (1994)..
18. B. J. Schwartz and P. J. Rossky, J. Phys. Chem. (in press).

Vibrational Dynamics in Liquids and Glasses Probed with Infrared Photon Echoes Using a Free-Electron Laser

D. Zimdars, A. Tokmakoff, B. Sauter, R.S. Francis, A.S. Kwok, and M.D. Fayer

Department of Chemistry, Stanford University, Stanford, CA 94305, USA

Abstract. The first photon echo experiments conducted on molecular vibrations in liquids and glasses are described. The experiments were performed on the asymmetric CO vibration of tungsten hexacarbony in two glass forming liquids. Echo decays were measured from room temperature to 10 K. The experiments were performed with the Stanford superconducting-linac-pumped free electron laser.

1. Introduction

Molecular vibrations are involved in a vast number of physical, chemical, and biological processes in condensed phases. The coupling between individual molecular vibrations and the external mechanical degrees of freedom (heat bath) is responsible for fluctuations in the structure and energy levels of a molecule, as well as for the flow of energy into and out of molecules and thermally activated processes. Vibrational dynamics are intimately related to the reactivity of chemical systems. In spite of the importance of the coupling of molecular vibrations to a heat bath, relatively little is known about the temperature-dependent dynamics of molecular vibrations in liquids and glasses.

The fundamental quantity that describes the dynamics of a molecular vibration in liquids or other condensed matter systems is the dephasing time, T_2. In principle, T_2 can be obtained from analysis of vibrational line shapes. However, in liquids and glasses a vibrational absorption line can have a contribution from inhomogeneous broadening. Infrared vibrational photon echo experiments can be used to remove the inhomogenous contribution, and reveal the true homogeneous line shape. As will be shown below, even in room temperature liquids a significant contribution to a vibrational absorption line can come from inhomogeneous broadening.

In general, coupling of vibrations to a bath is substantially weaker than the coupling of electronic states. This is demonstrated by the fact that vibrational transition energies have much smaller percentage gas to "crystal" shifts than electronic transitions. The weaker coupling gives rise to longer dephasing times and little or no solvation dynamics. The weaker coupling makes it possible to study dephasing in liquids with pulses that have narrow enough bandwidths to coherently couple only a well defined pair of states to radiation field.

In a vibrational photon echo experiment, two ps IR pulses tuned to a molecular vibration transition frequency are crossed in a sample. The first pulse

Springer Series in Chemical Physics, Vol. 60 **Ultrafast Phenomena IX**
Editors: P. F. Barbara · W. H. Knox · G. A. Mourou · A. H. Zewail

induces a coherence that decays due to dephasing. A second pulse, delayed by time τ, begins a rephasing process that results in the emission of an echo pulse. The echo pulse is emitted in a unique direction determined by wave vector matching conditions. The intensity of the echo pulse is measured as a function of τ. The decay of the echo intensity measures the vibrational dephasing time. The decay of the photon echo is the Fourier transform of the homogeneous spectrum. For example, if the echo decay is an exponential, i. e.,

$$I(t) = I_0 \exp(-4\tau/T_2). \tag{1}$$

the Fourier transform is a Lorentzian homogeneous line having width, $1/\pi T_2$.

The photon echo experiments were performed with ~1.5 ps (~10 cm^{-1} bandwidth) IR pulses at 5.04 mm (1983 cm^{-1}) generated with the Stanford superconducting-linac-pumped free electron laser (FEL). The FEL emits a 2 ms macropulse at a 10 Hz repetition rate. Each macropulse consists of ~0.5 μJ micropulses at a repetition rate of 11.8 MHz. The micropulse repetition rate of 11.8 MHz was reduced to 50 kHz by a germanium AOM single pulse selector, yielding an effective experimental repetition rate of 1 kHz. The first pulse is ~8 nJ and the second pulse is ~50 nJ, both measured at the sample. The more intense pulse was chopped at 25 kHz by a second AOM. The signal and a reference were measured with two HgCdTe detectors sampled by two gated integrators. The reference detector was used for shot intensity windowing.

Measurements were made on the T_{1u} CO stretching mode of $W(CO)_6$ in 2-methylpentane (2-MP), and in 2-methyltetrahydrofuran (2-MTHF) [1], which are glass forming liquids. These experiments are the first vibrational photon echoes in liquids and glasses and the first to follow the evolution of the homogeneous vibrational line width from a room temperature liquid to a low temperature glass.

2. Results

Figure 1 displays photon echo data taken in the low temperature glass at 10 K. The inset shows a log plot of the data. The decay is exponential (except for a very fast initial transient that vanishes at lower power), indicating that the homogeneous line is a Lorentzian. At this temperature, the absorption line width is 10.5 cm^{-1} (310 GHz). In contrast $T_2 = 240$ ps, yielding a homogenous line width of 1.3 GHz. Thus the absorption line is massively inhomogeneously broadened. The photon echo decays are exponential for all the temperatures at which it is unnecessary to perform convolutions to analyze the data, i. e., to ~150 K. This is 60 K above the glass transition temperature (T_g), well into the liquid state. This demonstrates conclusively that the vibrational lines of this system in the liquid are inhomogeneously broadened.

Above 150 K, the echo decay time approaches the pulse width. However, by analyzing the data with convolutions and observing the data over a large dynamic range, the homogeneous T_2 can be readily obtained. Figure 2 shows a semi-log plot of the room temperature echo decay of $W(CO)_6$ in 2-MP. A fit of a

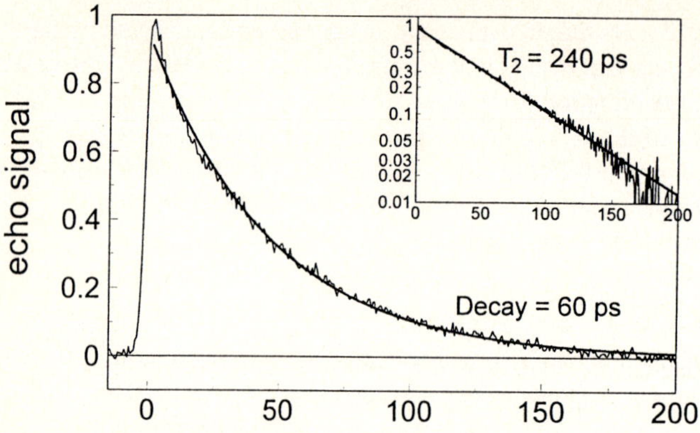

Figure 1 - Photon echo decay data for the CO asymmetric stretching mode (T_{1u}) of $W(CO)_6$ in 2-methylpentane glass at 10 K. The homogenous line width determined by the echo decay is 1.3 GHz (0.043 cm^{-1}) in contrast to the absorption line which is inhomogeneously broadened to 310 GHz (10.5 cm^{-1}).

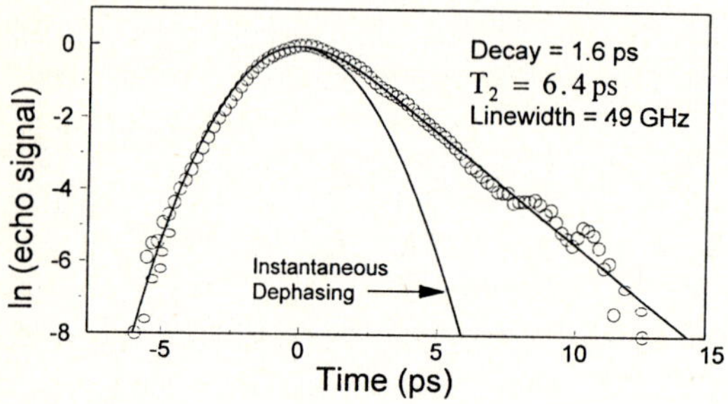

Figure 2 - Semilog plot of the room temperature echo decay, with a single exponential fit, and the instantaneous dephasing response function. The data decays in an exponential manner over >5 factors of e with a decay time of 1.6 ps, corresponding to a homogeneous linewidth of 49 GHz (1.6 cm^{-1}). The absorption linewidth at room temperature is 110 GHz (3.7 cm^{-1}).

single exponential with convolution is shown through the data, demonstrating that the echo decay is exponential over at least 5 factors of e. Also shown is the convolution with an instantaneous material response function. The data demonstrates that the echo formation in the room temperature liquid is in the $4/T_2$ limit, and that the absorption line is truly inhomogeneously broadened. The echo decays in 1.6 ps corresponding to a homogeneous linewidth of 49 GHz, in contrast to the 110 GHz absorption linewidth at room temperature.

The vibrational dephasing dynamics observed here are different than *electronic* photon echoes conducted in room temperature liquids on a 10 fs time

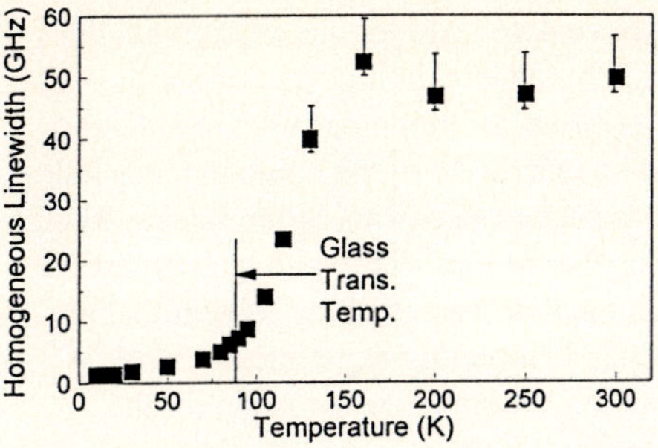

Figure 3 - Temperature dependent vibrational homogeneous linewidths, $1/\pi T_2$ obtained from the photon echo data. The glass transition temperature is 88 K.

scale. Because of the strong coupling between electronic states and nuclear motion, electronic dephasing occurs on an ultrashort time scale, corresponding to ballistic translational motions of the solvent molecules. The non-Markovian dynamics yield a short time Gaussian decay [4, 5]. Unlike electronic states, vibrational modes couple far more weakly to the solvent motions, so that vibrational dephasing occurs on a time scale slow compared to the ballistic motions of the bath. Since the data show the line is inhomogeneously broadened, some structural relaxation of the liquid is essentially static on the time scale of the vibrational dephasing.

The temperature dependent homogeneous line width for $W(CO)_6$ in 2-MP is shown in Figure 3. The linewidths, $1/\pi T_2$, were calculated from photon echo data using eq. 1. The line width increases gradually with temperature in the glass. The homogeneous line width is < 10 GHz thoughout the glassy region, compared to the inhomogeneous line width of ~ 300 GHz. The rate of increase of the homogeneous width becomes much steeper after the glass transition, $T_g = 88$ K. However, above 150 K, the temperature dependence of the line width becomes more gradual. As the temperature approaches 300 K, the homogeneous line slowly approaches the total line width. At room temperature, the total line width is 3.7 cm^{-1} while the homogeneous line width is 1.6 cm^{-1}. The gradual increase in the line width as room temperature is approached may be caused by motional narrowing.

Photon echo experiments were performed in the solvent, 2-MTHF [1] between 16 K and 140 K. The temperature dependence is qualitatively similar to that observed in the 2-MP solvent. Below the 2-MTHF glass transition ($T_g = 86$ K) the homogenous line width increases gradually with increasing temperature, becoming much steeper above T_g. However, the homogeneous widths are consistently wider in 2-MTHF. The wider homogeneous lines in 2-MTHF made it impossible to measure the echo decay above 140 K (~50 K above T_g) because of inadequate time resolution. The line is almost certainly still inhomogeneously broadened above this temperature since $W(CO)_6$ in 2-MTHF liquid has an absorption line width of 15 cm^{-1}, and the homogeneous width measured at 140 K is ~3 cm^{-1}. The narrower homogeneous widths observed in 2-MP at all temperatures made it possible to perform the experiments up to room temperature.

The photon echo data presented above demonstrates that vibrational lines of solute molecules can be inhomogeneously broadened in liquid solutions as well as glassy solutions. The data indicate the existence of inhomogenous broadening even at room temperature. The temperature dependence of the homogeneous line width reflects the wide variety of dynamic processes occurring in the liquid. The functional form of the increase in the homogeneous line width with temperature is due to the changes in the nature of the dynamical processes of the solvent that are coupled to the internal modes of the $W(CO)_6$. Continuing experimental work is aimed at elucidating the nature of the dynamic phenomena responsible for the vibrational dephasing.

Acknowledgment

The authors acknowledge Professor Alan Schwettman and Professor Todd Smith and their groups at the Stanford FEL whose continuous efforts made these experiments possible. This work was supported by the Medical Free Electron Laser Program (N00014-91-C-0170), the National Science Foundation (DMR90-22675), and the Office of Naval Research (N00014-92-J-1227-P02). B. S. thanks the Alexander von Humboldt Foundation for a Feodor Lynen Fellowship.

References

1. D. Zimdars, A. Tokmakoff, S. Chen, S. R. Greenfield and M. D. Fayer, Phys. Rev. Lett. **70**, 2718 (1993).
2. S. De Silvestri, A. M. Weiner, J. G. Fujimoto and E. P. Ippen, Chem. Phys. Lett. **112**, 195 (1984).
3. E. T. J. Nibbering, D. A. Wiersma and K. Duppen, Phys. Rev. Lett. **66**, 2464 (1991).
4. L. E. Fried and S. Mukamel, Adv. Chem. Phys. **84**, 435 (1993).

Ultrafast Optical and Vibrational Studies of Some Simple Chemical Reactions in Solutions

N. Pugliano, D. Raftery, E. Gooding, A. Szarka, and R.M. Hochstrasser

Department of Chemistry, The University of Pennsylvania,
Philadelphia, PA 19104, USA

Abstract:The dynamics of a few known gas phase reactions are different in solution as evidenced by rates, solvent sensitivity, state distributions, phase propagation and properties of the transition state.

1. Introduction

The influence of the solvent should not be underestimated in its ability to cause dissipation and to mold the reaction potentials and hence the dynamics and energy partitioning of condensed phase reactions. One approach towards characterizing the parameters of solution phase molecular dynamics is to contrast them with gas phase reactions that are to some extent understood. To accomplish this it is necessary to measure in solution some of the traditional parameters of gas phase dynamics, such as state-to-state rate constants, product energy distributions, isotope effects and other properties of the transition state. We have carried out such experiments for a number of reactions demonstrating in each case that ultrafast laser methods are adequate to access the primary processes, nascent distributions and transition states before the system reaches equilibrium. In this paper three examples are discussed: hydrogen abstraction reactions involv-ing the CN radical (1); the ground state unimolecular photodissociation of s-tetrazine(2); and the transition state dynamics involving the dissociation of I from HgI_2(3).

There are some essential differences between solution and gas phase reactions. The most obvious is the modification of potential surfaces, especially for low frequency motions and for polar or ionic molecules in polar or charged solvents. The forces exerted by solvent molecules cause energy relaxation involving both the internal and bath modes. This relaxation is sensitive to the total energy content of the reactant or product. Solution phase reactions may be in the spatial diffusion regime (4) where becoming energized is neither a rate controlling step nor a sufficient condition for reaction. In addition the internal and overall motions may be coupled in the solution so the reaction path may actually involve the motions of the solvent molecules (5). The solvent collisions may influence the dynamics near the transition state during the time that the internal forces which normally determine the reaction path are still operating. So the time spent in the transition state region and the reaction path may be modified by the solvent. The examples given below illustrate the occurrence of all these effects.

Springer Series in Chemical Physics, Vol. 60 **Ultrafast Phenomena IX**
Editors: P. F. Barbara · W. H. Knox · G. A. Mourou · A. H. Zewail

2. Hydrogen abstraction reactions with CN radicals

In order to study bimolecular reactions of A with BC in solution without interference from conventional diffusion control of the rate we generate the reactive species (A could be Cl (6), CN (1) or F) by a photodissociation (from Cl_2, ICN or XeF_2) carried out in the medium BC. In the present case A is CN, BC is $HCCl_3$ and the products of interest are HCN and CCl_3. The vibrational states of the HCN are interrogated by transient infrared spectroscopy and the rates, temperature effects and product vibrational state distributions are measured. Figure 1 shows the energetics of the experiment.

The CN is released with as much as 83% of the 33Kcal/mol excess as kinetic energy and a further 28Kcal/mol is released as reaction exoergicity. The detailed results of the experiments given elsewhere(1)include values for the reaction rates, isotope effects, vibrational energy distributions, temperature and viscosity dependence of rates and measurements of the vibrational relaxation time for DCN.

The results are very different from those found in the gas phase. The HCN formation constant of 177 ps^{-1} is much slower than in the gas phase where the reaction occurs on almost every collision (7). There is no reaction until the CN cools by nonreactive collisions. The nascent vibrational distribution (P_0:P_1:P_2= 0.81:0.19:<0.02) is totally different from the gas where large (inverted) populations are seen in v=1 and v=2 (7). The temperature dependence is close to that of the viscosity of chloroform indicating solvent control of the reaction. These results are consistent with a spatially diffusion controlled reaction where energization is not rate determining but reactive trajectories are controlled by the liquid state structure. The potential energy surface for the H-abstraction reaction is thereby modified to one with a more product-like barrier accounting for the slow rate and the more thermalized vibrational distribution than found in the gas phase. The achievement of a transition state configuration seems to involve considerable solvent reorganization.

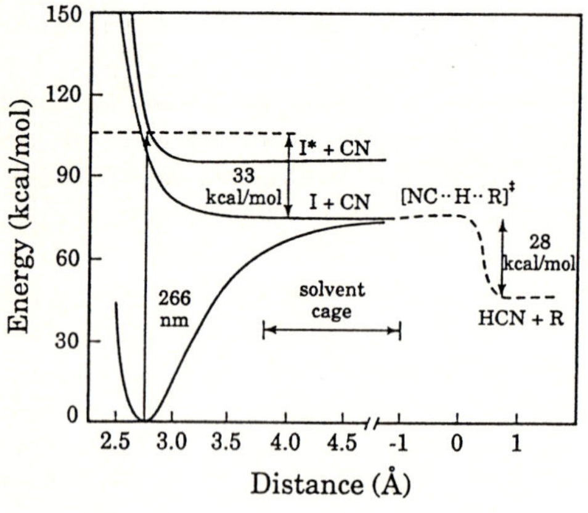

Fig. 1: PES for ICN and CN+ $HCCl_3$.

3. Unimolecular ground state dissociation of s-tetrazine

S-tetrazine dissociates into HCN and N_2 from high vibrational levels of its ground state reached after internal conversion from excited electronic states (8). We excite tetrazine by two-photon absorption (see Figure 2) and examine the generated HCN or DCN by transient infrared spectroscopy. The vibrational distribution we observe is inverted in the C-H stretching mode ($P_0<P_1$). The gas phase distribution on the other hand is close to being thermal. The basic cause of this disparity probably arises because the unimolecular rate is still slower than the solvent induced energy relaxation. Why this leads to a greatly modified product state distribution is not known yet but one possibility is that the relaxation leads to a new pathway that is not present in the gas phase such as a stepwise dissociation through a solvent stabilized intermediate structure.

The observation of an inverted distribution gave the opportunity to obtain measurements of the relaxation times of a number of vibrational states of DCN. The results shown in Figure 3 demonstrate the non-linear dependence of the relaxation rate on the vibrational quantum number. Two possibilities could account for this result. One, which seems unlikely but cannot yet be eliminated, is that the coupling to the bath is quadratic in the C-D stretch coordinate. The other is that the solvent induces internal energy transfer between the stretches and the bending modes (9) . from these results, the relaxation occurring high in the C-H stretch manifoldis expected to be ultrafast, which will make it challenging to obtain truly nascent vibrational distributions in such exoergic reactions.

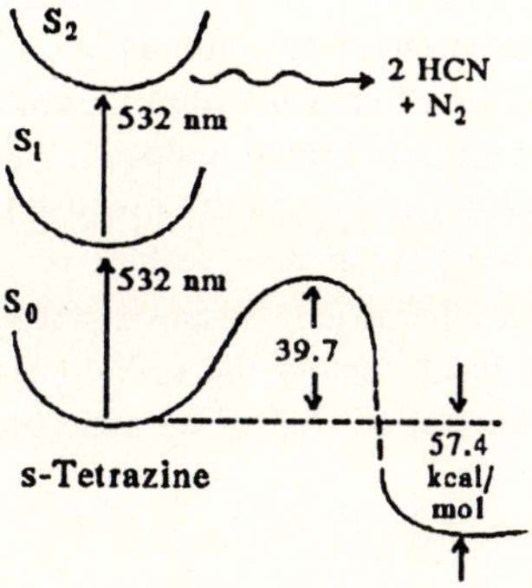

Fig.2: PES for s-tetrazine.

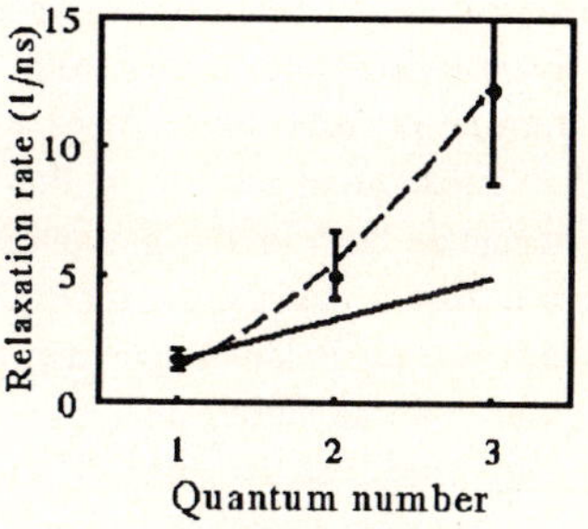

Fig.3: Vibrational relaxation of HCN versus quantum number.

4. Impulsive dissociation of HgI_2; bimolecular transition states

The photodissociation of HgI_2 presents the opportunity to examine the effects of the solvation dynamics near the transition state in solutions, since the electronically excited state that is reached by absorption of UV photons is a saddle

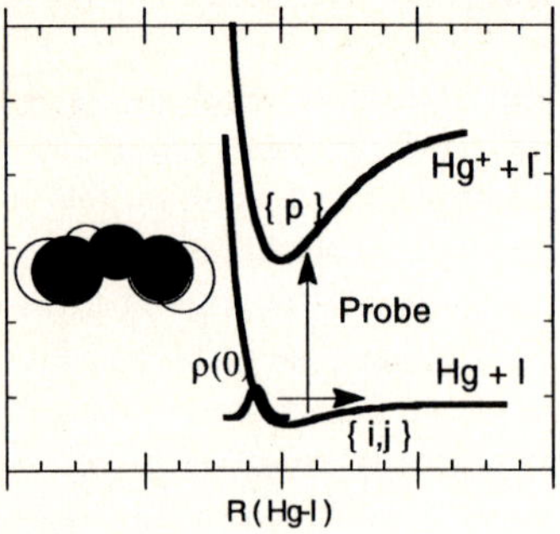

Fig. 4: PES for HgI and to scale sketch of HgI_2 photodissociation.

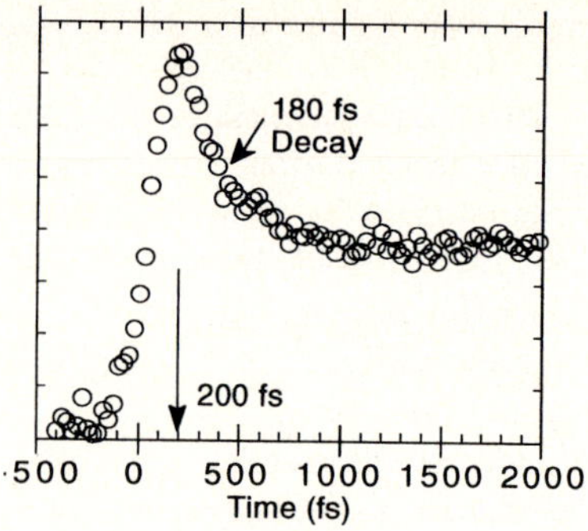

Fig. 5: HgI signal from HgI_2 pumped at 310 nm probed 600 nm.

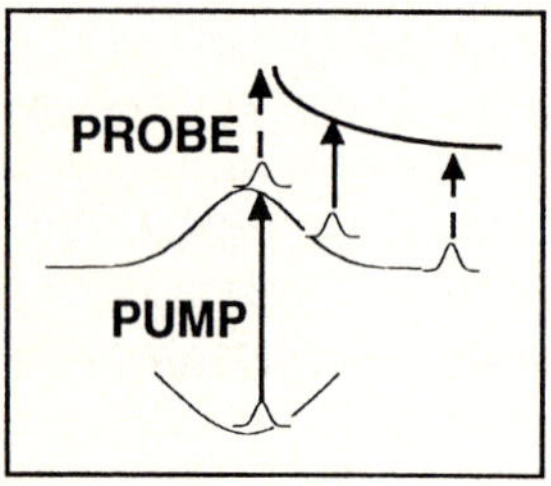

Fig. 6: HgI_2 transition state region.

point region of the surface. After excitation there is a strong force tending to compress one of the HgI bonds. As a result, the HgI product is generated by an "impulsive compression." This leads to product HgI with a relatively narrow distribution of position eigenstates which is a wavepacket in the energy eigenstate basis. The subsequent motion of this wavepacket on the ground state surface of HgI is observed both in the gas phase (10) and in solutions (3). The nascent states therefore define a density matrix $\rho(0)$ which evolves further under the influence of the solvent forces. In our experiments the ground state well is probed by absorption to an upper state as shown in Figure 4.

We find that the absorption spectra and hence the PES of HgI are extremely sensitive to solvent. There is not much frictional torque on the reaction as can be inferred from the sketch on Figure 4 in which an I atom has moved away one bond length without moving the center of mass. The signal at 600nm in Figure 5 shows the probe intercepting the HgI fragment while it is still under the influence of the recoil forces of the reaction. Here we believe we are seeing the effect of solvent collisions on the motion near the top of a barrier. The interpretation of the experiment is sketched in Figure 6.

The 600nm probe can only be absorbed significantly when the dissociation is incomplete. The 180fs decay of the signal indicates that the solvent molecules have slowed down the release of HgI from the transition state region. In this picture the delay of the peak signal to 200fs occurs because the transition state does not absorb this frequency until some HgI bond compression has occurred. Thus the solvent must drastically influence the transition state dynamics by cooling the fragments even while they are accelerating out of the reactive region.

5. Summary

These examples show clearly that there are profound differences between gas and solution phase reactions that are not accounted for by simple changes in the energetics. This is because the solvent can be involved in the reactive motion and thereby determine the location of the "point of no return" on the reaction paths.

HgI_2 photolysis presents a prototype example of the generation of a coherent state by a chemical reaction. We believe that the observation of such coherences in reactions will be quite common although in general one must distinguish between the "trivial" coherence introduced by the field of the light pulse and the impulsive generation of new structures by the internal forces acting on the atoms. In the latter case, of which HgI_2 is an example, the study of the resultant coherent transients leads to information about the transition states and the participation of solvent in the reaction.

6. Acknowledgements This research was supported by grants from NSF and NIH.

7. References

(1) D. Raftery, E.Gooding, A.Romanowsky and R.M.Hochstrasser, "Vibrational Product State Dynamics in Solution Phase Bimolecular Reactions: Transient IR Study of CN Radical Reactions," J. Chem. Phys. (submitted); see also paper in this volume.

(2) D. Raftery, E.Gooding and R.M.Hochstraser, "Vibrationally Excited Product State Distributions and Relaxation in the Unimolecular Dissociation of s-Tetrazine,"in preparation.

(3) N. Pugliano, D.Palit, A.Szarka and R.M.Hochstrasser, J. Chem. Phys. 99,7273(1993); see also paper in this volume.

(4) P. Hanggi, P.Talkner and N.Borkevec Rev. Mod. Phys. 62, 251 (1990)

(5) See for example: D.Raftery, R.J.Sension and R.M. Hochstrasser, "Chemical Aspects of Solution Phase Reaction Dynamics," in The Barrier Crossing Problem eds. G.R. Fleming and P.Hanggi (World Scientific),1993.

(6) D.Raftery, M.Iannone,C.M.Phillips and R.M. Hochstrasser, Chem.Phys.Lett. 201,513 (1993).

(7) V.R.Morris and W.M.Jackson, private communication.

(8) V.L.Windisch, A.B.Smith and R.M.Hochstrasser J. Phys. Chem. 92, 5366 (1988) and references therein.

(9) J.Owrutsky, D. Raftery and R.M. Hochstrasser "Vibrational Relaxation Dynamics in Liquids" Ann. Rev. Phys. Chem. (1994).

(10) M. Greubele, G. Roberts, and A. H. Zewail, Philos. Trans. R. Soc. London, Ser. A 332, 223 (1990).

Transient Resonance Impulsive Stimulated Raman-Scattering Study of Nascent Diiodide Ions: Experiment and Simulation

U. Banin, R. Kosloff, and S. Ruhman

Department of Physical Chemistry and the Fritz Haber and Farkas Research Centers, The Hebrew University, Jerusalem 91904, Israel

A novel coherent spectroscopy is applied to measure ultrafast vibrational relaxation. Its performance is quantitatively evaluated through computer simulations. Transient Resonant Impulsive Stimulated Raman Scattering (TRISRS), is used to record vibrational relaxation in rapidly evolving nascent I_2^-, following photodissociation of I_3^- in solution. The TRISRS experiment involves a sequence of three pulses. A primary UV photolyzing pulse at 308 nm initiates triiodide dissociation. At a certain delay with respect to the photolysis a secondary intense 'push' pulse at 616 nm sets in motion ground state coherent vibration. Its decay is followed via periodic transmission modulations of a continuously delayed tunable probing pulse[1]. In essence these modulations record a Raman Free Induction Decay (FID) whose duration defines the minimum time that a dynamic variable must be followed in order to reconstruct its power spectrum.

Qualitative interpretation of these trends is straight forward. At early push delays the nascent I_2^- is vibrationally excited, and the coherence built up in the high lying levels, oscillates at a reduced frequency due to the substantial anharmonicity of the potential. At these high levels of excitation efficient mechanisms of vibrational dephasing are manifest. As the delay increases vibrational energy relaxes rapidly, leading to an increase in the modulation frequency for coherences near the bottom of the potential, and the efficiency of vibronic dephasing is reduced. No further evolution is observed after a push delay of 4 psec, corroborating the conclusion that I_2^- relaxation is very rapid, derived earlier from transient absorption results[2], and from I_2^- dissociation-recombination experiments[3].

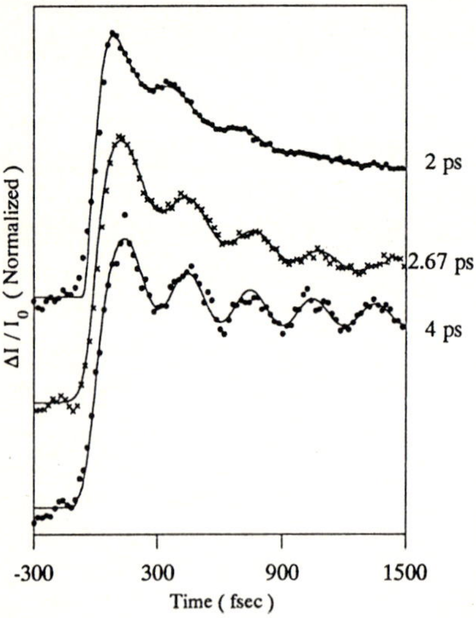

FIGURE 1. Experimental TRISRS bleaches at push delays of 2, 2.67 and 4 psec. Notice that the zero of time is defined as the center of the push pulse interaction. A marked evolution, both in the frequency and dephasing rate of the observed modulations is manifest. The modulations are fit to an exponentially decaying sinusoid. Modulation frequencies are 102, 105 and 112 cm^{-1}, and the dephasing times .5, .9 and 1.2 psec at the three delays respectively. In addition, 112 cm^{-1} is the frequency of TRISRS signals recorded at all subsequent pushing delays. Changing the probing pulse wavelength results in a phase shift of observed spectral modulations. This has been experimentally observed robing at 840 and 580 nm following the same push interaction.

Springer Series in Chemical Physics, Vol. 60 **Ultrafast Phenomena IX**
Editors: P. F. Barbara · W. H. Knox · G. A. Mourou · A. H. Zewail

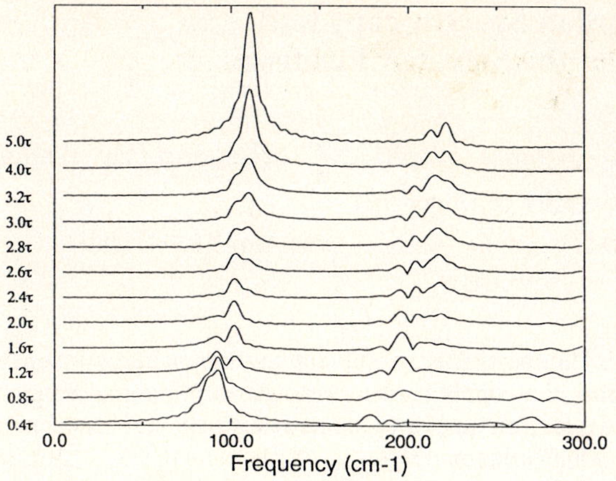

FIGURE 2. Fourier transforms of the simulated TRISRS signals as a function of the delay in reduced units. The fundamental peak in the transforms evolves nearly linearly up to a delay of 3τ, and then saturates at the asymptotic frequency of 112 cm^{-1}, despite the fact that substantial spectral relaxation is taking place. Thus TRISRS compliments transient absorption and is best adapted to following vibrational relaxation of highly excited molecules.

For more quantitative interpretation transient vibrational distributions are generated using a Master equation model[3]. Integrating the Master equation begins with 40 vibrational quanta, and energy relaxes exponentially with a rate constant of 0.43/τ. A grid based wave packet propagation technique is used, in conjunction with a model for I_2^- potentials, in order to represent the light matter interactions[4]. Push-probe transient transmission scans are then simulated for the various incoherent vibrational distributions obtained from the rate equations. First, push-probe cycles starting with a single vibronic level v are simulated. Then spectral modulations from the different v's are combined, according to their probability in a certain transient distribution, in order to produce a simulated TRISRS signal. Homogeneous mechanisms of dephasing are disregarded.

The simulations demonstrate that the push interaction at 620 nm is capable of instilling vibrational coherence in all chemically excited vibronic levels. In addition, the frequency of simulated spectral modulations after pushing the individual eigenlevels closely follows the vibronic Bohr frequency for the initially populated v's. Thus, as previously assumed, this aspect earmarks the levels populated at the instant of the push interaction. The simulations do however exhibit a nonuniform sensitivity to initial populations in different vibronic levels, with preference for detecting low lying levels.

[1] U. Banin and S. Ruhman, *J. Chem. Phys.*, **99**, 9318 (1993).
[2] U. Banin, and S. Ruhman, *J. Chem. Phys.*, **98**, 4391 (1993).
[3] J. C.Alfano, Y. Kimura, and P.F.Barbara, *Chem. Phys.*, **175**, 147 (1993)
[4] B.Hartke, R.Kosloff & S.Ruhman, *Chem. Phys. Lett.* **158**, 238 (1988).

Direct Transient-Absorption Spectroscopy and Absorption Anisotropy of the Solvated Electron in Water and Alcohols

P.F. Barbara, J.C. Alfano, Y. Kimura, P.J. Reid, C. Silva, and P.K. Walhout

Department of Chemistry, University of Minnesota,
Minneapolis, MN 55455, USA

Abstract. The participation of slow, diffusive solvent motion in the spectral dynamics of the solvated electron in alcohols is observed. Absorption anisotropy measurements on the solvated electron in water demonstrate the presence of anisotropic solvent fluctuations which relax in ~3 ps.

In this paper, we report the first direct femtosecond infrared-pump/variable probe measurements on the solvated electron in normal alcohols and on the absorption anisotropy of the solvated electron in water. The experiment are performed with a three pulse sequence.[1] First, an ultraviolet (UV) synthesis pulse (390 nm, 25 μJ, 2 kHz) generates electrons by two-photon detachment from I^-. After ~4 ns, the ground-state, equilibrated electrons were promoted to the electronically excited p-states by the pump pulse (780 nm, 4 μJ). The resulting spectral dynamics were observed by a variable wavelength probe pulse (450-1100 nm).

Transient absorption spectra of the solvated electron in ethanol are presented in Figure 1. Transients corresponding to probe wavelengths on the blue edge of the ground-state absorption (600-740 nm) show a fast, instrument-limited reduction in optical density (bleach) followed by a biphasic recovery with time-constants of ~0.5 ps and 5-10 ps. Transients probing on the red edge of the ground-state absorption (950 nm) show a transient increase in optical density which decays on the 20 ps timescale. Finally, transients probing at intermediate wavelengths (860 nm) show complicated non-monotonic behavior where the initial bleach recovery "overshoots" to yield a transient increase in absorption which decays in ~10 ps. At all wavelengths, a permanent bleach caused by reaction of the excited-state electrons with the solvent is observed.

Experiments have been performed on a series of normal alcohols from methanol to butanol. All solvents display a fast, ~0.5 ps component of the bleach recovery which is assigned to electronic relaxation.[2] In contrast, the slower component of the bleach recovery becomes longer in slower relaxing solvents, roughly following the solvent longitudinal relaxation time.[3] Therefore, the slow component of the bleach recovery corresponds to diffusional solvent motion. This observation demonstrates that although the excited-state lifetime of the electron is much shorter than the timescale for diffusive solvation, internal conversion from the excited state results in energy deposition into these slow relaxing solvent modes.

The transient absorption anisotropy of the solvated electron in water is presented in Figure 2.[4] These data clearly demonstrate the anisotropic response of the solvated electron. Since the solvent cavity surrounding the electron defines the energetics of the three orthogonal s-p electronic transitions, the observation of an anisotropy demonstrates the geometry of the solvent cavity is asymmetric (assuming

Springer Series in Chemical Physics, Vol. 60 **Ultrafast Phenomena IX**
Editors: P. F. Barbara · W. H. Knox · G. A. Mourou · A. H. Zewail

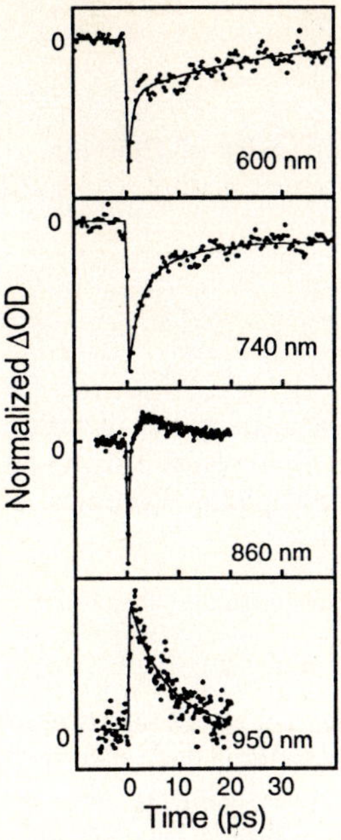

Figure 1. Transient absorption spectra of the solvated electron in ethanol. The probe wavelength is given in the figure.

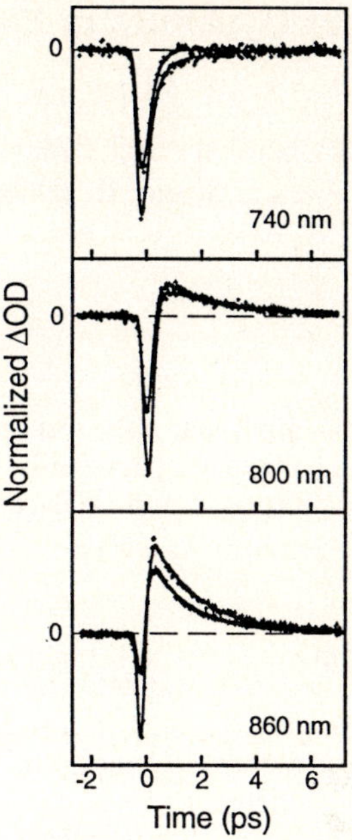

Figure 2. Diagonal and off-diagonal spectra of the solvated electron in water. The diagonal spectra is that with the largest decrease in optical density at all wavelengths.

that electronic dephasing is complete).[5] Also, a negative anisotropy is observed at 800 nm demonstrating the participation of transitons orthogonal to that defined by the pump. The anisotropy decays in ~3 ps indicating that anisotropic fluctuations of the solvent relax on a slow timescale relative to inertial and diffusional water solvation dynamics.[3] This result is consistent with recent molecular dynamics studies on aqueous electron solvation dynamics although the magnitude of the early-time anisotropy (~0.05 at 860 nm) is considerably smaller than theoretical predictions.[6] The corespondence in timescale between experiment and theory demonstrates that slow, anisotropic solvent fluctuations are operative in the solvation dynamics of the electron in water. This work was supported by a grant from the Basic Energy Sciences program of the DOE.

References.

1. Y. Kimura, J. C. Alfano, P. K. Walhout, and P. F. Barbara, J. Phys. Chem. 98, 3450 (1994).
2. J. C. Alfano, P. K. Walhout, Y. Kimura, and P. F. Barbara, Phys. Rev. Lett., submitted.
3. M. Maroncelli, J. Molec. Liquids 37, 1 (1993).
4. P. J. Reid, C. Silva, P. K. Walhout, and P. F. Barbara, Chem. Phys. Lett., submitted.
5. K. Wynne and R. M. Hochstrasser, Chem. Phys. 171, 179 (1993).
6. B. J. Schwartz and P. J. Rossky, Phys. Rev. Lett. (1994) in press.

Femtosecond Dynamics in Polar Liquids

T. Joo, M. Cho, J. Yu, Y. Jia, and G.R. Fleming

Department of Chemistry and James Franck Institute, University of Chicago, 5735 S. Ellis Ave., Chicago, IL 60637, USA

Abstract. New nonlinear spectroscopic methods for characterizing the dynamics of solvents and solutes are described. The nature of signals observed at 3rd and 5th order nonlinear response and their relationships to the liquid phase dynamics are discussed.

1. Introduction

Electronic spectra in liquids are often broad and featureless, a result of complex solvent-solute dynamics. In the simplest view (optical Bloch limit) each electronic transition is considered to be a convolution of homogeneous (infinitely fast) and inhomogeneous (static, usually Gaussian) lineshapes. This simple picture is not supported by recent simulations[1-4] and experiments such as photon echo[5-7], optical Kerr[8], and fluorescence Stokes shift[9] measurements. In highly polar liquids such as water and acetonitrile the initial response of the solvent is mostly inertial (Gaussian like) rather than diffusive (exponential) and occurs on a 20-100 fs timescale. From two pulse photon echo studies, Nibbering et al.[6] argued that the dynamics in polar liquids is well described by a single stochastic process captured in its intermediate modulation regime. In other studies, a very fast response as well as a relatively slow response has been proposed[5,10].

In the conventional 2 and 3 pulse photon echo measurements, the time integrated echo signal rather than the actual echo profile is measured. Even though the actual echo is very sensitive to the detailed nature of the dynamics, the time integrated echo signal is not very sensitive[11]. In addition when the inhomogeneous broadening is not large compared to the homogeneous broadening the photon echo technique can produce ambiguous results. Recently, a new technique based on the fifth order nonlinear response has been proposed[12]. In the Markovian limit this technique is completely insensitive to the ratio of homogeneous and inhomogeneous broadening. Here we report an electronic dephasing study on a dye (HITCI) in polar liquids using a nonlinear technique based on this fifth order three pulse scattering[12] as well as conventional 2 and 3 pulse photon echoes. The combination of all three types of measurements provides a detailed picture of the system-bath interaction.

2. Brief account of various 'echo' methods

The pulse sequences with their relative time orderings for the two-pulse photon echo (2PE), three-pulse stimulated photon echo (3PE), and the fifth order three-pulse photon echo (F3PE), called FOTS in ref. 12, are shown in Fig. 1. In the 2PE,

Springer Series in Chemical Physics, Vol. 60 **Ultrafast Phenomena IX**
Editors: P. F. Barbara · W. H. Knox · G. A. Mourou · A. H. Zewail

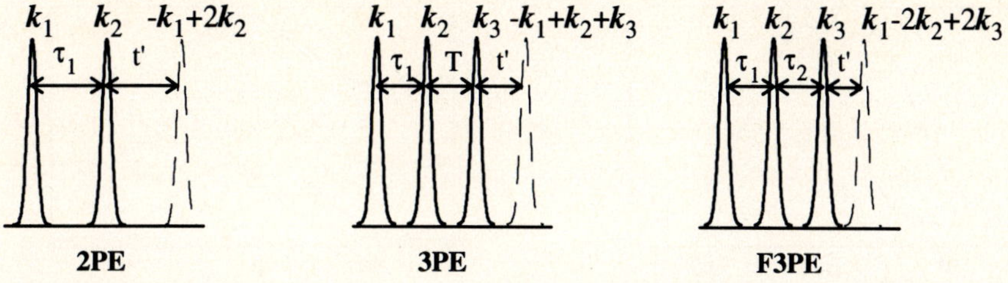

Figure 1. Pulse sequences for 2PE, 3PE, and F3PE. Time increases from left to right. $\boldsymbol{k}_i$ denotes the wave vector of the i-th pulse. Dashed curves represent the 'echo' signal.

the system is in the electronic coherence state during τ_1 and it undergoes rephasing during t' provided the phase memory is not lost. The 3PE is similar to 2PE except that there is a dwell time, T, in a population state. During T spectral cross relaxation (spectral diffusion) can occur. In the 2PE and 3PE measurements the signal integrated over t' is measured as a function of τ_1. In the F3PE the system is placed in an electronic coherence state by the first pulse and evolves during τ_1. The second pulse involves two interactions which leave the system in an electronic coherence which is the complex conjugate of the first. During τ_2 the system experiences rephasing. The third pulse also involves two interactions leaving the system back into a coherence state with the same phase sign as in the first period. Thus with large inhomogeneous broadening the system will fully rephase when $\tau_2 = \tau_1+t'$. Since both τ_1 and τ_2 are controllable, this technique provides the possibility of controlling the rephasing process in contrast to the conventional 2PE and 3PE where the rephasing cannot be controlled. In the F3PE when the signal integrated over t' is measured with τ_1 set equal to τ_2 ('diagonal'), the inhomogeneous contribution can be removed completely in the Markov limit[12]. The 'off-diagonal' case ($\tau_1 \neq \tau_2$) also provides useful information. Note that it is not possible for the full rephasing to occur when $\tau_1 > \tau_2$. When τ_2 is scanned with τ_1 fixed there will be no signal until τ_2 is equal to τ_1. That is, the signal will be delayed as much as τ_1 when full rephasing is possible. Thus the 'off-diagonal' measurement allows one to look at the rephasing more directly.

3. Experimental Results

Fig. 2 and 3 show 2PE and 3PE measurements on HITCI in Glycerol/water (70/30 by weight). The data for the same molecule but in pure Ethylene glycol are qualitatively similar. 20 fs transform-limited pulses from a home-built Kerr Lens Mode-locked Ti:Sapphire laser[13] are used. The pulse autocorrelation shows that the pulse is very close to Gaussian in shape. In all the measurements, the accuracy of zero time is better than 1 fs. This is achieved by simultaneous measurements of the two signals in two different phase matching directions ($2\boldsymbol{k}_1-\boldsymbol{k}_2$ and $-\boldsymbol{k}_1+2\boldsymbol{k}_2$ in the 2PE and $\boldsymbol{k}_1-\boldsymbol{k}_2+\boldsymbol{k}_3$ and $-\boldsymbol{k}_1+\boldsymbol{k}_2+\boldsymbol{k}_3$ in the 3PE), which are mirror images to each other. The 2PE signal clearly shows a decay yielding a decay time of ~20 fs after deconvolution, as well as a prominent shift to the positive delay, which is in part due to the finite pulse duration. The 2PE signal, however, is not very sensitive to

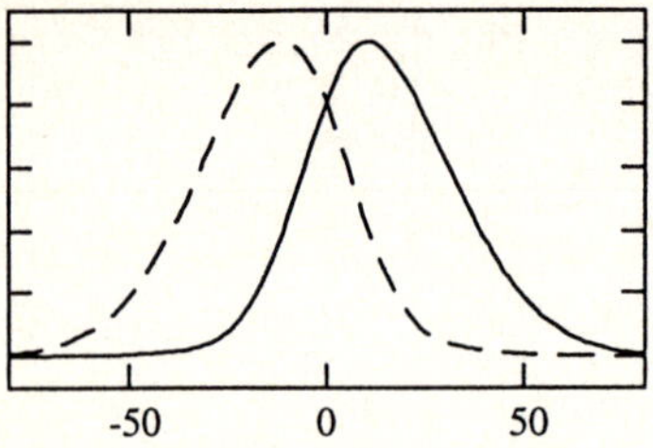

Figure 2. 2PE signal of HITCI in glycerol/water. Solid and dashed lines represent the signals in $2\boldsymbol{k}_1$-$\boldsymbol{k}_2$ and -$\boldsymbol{k}_1$+$2\boldsymbol{k}_2$ wave vector directions, respectively.

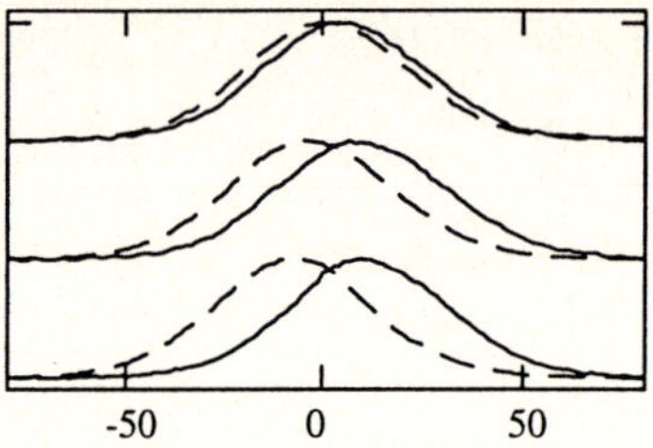

Figure 3. 3PE signal in $\boldsymbol{k}_1$-$\boldsymbol{k}_2$+$\boldsymbol{k}_3$ and and -$\boldsymbol{k}_1$+$\boldsymbol{k}_2$+$\boldsymbol{k}_3$ wave vectors. From top to bottom T = 200 ps, 300 fs, and 20 fs.

the different functional forms arising from different dynamic models[11]. Before drawing conclusions we need to consider the other measurements.

Fig. 3 shows 3PE signals for several different values of T. The signals do not peak at zero delay indicating that there is a rephasing process involved in the signal though the individual signals are quite symmetric. Moreover, the 'shift', i.e. the delay of the peak from zero, decreases as T increases. Fig. 4 shows the shift as a function of T. A minimum of three exponentials is needed to give a good fit. For glycerol/water we found 46 fs, 6.7 ps, and 230 ps components with their relative amplitudes of 1, 0.32, and 1, respectively. Simulations show that the time scale of the 'shift' decrease represents roughly the time scale of the process involved[5]. As evident in the 3PE data, the 46 fs process can cause rephasing. This capability of rephasing in turn suggests that the physical process is an inertial (Gaussian) process as opposed to a diffusive non-recoverable process. Within a 100 fs, however, the relaxation becomes diffusive and can no longer cause rephasing. This will be shown clearly by the F3PE measurements described below. The two much slower (ps) processes must be in their slow modulation limit and act as inhomogeneous broadening. These processes also change their nature from inhomogeneous to homogeneous at long enough times.

The dynamics discussed in the previous paragraph can be more clearly seen in the F3PE. Fig. 5 and Fig. 6 shows F3PE measurements. The 'Diagonal' scan (Fig.

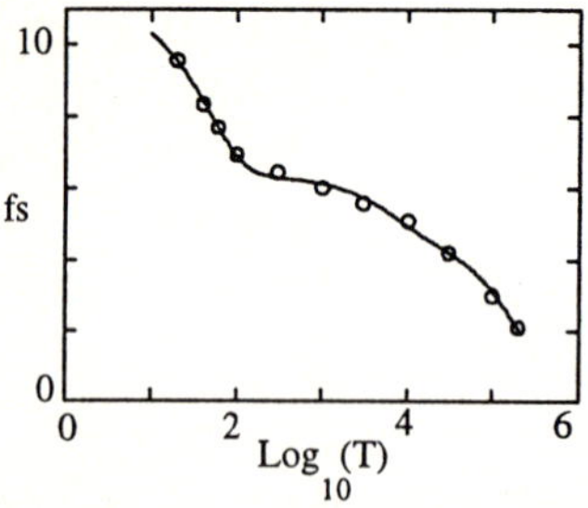

Figure 4. The peak 'Shift' vs T(fs) in 3PE. Solid line represents three exponential fit.

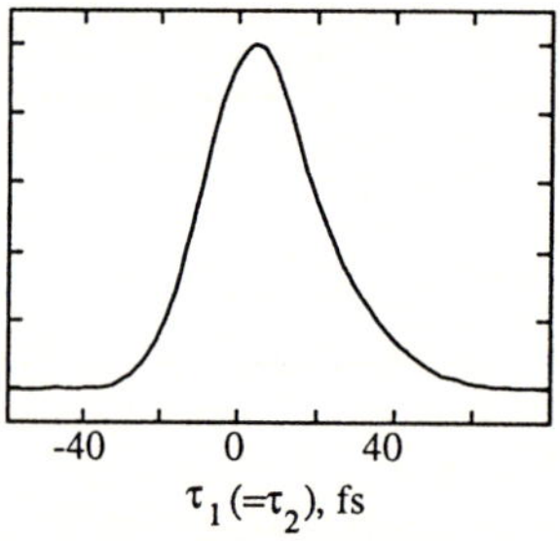

Figure 5. F3PE 'diagonal' signal of HITCI in glycerol/water.

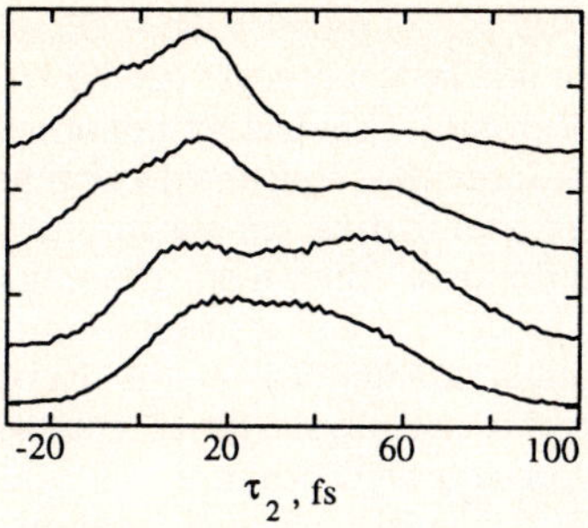

Figure 6. F3PE 'off-diagonal' of HITCI. From top to bottom, τ_1 = 80, 60, 50, and 20 fs.

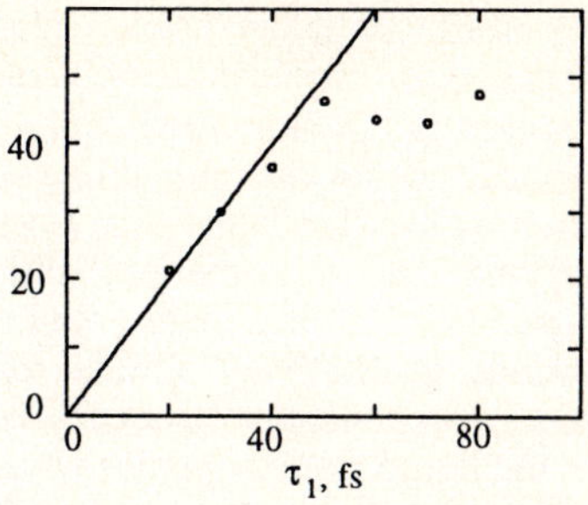

Figure 7. F3PE 'off-diagonal' delay (fs) vs. τ_1. Solid line represents Bloch limit.

5) shows a prominent asymmetry. This decay represents pure homogeneous dephasing, free of any inhomogeneity, in the Markov limit. The 3PE data show, however, that the dynamics is hardly in the Markovian limit and that several (at least three) processes must be included. Nonetheless, the prominent asymmetry of the signal indicates the ability of the technique to remove contributions from slow dynamics. When the signal is examined more closely, it actually consists of at least two components: one peaked around ~5 fs; and a decaying component which appears as a shoulder. The signal peaked around 5 fs is results from the 'coherent artifact' which arises from the Feynman diagrams[12] that are not properly time-ordered. The decaying part can be identified as an F3PE component. It is obvious that to get detailed information on the dynamics a full numerical simulation considering all Feynman diagrams with explicit treatment of finite pulse duration is necessary. A numerical calculation of the 5-th order signal is underway.

The 'off-diagonal' scan (Fig. 6) also provides valuable information on the dynamics. As discussed in section 2, when τ_2 is scanned while τ_1 is fixed, the actual echo profile can be seen. For example, for τ_1 = 50 fs, the data shows a peak near 50 fs. It has an additional peak, the 'coherent artifact', near zero delay where $\boldsymbol{k}_2$ and $\boldsymbol{k}_3$ overlap in time. The peak appearing near τ_1 (= 50 fs) shows that almost full rephasing can be achieved for times 50 fs or less. That is, the electronic memory is almost fully retained on this time scale. The delay of the F3PE signal increases as τ_1 increases. The delay as a function of τ_1, after deconvolving the finite pulse duration effect, is plotted in Fig. 7. It increases linearly for small τ_1 and reaches a plateau. If this rephasing is caused by some large inhomogeneous (static) broadening, the delay will increase linearly with τ_1 (solid line in Fig. 7). This 'off-diagonal' results together with the 3PE unambiguously proves that the initial fast response in liquid is an inertial response of the solvent molecules.

4. Conclusions and Future Work

A new nonlinear spectroscopic method based on 5-th order nonlinearity is presented. It is shown that this new technique can provide critical information on the dynamics on ultrashort time scales. This new information together with accurate three pulse photon echo measurement enable us to identify at least three processes with distinct time scales. The fastest process occurs on the time scale of 50 fs and the physical nature is mostly inertial. Even this fastest process can be captured in the slow modulation limit allowing rephasing. This has an important implications for the optical line shape. That is, each individual line should also be close to Gaussian rather than Lorentzian. To be more quantitative, we are currently carrying out a numerical simulations using multimode Brownian oscillator model[14]. Even though popular Kubo stochastic model[15] can also be used to produce qualitatively similar results, the fundamental assumption in the Kubo model - exponential frequency correlation function - is not physically acceptable for the liquid state dynamics where an inertial component is significant.

References.

1. M. Maroncelli and G.R. Fleming, J. Chem. Phys., 89, 5044 (1988).
2. E.A. Carter and J.T. Hynes, J. Chem. Phys., 94, 5961 (1991)
3. M. Cho, G.R. Fleming, S.Saito, I. Ohmine and R. Stratt, J. Chem. Phys., in press.
4. N.E. Shemetulskis and R.F. Loring, J. Chem. Phys., 97, 1217 (1992).
5. T. Joo and A.C. Albrecht, Chem. Phys., 176, 233 (1993).
6. E.T.J.Nibbering, D.A.Wiersma and K.Duppen, Phys. Rev. Lett., 66, 2464 (1991).
7. J.-Y. Bigot, M.T. Portella, R.W. Schoenlein, C.J. Bardeen, A. Migus, and C.V. Shank, Phys. Rev. Letters, 66, 1138 (1991).
8. See, for example, S. Ruhman and K.A. Nelson, J. Chem. Phys., 94, 859 (1991).
9. S.J. Rosenthal, R. Jimenez, G.R. Fleming, P.V. Kumar, and M. Maroncelli, J. Mol. Struct., in press.
10. M.K. Lawless and R.A. Mathies, J. Chem. Phys., 96, 8037 (1992).
11. M. Cho and G.R. Fleming, J. Chem. Phys., 98, 2848 (1993).
12. M. Cho and G.R. Fleming, J. Phys. Chem., 98, 3478 (1994).
13. M.T. Asaki, C.-P. Huang, D. Garvey, J. Zhou, H.C. Kapteyn, and M.M. Murnane, Opt. Letters, 18,, 977 (1993).
14. Y.J. Yan and S. Mukamel, J. Chem. Phys., 94, 179 (1991).
15. R. Kubo, Adv. Chem. Phys., 15, 101 (1969).

Femtosecond Molecular Dynamics in Gases and Liquids Through Photon Echoes with Incoherent Light

R. Zhang, T.-S. Yang, and A.B. Myers

Department of Chemistry, University of Rochester,
Rochester, NY 14627-0216, USA

Abstract. Photon echo experiments with temporally "incoherent" light are used to explore rovibrational relaxation in the B state of I_2 vapor and electronic spectral broadening in I_2 and azulene in solution.

Spectrally broadband, temporally "incoherent" light has been employed to obtain femtosecond time resolution in various nonlinear optical experiments by exploiting the fact that in many such processes it is the coherence time of the light, not its pulse duration, that dictates the time resolution of the experiment.[1] One such process is the photon echo,[2-5] which can distinguish the contributions to molecular electronic spectral line-widths from "homogeneous" and "inhomogeneous" broadening if the latter is static on the time scale of the entire matter-radiation interaction sequence. Unfortunately, in most liquids, spectral diffusion during the incoherent light source's pulse duration is likely to wash out any signature of inhomogeneous broadening. We have approached this problem by realizing that the time scale on which spectral diffusion can act may be shortened by making the first two fields orthogonally polarized such that they generate an orientational grating, which decays with the rotational relaxation time. Only processes faster than rotational relaxation should then contribute to the homogeneous dephasing. This idea is applied here to I_2 in vapor and solution phases and azulene in solution.

The broadband pump pulses (centered at 560-562 nm with 175 cm^{-1} width for I_2 and at 680-700 nm with 400-1400 cm^{-1} width for azulene) are produced by a cavityless Q-switched YAG-pumped dye "laser" containing no output coupler or tuning element. The spontaneous emission is amplified by passage through a second dye cell. The pulse duration is about 8 ns. The probe is obtained either by beamsplitting the pump or from the 532 nm YAG output. Signals are recorded as a function of delay τ between the two pump pulses at constant probe delays of $\approx$1 ns for the solution phase experiments and >10 ns for the vapor phase studies.

Our theoretical description of time-delayed four-wave mixing with incoherent light in a multilevel system is presented in detail elsewhere.[6] In I_2 vapor, parallel polarized pulses are used to form a long-lived population grating, and a long probe delay insures that the probe field interacts last. We calculate the signal by summation over rovibrational levels of the ground and B excited states, using literature values for the spectroscopic constants. Expressions are derived both for the case in which no energy relaxation occurs during the "waiting time" for the probe pulse and for the case where rotational relaxation is complete, but no vibrational relaxation occurs. For the solution phase experiments, perpendicularly polarized pump pulses form an orientational grating. To probe this short-lived grating with our long pulses the probe must temporally overlap the pumps, requiring consideration of time orderings of the field interactions in which the "probe" is not last. The theoretical signals are calculated by summation over vibronic eigenstates with varying amounts of homogeneous and inhomogeneous broadening assumed.

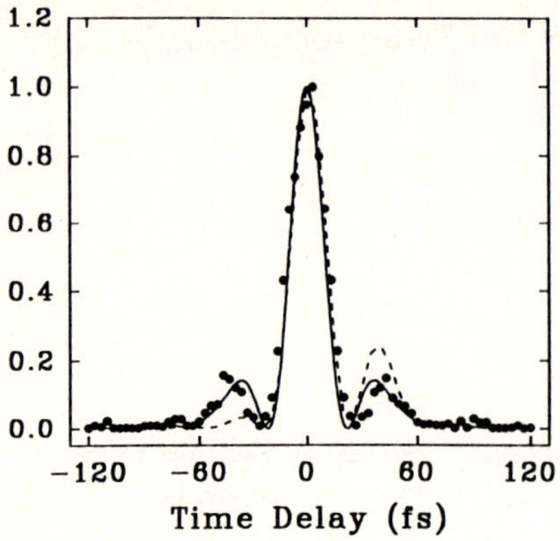

Figure 1. Points: data for 1 mM azulene in cyclohexane, orthogonally polarized pumps at 680 nm, probed at 532 nm. Lines: calculations for purely homogeneous broadening (solid) and for 30 cm^{-1} homogeneous, 285 cm^{-1} inhomogeneous widths (dashed).

The experimental result for I_2 vapor shows beats corresponding to a 91 cm^{-1} frequency, consistent with the known vibrational spacings in the B state at our pump wavelength. The experimental signal behavior falls between the calculated curves for no energy relaxation and for complete rotational energy randomization during the ≈10 ns waiting time.[6] The experimental signal from I_2 in hexadecane appears symmetric and exhibits no evident beats. Detailed numerical calculations demonstrate that it is very difficult to distinguish between homogeneous and inhomogeneous broadening in this system.[6] However, reducing the homogeneous width below about 25 cm^{-1} does produce a definite beat at positive times, inconsistent with experiment. This suggests an upper limit of about 200 fs for the B state dephasing time in solution.

Fig. 1 shows data for azulene in cyclohexane at a wavelength where the broadband pump pulses overlap both the electronic origin and the first strong vibronic band, generating clear beats. The calculated curves assuming purely homogeneous (dashed) and mostly inhomogeneous broadening (solid) are quite different in this case, and indicate that the electronic transition is mostly homogeneously broadened on the time scale of the orientational relaxation. Data for azulene excited near the electronic origin (not shown) also agree well with the calculation for pure homogeneous broadening.

In summary, our solution phase results provide no evidence for inhomogeneous broadening persisting on the 4-8 ps reorientational time. While light sources having "long" pulse durations and femtosecond coherence times are clearly useful for distinguishing between "static" and homogeneous contributions to electronic spectral linewidths in gases and low-temperature solids, true femtosecond pulses are probably required to reveal time scales for spectral diffusion in most liquids. In either case, full interpretation of such data still requires a proper multilevel model for the vibrational structure of the electronic transition.

References

1. N. Morita and T. Yajima, *Phys. Rev. A* **30**, 2525 (1984)
2. A.M. Weiner, S. De Silvestri and E.P. Ippen, *J. Opt. Soc. Am. B* **2**, 654 (1985)
3. Y.S. Bai and M.D. Fayer, *Chem. Phys.* **128**, 135 (1988)
4. J.-Y. Bigot, M.T. Portella, R.W. Schoenlein, C.J. Bardeen, A. Migus and C.V. Shank, *Phys. Rev. Lett.* **66**, 1138 (1991)
5. E.T.J. Nibbering, D.A. Wiersma and K. Duppen, *Phys. Rev. Lett.* **66**, 2464 (1991)
6. T.-S. Yang, R. Zhang and A.B. Myers, *J. Chem. Phys.*, in press

Coherence in Photodissociation–Recombination Dynamics of I_2 Isolated in Rare-Gas Solids

R. Zadoyan, Z. Li, C.C. Martens, and V.A. Apkarian

Department of Chemistry, University of California, Irvine, CA 92717, USA

Abstract. Pump-probe measurements (with time resolution of ~150fs), together with molecular dynamics simulations, yield a detailed description of the photodynamics of I_2 isolated in rare gas solids (RGS).

1. Introduction

I_2 doped RGS are studied as prototypes of condensed phase reactive dynamics: the breaking and remaking of bonds. Upon dissociative excitation of I_2 in solid Ar and Kr, the I atoms recombine due to a complete cage effect. Description of the process requires following of the dynamics for many picoseconds, at molecular configurations that are far removed from the vertically accessible Franck-Condon region. Due to the large number of degrees of freedom in condensed media, the time dependent amplitudes of an initially prepared superposition state will evolve irreversibly prior to reaching such configurations, limiting the utility of spectroscopic techniques that rely on amplitude coherence. In contrast, in systems with large density of states, pump-probe measurements enable classical descriptions of both dynamics and spectroscopy. Taking advantage of these considerations, we provide a detailed description of the solid state many-body dynamics and compare the results to equivalent measurements in other media.

2. Results and Discussion

In the 640-735 nm spectral range, $I_2(A^3\Pi_{1u})$ is prepared above its dissociation limit and probed by laser induced fluorescence from the ion-pair states using the second harmonic of the pump.[1] The relevant potential energy surfaces are shown in figure 1a. Due to the double valued nature of the difference potentials, for a given probe wavelength, two molecular configurations are simultaneously interrogated. The outer resonance probes the cage wall directly. Transients obtained in Kr are shown in 1b, along with the theoretical reproduction of the data. The signals can be understood by reference to the ensemble of trajectories shown in 1c. Upon creation on the repulsive wall of the A state, the I-I bond stretches to ~ 4.6Å, pushing the cage atoms away. The initial stretch over the inner window, at 3.2Å, is captured as the response limited first peak in the signal. The collision with the cage, at t ~400fs, is captured by the outer window. The first recursion to the inner window occurs at a delay of ~1.1ps, with atoms recoiling after losing nearly all of their excess energy.

Springer Series in Chemical Physics, Vol. 60 **Ultrafast Phenomena IX**
Editors: P. F. Barbara · W. H. Knox · G. A. Mourou · A. H. Zewail

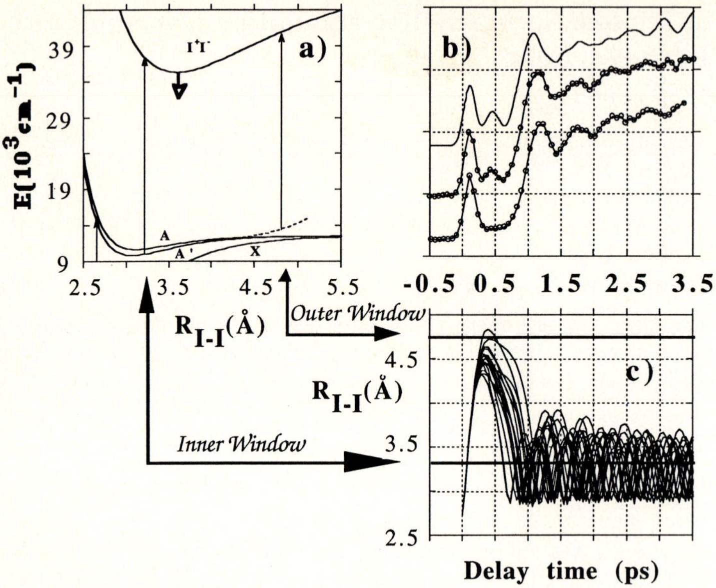

Fig.1: a)Potential energy curves of I_2 including sudden cage wall, b) Experimental (λ_{pump}=704 nm and 726 nm) and theoretical (top trace) pump-probe signals, c)Ensemble of trajectories from classical simulations. The features in b) are understood by the crossings of the trajectories with the horizontal lines in c).

The subsequent vibrational relaxation occurs in the bound part of the molecular potential, with partial retention of coherence, as evidenced by the oscillations in the signal which last for ~4ps. To reproduce the modulations, the zero-point motion of the lattice is incorporated in the simulations by temperature scaling [2]. The coherence is more extensive in Ar, and persists for t~12ps. Frequency analysis of the oscillatory signal in Ar reveals the lattice modes that are directly coupled to the molecular motion. The details of the initial atom-cage collision govern the extent of coherence in subsequent dynamics.

In Xe the ion-pair state is reactive and leads to dissociation of the molecule via ionic potentials. This reactive dynamics is akin to the condensed phase harpoon reactions, which show an inverse cage-effect [3,4].

Preparation of I_2 in the $B(^3\Pi_0)$ or $^1\Pi_{1u}$ states in RGS, leads to fast vibronic relaxation, predissociation, and electronic dephasing, precluding the observation of any coherent dynamics in real-time. Resonant Raman measurements reinforce the prior studies [5], which conclude that dephasing times range between 50 and 200 fs. Vibronic relaxation in the B state is complete in 6 ps, and predissociation is complete in ~ 1ps. Nevertheless, caging of the predissociated molecule is complete in both Ar and Kr, even when the molecule is prepared at excess energies of ~1eV. Although recombination of the caged fragments on the A/A' surfaces is observed, all coherences are lost.

References

1. R. Zadoyan, Z. Li, P. Ashjian, C. C. Martens, and V. A. Apkarian, Chem. Phys. Lett. **218**, 504 (1994).
2. J. P. Bergsma, P. B. Berens, K. R. Wilson, D. R. Fredkin, and E. Heller, J. Phys. Chem. **88**, 612 (1984).
3. R. Zadoyan and V. A. Apkarian Chem. Phys. Lett. **206**, 475 (1993).
4. M. E. Fajardo, R. Withnall, J. Feld, F. Okada, W. G. Lawrence, L. Wiedeman, and V. A. Apkarian, Laser Chem., **9**, 1(1988).
5. H. Kono and S. H. Lin, J. Chem. Phys. **84**, 1071 (1986).

Chemical-Substitution and Deuterium-Isotope Effects on Ultrafast Intermolecular Electron Transfer: Possible Role of Molecular Vibrations

Y. Nagasawa [1], *A.P. Yartsev* [2], *K. Tominaga* [2], *and K. Yoshihara* [1,2]

[1]The Graduate University for Advanced Studies, Myodaiji, Okazaki 444, Japan
[2]Institute for Molecular Science, Myodaiji, Okazaki 444, Japan

Deuterium isotope effect was observed for the ultrafast intermolecular electron transfer (ET) between coumarin 152 (C152) and aniline (AN) in the time region of 3 ~ 13 ps. The ET was ~10 % slower for the deuterated AN compared to the normal one. Hindered rotation of the amino group may be contributing to the fast part of the solvation process. Deuterium effect was minor for the ET occurring in the region of < 2 ps, which indicates other ultrafast dynamics contributing to the ET.

We have observed ultrafast non-exponential fluorescence quenching of 7-aminocoumarin dyes in electron donating solvents by means of femtosecond fluorescence up-conversion technique [1]. We regarded this fluorescence quenching as an ultrafast intermolecular electron transfer (ET) from the solvent to the excited dye. Some of the fast reactions occur much faster than the diffusive solvation process.

To examine the role of vibrations on such an ET, experiments on deuterium isotope effects were carried out. C152 in AN has a lifetimes of 17 ps (59 %) and 285 ps (41 %) and did not show any deuterium effect. These lifetimes are longer than the solvation times of AN, which are 6.7 ps (81 %) and 13.3 ps (19 %). When the lifetimes become somewhat shorter than the solvation time, deuteration effect was observed. In Figure 1 (a) the fluorescence decays of C152 in AN and in AN-d7 (aniline with all the seven hydrogens deuterated) are shown. The lifetimes of C152 in AN (dashed line) are 3.7 ps (41 %) and 12.4 ps (59 %), and in AN-d7 (solid line) they become about 10 % longer. However, when the reaction becomes much faster, the isotope effect becomes minor again. The fluorescence decay of C151-d2 (C151 with deuterated amino group) in AN-d7 is shown in Figure 1 (b) which shows only a slight difference compared to that of C151 in AN (590 fs (50 %) and 1.9 ps (50 %)).

We also carried out an experiment in AN-d2 (AN with only the amino group dueterated) and observed similar isotope effect for C152. The deuteration of the amino hydrogens is more important than that of the phenyl hydrogens. An isotope effect of alcoholic solvent, glycerol, has been examined for intramolecular ET of metal-metal charge transfer complex [2]. In this case deuteration of hydroxyl group was carried out. The inertial component of the solvation process [3] caused by the hindered rotation of the hydroxyl group was important for the ultrafast ET. In the present intermolecular ET of C152 and AN, hindered rotation of the amino group can be important. However, when the reaction becomes much faster (C151 in AN), the isotope effect becomes minor. This indicates that much faster nuclear dynamics is also contributing to the ultrafast ET [4].

Springer Series in Chemical Physics, Vol. 60 **Ultrafast Phenomena IX**
Editors: P. F. Barbara · W. H. Knox · G. A. Mourou · A. H. Zewail

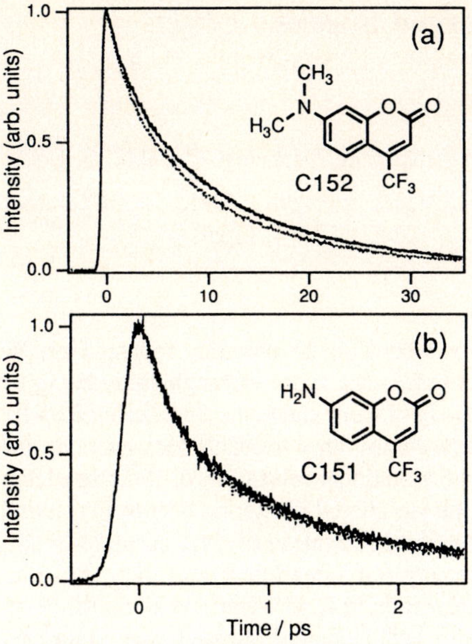

Figure 1. (a) Fluorescence decays of C152 in normal aniline (AN) (dashed line) and in deuterated AN-d7 (solid line).
(b) C151 in AN (dashed line) and C151-d2 in AN-d7 (solid line).

References

[1] Y. Nagasawa, A. P. Yartsev, K. Tominaga, A. E. Johnson, and K. Yoshihara, *J. Am. Chem. Soc.*, **115**, 7922, (1993).
[2] K. Tominaga, D. A. V. Kliner, A. E. Johnson, N. E. Levinger, and P. F. Barbara, *J. Chem. Phys.*, **98**, 1228, (1993).
[3] S. J. Rosenthal, X. Xie, M. Du, and G. R. Fleming, *J. Chem. Phys.*, **95**, 4715, (1991).
[4] H. Sumi and R. A. Marcus, *J. Chem. Phys.*, **84**, 4894, (1986).

Vibrational Relaxation of a Charged Species in Non-Polar Solvent

J.C. King, M.C. Asplund, and C.B. Harris

Department of Chemistry, University of California, Berkeley, CA 94720, USA

Early studies of vibrational energy relaxation in solution focussed on the relaxation of small, nonpolar, non-ionic molecules - the most thorough study being the dissociation and recombination of I_2.[1] This body of work showed a time scale of 10-100 ps for vibrational relaxation of ***non-ionic species in solution*** is usual. However one area that has not been studied until recently is the vibrational relaxation of ionic species in solution. In one of the first experiments on the vibrational relaxation of ions in solution, Hochstrasser and coworkers[2] found that vibrational relaxation of N_3^-, occurred in only 3 ps in polar solvents. This trend of extremely fast vibrational relaxation has been further supported by studies of vibrational relaxation in I_2^- in water and alcohols. Work by both Barbara and coworkers[3] and Banin and Ruhman[4] showed that vibrational relaxation occurs in 2-5 ps. These studies show that vibrational relaxation is an extremely rapid process for an ion in a polar solvent. However the reason for this increase in cooling rate and whether the same factors govern the rate of cooling as in neutral species is unclear. To more closely connect this work to the previous body of work on non-ionic species, ***we have performed a study of the vibrational relaxation of an ionic species (I_2^-) in a nonpolar solvent (CCl_4).*** These results are directly comparable to the vibrational relaxation of I_2, a neutral, non-ionic equivalent, which has been previously studied in the same solvent.[5,6]

To study the vibrational relaxation of I_2^-, I_3^- is photolyzed to produce I and vibrationally excited I_2^- in the manner used by Banin and Ruhman. Their work showed that the I_2^- was excited to the v = 25 level and had ~2500 cm^{-1} of excess energy. We are probing the I_2^- near its absorption at 750 nm with a standard pump/probe scheme.

Figure 1 shows the cooling at three wavelengths. As shown in simulations by Barbara and co-workers[3], the absorption at 600 nm represents a vibrationally excited I_2^-, at 670 nm an I_2^- with a lower amount of vibrational energy and at 760 nm, an I_2^- near (or at) the bottom of the well. What this data shows is faster cooling (~5 ps) near the top of the well at 600 nm, a mixture of cooling times (5 ps and 40 ps) at intermediate vibrational excitation and slower cooling (~55 ps) near the bottom of the well. Surprisingly, in this case the relaxation of the I_2^- is only a factor of 2 to 3 faster than for I_2 in the same solvent.

To find the reason for this change in vibrational relaxation times, it is important to identify the differences between I_2 and I_2^-. The most obvious difference is the charged nature of the I_2^- and thus the charge/induced dipole interactions with the nonpolar solvent. These additional longer range forces may play an important role in reducing the relaxation time. Another important difference between I_2 and I_2^- is the intramolecular potentials. The I_2^- potential has a frequency of ~115 cm^{-1} while iodine has a fundamental frequency of ~215 cm^{-1}. Previous work has shown that the amplitude of the solvent's friction at the frequency of the oscillator is an important indicator of how fast relaxation

Springer Series in Chemical Physics, Vol. 60 **Ultrafast Phenomena IX**
Editors: P. F. Barbara · W. H. Knox · G. A. Mourou · A. H. Zewail

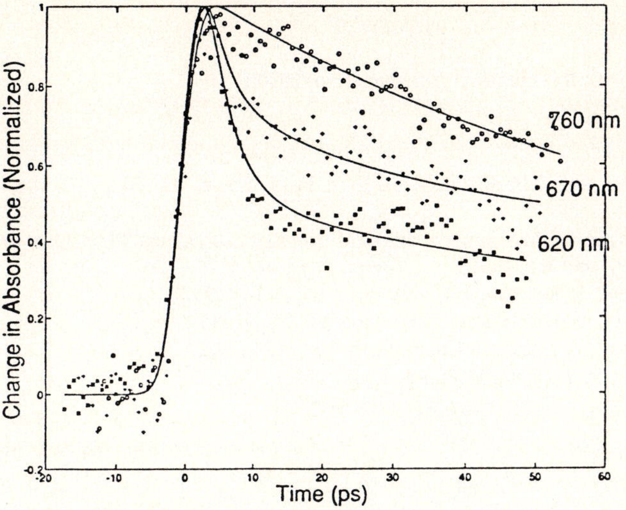

Figure 1:
Transient Absorption curves at three wavelengths, showing red shifting of the absorption maximum due to vibrational cooling.

will occur and most solvents do have a higher friction at lower frequencies. This factor would also increase the rate of relaxation of I_2^- as compared to I_2. A priori it is not clear how much of the increase in relaxation rate is due to the charge/induced dipole interactions and how much is due to the change in oscillator frequency.

To understand the contributions from these two different factors, it is interesting to compare our results to what one would expect just considering the change in oscillator frequency. From a very simple perspective, theory predicts[7,8] that the relaxation rate is proportional to the amplitude of the friction of the solvent at the frequency of the oscillator. If we look at the friction or equivalently the I_2/CCl_4 force autocorrelation function (FAF) of the solvent at the frequencies of the two oscillators (200 cm^{-1} for I_2 and 100 cm^{-1} for I_2^-) it should give an indication of what the increase in the rate of relaxation will be. The FAF of I_2 in Xe was previously determined[7,8], and neglecting the internal degrees of freedom, the force autocorrelation function should be qualitatively similar to CCl_4, which has nearly the same mass as and polarizability. Comparing the FAF at these two frequencies, the amplitude difference is of the same order as the observed change in relaxation rates. *Therefore the increase in relaxation rate is just what is expected for this change in oscillator frequencies.* The fact that the relaxation of I_2^- is well described by this simple theory indicates that the charge on the I_2^- does not couple well to any of the modes in the solvent.

In summary, our results show that relaxation of I_2^- in CCl_4 occurs in about 55 ps. This is a factor of ten slower than has been found for I_2^- in polar solvents. It is however a factor of three times faster than I_2 in the same solvent. This is approximately the change that would be expected for just the difference in oscillator frequency between the two species and indicates that the charge on the I_2^- does not couple well to modes in the solvent. We are currently working on extending the wavelength range for probing the I_2^- to get a more complete picture of the dynamics.

Acknowledgements

This work was supported by the National Science Foundation

1. A.L. Harris, J.K. Brown and C.B. Harris, Ann. Rev. Phys. Chem. **39**, 341, (1988) and references therein.
2. J.C. Owrutsky, Y.R. Kim, M. Li, M.J. Sarisky, and R.M. Hochstrasser, Chem. Phys. Lett. **184**
3. D.A.V. Kliner, J.C. Alfano and P.F. Barbara, J. Chem. Phys. **98**, 5375 (1993).
4. U. Banin, A. Waldman and S. Ruhman, J. Chem. Phys. **96**, 2416 (1992).
5. M. Berg, A.L. Harris and C.B. Harris, Phys. Rev. Lett. **54**, 951 (1985).
6. A.L. Harris, M. Berg, and C.B. Harris, J. Chem. Phys. **84**, 9788 (1986).
7. D. E. Smith and C.B. Harris, J. Chem. Phys. **92**, 1304 (1990).
8. D.E. Smith and C.B. Harris, J. Chem. Phys. **92**, 1320 (1990).

Femtosecond Dynamics of Aggregate Excitons in Liquids

M. v. Burgel, D.A. Wiersma, and K. Duppen

Ultrafast Laser and Spectroscopy Laboratory, Department of Chemistry, University of Groningen, Nijenborgh 4, 9747 AG Groningen, The Netherlands

Abstract. By photon echo experiments and pump-probe spectroscopy on aggregates of molecules in liquids, it is shown that delocalization of excitation considerably slows down optical dephasing, due to motional narrowing over ultrafast solvent fluctuations. An expected consequence is that the optical dynamics occurs closer to the Markovian limit than for non-aggregated single molecules in the same liquid.

When solvated molecules aggregate, i.e. form ordered one-dimensional molecular chains of considerable length, the optical properties are clearly different from those of single molecules in liquids. The excited electronic states of these aggregates are best described as excitons, which are delocalized over several molecules. In Fig. (1) the absorption spectrum of a carbocyanine dye in water is shown, together with its chemical structure. The formation of aggregates leads to the appearance of a new absorption band at sufficiently high concentrations ($c > 10^{-5}$ M), narrowed and red-shifted compared to the monomer absorption. Both these features are signatures of exciton behaviour, and are due to some degree of motional narrowing over single-molecule properties. The relevant level structure is shown in Fig. (2). In the lowest exciton band, one single-molecule excitation is shared by a certain number of molecules, in the next highest exciton band two single-molecule excitations are present on the same sites, etc. Due to the solvent perturbations, there is dynamical disorder among the molecules of the aggregate, and consequently localization of the excited state wavefunction occurs on some finite section of the molecular chain. In

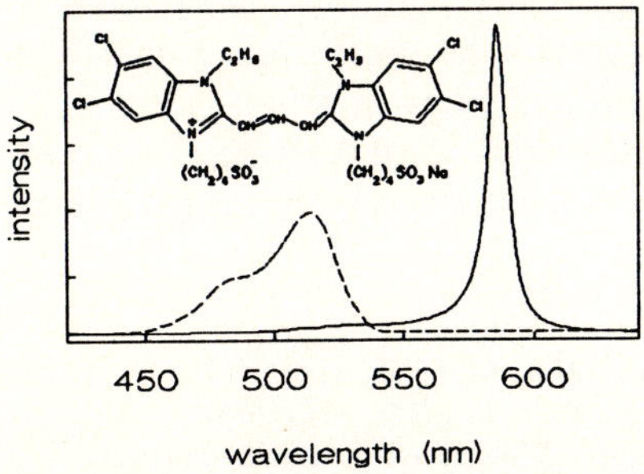

Figure 1 Absorption spectrum of TDBC in water. Dashed line: monomer; solid line: aggregate.

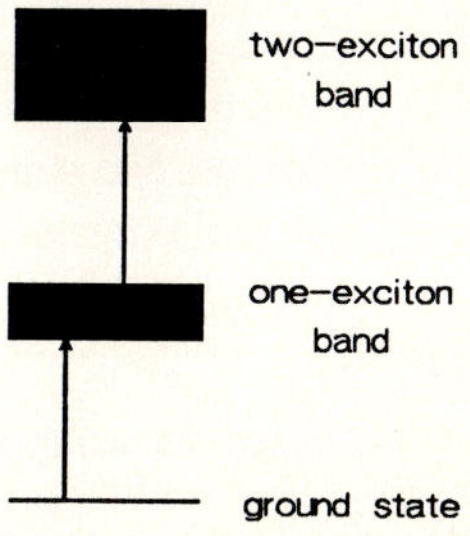

Figure 2 Level scheme for the two lowest exciton bands of TDBC in water.

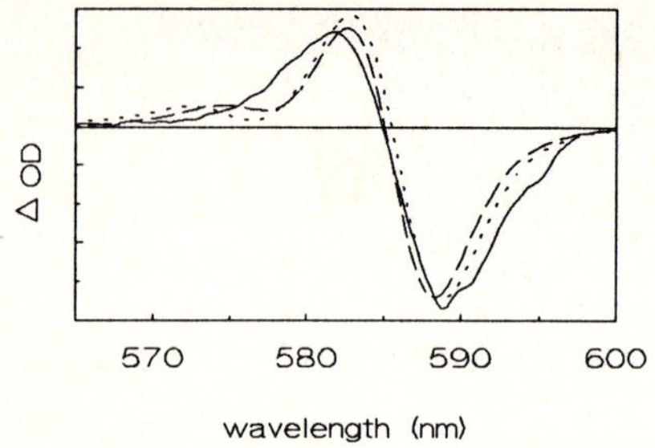

Figure 3 Pump-probe signal for TDBC in water. A bleach (588 nm) and induced absorption (582 nm) are observed.

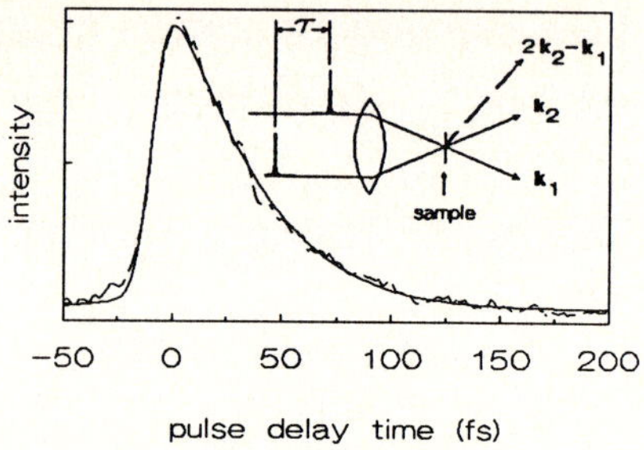

Figure 4 Two-pulse photon echo decay. Inset: lay-out of the experiment. Dashed line is measured, solid line is calculated.

addition, rapid dephasing of the optical transitions of the aggregate is induced. It can therefore be expected that delocalization and dephasing are somehow connected.

The delocalization length can be determined by ultrafast pump-probe spectroscopy. In such an experiment, a bleaching of the entire absorption lineshape is induced by an ultrashort pump pulse. This can be measured by a delayed probe pulse. In addition, absorption is induced on the one- to two-exciton transition [1-3]. This absorption is blue-shifted compared to the bleach, since two excitations cannot be at the same time at the same single-molecule position. For small delocalization lengths this effect is more important than for large ones. Hence, the frequency difference between the two transitions depends directly on the delocalization length. The result of a pump-probe experiment is shown by the solid line in Fig. (3). When this line shape is fitted using conventional exciton band theory, a delocalization length in the order of 16 (dotted line) to 18 (dashed line) molecules is found.

The optical dephasing time can be determined by ultrafast photon echo experiments. A result of such an experiment with 10 fs optical pulses is shown in Fig. (4). The mono-exponential decay suggests that the dynamics occurs in the Markov limit, which gives for the dephasing time: $T_2=160$ fs. This is by far the longest coherence decay ever measured in a room temperature liquid. Closer inspection of the echo decay and the observed absorption spectrum reveals that the Markovian T_2-concept is, just as for single molecules [4,5], too simple for these aggregate excitons in liquids [6]. The theoretical echo decay, based upon non-Markovian fluctuations on two time scales (5 fs and 10 ps) is shown in Fig. (4).

In conclusion we can state that delocalization and optical dephasing are closely connected due to motional narrowing processes over ultrafast dynamic perturbations.

References:

1. H. Fidder, J. Knoester and D.A. Wiersma, J. Chem. Phys. **98**, 6564 (1993).
2. A.E Johnson, S. Kumazaki and K. Yoshihara, Chem. Phys. Lett. **211**, 511 (1993).
3. K. Minoshima, M. Taiji, K. Misawa and T. Kobayashi, Chem. Phys. Lett. **218**, 67 (1994).
4. J.-Y. Bigot, M.T. Portella, R.W. Schoenlein, C.J. Bardeen, A. Migus and C.V. Shank, Phys. Rev. Lett. **66**, 1138 (1991).
5. E.T.J. Nibbering, D.A. Wiersma and K. Duppen, Phys. Rev. Lett. **66**, 2464 (1991).
6. M. van Burgel, D.A. Wiersma and K. Duppen, to be published.

Locally Ordered Structures in Molecular Liquids Revealed with Femtosecond Fourier-Transform Raman Spectroscopy

W.T. Lotshaw[1], *P.R. Staver*[1], *D. McMorrow*[2], *N. Thantu*[2], and *J.S. Melinger*[2]

[1]G.E. Research, Rm. KWC 627, 1 River Road, Schenectady, NY 12309, USA
[2]Naval Research Laboratory, Code 6613, Washington, DC 20375, USA

The issue of local order in molecular liquids, and the influence of such order on molecular dynamics and dielectric relaxations in liquids and solutions, is a topic of substantial importance to theoretical representations of the liquid state and interpretations of solvent effects in solute relaxations and chemical reactivity in solution. Previously [1] we have used time and frequency domain analyses of the optical heterodyne detected optical Kerr effect (OHD OKE) to examine the complicated vibrational and Brownian molecular motions inpure liquids, and identified the intermolecular origin of the low-frequency Raman band in organic liquids (typically centered in the range 10-100 cm^{-1}). These bands are attributed to Raman-active librational motions in non-associated molecular liquids, with bandwidths in the range from 20-100 cm^{-1}.

We address the question of local order and structure in molecular liquids through the dependence of the measured intermolecular vibrational dynamics and spectra on the number density of benzene in the solvent carbon tetrachloride (CCl_4). Previous studies [2] have shown a bimodal structure in the intermolecular vibrational Raman spectrum below 150 cm^{-1}. Considering the symmetry of this molecule, it was suggested [2] that the low frequency part of the spectrum (< 20 cm^{-1}) could be due to the existence of locally ordered structures (dimers or other aggregates). The results of this study support this interpretation, and provide a means for interrogation of the concentration dependence of these specific intermolecular interactions.

Figure 1 (*left*) shows the OHD OKE transient measured for 9.1% volume fraction of benzene in CCl_4 with $\leq$ 40 fs pulses from a continuously tunable, synchronously pumped Ti:Sapphire laser [3]. The purely electronic hyperpolarizability dominates the response near $\tau = 0$, while significant contributions from both benzene and CCl_4 are present for $\tau > 0$. The prominent oscillations are due to the *intra*molecular Raman modes of CCl_4 at 217 and 314 cm^{-1}. Analysis of this very complex transient is facilitated by the use of discreet Fourier transform (FT) methods, which are described in detail elsewhere [1]. The frequency response function of the OKE transient shown in fig. 1, deconvoluted from the spectral density of the laser pulse correlation function, is also shown in figure 1 (*right*). The intramolecular Raman modes of CCl_4 and a broad, low frequency band due to the intermolecular dynamics of the solution are evident.

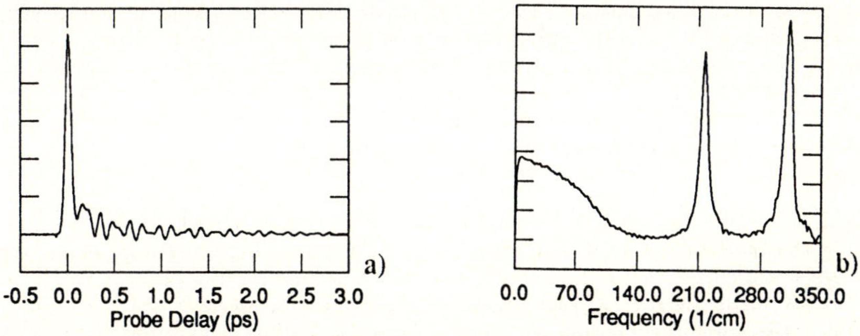

Figure 1. (a) Femtosecond OHD OKE transient measured for a 9.1% (v/v) solution of benzene in carbon tetrachloride; (b) spectral density representation of a) (amplitude arb. units).

Editors: P. F. Barbara · W. H. Knox · G. A. Mourou · A. H. Zewail

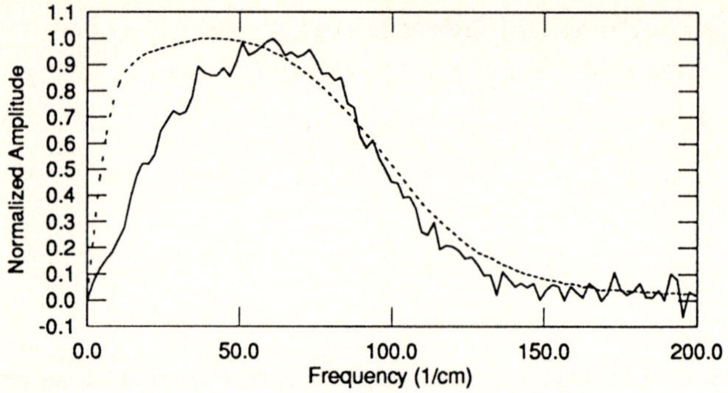

Figure 2. Vibrational part of the intermolecular Raman spectra for pure benzene (*dashed*) and the benzene contribution to the 9.1% solution in carbon tetrachloride (*solid*) illustrating the spectral evolution that occurs upon dilution.

The data of figure 1 contain several overlapping contributions that must be separated before a detailed interpretation of the intermolecular dynamics is possible. Since we are interested in the intermolecular vibrational part of the response, the Brownian relaxation due to diffusive reorientation of benzene is removed by a subtraction procedure described previously [1]. For the 9.1% benzene solution, contributions to the intermolecular dynamics by CCl_4 are significant. These contributions are removed by subtraction of the pure CCl_4 spectrum from that of the solution, which assumes that the intermolecular and intramolecular parts of the CCl_4 spectrum are affected commensurably by the dilution. This assumption is reasonable, and supported by the unaltered lineshapes of the intramolecular resonances on dilution.

The dashed curve in figure 2 shows the intermolecular vibrational Raman part of pure benzene deduced from the FT analysis. We had previously attributed this lineshape to a bimodal vibrational response in which the low frequency part was attributed to dimers or higher aggregates, and the high frequency part was attributed to the librational motions of non-aggregated molecules [2]. Dilution is expected to shift the dimer-monomer equilibrium in favor of the monomer, thereby reducing the low-frequency amplitude of the bimodal vibrational distribution. The solid curve in figure 2 shows the intermolecular vibrational Raman part of the 9.1% benzene/CCl_4 solution with the CCl_4 contribution subtracted off as described above. The absence of the low-frequency shoulder in the 9.1% solution which is evident in the spectrum of pure liquid benzene is consistent with a dimer/aggregate origin of this feature in the pure liquid.

The changes we observe occur at frequencies less than 50 cm^{-1}, and are nearly impossible to address by spontaneous light scattering due to Rayleigh scattering and background problems. The femtosecond OHD OKE methods are ideally suited to the analyses presented here because the attributes of: *i*) background suppression and phase sensitive signal detection permit isolation of the spectroscopic signal of interest; *ii*) high sensitivity and signal to noise ratio result in the precise characterization of the picosecond Brownian dynamics; and *iii*) a wide dynamic range and order reduction in the nonlinear response function permit a highly accurate, discreet FT analysis of the more complex elements in the measured transients.

1. D. McMorrow and W.T. Lotshaw, Chem. Phys. Lett. **174**, 85 (1990); J. Phys Chem. **95**, 2028 (1991).
2. D. McMorrow and W.T. Lotshaw, Chem. Phys. Lett. **201**, 369, (1993).
3. P.R. Staver and W.T. Lotshaw, manuscript in preparation, 1994.
4. T.R. Dickson, PhD Thesis, University of Toronto, 1991.

Femtosecond Studies of Charge Separation in the Fluorescent State of DCM

H. Zhang, A.M. Jonkman, P. van der Meulen, and M. Glasbeek

Laboratory for Physical Chemistry, University of Amsterdam,
Nieuwe Achtergracht 127, 1018 WS Amsterdam, The Netherlands

Abstract. Fluorescence decays of the laser dye DCM were measured using fluorescence upconversion with a resolution of 150 fs. The results reveal solvent controlled intramolecular electron transfer in the adiabatic limit.

Solvation dynamics has long been recognized to be of great importance for the understanding of the rate of intramolecular electron transfer (ET) reactions[1]. Especially when the ET reaction takes place in the so-called strong adiabatic coupling limit, solvation dynamics may predominantly determine the ET rate constant[2]. Recently, for a number of molecules solvent controlled ET in the photoexcited state has been considered in the strong adiabatic coupling regime[3]. In this report, we present results of a femtosecond pump-probe study of the charge separation reaction in the photoexcited state of a large laser dye molecule, 4-(dicyanomethylene)-2-methyl-6-(*p*-dimethylaminostyryl)-4-H-pyran (DCM). We have performed a subpicosecond study of the solvatochromism of the DCM molecule. A major finding is the strong mixing of LE and the CT excited states for DCM (i.e., the adiabatic coupling limit applies) so that the solvation dynamics constitutes the rate determining step in the ET process. In our experiments we have applied the fluorescent upconversion technique. Pulses from a femtosecond Tsunami Ti:sapphire laser were frequency doubled before exciting the DCM sample near 420 nm. The emission of DCM dissolved in methanol at ambient temperature was gated with the fundamental beam and the sum frequency was photodetected. The time resolution in our experiment is approximately 150 fs. Fluorescent decays were observed for DCM dissolved in methanol at a few detection wavelengths. After deconvolution of the experimental decay signals using the instrumental response function, the signal intensity, $I(\lambda,t)$, could in all cases be fitted to a sum function of three exponential functions. In reconstructing the time-resolved spectra we followed the procedure outlined in Ref. 4. In Fig. 1 we plot the time-dependent Stokes shift C(t), where $C(t)=(\nu(t)-\nu(0))/(\nu(\infty)-\nu(0))$ and $\nu(t)$ corresponds to the frequency of the maximum of the fluorescence band, of DCM in methanol. The data points of C(t) appear to be equal to the solvent relaxation times for methanol[5]. It is concluded that the time dependence of the Stokes shift for DCM in methanol is fully determined by the solvation dynamics. Fig. 1 also presents the evolution with time of the integrated fluorescence intensity of DCM in methanol. It is found that *the integrated intensity shows a rapid decrease to about 50 % of its maximum value on the (picosecond) time*

Springer Series in Chemical Physics, Vol. 60 **Ultrafast Phenomena IX**
Editors: P. F. Barbara · W. H. Knox · G. A. Mourou · A. H. Zewail

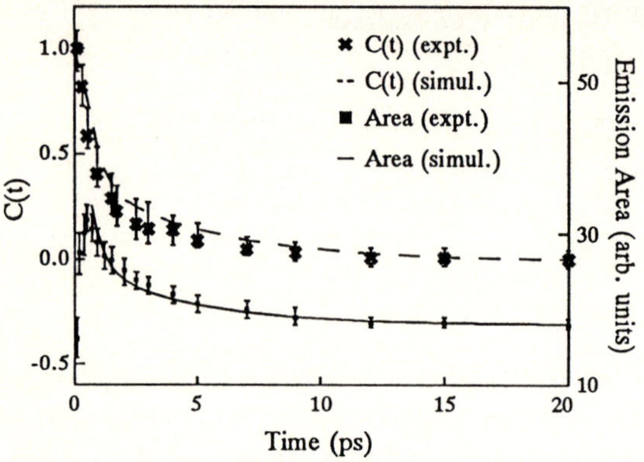

Fig.1 Experimental and computer simulated time-dependent Stokes shift, C(t), and integrated fluorescence intensity, I(t), of DCM in methanol.

scale of the solvation process. Recently, time-resolved Stokes shift measurements of DCM in a few alcoholic solvents have also been studied using stimulated emission as a probe[6]. However, in the latter work the integrated intensity was not considered explicitly. It is also remarked that the ν^3 dependence effect on the emission intensity is readily verified to be less than 25%. As has been discussed in a few recent papers[3], in the limit of strong adiabatic coupling of the LE and CT states solvation will appreciably affect the electronic structure, and concomitantly, the radiative character of the excited state will also be changed as time progresses. Assuming linear coupling of the electronic states with the surrounding dielectric medium, the generalized Smoluchowsky equation for the excited state population density distribution $\rho(t)$, was solved for suitable values of the optical parameters, and we simulate the time dependence for the function, C(t), and the integrated intensity for the spontaneous emission, as represented by the plotted curves in Fig.1. We conclude that *the intramolecular ET process for DCM in polar solvents is appropriately discussed within the strong adiabatic coupling model in the inverted regime.*

References

1. R.A. Marcus and N. Sutin, Biochim.Bioph.Acta **811**, 265 (1985); J.Ulstrup, *Charge Transfer Processes in Condensed Media*, Springer-Verlag, Berlin, 1979.
2. I. Rips and J.Jortner, J.Chem.Phys., **87**, 6513 (1987).
3. K. Tominaga, G.C. Walker, W. Jarzeba, and P.F. Barbara, J.Phys.Chem., **95**, 1075 (1991); T.J. Kang, W. Jarzeba, and P.F. Barbara, Chem.Phys., **19**, 81 (1990).
4. M. Maroncelli and G.R. Fleming, J.Chem.Phys., **86**, 6221 (1987).
5. W. Jarzeba, G.C. Walker, A.E. Johnson, and P.F. Barbara, Chem.Phys., **152**, 57 (1991).
6. D.C. Easter and A.P. Baronavski, Chem.Phys.Letters **201**, 153 (1993).

Coherent Optical Polarization of Large Coherence-Length Frenkel Excitons in Pseudoisocyanine *J* Aggregates

S. Kobayashi and F. Sasaki

Electrotechnical Laboratory, 1-1-4 Umezono, Tsukuba, Ibaraki 305, Japan

Abstract. An unusually slow free induction decay is observed in excitons of pseudoisocyanine J aggregates. The dephasing mechanisms of the coherent optical polarization of large coherence length Frenkel excitons are discussed on the bases of exciton energy fluctuation.

One dimensional molecular aggregates called J aggregates show a sharp exciton band (J band) and a superradiant emission as a result of collective excitation [1]. We have reported that the J aggregates of pseudoisocyanine (PIC) show an extraordinary large nonlinear optical (NLO) susceptibility [2] and also fast switching–off time (1.3 ps) at 77 K [3]. Furthermore PIC J aggregates show a novel excitation density dependent NLO response which could be useful as an optical logic element [4]. These unique NLO properties are attributed to the one–dimensional Frenkel excitons with the large coherence length which is extended over thousand molecules at 77 K [2].

In this report we present experimental results in which the coherent interaction between Frenkel excitons and a radiation field measuring a perturbed free induction decay (FID) by using a standard pump–probe method with low density (1 to 10 $\mu J/cm^2$) 200 fs pulse. The pump field produces any substantial changes in the probe field through materials polarization. Important information on the exciton–phonon and the exciton–exciton interactions are contained in the temporal line shape of optical response under ultrashort pulse excitation [5].

Samples were made by dissolving PIC bromide to a concentration of 4 mM in a 1:1 mixture of water and ethyleneglycol (WEG). The pulses are the amplified output of a hybrid mode–locked dye laser with 1 kHz repetition. For the pump and probe experiments pulses are obtained through the generation of a white light continuum in a water cell.

Figure 1 shows the oscillatory structures at the J band in the differential transmission spectra ($\Delta T/T$) of the probe pulse at the very early time when the probe pulse proceeds the pump pulse. The oscillatory structure is assigned to the perturbed FID and has been reported in dye solutions and semiconductors [5] but there has been no report about Frenkel excitons. When PIC J aggregates were excited with relatively high density ultrashort pulse (a few mJ/cm^2) such oscillatory structures were disappeared [6].

The rise–up of the oscillatory structures indicates the dephasing time (T_2) of polarization and one can estimate T_2 between 300 fs and 600 fs depending on the models applied. The fluctuation of the exciton energy level causes the FID. The dephasing time estimated from the width of the J band (30 cm^{-1} in WEG at 77 K) is

ments. At sufficiently high temperature, when the critical correlation length ξ falls below $3\xi_0$ ($\xi_0 \equiv$ a characteristic molecular length), both 5CB and MBBA begin to make a transition from an ordered liquid to a simple liquid; the slow dynamics start deviating from LdG theory and the intradomain dynamics become temperature dependent. The dynamics within the domains are temperature independent because the local domain structures are preserved during the relaxation. The time dependence of the intradomain dynamics in both MBBA and 5CB is a power law decay with an identical power law exponent of 0.63 [3]. These identical features of the two systems suggest a universality of the fast time scale dynamical behavior in the isotropic phase of liquid crystals. A calculation based on extended LdG theory gives a power law with an exponent of 0.5.

In polymers, local order is created by the restricted geometry, connectivity of chemical bonds and local steric interactions. These restrictions create extended orientational correlations among side groups of the polymer. The side group dynamics in two very different polymers poly(2-vinylnaphthalene) (P2VN) [4a] and poly(methylphenylsiloxane) (PMPS) [4b] were studied. The TG-OKE experiments are able to observe solely the side group dynamics in P2VN and PMPS. Local structures cause the orientational relaxation of side groups of both P2VN and PMPS to be viscosity and temperature independent. The TG-OKE experiments on PMPS melt indicated that the orientational dynamics of the phenyl side groups are independent of temperature over a broad temperature range, 25 °C to 143 °C. Over this broad temperature range, the ratio of the viscosity to the temperature, η/T, changes by a factor of 40. In glass forming liquids, a dynamic critical transition occurs at T_c ($T_c > T_g$, the glass transition temperature), and the density correlations on fast time scales spontaneously arrest below T_c. The resulting local inhomogeneity causes a bifurcation of the relaxation processes into α and β relaxations below T_c. The temperature independent nature of the dynamics and the local character of the dynamics are distinctive features of β relaxation [4b]. In the PMPS melts, following the ultrafast librational dynamics, we observe an exponential decay (τ = 1.95 ps) at short time that merges (~10 ps) into a power law decay with an exponent of 0.82. Relaxations involving a distribution of local inhomogeneities or nonergodic constrained release dynamics can give rise to power laws.

References

(1) F. W. Deeg, and M. D. Fayer, J. Chem. Phys. **91**, 2269 (1989)
(2) S. R. Greenfield, A. Sengupta, J. J. Stankus, M. Terazima, and M. D. Fayer, J. Phys. Chem. **98**, 313 (1994).
(3) J. J. Stankus, R. Torre, and M. D. Fayer, J. Phys. Chem. **97**, 9478 (1993).
(4) (a) A. Sengupta, M. Terazima, and M. D. Fayer, J. Phys. Chem. **96**, 8619 (1992).
(b) A. Sengupta, and M. D. Fayer, J. Chem. Phys. **100**, 1673 (1994).

Femtosecond Dynamics in Hydrogen-Bonding Solvents: Temperature Dependence of the Inertial and Diffusive Relaxation of Liquid Acetic Acid

Y.J. Chang and E.W. Castner, Jr.

Brookhaven National Laboratory, Chemistry Department, Upton, NY 11973, USA

Abstract. Ultrafast solvent dynamics of acetic acid was measured using femtosecond Raman-induced Kerr effect spectroscopy at 292 K and 381 K, near the melting and boiling points of the liquid, respectively.

Femtosecond nonlinear optical spectroscopy has proven to be a powerful tool with which to probe both the inertial and diffusive dynamics of molecular liquids [1]. Only very recently have these techniques been applied to the study of dynamics in complex, hydrogen-bonding (H-bonding) liquids [2]. We present the results of our study on the solvent dynamics of neat liquid acetic acid, CH_3CO_2H at 292 and 381 K. The measurements of the inertial and diffusive dynamics were made using the off-resonant, optical-heterodyne-detected, Raman-induced Kerr effect (OHD-RIKE) technique which provides a signal linear in the third-order nonlinear-optical response [1a]. Because the melting and boiling points of acetic acid are *ca.* 289.4 and 390 K, respectively, our measurements have probed the extremes of the liquid phase. One of the most important uses for such ultrafast solvent relaxation data is to predict the rate of solvent-limited chemical reactions in solution, particularly adiabatic electron-transfer reactions.

Acetic acid forms a cyclic dimer of C_2 symmetry in the crystalline, liquid, and gas phases because of a two-fold H-bond [2]. Unlike other complex H-bonding liquids such as the amides [3a] or water [3b,3c], which may form linear chains and tetrahedrally coordinated networks, liquid acetic acid is only observed as a monomer or dimer. Turi and Dannenberg [2] used high level *ab initio* and semi-empirical molecular-orbital calculations to explain this observation as resulting from the increased stability of 3.1 kcal/mole, of the doubly H-bonded configuration relative to the singly H-bonded dimer. Thus, the acetic acid liquid is a simpler system in which to study the dynamics of H-bonding liquid interactions.

Our raw transient birefringence data for acetic acid are shown in Figure 1(A). The delayed response following the near instantaneous electronic hyperpolarizability response contains both the inertial ($t \leq 1$ ps) and diffusive ($t \geq 1$ ps) solvent relaxation. For both temperatures, the rotational diffusion response fits best to a triple exponential decay function with the longest time constant of 22 ps (292 K) and 7.5 ps (381 K). The dynamics occurring in the inertial regime is best studied in the frequency domain representation of our data. By doing a Fourier-transform deconvolution on the birefringence transients shown in Figure 1(A) [1a], we obtain a frequency domain representation of the nonlinear-optical susceptibility, $\mathrm{Im}[R_{OHD\text{-}RIKE}(\omega)]$, shown in Figure 1(B). Lineshape analysis resolves the low-frequency spectra into three components: The lowest-frequency multi-Lorentzian feature corresponding to rotational diffusion, three Gaussian-shaped intermolecular vibrational or librational bands centered at 36.5, 105.3 and 162.6 cm^{-1} for 292 K spectrum and 33.9, 99.9 and 204.1

Springer Series in Chemical Physics, Vol. 60 **Ultrafast Phenomena IX**
Editors: P. F. Barbara · W. H. Knox · G. A. Mourou · A. H. Zewail

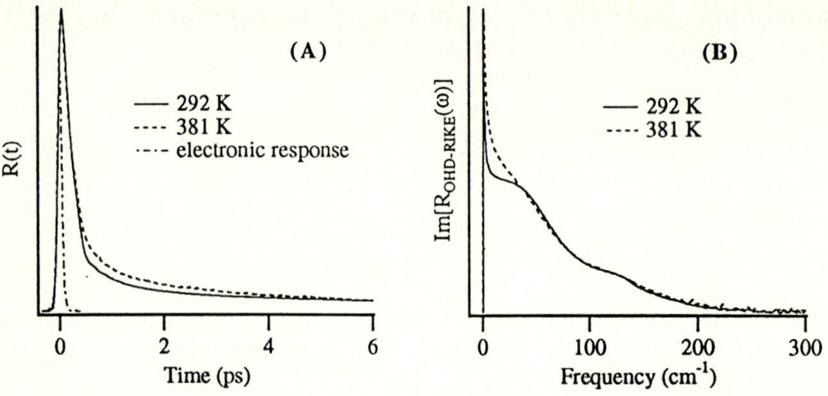

Fig. 1 (A) Time-domain OHD-RIKE data. (B) Frequency-domain OHD-RIKE data.

cm^{-1} for 381 K spectrum, and an asymmetric band centered at 9.9 cm^{-1} for both spectra which may be assigned to interaction-induced polarizability anisotropy. The low-frequency Raman work of Faurskov-Nielsen and Lund shows that the three intermolecular vibrational (or librational) frequencies of liquid acetic acid have peaks at 55, 115, and 165 cm^{-1} in the R(ω) representation of their data. They have assigned these bands to a symmetric stretch, an out-of-plane bend, and a librational motion involving the acetic acid dimer, respectively [4]. We have recently discussed the differences in the observed spectral features between R(ω) representation and $Im[R_{OHD\text{-}RIKE}(\omega)]$, particularly for those features below ~100 cm^{-1} [3(d)]. Although the intermolecular vibrational bands broaden somewhat in going from 292 to 381 K, Fig. 1(B) shows that the acetic acid C_2 dimers persist to a large extent close to the boiling point of this liquid, which is consistent with the predominating four-coordinate H-bond structure of water at 373 K [5].

This research was carried out at Brookhaven National Laboratory under contract DE-AC02-76CH00016 with the U. S. Department of Energy and supported by its Division of Chemical Sciences, Office of Basic Energy Sciences.

REFERENCES

[1] (a) D. McMorrow and W. T. Lotshaw, J. Phys. Chem. **95**, 10395 (1991); (b) Y.-X. Yan, L.-T. Cheng, and K. A. Nelson, *Advances in Nonlinear Spectroscopy*, eds. R. J. H. Clark and R. E. Hester (Wiley, Chichester, 1988), p. 299.
[2] L. Turi and J. J. Dannenberg, J. Phys. Chem. **97**, 12197 (1993).
[3] (a) K. Wynne, C. Galli and R. M. Hochstrasser, Chem. Phys. Lett. **193**, 17 (1992); (b) Y. J. Chang and E. W. Castner, Jr., J. Chem. Phys. **99**, 7289 (1993); *ibid*, **99**, 113 (1993); (c) S. Palese, L. Schilling, R. J. D. Miller, and W. T. Lotshaw, J. Phys. Chem., in press; (d) Y. J. Chang and E. W. Castner, Jr., J. Phys. Chem. submitted February, 1994.
[4] O. Faurskov Nielsen and P.-A. Lund, J. Chem. Phys. **78**, 652 (1983).
[5] D. Eisenberg and W. Kauzmann, *The Structure and Properties of Water*, (Oxford University Press, Oxford, 1969), p. 178.

Optical Dephasing Dynamics in Room-Temperature Liquids

P. Cong, Y.J. Yan, H. Deuel, and J.D. Simon

Department of Chemistry, University of California at San Diego,
9500 Gilman Drive, La Jolla, CA 92093-0341, USA

Abstract Experimental measurements of optical dephasing dynamics using femtosecond four-wave mixing spectroscopy are presented. The results are compared to non-Markovian calculations using a Brownian oscillator model for the solvent.

The dynamic aspects of electronic relaxation of molecules in solution are central to the understanding of a wide range of condensed phase photophysical and photochemical processes. While the population relaxation rate of electronically excited species can be measured by a number of spectroscopic methods, the evolution of the phase coherence in liquids has been difficult to characterize due to its ultrafast nature. With advances in the technology of generating ultrashort laser pulses, the relaxation of electronic coherence can finally be studied in real time. In the past decade, femtosecond photon echo measurements have provided information on the electronic dephasing dynamics of molecules in room-temperature solution [1-3]. A prerequisite for using this method to address such dynamics is that the laser pulses must be much shorter than the time scale of the dephasing process. It has been demonstrated by Shank and co-workers [2] and Wiersma and co-workers [3] that ~ 10 fs pulses are needed to resolve the electronic dephasing dynamics of dye molecules in room-temperature solution. In this paper, a new approach to study the ultrafast dephasing dynamics in liquids is presented: degenerate femtosecond pump-probe spectroscopy that utilizes variable laser pulse-widths and tunable wavelengths [4]. A comprehensive theoretical treatment based on a multimode Brownian oscillator model of solvation [5] is developed and quantitative agreement between theory and experiment is obtained.

The experimental setup for transient absorption has been described elsewhere [4, 6]. The signal observed around zero time delay in a degenerate four-wave-mixing (4WM) experiment is composed of both population bleaching kinetics and coherent coupling between the pump and probe laser beams. Fixing the central wavelength of the laser, the relative contribution of the coherent coupling term increases with increasing laser pulsewidth. In the case of HITCI in ethylene glycol, this dependence is shown in Figure 1 and has been explained in ref. [4].

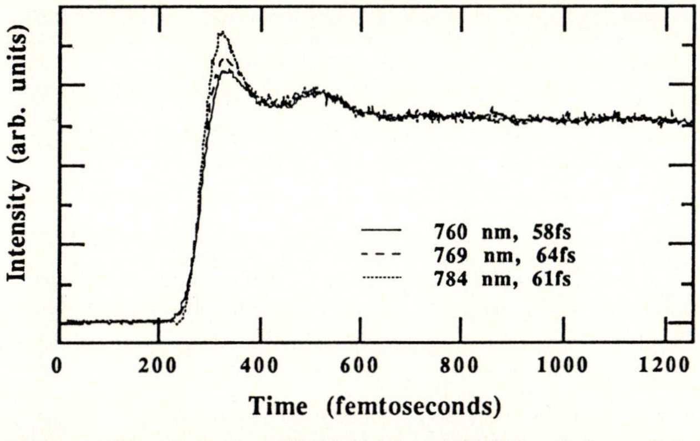

Figure 1. Time-dependent 4WM signal on HITCI in ethylene glycol as a function of excitation wavelength.

Springer Series in Chemical Physics, Vol. 60 **Ultrafast Phenomena IX**
Editors: P. F. Barbara · W. H. Knox · G. A. Mourou · A. H. Zewail

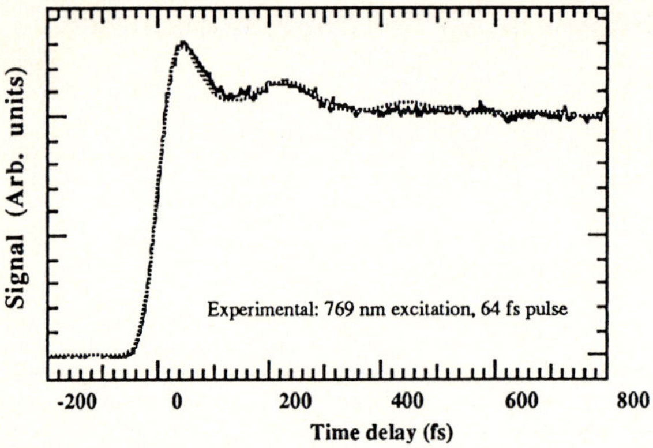

Figure 2. Comparison of experimental data on HITCI with the calculations of the non-Markovian theoretical model.

Clearly, the coherent spike amplitude increases when the central frequency of the laser is tuned away from the resonance wavelength (750 nm, for HITCI in ethylene glycol, see Figure 1). It is rather straightforward to understand this phenomenon. The population term is proportional to $1/\Delta\omega^2$, ($\Delta\omega = |\omega(\text{laser}) - \omega(\text{mol})|$), and the coherent contribution decays as $1/\Delta\omega$. The larger the detuning, the more dominant the coherent term becomes near zero time delay (see ref. [4] and [6] for more details). Therefore, in addition to the laser pulsewidth, the laser wavelength also affects the relative contribution of coherence and population dynamics to the overall absorption signal. This provides a second experimental parameter whose evolution in a degenerate 4WM experiment depends on the material dephasing time. A quantum beat pattern is also present in both curves. These clearly overdamped oscillations are caused by the coherent excitation and propagation of vibrational states of the dye molecule accompanying the electronic transition.

Electronic dephasing dynamics can be described by a variety of theoretical models, ranging from the Bloch equation approach [4, 7], to non-Markovian descriptions such as the Kubo model [2,3] and multi-mode Brownian oscillator model [5]. We have developed a comprehensive theoretical model [6] that is based on a multi-mode Brownian oscillator description of solvation [6]. Unlike earlier work, the pump and probe laser pulsewidths and wavelengths are explicitly treated. The experimental data shown in Figure 1 can be quantitatively described by this model. The fit requires a spectral fluctuation of 150 wavenumbers, a solvent correlation time of 300 fs, an intramolecular vibrational frequency of ~ 150 wavenumbers, and an inhomogeneous linewidth of 550 wavenumbers. Figure 2 shows a comparison between theory and experiment. This result demonstrates the inhomogeneous nature of the absorption spectrum and the non-Markovian aspects of the dynamics.

This work is supported by the National Science Foundation.

1. K. A. Nelson and E. P. Ippen, *Adv. Chem. Phys.* 75 (1989) 1, and references therein.
2. E. T. J. Nibbering, D. A. Wiersma, and K. Duppen, *Phys. Rev. Lett.* 68 (1991) 2464.
3. J.-Y. Bigot, M. T. Portella, R. W. Schoenlein, C. J. Bardeen, A. Migus, and C. V. Shank, Phys. Rev. Lett. 66 (1991) 1138.
4. P. Cong, H. P. Deuel, and J. D. Simon, *Chem. Phys. Lett.* 212, 367 (1993).
5. Y.-J. Yan and S. Mukamel, *J. Chem. Phys.* 89 (1988) 5160.
6. P. Cong, Y. J. Yan, H. P. Deuel, and J. D. Simon, *J. Chem. Phys.* in press.
7. M. W. Balk and G. R. Fleming, *J. Chem. Phys.* 83 (1985) 4300.

Electronic Dephasing in Mixed Crystals and Glasses at High Pressure

B.J. Baer and E.L. Chronister

Department of Chemistry, University of California, Riverside, CA 92521, USA

Abstract. We compare the effect of pressure on the homogeneous dephasing of a mixed crystal of pentacene in p-terphenyl with a Rhodamine 640 doped PMMA glass. Very different pressure effects on the dynamics are observed for the crystalline and glassy solids.

1. Introduction

Photon echo spectroscopy is used to investigate the effect of temperature and high pressure on homogeneous dephasing in mixed crystals and glasses. Optical dephasing in crystals can be dominated by interaction of a pseudo-local phonon with an excited chromophore,[1] while optical dynamics in glasses can be attributed to tunneling associated with a distribution of two level systems (TLS).[2]

Pressure is a useful experimental parameter which can be used to increase the density of a crystal or glass, with a resulting increase in the coupling between an impurity molecule and the host matrix. In a crystal, a pressure increase typically results in increased phonon frequencies (including pseudo-local phonon modes), however, it is less clear how pressure affects the potential energy surface of a glass.

2. Results and Discussion

In figure 1 we show the effect of pressure on the optical dephasing times of a dilute mixed crystal of pentacene in p-terphenyl as a function of temperature. The data at various pressures are fit to the equation $T_2^* = T(\infty)\exp(\Delta E/kT)$ where $T(\infty) = 3$ ps for this system.[3] The Arrhenius type plots clearly show the increased dephasing rates at higher temperatures as well as the increased activation energy ΔE associated with the pseudolocal phonon at higher

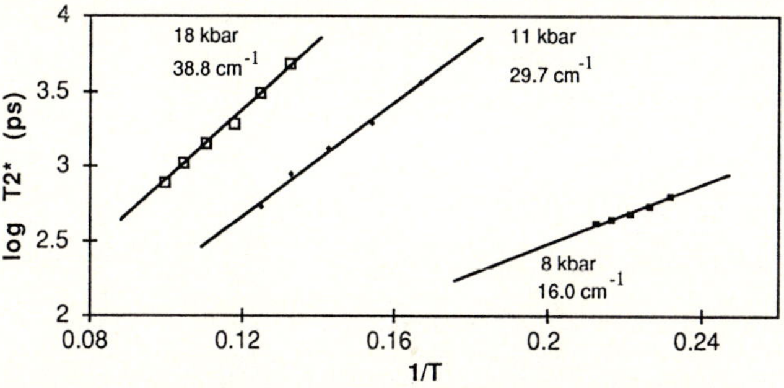

Figure 1. Arrhenius plots of the dephasing time for pentacene in p-terphenyl at different pressures.

Springer Series in Chemical Physics, Vol. 60 **Ultrafast Phenomena IX**
Editors: P. F. Barbara · W. H. Knox · G. A. Mourou · A. H. Zewail

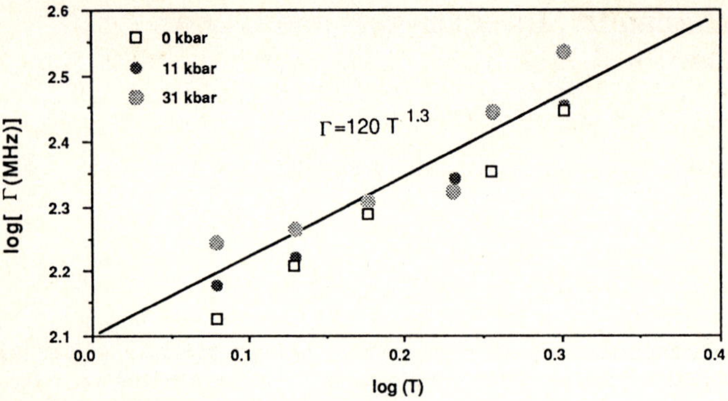

Figure 2. Log-log plot of the dephasing rate versus temperature for Rhodamine 640 in PMMA at different pressures

pressures. The slope of the Arrhenius fits yield the pressure induced change in the energy of the pseudo-local phonon, as shown in figure 1.

In figure 2 we show the effect of pressure on the homogeneous linewidth (dephasing time) of Rhodamine 640 dissolved in a PMMA glass. The data are fit to the equation $T_2^* = aT^b$ (with a=120±10MHz and b=1.3±0.07). Although increased density clearly alters the intermolecular interactions between the impurity molecule and the matrix, the electronic dephasing rate of Rh 640 in PMMA appears to be pressure independent over the temperature range investigated. This suggests that even though the tunneling coordinate is compressed, the tunneling dynamics for this system are unaffected by the density change (e.g. >15% volume compression at 31 kbar), possibly due to increased tunneling barrier heights.

3. Summary

A clear difference is observed in the effect of pressure on optical dephasing in mixed crystals versus the effect of pressure on spectral diffusion in a PMMA glass. Specifically, increased pressure shows little effect on the optical linewidth an impurity chromophore in a PMMA glass (over the temperature range studied), in clear contrast to the significant pressure effects observed in mixed crystals [3,4]. The contrasting effect of pressure on optical dephasing in crystals and glasses may provide a new tool for investigating mechanisms of optical dephasing.

4. Acknowledgment We acknowledge the National Science Foundation (CHE-9008551) for financial support of this research.

References

1. W. H. Hesselink and D. A. Wiersma, J. Chem. Phys. 73 (1980) 648.
2. L. R. Narasimhan, Y. S. Bai, M. A. Dugan, and M. D. Fayer, Chem. Phys. Lett. 176 (1981) 335.
3. B. J. Baer and E. L. Chronister, J. Phys. Chem. submitted.
4. R. A. Crowell and E. L. Chronister, Chem. Phys. Lett. 216 (1993) 293.

On the Role of Quantum Interference in the Free-Induction Decay from Wave-Packet Systems

J.N. Sweetser, T.J. Dunn, and I.A. Walmsley

The Institute of Optics, University of Rochester, Rochester, NY 14627, USA

Abstract. The role of vibrational wave packets in the ground and excited electronic states of a molecule in determining the temporal structure of the free induction decay is examined. Interference between wave packets on different surfaces causes a temporal shortening of the emitted light pulses, and modifies the dispersion of the resonant transition.

The emission of light from atoms and molecules in which the internal degrees of freedom are in wave packet states has been suggested as a possible source of coherent radiation with certain desirable properties, such as short duration or specific pulse shapes.[1,2] We present in this paper a theoretical and experimental examination of some of the issues associated with such a source. In particular, we address the case of free induction decay following the creation of a vibrational wave packet in a molecular sample by the application of an ultrashort optical pulse near an electronic resonance. In this case ultrashort means that the duration is such that not only is dephasing negligible, but also the area theorem does not hold.

The experimental apparatus is the same as that used in Ref.[3] A pulse of mean wavelength 630 nm , with about 1mJ energy is incident upon a vapor of sodium contained in a heat pipe held at a temperature of 650 K. The dimer density is about $10^{15}cm^{-3}$, and the interaction region is about 8cm long. The transmitted radiation is detected with 60fs temporal resolution by upconverting it in a nonlinear crystal. The free induction decay from the sample is characterized by its high degree of spatial coherence --it is emitted in the direction of the pump pulse, and is very intense compared to the fluorescence. Fig. 1 shows the results of a typical measurement. The complicated temporal structure is quite unlike that measured for the fluorescence from the same system under the same conditions. There are several features worthy of note. First, the fastest structure has a periodicity that is not that of the excited state ($A^1\Sigma_u^+$) wave packet alone. Second, the radiation consists of pulses that are shorter than those measured in the fluorescence from the outer turning point of the wave packet's trajectory. Third, there is an overall modulation envelope to the radiation that is not observed in the fluorescence. These features may be explained in terms of two principles: the interference of wave packets in different electronic states, and the modification of dispersion of the sample by quantum interference.

The emission consists in general of two parts, an incoherent part that depends only on the wave packet in the excited state, and a coherent part, that depends on the wave packets in both excited and ground electronic states (of a two-state molecule). In terms of the state vectors of the wave packets in the ground state ($|\Psi_g(t)\rangle$) and excited state ($|\Psi_e(t)\rangle$), the detected signal from the coherent emission is proportional to the two-time field correlation function:

$$\left\langle \hat{E}^{(+)}(t)\hat{E}^{(-)}(t')\right\rangle_{Coh} \propto \left\langle \Psi_e(t)\middle|\Psi_g(t)\right\rangle\left\langle \Psi_g(t')\middle|\Psi_e(t')\right\rangle. \qquad (1)$$

Thus, in contrast to the incoherent emission, it is clear that the coherent signal will reflect the modulations arising from the interference of the ground and excited state wave packets that are generated by absorption and intrapulse Raman scattering respectively. Thus the free induction decay consists of pulses that are spaced by the commensurate period of the ground and excited state vibrational modes (approximately 200fs and 310fs for our experiment) leading to the emission of pulses of about 100fs duration spaced by 600fs. Because the emission depends on the correlations between the two wave packets, the emitted pulses are of significantly shorter duration than found

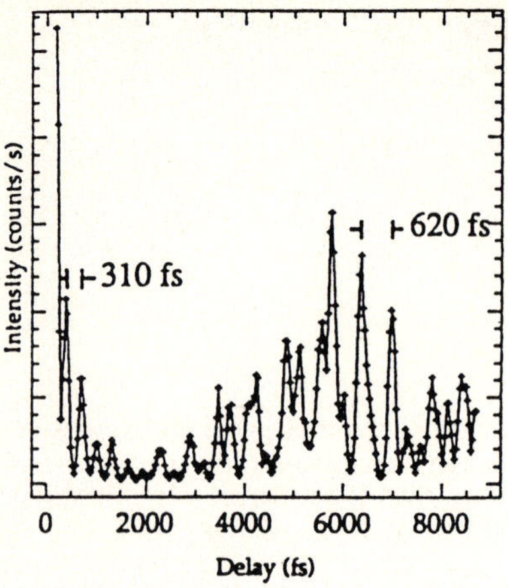

Fig 1. Measured free induction decay from sodium dimers. The pulse repetition period is the commensurate of the ground and excited vibrational wave packets, indicating that they interfere to completely suppress emission at certain times.

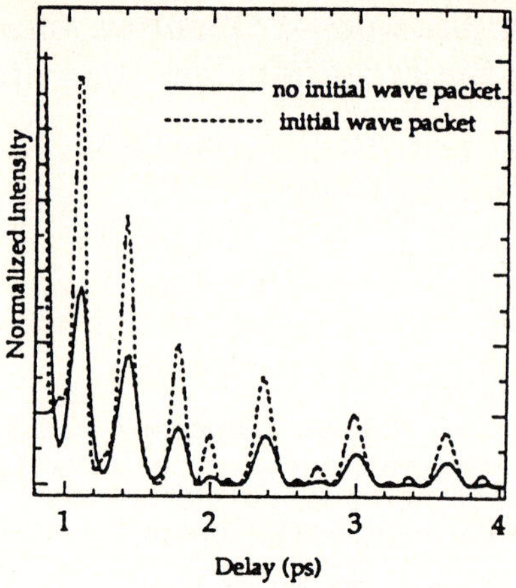

Fig. 2. A theoretical plot showing the transmission of a probe pulse when (i) a wave packet exists in the excited state, and (ii) there is no wave packet. The modification of the dispersion is clearly seen.

in the incoherent radiation. Furthermore, if the excited state wave packet is time-stationary, then any signal modulation will be due solely to the motion of the ground state wave packet. It is this feature that leads to significant temporal modulation of the coherent emission even during periods when the excited state wave packet has "decayed" or become diffuse, due to Hamiltonian evolution on the anharmonic potential surface of the excited state.

The slow temporal modulations that appear on the signal arise because the input pulse is distorted upon propagation through a resonant medium. A novel feature of coherent propagation in the weak field limit through multi-level system is that the pulse area decays to zero much more quickly than in a two-level system, even though the transmitted field envelopes are quite similar, because of the interference of the different radiating transitions. In the strong-field limit we do not find any evidence of self-induced transparency. We believe that this is because we are in a extreme regime of transiency in which the McCall-Hahn area theorem is invalid. The presence of a coherent superposition of vibrational states can also alter the transmitted pulse shapes, since the absorption and dispersion of the sample undergoes a temporal modulation. We have observed the effects of this modulation on a delayed probe pulse, and the results of a calculation are shown in Fig. 2.

In conclusion, we note that the fluorescence from an excited molecule reveals effects of wave packet dynamics in different electronic states , through the coherent emission, and to the wave packet dynamics in a single electronic state , through the incoherent emission. The free induction decay from a wave packet system exhibits further temporal modulation because of the modification of the absorption and dispersion of the sample by the wave packets.

References

[1] J. Parker and C. R. Stroud, Jr., Phys. Rev. A, 41, 1602 (1990); L. D. Noordam, H. G. Muller, A. ten Wolde and H. B. van Linden van den Heuvell, J. Phys. B 23, L115 (1990)
[2] I. Averbukh and M. Shapiro, Phys. Rev. A, 47, 5086 (1993)
[3] T. J. Dunn, J. N. Sweetser, I. A. Walmsley, and C. Radzewicz, Phys. Rev. Lett., 70, 3388 (1993)

Femtosecond Two-Dimensional Raman Spectroscopy of Liquid Water

S. Palese[1], *J.T. Buontempo*[1], *Y. Tanimura*[1], *S. Mukamel*[1], *R.J.D. Miller*[1], *and W.T. Lotshaw*[2]

[1]Department of Chemistry and Institute of Optics, University of Rochester, Rochester, NY 14627-0216, USA

[2]General Electric Research and Development Center, P.O. Box 8, Schenectady, NY 12301, USA

Abstract. Higher order Raman spectra for water have been calculated, within the Brownian oscillator model, using femtosecond optical Kerr effect (OKE) results to demonstrate the distinctions between the homogeneous and inhomogeneous limits. Experimental results suggest this distribution is inhomogeneously broadened.

Critical to the understanding of condensed phase reaction dynamics is the understanding of a solvent's response to and influence on the system dynamics. Although models of a solvent's vibrational and orientational dynamics can be tested against infrared and Raman studies of the neat solvent, there are ambiguities as the spectral density is usually masked by inhomogeneous broadening. However, valuable information about the solvent nuclear dynamics can be extracted from nonlinear spectroscopies such as infrared and Raman echoes. Using the formalism of Tanimura and Mukamel [1] we have calculated the 3rd, 5th, and 7th order Raman spectra for liquid water employing a multimode Brownian oscillator model to represent the water modes. This model is expected to be appropriate for liquid water because it accounts for nuclear dynamics occurring on a finite time scale, extrapolating continuously between homogeneous and inhomogeneous limits.

The $\chi^{(3)}$ response function of water was experimentally determined using OKE spectroscopy with pulse widths of 38 fs to provide the spectral density between 1-800 cm^{-1}. The analytical fit to these results and calculated free induction decay is shown in fig. 1. Using the fit, the higher order Raman responses were calculated in the

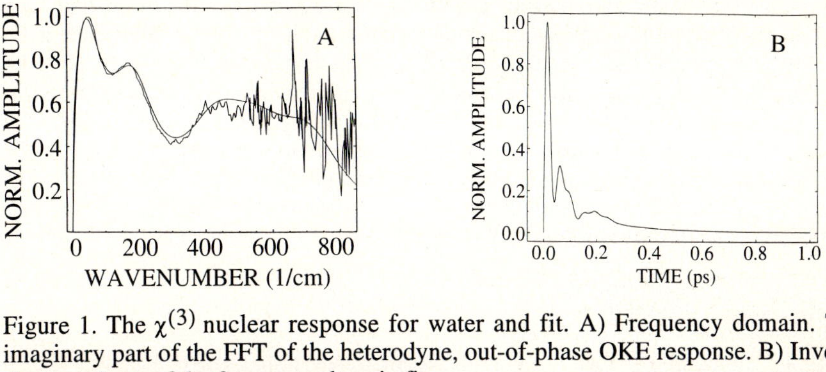

Figure 1. The $\chi^{(3)}$ nuclear response for water and fit. A) Frequency domain. The imaginary part of the FFT of the heterodyne, out-of-phase OKE response. B) Inverse sine transform of the frequency domain fit.

Springer Series in Chemical Physics, Vol. 60 **Ultrafast Phenomena IX**
Editors: P. F. Barbara · W. H. Knox · G. A. Mourou · A. H. Zewail

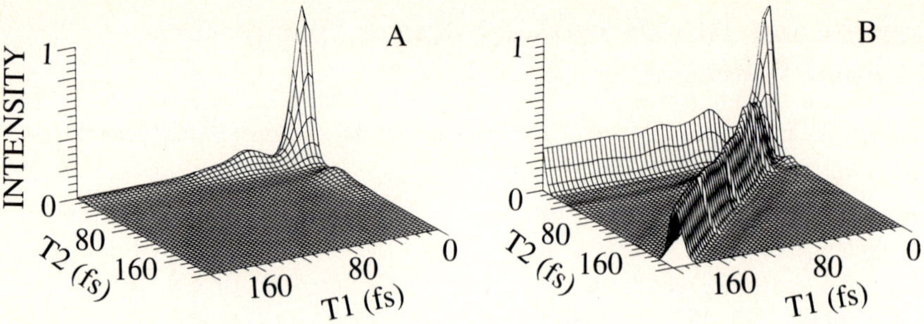

Figure 2. The calculated $|\chi^{(5)}|^2$ response. A) Homogeneous limit. B) Inhomogeneous limit. A) T1- Excitation delay, T2- Probe delay from second excitation.

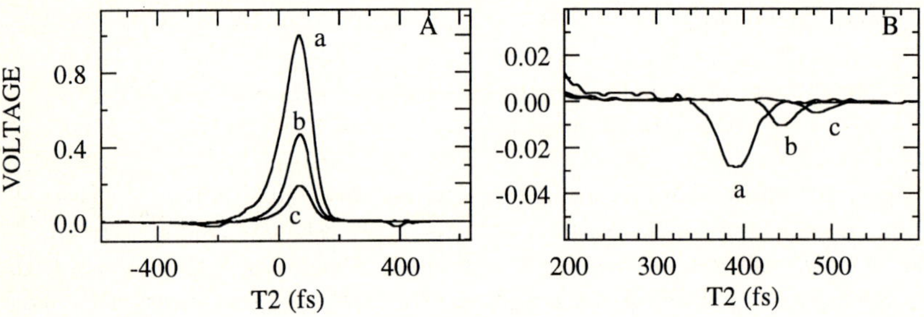

Figure 3. The correlation response collected along the diffracted beam direction (A). T1 = a) 253 fs, b) 303 fs, c) 353 fs. B) Recurrence at $\approx$ 2(T1) expanded.

homogeneous and inhomogeneous limits. These calculations are shown in fig. 2 for the 5th order ($\propto \chi^{(5)}$) response. The 7th order ($\propto \chi^{(7)}$) response, which is analogous to the Raman echo, shows similar behavior. In 5th order there are clear distinctions between these limits even for a multimode system as complex as the intermolecular modes of water. The most valuable diagnostic, in this regard, is the rephasing of the coherence in the nuclear polarization for the inhomogeneous limit.

Preliminary three pulse grating studies ($\chi^{(3)}$ $k_D = (k_1 - k_2) + k_{probe}$, $\chi^{(5)}$ $k_D = (k_1 - k_{1'}) + (k_{1''} - k_2) + k_{probe}$) conducted using amplified (300 nJ/pulse) 60 fs pulses suggest that the water distribution is inhomogeneously broadened (fig. 3). The large peak is the $\chi^{(3)}$ contribution. In this configuration the decay of the recurrence is not due to the homogeneous relaxation time of liquid water, but instead to the temporal overlap of the 1st and 2nd pulses which form the 2nd Raman excitation. However this configuration has superior S/N in discriminating the recurrence relative to collection along the probe direction, due to the presence of a strong background from the two pulse homodyne signals. Even for signal collection along the probe direction mixing between the various orders, because of the square law detector, will lead to an observed decay of the recurrence which is not related to the homogeneous lifetime.

References

1. Y. Tanimura and S. Mukamel, J. Chem. Phys. **99**, 9496 (1993).

Solvent Dependence of the Femtosecond Three-Pulse Photon Echo for LD690 in n-Alcohols

C.J. Bardeen and C.V. Shank

Department of Chemistry, University of California, Berkeley, CA 94720, USA and Materials Sciences Division, MS 70-193A, Lawrence Berkeley Laboratory, Berkeley, CA 94720, USA

Abstract: The fast dynamics of the oxazine dye LD690 in a series of n-alcohols are investigated using several techniques, including linear absorption, resonance raman, and the femtosecond three pulse photon echo.

The coupling of solvent nuclear motion to the solute's electronic degrees of freedom strongly affects the electronic absorption spectrum of the solute, usually as a source of broadening. In this work the molecule LD690 is used as a probe of the solvent-solute interaction in n-alcohols and we use a combination of methods (linear absorption, resonance raman, and the femtosecond three pulse photon echo (3PPE)) to determine the origins of its spectral broadening in solution.

To describe these experiments we use the formalism developed by Mukamel and others [1]. The molecule is considered to have two electronic levels, each coupled to a set of normal modes. For each mode, the ground and excited state minima are displaced from each other by an amount D. Intramolecular vibrational modes are assumed to execute oscillatory motion along their respective nuclear coordinates, while modes that describe the interaction of the chromophore with the solvent undergo damped motion as they experience dissipation due to the environment. The solvent-solute interaction consists of two modes, one of which has a very slow relaxation time and contributes to spectral broadening as a gaussian distribution of transition frequencies. The other mode is characterized by a correlation function $M(t)=\Delta^2 \mathrm{sech}^2(t/\tau_c)$ and describes the femtosecond dephasing.

The absorption spectrum of LD690 is observed to redshift and narrow as the solvent is varied from methanol along the series of n-alcohols to 1-decanol, indicating that the solvent has a significant effect on the electronic transition. The resonance raman spectrum shows 18 peaks with significant intensity, with a dominant peak at 587 cm^{-1}. In the limit of fast dephasing, the resonance raman intensity is proportional to $D^2\omega^2$, and in this way we can estimate the relative couplings of the modes to the electronic transition. Using parameters derived from the resonance raman data and the time-resolved experiments described below, we were able to reproduce the linear absorption spectrum fairly accurately.

In the 3PPE experiment, if we set t_{13}, the delay of the third pulse, to a fixed value and scan t_{12}, the delay of the second pulse, we measure the loss of coherence between the ground and excited state wavefunctions. Fig. 1 shows this experiment done with 10 fs pulses for LD690 in methanol and in 1-pentanol. Here t_{13}=120 fs in order to suppress the 587 cm^{-1} mode oscillations [2]. Also shown are calculated signals based on parameters derived from the resonance raman, the experimental pulse shape, and the broadening parameters discussed below. The strong asymmetry in these signals indicates that in both solvents there is significant inhomogeneity on the femtosecond timescale. Also, for $t_{12}>0$, the signal in 1-pentanol decays more slowly than that in methanol, indicating that the fast component (homogeneous broadening) is changing from solvent to solvent.

Springer Series in Chemical Physics, Vol. 60 **Ultrafast Phenomena IX**
Editors: P. F. Barbara · W. H. Knox · G. A. Mourou · A. H. Zewail

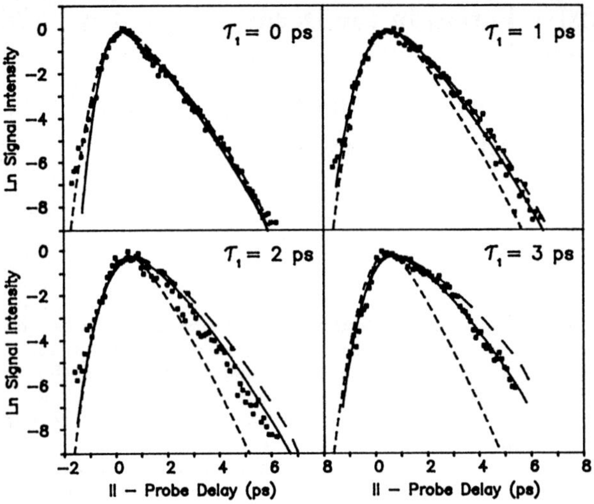

Fig. 1. **Points**: Raman echo data from 1:1 $CH_3Cl:CDCl_3$. **Short Dashed**: Signal decay in the absence of a rephasing pulse. **Long Dashed**: Best fit to Raman echo with a static inhomogeneity. **Solid**: Fit to Raman echo data including a spectral diffusion time of 5 ps.

References.

1. R.W. Schoenlein, L.A. Peteanu, R.A. Mathies and C.V. Shank, Science **254**, 412 (1991).
2. A.H. Zewail, M. Dantus, R.M. Bowman, and A. Moktari, J. Photochem. Photobiol. A: Chem. **62/3**, 301 (1992).
3. N.F. Scherer, L.D. Ziegler, and G.R. Fleming, J. Chem. Phys. **96**, 5544 (1992).
4. U. Banin and S. Ruhman, J. Chem. Phys. **98**, 4391 (1993).
5. K.S. Schweizer and D. Chandler, J. Chem. Phys. **76**, 2296 (1982).
6. Y.-X. Yan, L.T. Cheng, and K.A. Nelson, in *Advances in Nonlinear Spectroscopy*, edited by R.G.H. Clarke and R.E. Hester (Wiley, New York, 1987).
7. D. Vanden Bout, L. Muller and M. Berg, Phys. Rev. Lett. **67**, 3700 (1991).
8. D. Zimdars, A. Tokmakoff, S. Chen, S.R. Greenfield, M.D. Fayer, T.I. Smith and H.A. Schwettman, Phys. Rev. Lett. **70**, 2718 (1993).
9. L.J. Muller, D. Vanden Bout, and M. Berg, J. Chem. Phys. **99**, 810 (1993).
10. G. Döge, R. Arndt, H. Buhl and G. Bettermann, Z. Naturforsch. Teil A **35**, 468 (1980).

Part III

Spectroscopy and Advances in Measurements

Interferometric Measurements of Femtosecond Laser-Cavity Dispersion

K. Naganuma

NTT Opto-electronics Laboratories, 3-1 Morinosato-wakamiya, Atsugi 243-01, Japan

Abstract. We present the cavity Fourier-transformed crosscorrelation, a quick and universal way to measure cavity dispersion of lasers using their amplified spontaneous emission.

1. Introduction

Wavelength dispersion measurement method has been drawing much attention as an indispensable tool for femtosecond laser development. For individual optical element, these methods have been well established using interferometers[1-3].

W. H. Knox developed a method called frequency-domain dispersion (FDD)[4] and pioneered the cavity dispersion measurements of live mode-locked lasers. FDD requests two aspects of target lasers; spontaneous pulsation without external modulation and an intra-cavity tuning element keeping the cavity length constant. For semiconductor lasers, femtosecond time-of-flight measurements were conducted to detect wavelength-dependent time separation between internally reflected echo-pulses[5].

Here we describe an extension of the interferometric method to measure the laser cavity dispersion. Using this method, any laser, irrespective of lasing mode, can be measured for the entire gain band wavelengths without any external light source.

2. Principle of Cavity Fourier-Transformed Crosscorrelation (FTC)

In the interferometric dispersion measurements, a white light electric field E_i is crosscorrelated to a convoluted field $E_s = E_i \otimes t_s$, where t_s is the transfer function of sample. To obtain the convoluted field, ordinary optics needs to be inserted in one of the arms of an interferometer. In contrast, when the sample is a cavity, the convoluted field is automatically generated while a light is circulating inside the cavity, which enables dispersion measurements on the cavity placed outside the interferometer. The path difference of the now empty interferometer should be equal to the sample cavity length to allow E_s to catch up to E_i. If the sample cavity is light emitting, *i.e.* a laser or an optical amplifier, its own amplified spontaneous emission (ASE) serves as the light source. At last, the basic measurement setup is just an imbalanced Michelson interferometer to crosscorrelates ASE from a target laser kept under the lasing threshold.

For instance, a long-range scan of interferometric correlation for a 1.55-μm laser diode is depicted in Fig. 1. Ordinary interferogram, or electric field autocorrelation

Springer Series in Chemical Physics, Vol. 60 **Ultrafast Phenomena IX**
Editors: P. F. Barbara · W. H. Knox · G. A. Mourou · A. H. Zewail

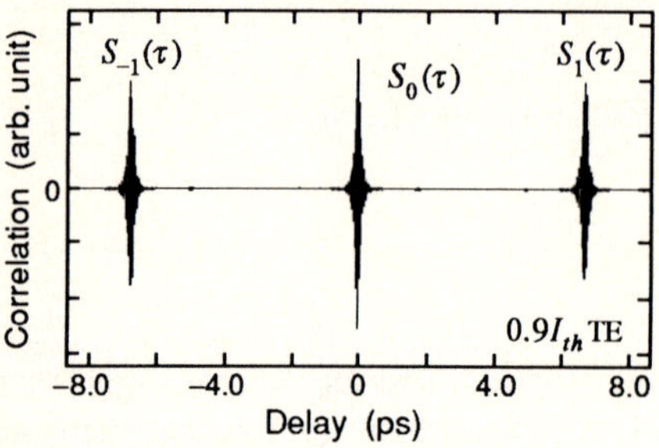

Fig. 1 Typical interferometric correlation for a semiconductor laser diode. In addition to the ordinary interferogram S_0, subfringe S_1 carrying round-trip transfer function information appears. In the opposite time-delay direction, its mirror image S_{-1} comes out with functions of interferometer arms exchanged.

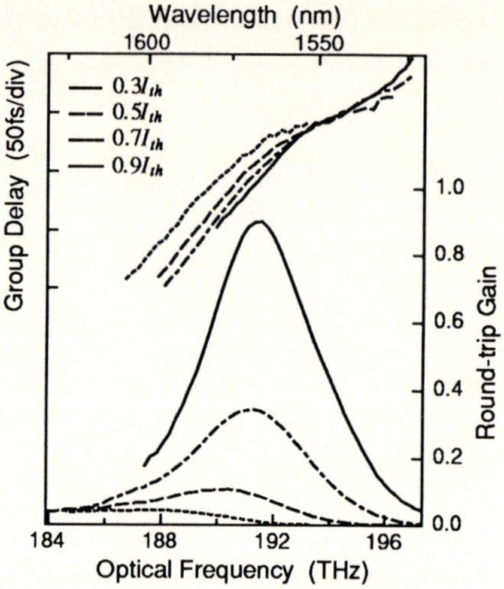

Fig. 2 Measurement results on a 1.55-μm strained-layer quantum well BH laser diode. When the current is raised to I_{th}, the gain approaches unity as expected.

$S_0(\tau)$ appears near the zero-delay position. When time delay τ is increased, an interference signal reappears near the cavity round-trip time (≈ 7 ps). This subfringe $S_1(\tau)$ should contain information on the transfer function of the cavity. To extract the information, we exploit Fourier transform of interference signals which is expressed as $\tilde{S}_0(\omega) = t_{\text{bias}}(\omega)U(\omega)$ and $\tilde{S}_1(\omega) = t_{\text{RT}}(\omega)t_{\text{bias}}(\omega)U(\omega)$. By calculating the quotient, $\tilde{S}_1(\omega)/\tilde{S}_0(\omega)$, pure round-trip transfer function $t_{\text{RT}}(\omega)$ can be separated with the apparatus-specific bias $t_{\text{bias}}(\omega)$ and the ASE spectrum $U(\omega)$ completely eliminated. Then the round-trip gain is calculated as $|t_{\text{RT}}(\omega)|^2$, while the group delay spectrum is obtained by $d\arg[t_{\text{RT}}(\omega)]/d\omega$.

In experiment, we used a Michelson interferometer whose path difference is calibrated by a He-Ne laser. ASE from a sample cavity was fed to the interferometer through a signal-mode optical fiber, to reduce beam divergence. Semiconductor sample cavities were temperature-controlled during measurements to prevent cavity length drift. Results for a 1.55-μm strained-layer quantum well BH laser diode[6] are shown in Fig. 2. On the longer wavelength side, group delay decreases for larger injection current due to the α-parameter effect. In the case of this laser, the dispersion at the gain peak wavelength remains almost constant across different injection currents, allowing extrapolation of these below-threshold measurements to the lasing state.

3. Dual-interferometer scheme

With typical femtosecond lasers where discrete optics are assembled on a chassis, the cavity length fluctuates ceaselessly (ca. $\Delta L \approx 0.5$ μm) because of air flow and ambient

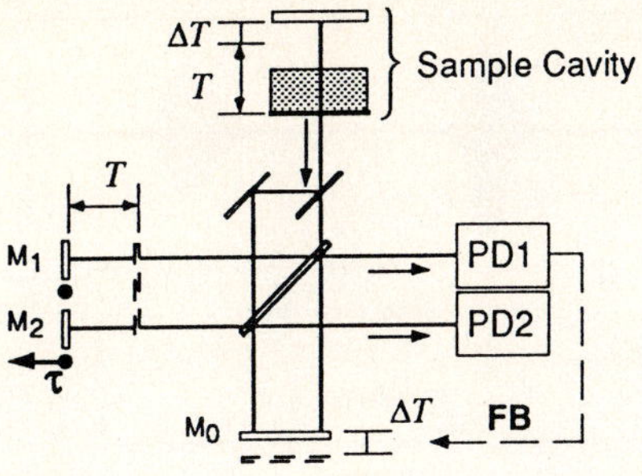

Fig. 3 Dual-interferometer scheme. Interferometer #1 which is kept still detects cavity length fluctuations and shifts the time-delay origin accordingly by moving the PZT-attached mirror M_0. FB, feedback; PD, photo-detector.

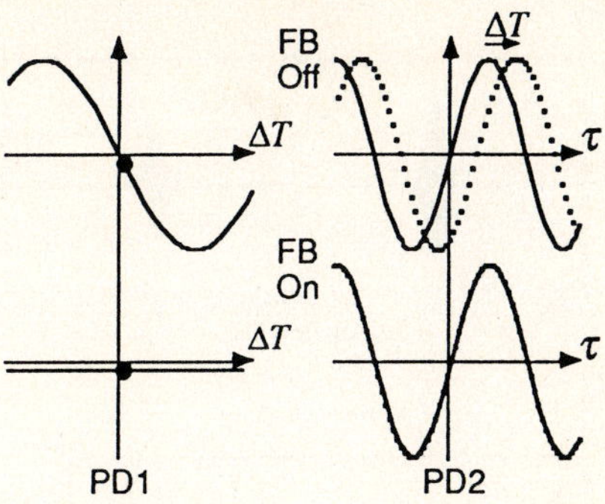

Fig. 4 Stabilization operation. The time-delay τ of the scanning interferometer is defined based on the displacement between M_1 and M_2, which is measured by a third local interferometer using a He-Ne laser (not shown in Fig. 3).

vibration, which hindered measurement efforts. To address this problem, we devised a dual-interferometer scheme[7] shown in Fig. 3 in which one of the interferometers is dedicated to detecting length change in the test cavity and enabling the influence of fluctuations to be nullified.

Figure 4 shows the operation principle behind the scheme. Without feedback to a shared mirror M_0, the output of interferometer #1 (PD1) follows the round-trip time change ΔT as shown in the top-left figure, while the crosscorrelation signal on the output of interferometer #2 (PD2) suffers a shift corresponding to ΔT (top right). Once the feedback is activated (PD1 output is fed to M_0 so that it is kept constant as shown in the bottom-left figure), interferometer #1 tracks the sample cavity continuously and provides a reference point anchored to the changing round-trip time. The crosscorrelation signal is taken on the time-delay axis relative to this reference point and is thus free from the unwanted shift (bottom right).

4. Femtosecond laser measurements using dual-interferometer

By now two femtosecond laser cavities have been measured using this method. Figure 5 shows measurement results on a NaCl:OH^- color-center laser cavity. The laser was a Burleigh Instruments Model FCL-230 with a 1.5-m cavity length ($L/2$), and generated 130-fs pulses under additive-pulse modelocking (APM) operation. An auxiliary cavity for the APM was blocked to measure the main cavity characteristics. Under a cw (non-pulsed) pumping by a Nd:YAG laser, correlation signals were measured for an ASE light with a power of 48 nW out of the fiber to the interferometer. In Fig. 5(a), small difference between $S_0(\tau)$ and $S_1(\tau)$ implies that intracavity dispersion is tiny.

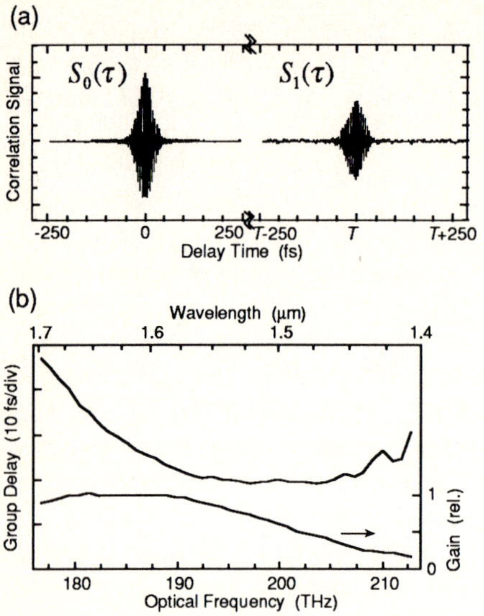

Fig. 5 Measurement results for a femtosecond color-center laser cavity. (a) correlation signals. (b) group delay and power gain spectrum. Tiny dispersion in a femtosecond domain has been clearly detected.

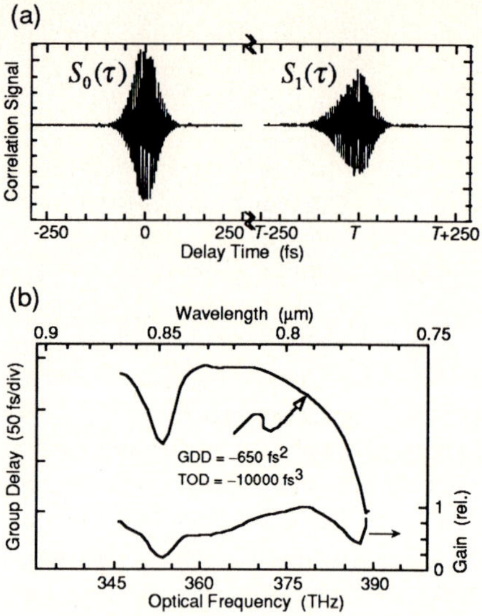

Fig. 6 Measurement results for a mode-lock Ti:sapphire laser cavity having an SF6 prism pair. This laser operates under negative GDD. It also shows a large negative TOD due to the overcompensation by the prism pair.

Figure 5(b) shows the obtained spectrum for group delay and power gain. Accuracy for these measurements is estimated to be 0.59 fs for group delay and 3.4% for gain with a 3.7-THz (28-nm) resolution. The group delay spectrum shows that this laser operates in wavelengths where group-delay dispersion (GDD) is very close to zero. For a NaCl:OH^- color-center amplifier, Sucha *et al.*[8] pointed out that negative GDD of glass partially compensates positive GDD accompanied by the crystal. Similar circumstances hold true for this laser cavity. Yet regarding to the third-order dispersion (TOD), both components have the same sign, which is responsible to a positive TOD observed here.

Figure 6 shows measurement results on a self-modelock Ti:sapphire laser cavity. Besides a laser rod (20-mm long), the cavity contains a dye (HITCI) jet for the modelocking initiation, as well as a prism pair (SF6, apex distance $l = 250\,\mathrm{mm}$) for the dispersion control. Total cavity length amounts to 1.83 m ($L/2$). Typical pulsewidth from the laser is 100 fs. In Fig. 6(a), a rather broad interferogram $S_0(\tau)$ implies a spectral limitation, probably resulting from an unoptimized combination of mirrors. The asymmetric contour of the subfringe $S_1(\tau)$ is a quick indication of large TOD in the cavity.

In Fig. 6(b), the group delay spectrum carries downward spikes on both sides. Together with dips on the gain, they reiterate the mirror problem. Between the spikes, the group delay is approximated by a parabola from which the cavity dispersion at the pulse center wavelength is read GDD=-650 fs^2 and TOD=-1×10^4 fs^3.

In terms of dispersion, dominant components in the cavity are the rod and the prism pair. At an 800-nm wavelength, dispersion of the latter is expressed as [9], $D_2(\mathrm{fs}^2) = -25.7l + 674x$ and $D_3(\mathrm{fs}^3) = -63.2l + 452x$, where both l (apex distance) and x (prism insertion) are in mm. Difficulty in the dispersion estimation with prism pair arises from its large x-dependence. Once the D_2 is given by the measurement, combining with a measured dispersion of the rod (D_2 =2560 fs^2, D_3 =2040 fs^3), the formula predicts an accompanying cavity TOD of -11500 fs^3, which is in satisfactory agreement with the measured TOD. Thus the measurement results are highly consistent with the now common dispersion control prescription.

In conclusion, owing to the universality, the cavity FTC method might be especially powerful when applied to such new solid state lasers with which achievement of modelock operation is being endeavored. It can also serve effortless measurements of passive waveguides having parallel facets by injecting a white light externally.

References

1. W. H. Knox, N. M. Pearson, K. D. Li, and C. A. Hirlimann, Opt. Lett. **13**, 574 (1988).
2. K. Naganuma, K. Mogi, and H. Yamada, Opt. Lett. **15**, 393 (1990).
3. M. Beck, and I. A. Walmsley, Opt. Lett. **15**, 492 (1990).
4. W. H. Knox, Opt. Lett. **17**, 514 (1992); W. H. Knox, and J. P. Gordon, J. Opt. Soc. Am. B **10**, 2071 (1993).
5. K. L. Hall, G. Lenz, and E. P. Ippen, IEEE J. Lightwave Technol. **LT-10**, 616 (1992).
6. K. Naganuma, Appl. Phys. Lett. **64**, 261 (1994).
7. K. Naganuma, and Y. Sakai, Opt. Lett. **19**, 489 (1994).
8. G. Sucha, S. R. Bolton, and D. S. Chemla, IEEE J. Quantum Electron. **QE-28**, 2163 (1992).
9. R. L. Fork, C. H. Brito Cruz, P. C. Becker, and C. V. Shank, Opt. Lett. **12**, 483 (1987); We used the refractive index data for SF-6 glass found in HOYA Optical Glass catalog, HOYA Co.

Near-Field Time-Resolved Fluorescence Spectroscopy on Single Molecules and Photosynthetic Membranes

X.S. Xie[1], *G.R. Holtom*[1], *L. Mets*[2], *and R.C. Dunn*[1]

[1]Pacific Northwest Laboratory, P.O. Box 999, Richland, WA 99352, USA
[2]Department of Molecular Genetics and Cell Biology, University of Chicago, 1101 East 57th Street, Chicago, IL 60637, USA

Abstract Fluorescence lifetime measurements on single dye molecules and photosynthetic membranes with nanometer spatial resolution are achieved by incorporating time-correlated single photon counting with a near-field fluorescence microscope.

Near-field Fluorescence Microscopy

Near-field Fluorescence Microscopy [1] permits the fluorescence imaging and spectroscopy of single chromophores in an ambient environment with a spatial resolution beyond the diffraction limit [2,3]. Time resolved spectroscopy combined with near-field optics [3,4] will allow one to probe molecular motions and chemical reactions on a single molecule basis.

Probing Single Molecule Dynamics

Figure 1A shows a fluorescence image of single sulforhodamine 101 molecules dispersed on a glass surface. The image was obtained by raster scanning the sample while keeping the aluminum coated fiber tip [1] approximately 7nm above the suraface, as shown in Fig. 1B. The sub-diffraction intensity profile (dots in Fig. 1B) has a FWHM of 125nm, limited by the tip aperture. We have implemented time correlated photon counting with the near-field optics to measure fluorescence lifetimes [3,4]. A three dimensional histogram (x,y,t) of emission counts was accumulated in 20min as the sample was raster scanned, resulting in a six nanosecond long

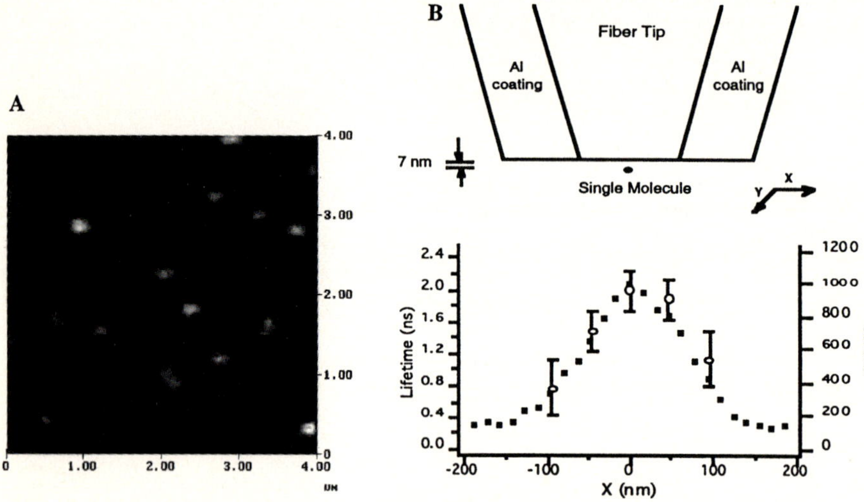

Figure 1. A. Near-field fluorescence image of single sulforhodamine 101 molecules dispersed on a glass surface. B. Single molecule emission lifetimes (circles) and emission intensities (dots) are plotted as a function of x displacement across a single molecule.

Springer Series in Chemical Physics, Vol. 60 **Ultrafast Phenomena IX**
Editors: P. F. Barbara · W. H. Knox · G. A. Mourou · A. H. Zewail

movie of the fluorescence decays of a 2μmX2μm field of single molecules. From this movie, we extracted the single molecule emission lifetimes (circles in Fig. 1B) as a function of the lateral displacement of the tip relative to the molecule. Shortening of fluorescence lifetimes near the tip edges is due to excited state energy transfer to the aluminum coating, which can be minimized by centering the tip on the molecule. This demonstrates the feasibility for detailed studies of photoinduced chemical reactions of single molecules in specific local environments.

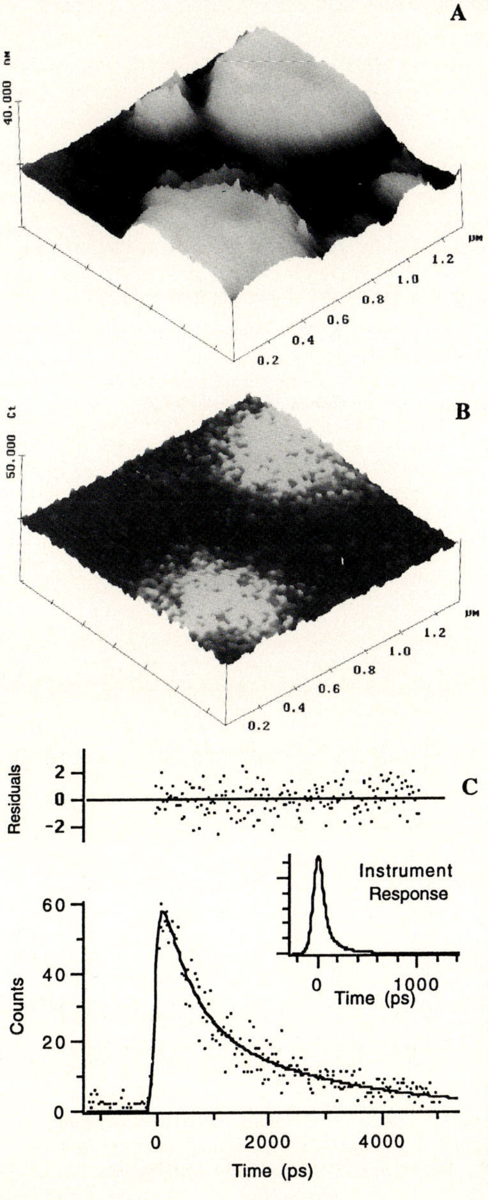

Figure 2. A. Shear force image of four intact photosynthetic membrane fragments, 7nm thick, laying flat on a mica surface. B. Near-field fluorescence image taken simultaneously with A, which shows only the two big pieces contain fluorescent chlorophyll molecules. C. Fluorescence decay measured by holding the tip above a single spot of the membrane. The decay is fit with a sum of two exponentials with lifetimes of 450ps (65%) and 2.7ns(35%). The instrumental response shown in the inset has a FWHM of 120ps, limited by the avalanche photodiode detector used.

Photosynthetic Membranes

The near-field technique offers the possibility of a spectroscopic mapping of various proteins in the photosynthetic membrane[4]. Figures 2A and 2B show the simultaneous shear force image[1] and near-field fluorescence image of intact thylakoid membrane fragments from green alge containing only light harvesting complexes. These images enable us to correlate topographic and spectroscopic information. The four bilayer membrane pieces seen in Fig. 2A lay down flat on a mica surface with only the two big pieces containing fluorescent chlorophyll molecules as seen in Fig. 2B. The emission decay shown in Fig. 2C was measured at a single spot on the membrane. This demonstrates that electron and energy transfer dynamics can be studied on a single membrane protein in the native state with this nonperturbative approach.

Acknowledgements

We acknowledge the contributions of Vey Allen, Gordon Anderson and Aida Pascual and the funding from the U. S. Department of Energy under Contract No.DE-AC06-76RLO 1830.

References

1. E. Betzig, J. K. Trautman, *Science* **257**, 189 (1992).
2. E. Betzig, R. J. Chichester, Science **262**, 1422 (1993); W. P. Ambrose, P. M. Goodwin, J. C. Martin, R. A. Keller, *Phys. Rev. Lett.*, **72**, 132 (1994).
3. X.S . Xie, R. C. Dunn, *Science* in press.
4. R. C. Dunn, G.R. Holtom, L. Mets, X. S. Xie, *J. Phys. Chem.* **98**, 3094 (1994).

Frequency-Resolved Optical Gating: Measuring the Intensity and Phase of an Ultrashort Laser Pulse

K.W. DeLong[1], *D.N. Fittinghoff*[1], *R. Trebino*[1], *A. Sullivan*[2], *J. Hunter*[2], *W.E. White*[2], *and D.J. Kane*[3]

[1]Sandia National Laboratories, MS 9057, Livermore, CA 94551-0969, USA
[2]Lawrence Livermore National Laboratories, Livermore, CA 94550, USA
[3]Southwest Sciences, Inc., 1570 Pacheco Street, Suite E-11, Santa Fe, NM 87501, USA

Frequency-Resolved Optical Gating, (FROG) has achieved considerable success in measuring the complete intensity and phase evolution of an ultrashort laser pulse in a single shot without any assumptions on the form of the pulse. It combines ideas from acoustics, astronomy, and image science, and uses a simple experimental apparatus along with a numerical algorithm in order to reliably solve this long-standing problem for pulses in the infrared, visible and ultraviolet.

1. Introduction

The measurement of the intensity and phase of an arbitrary ultrashort optical pulse has been the subject of vigorous research activity for many years. Traditionally, the intensity autocorrelation and the spectrum, and more recently the interferometric autocorrelation, have been the main diagnostic techniques used for characterizing ultrashort pulses. Recently, several methods based on time-frequency techniques have been proposed. So far only one of these, Frequency-Resolved Optical Gating, or FROG [1-4], can operate on a single-shot basis. FROG has been demonstrated in the visible, ultraviolet, and infrared, on pulses with energies varying from 5 nJ to 300 μJ.

The FROG technique consists of a fairly simple experimental apparatus, combined with a numerical data analysis algorithm. Experimentally, the pulse to be measured is split into two beams, and these are then mixed in a (nearly) instantaneously responding nonlinear medium in an autocorrelation-type geometry. The mixing signal from this process is simply frequency resolved (using a spectrometer). The nonlinear optical process can be Kerr polarization rotation, self-diffraction, second- or third-harmonic generation, cascaded second-order effects, etc. In recording the spectrum of the mixing signal for all relevant time delays of the two input beams, one generates what is known as the FROG trace. This FROG trace is then used as the input to a numerical algorithm. This algorithm, which is based on phase-retrieval techniques, inverts the FROG trace to give the electric field of the pulse (intensity and phase) as a function of time uniquely and reliably.

Springer Series in Chemical Physics, Vol. 60 **Ultrafast Phenomena IX**
Editors: P. F. Barbara · W. H. Knox · G. A. Mourou · A. H. Zewail

2. Theory

Polarization-gate (PG) FROG is the most intuitive version of FROG [1, 2]. In PG FROG, the probe pulse passes through two crossed polarizers. The gate pulse, which is polarized at a 45° angle to the probe, is focused along with the probe into a material with a third-order nonlinear optical response (e.g. glass). As a result of the nonlinear Kerr effect, a birefringence is induced in the material, and, in the limit of small induced birefringence, a signal field of the form

$$E_{sig}(t,\tau) = E(t)|E(t-\tau)|^2$$

is transmitted through the second polarizer, where τ is the time delay between the probe and gate pulses. This signal field is then frequency-resolved by a spectrometer, and the intensity of the spectrum detected by a detector array. The spectrum of this signal is the FROG trace, defined by

$$I_{FROG}(\omega,\tau) = \left| \int_{-\infty}^{\infty} dt\, E_{sig}(t,\tau)\exp(i\omega t) \right|^2 .$$

The delay between the two beams can be varied either by a standard delay line (multi-shot version) or by crossing the beams at finite angle (as in a single-shot autocorrelator). In the single-shot version of FROG, an imaging spectrometer, or a simple lens-grating-lens combination, should be used to eliminate astigmatic effects present in non-imaging spectrometers. This single-shot version of FROG allows one to obtain FROG traces in real time, making FROG a valuable real-time monitor of pulse quality.

The FROG trace is a type of *spectrogram* [5] of the pulse. Due to their visual appeal and intuitiveness, the spectrogram (a time-frequency quantity) is the preferred representation of complex waveforms in fields such as acoustics. The spectrogram contains all the information about the waveform, except for trivial ambiguities such as overall phase and absolute time of arrival.

3. FROG Geometries

As mentioned above, there are various beam geometries which can be used to generate FROG traces [6]. The polarization-gate geometry, discussed above, is an automatically phase-matched process, and therefore can be used to measure longer pulses in the single-shot geometry (where large crossing angles would be necessary). PG FROG requires propagation of the probe through a polarizer, and so may not be suitable for very short (less than ~70 fs) or short-wavelength pulses (although one could insert a similar polarizer in the gate beam, and analytically remove the effect of the polarizers).

In self-diffraction (SD) FROG [3], the mixing signal at $2\mathbf{k}_{probe} - \mathbf{k}_{gate}$ is input into the spectrometer, where $\mathbf{k}$ is the spatial wavevector. This process is not phase-matched, so that the single-shot version cannot obtain large temporal delays due to the small crossing angles required. SD FROG does not require propagation through external optics, however, so that it may be the most suitable method for ultraviolet or extremely short pulses.

For low-power pulses, less than roughly 100 kW peak power, it becomes necessary to use second-harmonic generation (SHG) as the FROG nonlinearity [7]. Despite its experimental simplicity, SHG FROG has several disadvantages. Due to the equivalence of the probe and gate beams, SHG FROG traces are always symmetric with respect to delay time, just as usual second-order autocorrelations are. This feature significantly reduces the utility of the SHG FROG trace, making it difficult to use SHG FROG as a real-time monitor. It also introduces a temporal ambiguity into the retrieved field, so that the direction of time is indeterminate. For example, equal magnitudes of positive and negative chirp are indistinguishable. This in turn can lead to difficulties for the pulse-retrieval algorithm, which, although quite reliable, is not as robust as the algorithm for PG FROG.

Several geometries which can in principle be used for FROG have not yet been demonstrated experimentally. Third-harmonic generation, while perhaps useful for long wavelength sources, is clearly limited in its wavelength range. Using cascaded second-order effects (frequency doubling one beam, and then using the other beam to parametrically up- or down-convert the signal), may provide the advantages of third-order FROG along with the low-power sensitivity of SHG FROG, although they too are limited in their wavelength range.

4. Pulse-Retrieval Algorithm

The problem of the inversion of the FROG trace to obtain the pulse intensity and phase is equivalent to the well-known two-dimensional phase-retrieval problem in imge science. The algorithm we use is based on the iterative-Fourier-transform type of phase-retrieval algorithm developed by the astronomy and image-science communities [2, 8].

The details of the algorithm have been described elsewhere [9]. In its basic form, the FROG algorithm consists of Fourier transforming a guess field back and forth between the time and frequency domains. In the frequency domain, the magnitude of the guess field is replaced with the square root of the FROG trace intensity, while the phase of the guess field is left unchanged. The algorithm itself determines the proper phase; hence the name "phase retrieval." In the time domain, we enforce the constraint that the signal field must be generatable from a physically realizable electric field, according to the equations which govern the particular nonlinear interaction geometry used.

In the basic FROG algorithm as originally published, a simple integration over the delay variable τ was used to generate a new guess for the pulse electric field. This *ad hoc* technique was quite effective for most pulses in the PG version of FROG. It could not, however, successfully invert the traces of pulses with complicated intensity profiles, nor did it perform as well as hoped for in other

FROG geometries, such as SD and SHG. Instead, we apply the method of generalized projections in these cases. Briefly, this involves minimizing an error function which ensures adherence to the governing nonlinear optical equations. This technique is very effective, and gives excellent results in the cases where the basic algorithm fails. Generalized projections are discussed in detail elsewhere [7].

5. Measurement of Ti:Sapphire Laser Pulses

We used FROG to measure pulses from a Spectra-Physics Tsunami Ti:Sapphire laser. Due to the low pulse energies available from the oscillator, it was necessary to use SHG as the nonlinear optical interaction. In general, the Tsunami outputs near-transform-limited pulses over most of its normal lasing range. We were able, with some work, to obtain several unusual pulses from the laser.

Figure 1 shows the FROG trace, and Figure 2 the resulting electric field, of a pulse with significant cubic spectral phase distortion. This pulse was obtained by maximizing the amount of prismatic glass in the cavity. The amount of cubic phase distortion is sufficient to cause significant satellite pulses in the time domain. We believe that the excessive amount of glass in the cavity imparts this high-order phase characteristic to the pulse.

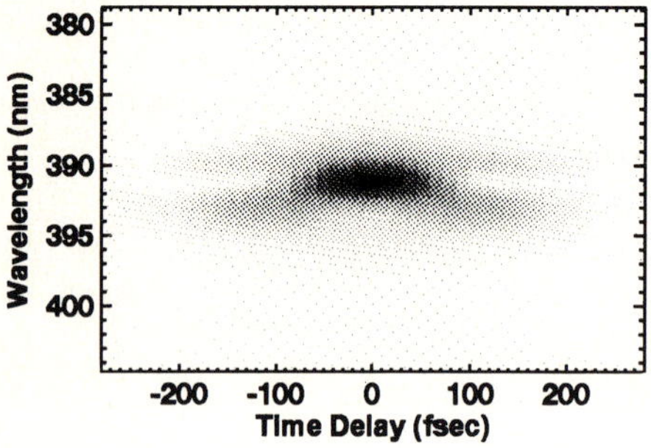

Figure 1. The SHG FROG trace of a pulse from a Ti:Sapphire laser with excess glass in the cavity. The presence of the horseshoe-shaped features indicates spectral cubic phase.

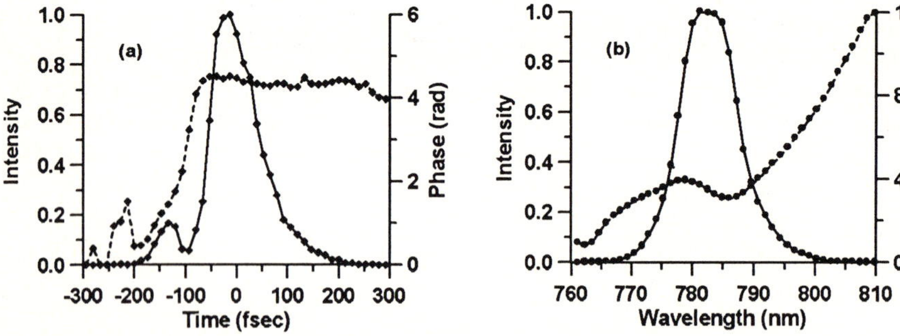

Figure 2. The intensity and phase in the (a) time and (b) frequency domains as derived from the data of Fig. 1.

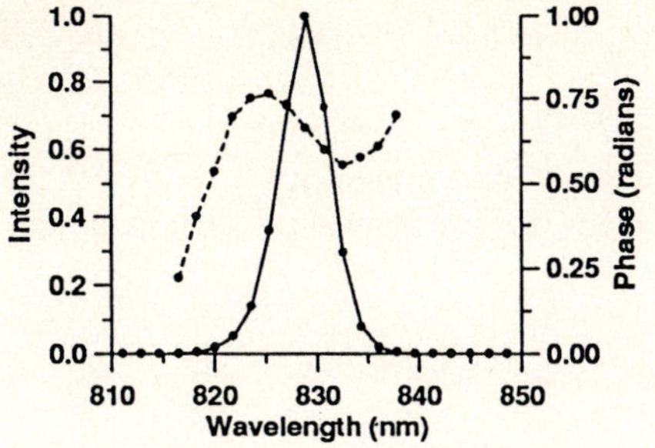

Figure 3. The intensity and phase as a function of wavelength of a pulse from a Ti:Sapphire laser which has been tuned outside its normal tuning range. The cubic phase present in the pulse is most likely a result of operation near the edge of the mirror coatings, where the dispersion of the mirrors become large. This pulse was also measured using SHG FROG.

Figure 3 shows another pulse with cubic spectral phase distortion. In this case, the laser was tuned to 830 nm, near the edge of the lasing range of the laser when the short-wavelength mirror set was installed. In this case, we believe that the pulse obtains its nonlinear chirp from the mirror coatings, which are being used outside their intended bandwidth. The amount of cubic phase in Fig. 3 is not sufficient to cause significant satellites in the time domain (not shown).

6. References

1. D. J. Kane and R. Trebino, Opt. Lett. **18**, 823 (1993).
2. R. Trebino and D. J. Kane, J. Opt. Soc. Am. A **10**, 1101 (1993).
3. D. J. Kane and R. Trebino, IEEE J. Quant. Electr. **29**, 571 (1993).
4. D. J. Kane, A. J. Taylor, R. Trebino and K. W. DeLong, Opt. Lett., (in press).
5. R. A. Altes, J. Acoust. Soc. Am. **67**, 1232 (1980).
6. K. W. DeLong, R. Trebino and D. J. Kane, J. Opt. Soc. Am. B, (in press).
7. K. W. DeLong, R. Trebino, J. Hunter and W. E. White, J. Opt. Soc. Am. B, (submitted).
8. J. R. Fienup, Appl. Opt. **21**, 2758 (1982).
9. K. W. DeLong and R. Trebino, J. Opt. Soc. Am. A, (in press).

Spatio-Temporal Holography

M.C. Nuss, M. Li, T.H. Chiu, and A. Partovi

AT&T Bell Laboratories, 101 Crawfords Corner Road, Holmdel, NJ 07733-3030, USA

Abstract. The mapping of temporal pulse shape information into spatial images is demonstrated using spatio-temporal holography. This is possible due to the fundamental equivalence of time and space in such an optical setup. Such time-space conversions have potential applications in multiplexing, demultiplexing, packet header recognition as well as in image processing.

There are quite a few analogies between spatial problems and their corresponding temporal counterparts in optics. Examples are the duality between diffraction and dispersion, which has resulted in the concept of a "time lens" [1], as well as the similarity between holographic and photon-echo optical memories. But the time and space versions of the above mentioned optical systems are quite different and do not allow to convert time into space and vice versa.

Here, we present an optical system in which the time-space duality is so complete that temporal and spatial information can be converted into each other [2]. Our experiment makes use of the fundamental equivalence of time and space in Fourier-transform holography with its two incarnations, conventional (spatial) holography and temporal (sometimes also called spectral) holography [3,4]. Fig. 1 illustrates the principle of our experiment, which contains elements of both spatial and temporal holography setups. A signal pulse carrying the temporal information and a short, featureless reference pulse are spectrally dispersed by a diffraction grating in the input plane, collimated by a lens, and interfere in a holographic medium placed at the spectrum plane.

The principle of the experiment is that once the hologram is written, it can fundamentally not be distinguished from a hologram written by spatial replicas of the temporal signal, with the scaling parameter given by $\lambda_0/c \cdot d\cos\theta$, where λ_0 is the center wavelength of the optical pulses, d the pitch of the grating, c the speed of light, and θ the diffraction angle [2]. Thus,

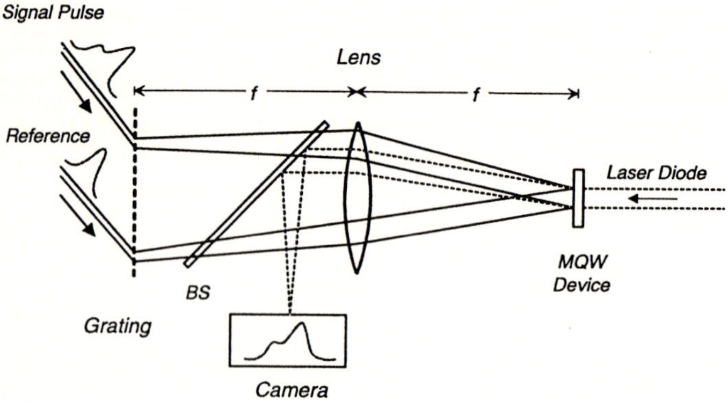

Fig. 1 Schematic view of the spatio-temporal holographic setup

Springer Series in Chemical Physics, Vol. 60 **Ultrafast Phenomena IX**
Editors: P. F. Barbara · W. H. Knox · G. A. Mourou · A. H. Zewail

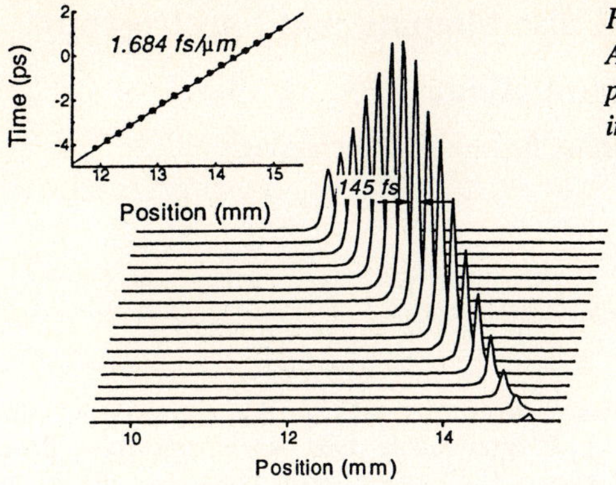

Fig. 2
Autocorrelation peaks of 100 fs pulses recorded by a CCD camera in the detector plane

this offers the opportunity to convert temporal data into spatial data and vice versa. For example, in the experiment described below, we accomplish time-to-space mapping by writing a temporal hologram with spectrally dispersed pulses, but then interpret the temporal hologram as a spatial hologram by reading it out with a monochromatic plane wave (Fig. 1).

In the experiment, we use a 600 lines/mm diffraction grating and a 200 mm focal length Fourier-transform lens. The center wavelength of the femtosecond laser is roughly 830 nm. A multi-quantum-well (MQW) photorefractive device, located at the spectrum plane, serves as a real-time holographic medium [5]. The MQW device is read out with a few mW from a collimated single-mode diode laser at 850 nm. The diffracted light travels back through the setup and is being picked off by a thin beam splitter *BS* and focused onto the detector plane.

Fig. 2 shows the correlation peaks recorded by a CCD camera in the detector plane for different time delays between the signal and reference pulses in the case where both signal and reference are identical 100 fs laser pulses. As the time delay between the two input pulses is varied, the correlation peaks move in the detector plane. We can obtain a calibration of the time-to-space mapping process by plotting the position of the correlation peaks versus the time delay between the two femtosecond input pulses (see inset in Fig. 2). We have also obtained the cross-correlation between a longer pulse and a short reference pulse by propagating one of the two pulses through 50 cm of single-mode fiber. In this case, the correlation peak is stretched out to 200 µm, corresponding to a 340 fs FWHM.

References

[1] B. H. Kolner, and M. Nazarathy, Opt. Lett. **14**, 630 (1989)
[2] M. C. Nuss, M. Li, T. H. Chiu, A. M. Weiner, A. Partovi, Opt. Lett. **19**, 664 (1994)
[3] A. M. Weiner et al., IEEE Journ. Quant. Electr. **J-QE 28**, 908 (1992)
[4] Y. T. Mazurenko, Appl. Phys. **B 50**, 101 (1990)
[5] A. Partovi et al., Appl. Phys. Lett. **62**, 464 (1993)

Programmable Femtosecond-Pulse Shaping and Spectroscopy

M.M. Wefers, H. Kawashima, and K.A. Nelson

Department of Chemistry, Massachusetts Institute of Technology,
Cambridge, MA 02139, USA

The ability to 'shape' ultrafast waveforms will profoundly expand the capabilities of optical spectroscopy [1]. Programmable fsec waveforms have been generated previously by using a liquid crystal (LC) spatial light modulator (SLM) to selectively attenuate and/or retard the spatially separated frequency components of an angularly dispersed ultrashort pulse [2,3]. Here we report the amplification and improved fidelity of ultrashort waveforms generated in this way, and their application to the coherent control of potassium vapor.

A pulse shaping device has been designed that incorporates two well registered LC SLMs in a single housing. By aligning the liquid crystals of the two masks at 45 and -45 degrees respectively and sandwiching the device between horizontal polarizers, complete manipulation of both spectral amplitude and phase is possible. The SLMs consist of 128 97-micron pixels separated by 3-micron gaps. Because the spotsize of the incident frequency components is focused to about 60 microns, the limitations arising from the gaps and the discreteness of the SLMs can easily be compensated for waveforms whose temporal range is up to 40 times as long as the incident pulse. Multiple pulse sequences with arbitrary delay, phase, and chirp have been generated using an incident 75-fsec optical pulse from a Ti:sapphire oscillator. Optical square pulses of up to 1.8 psec have also been generated.

Figure 1a shows a waveforms of four equal intensity, bandwidth limited, phase-locked pulses. For waveforms consisting of more than four pulses the agreement with the desired waveform (dotted) begins to diminish, as in fig. 1b. Such imperfections can be corrected by iteratively improving the experimentally generated waveform, preferably after amplification, by using pulse characterization techniques such as FROG [4].

We have amplified the shaped waveforms in a Ti:sapphire regenerative amplifier. Since the frequency operations of shaping and amplification are linear, the temporal profile of the shaped waveform is preserved. Figure 2a shows the cross-correlation of an unamplified 800-fsec square pulse (dotted) along with the auto-correlation of the experimental amplified square pulse (circles) and the anticipated auto-correlation based on the unamplified waveform (solid). Figure 2b shows the cross-correlations of an unamplified (dashed) and amplified (solid) phase-locked equal-intensity three pulse sequence. The broadening of the individual pulses in the amplified waveforms is similar to that of an unshaped single pulse. The amplification produces a uniform transverse spatial

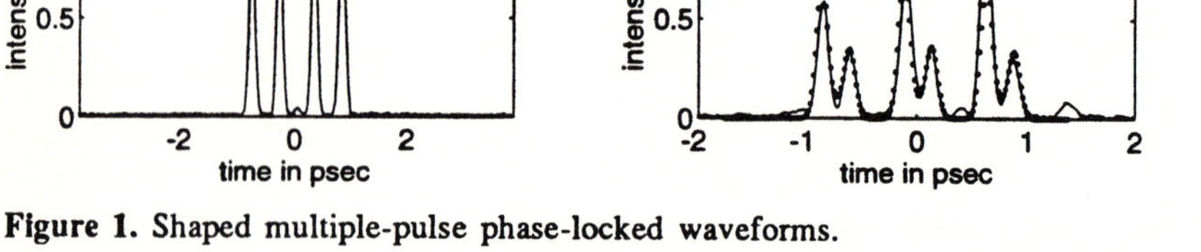

Figure 1. Shaped multiple-pulse phase-locked waveforms.

Springer Series in Chemical Physics, Vol. 60 **Ultrafast Phenomena IX**
Editors: P. F. Barbara · W. H. Knox · G. A. Mourou · A. H. Zewail

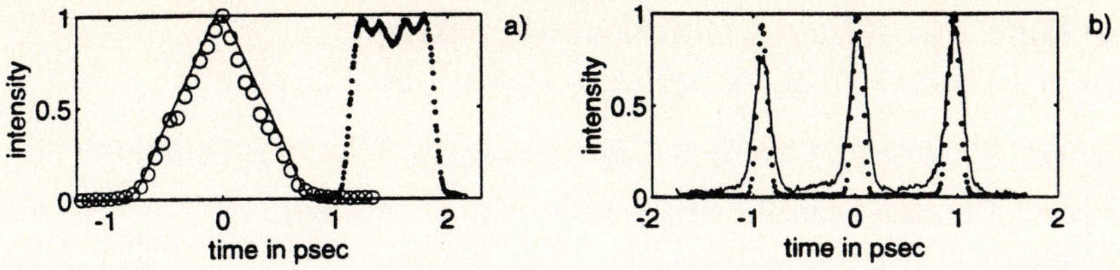

Figure 2. Auto-correlation (a) and cross-correlation (b) of shaped amplified and unamplified waveforms. See text for details.

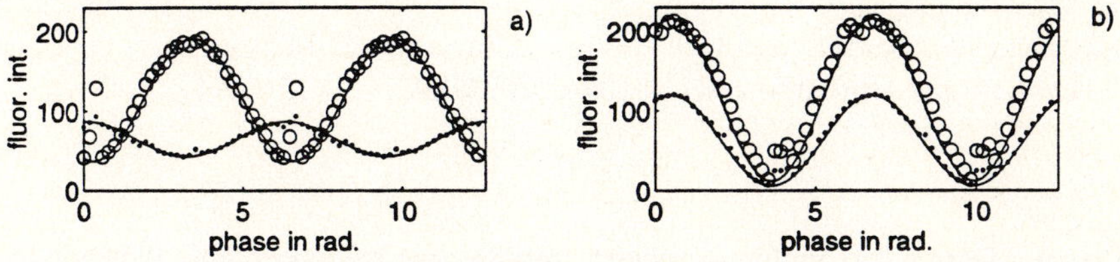

Figure 3. Frequency-resolved fluorescence from K vapor versus the phase difference between two phase locked pulses with time separation a) 288 fsec and b) 566 fsec.

profile unlike the unamplified shaped waveforms whose transverse profiles shifts slightly along the temporal profile as a result of diffraction off the SLMs [5].

Finally we have used pairs of equal intensity unamplified phase-locked pulses to control selectively the fluorescence from atomic K vapor. With resonances at 766.5 nm and 769.9 nm, arising from spin-orbit coupling, the sample behaves as a 3-level system for our 75-fsec incident pulses. By controlling the relative phase and delay of the pulses, fluorescence from the excited state levels can be enhanced or suppressed. Figure 3 shows the frequency-resolved fluorescence intensity versus the phase difference between the pulses for a pulse pair delayed by 288 fsec (fig. 3a) and 576 fsec (fig. 3b). The circular (dot) symbols denote fluorescence from the 766.5-nm (769.9-nm) resonance while the solid curves give the best sinusoidal fit to the data. The real data, with relative phases between 0 and 2-pi are duplicated in fig. 3a and fig. 3b to better show the phase-dependent trends. The 288-fsec delay is half the period associated with the spin-orbit splitting. In this case the fluorescence from one level can be enhanced while the fluorescence from the second level is suppressed. For the 566-fsec delay, corresponding to a full period of the splitting, the fluorescence intensity from the two excited state levels can be enhanced or suppressed simultaneously. These features are clearly demonstrated in fig. 3. The anomalous points near the minima of the sinusoidal curves are reproducible and are probably associated with the high repetition rate (76-MHz) of the unamplified light and the long-lived coherences of the potassium.

This research was funded in part by NSF grant CHE-8901722.

References

[1] W. S. Warren, H. Rabitz, and M. Dahleh, Science **259**, 2279 (1993)

[2] A. M. Weiner, D. E. Leaird, J. S. Patel, and J. R. Wullert, IEEE J. Quantum Electron. **28**, 908 (1992)

[3] M. M. Wefers and K. A. Nelson, Opt. Lett. **18**, 2032 (1993)

[4] D. J. Kane and R. Trebino, Opt. Lett. **18**, 823 (1993)

[5] M. M. Wefers and K. A. Nelson, submitted to J. Opt. Soc. Am. B (1994)

Real-Time Two-Photon Confocal Microscopy Using a Femtosecond, Amplified Ti:Sapphire System

J. Squier[1], *T. Norris*[1], *C. Bliton*[2], *G.J. Brakenhoff*[3], *and B. Athey*[4]

[1]Center for Ultrafast Optical Science, University of Michigan, 2200 Bonisteel Blvd., IST Bldg., Rm. 1006, Ann Arbor, MI 48109-2099, USA
[2]Meridian Instruments, Inc., 2310 Science Parkway, Okemos, MI 48864, USA
[3]Confocal Microscopy Group, Department of Molecular Cell Biology, University of Amsterdam, Plantage Muidergracht 14, 1018 TV Amsterdam, The Netherlands
[4]Department of Anatomy and Cell Biology, University of Michigan Medical School, Ann Arbor, MI 48109-0616, USA

Abstract. Using a Ti:sapphire kilohertz amplifier we have been able to produce the first real time 2-photon confocal microscope images.

Confocal microscopes improve resolution by creating optical sections with spatial filtering. These optical sections can be reconstructed to produce complete three-dimensional images suitable for studying a variety of important biological problems. However, over the time course of an image scan, confocal microscopes illuminate the full specimen volume with the excitation wavelength (as in standard fluorescence microscopy) which can limit the image quality. The use of a short optical pulse (from an ultra-short pulse laser) allows one to utilize two-photon excitation, which occurs only at the focus of the (high NA) objective, with several advantages following. Background scattered light is lower, and fluorophore bleaching only occurs at the objective focus instead of throughout the sample volume. Consequently, three-dimensional imaging experiments may be performed at higher excitation powers, greatly improving signal-to-noise, or lower excitation power could be used, and the number of image sets collected over time could be increased. Finally, since the excitation volume is lower with two-photon excitation, caged compounds may be released in a small (diffraction-limited) volume in a specimen. Thus, a new tool may be developed for the researcher: a miniature (micron-cubed) opto-chemical probe that can be used in a very non-invasive way.

2-photon laser scanning fluorescence microscopy was first demonstrated by Denk et al [1]. We have extended this technique to a confocal imaging system which uses a line focus and bilateral scanning technology to provide the fastest image capture available (Meridian Instruments, InSight Plus). Thus, we are able to produce real-time 2-photon images, to our knowledge, for the first time.

The laser source is a 450 kHz Ti:sapphire amplifier system [2]. The output of the system can be readily varied in repetition rate and energy. The chirp of the output pulse can also be varied positively or negatively, enabling us to compensate for quadratic and cubic phase errors which might occur as the pulse travels through the microscope. The flexibility of the laser enables us to study the various parameters which produce the most optimal images.

Springer Series in Chemical Physics, Vol. 60 **Ultrafast Phenomena IX**
Editors: P. F. Barbara · W. H. Knox · G. A. Mourou · A. H. Zewail

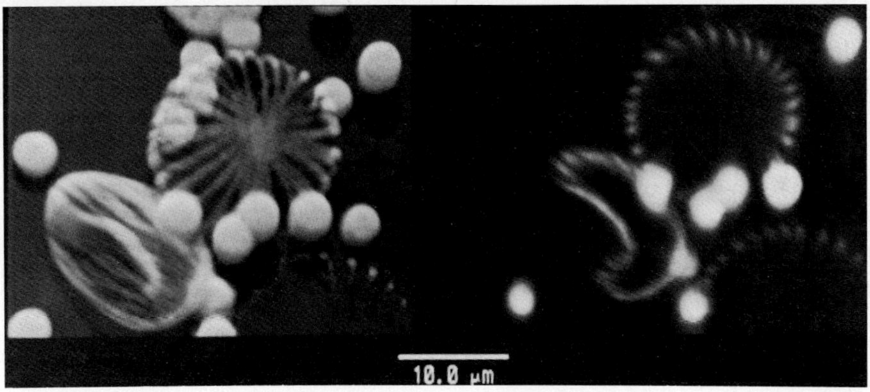

Figure 1. Image on the left is a three-dimensional image of spathaphyllum pollen, acridine orange label, and 4.7 µm beads, phycoerythrin label. The second image is one of the two-dimensional cross-sections which are used to reconstruct the full three-dimensional profile.

An example of the images produced by this technique are shown in figure 1.The laser pulse width was approximately 120 fs, at a center wavelength of 800 nm. The sample is spathaphyllum pollen, acridine orange label, and 4.7 µm beads, phycoerythrin label.

The first quantity of interest when attempting to make images with 2-photon processes, is a measure of the energy flux required to produce equivalent fluorescence intensities when compared to single photon excitation. Simple estimations based on the cross sections used in this series of experiments, places the equivalent 2-photon flux at 10-100 times that required in single photon excitation. This is notable from several respects. First, it is often thought that 2-photon excitation should be less damaging to a specimen. This is partially due to the fact that a longer wavelength is used, but also due to the fact that damage should be limited to essentially the focus, rather than throughout the entire focal volume. However, based on the fact that 10-100 times more flux is necessary to generate an equivalent fluorescent intensity, the damage threshold for the specimen must also be lowered by this equivalent amount in order to take advantage of this process [3].

Further, we have also found that there is an optimum repetition rate at which to run the laser based on the thermal loading of the specimen [3]. At high repetition rates the maximum energy E must be distributed during the thermal relaxation time of the specimen over a number of pulses n. The generated fluorescence intensity is then proportional to nE^2. However, by lowering the repetition rate so that there is only one pulse within the thermal relaxation time, the fluorescence intensity goes as n^2E^2. Thus for microsecond recovery times, we can expect 100 times greater fluorescence yield by using a repetition rate on the order of 1 MHz as opposed to 100 MHZ, with the same average 2-photon power in both cases.

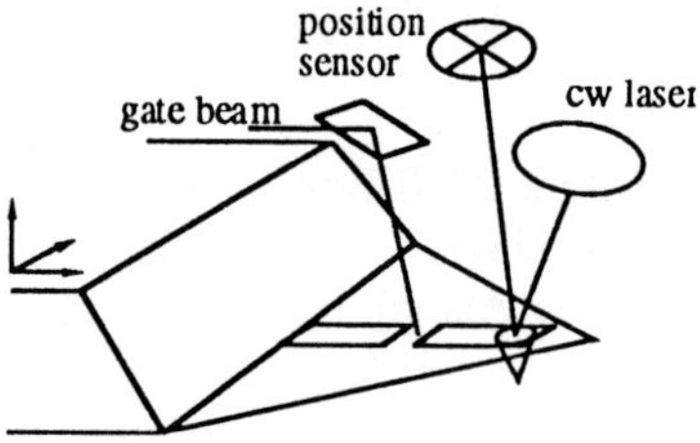

Figure 1 Diagram of ultrafast scanning force probe

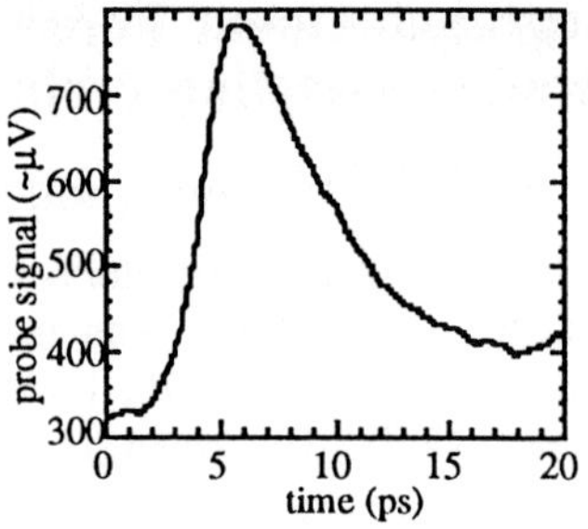

Figure 2 Waveform acquired by probe

switch for a period of about 1 ps (This samples the electronic waveform on the device-under-test by allowing an amount of current proportional to the waveform voltage to pass on to the detection electronics); The optical delay of the switching beam is scanned to perform equivalent time sampling of the repetitive waveform.

The experiment

In low-force contact, our probe has a dark conductance of 0.05 nS as measured using a Fluke 27 multimeter. The signal to be measured is generated by a 20 x 20-µm photoconductive switch on LT GaAs and carried on a coplanar stripline with 20-µm wide strips and spacing. The generating and sampling switchs are activated by two beams derived from a 100-fs Ti:Sapphire laser operating at 800 nm. Delay in the arrival of the two beams is scanned by a computer controlled optical delay line. The laser operates at a repetition rate of 100 MHz and produced 9 mW in the excitation beam and 0.8 mW in the probe beam.

The results

Images scanned by the probe show 10 to 100-nm features. We measured an impulse response of 2.5 ps (see figure 2). Signal-to-noise equal to unity was seen at a signal level on the order of 10 µV/√Hz.

We also constructed a probe with fiber illumination to integrate the ultrafast and ultrafine resolution into one instrument. With this fiber coupled probe we scanned an image of the transmission lines mentioned above and brought the probe into contact with one electrode to measure a picosecond transient. The temporal resolution of this result was limited to 4.5 ps by the dispersion of the 100 fs pulses in 30 cm of single-mode optical fiber. While an offset was present in the measurement, a signal to noise of unity was obtained at 1 µV with a 1-second time constant.

Conclusions and References

2.5 ps temporal resolution and 100-nm spatial resolution have been demonstrated using a single probe. This probe has been integrated into an instrument which can take advantage of the fine spatial resolution of scanning force imaging to aid in locating and placing the probe on fine devices for waveform measurements with µV/√Hz sensitivity. This probe not only provides a critical solution to the problem of internal circuit testing on VLSI, but it also opens the door to investigation individual mesoscopic devices.

D. R. Grischkowsky, *et al* IEEE J. Quantum Electron. **QE-24**, 221 (1988)

J. Kim, S. Williamson, J. Nees, S. Wakana, and J. Whitaker, Appl. Phys. Lett. **62** (18), 3 May 1993.

C.A. Spindt, L. Brodie, L. Humpfrey, and E. R. E. R. Westerberg, J. Appl. Phys. **47**, 5248 (1976).

Single-Shot, Pulse-Length-Limited Ultrafast Transient-Absorption Spectroscopy

J.T. Fourkas, L. Dhar, and K.A. Nelson

Department of Chemistry, Massachusetts Institute of Technology, Cambridge, MA 02139, USA

Irreversible ultrafast dynamics, such as photo-induced chemical reactions and high-intensity processes, have proven difficult to study spectroscopically in solids and viscous liquids. Although gases and low-viscosity liquids can be flowed fast enough that each set of laser pulses interrogates a new portion of the sample, such an approach is not practical in other samples. One way to apply ultrafast spectroscopic techniques to the study of irreversible processes in samples that cannot be flowed is to take an entire time-resolved data set in a single laser shot. Indeed, single-shot spectroscopic techniques have existed since almost the dawn of ultrafast lasers [1]. The impetus for using these methods, which revolve around the spatial encoding of temporal information, arose from the low repetition rates of early ultrafast lasers. However, as it became possible to produce shorter laser pulses at higher repetition rates, limited time resolution became an important constraint in these techniques. Thus, the prevalent practice in ultrafast spectroscopy is to delay variably the time between excitation and probe pulses, recording decays on a point-by-point basis along the time axis and averaging the signal over thousands of laser shots. Nevertheless, the attractions of single-shot time-resolved spectroscopy are undeniable, both for its insensitivity to shot-to-shot laser fluctuations and its fast data-collection rate.

We have demonstrated a new scheme for using the spatial encoding of temporal information to obtain ultrafast transient-absorption data in a single laser shot [2]. Pump and probe beams, which are separated by an angle that is generally much less than 90°, are cylindrically focussed such that their focal lines are spatially coincident within a sample. Because the pump and probe pulses reach different parts of the sample at different times, the inter-pulse delay is spatially encoded along the cylindrical axis. The decay information is then contained in the spatial intensity profile of the probe beam. As opposed to the geometries used in previous techniques, our geometry, which is similar ones used in single-shot autocorrelation schemes [3,4], is capable of providing pulse-length-limited resolution in samples of readily attainable thicknesses with laser pulses as short as tens of femtoseconds.

The experimental setup is shown in Fig. 1. The pump and probe beams are incident on a cylindrical lens that focusses the beams onto a common focal line in the sample. To normalize for shot-to-shot fluctuations in the intensity profile of the probe beam, a reference beam is brought into the lens parallel to and directly below the probe beam, such that it overlaps the other two beams in the sample, but arrives at an earlier time. The probe and reference beams are

Springer Series in Chemical Physics, Vol. 60 **Ultrafast Phenomena IX**
Editors: P. F. Barbara · W. H. Knox · G. A. Mourou · A. H. Zewail

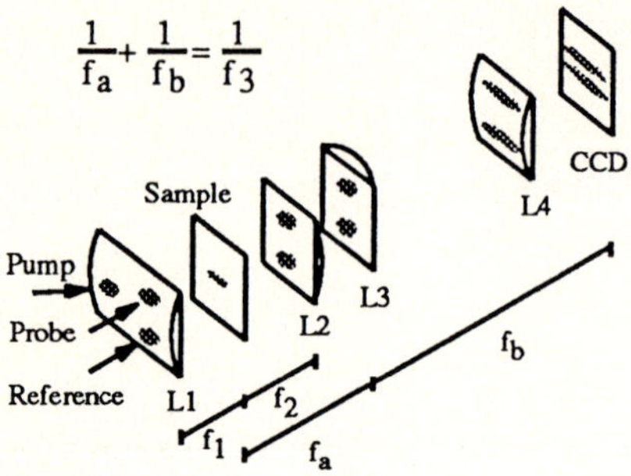

Figure 1. Setup for single-shot experiment. The L's are lenses and the f's focal lengths.

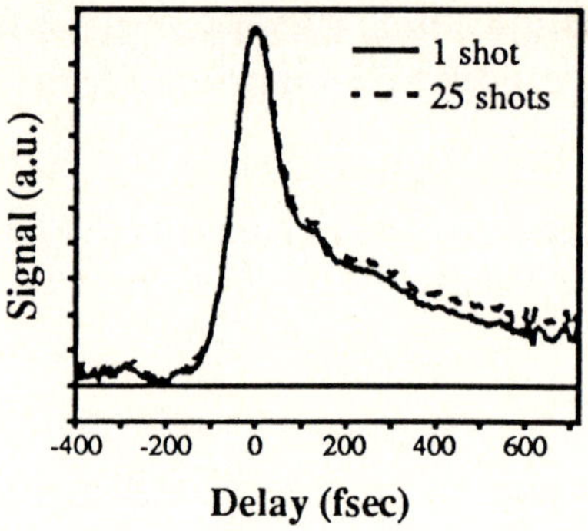

Figure 2. Single-shot data recorded from ethyl violet in methanol with one and 25 laser shots. Note the 155-fs oscillations.

then collimated and imaged onto a CCD, which records their spatial profiles on each laser shot. Data are recorded in a sample that has an essentially constant response on the time scale of the experiment (a few picoseconds) to normalize for the intensity profile of the excitation beam.

Shown in Fig. 2 are typical data, taken on a 100-μm-pathlength sample of ethyl violet in methanol. One set was taken in a single shot and one set with 25 shots. The 620-nm laser pulses were 72 fs in duration and were separated by an angle of 25°. A normalization scan for the pump-beam profile was taken in a similar cell using cresyl violet in methanol. The S/N of the single-shot data set is comparable to that of the 25-shot data set. Ethyl violet is known to display a 155-fs molecular-vibrational oscillation [5], which is clearly evident in the figure. Furthermore, a convolution of the known response of ethyl violet with the instrument response function expected for 72-fs pulses in a standard, collinear pump/probe experiment provides an excellent fit to the data.

These data demonstrate the great promise of single-shot ultrafast spectroscopy in studying irreversible phenomena on a femtosecond time scale. This and related techniques will open to investigation a wide range of samples that cannot be studied with conventional ultrafast techniques.

This research was supported by the ONR, grant N00014-90-J-4070.

References

[1] M. Malley and P. M. Rentzepis, Chem. Phys. Lett. **3** , 534 (1969).
[2] L. Dhar, J. T. Fourkas, and K. A. Nelson, Opt. Lett. **19,** 643 (1994).
[3] A. Brun, P. Georges, G. Le Saux, and F. Salin, J. Phys. D **24,** 1225 (1991).
[4] D. J. Kane and R. Trebino, Opt. Lett. **18,** 823 (1993).
[5] F. W. Wise, M. J. Rosker, and C. L. Tang, J. Chem. Phys. **86,**2827 (1987).

Observation of Fifth-Order Optical Response of Liquid CS_2 by Non-Resonant Six-Wave Mixing

K. Tominaga, Y. Naitoh, T.J. Kang, and K. Yoshihara*

Institute for Molecular Science, Myodaiji, Okazaki 444, Japan
*Permanent address: Department of Chemistry, Taegu University, Kyongsan 713-714, Korea

Abstract. We have observed a fifth order optical response of liquid CS_2 by non-resonant six-wave mixing which unambiguously determines the inhomogeneous contribution to intermolecular vibrational dephasing.

Vibrational dephasing in liquids has been a crucial probe of the interpretation of molecular dynamics in condensed phases. In the studies it is a key issue to investigate the relative importance of homogeneous and inhomogeneous contributions to the dephasing processes. Recent advances in nonlinear vibrational spectroscopies have allowed us to unambiguously determine the two contributions to a vibrational linewidth, and the techniques have been demonstrated for high frequency intramolecular vibrational modes [1-3]. In this report we have performed six-wave mixing spectroscopy to distinguish the homogeneous and inhomogeneous contributions to intermolecular vibrational modes.

Dephasing processes of low frequency modes have been observed by a femtosecond optical Kerr and impulsive stimulated light scattering (ISS) experiments using pulses whose durations are short compared to the vibrational oscillation period [4]. These experiments based on the third order susceptibility can not distinguish the two contributions. Recently Tanimura and Mukamel have proposed a fifth order experiment that is capable of distinguishing inhomogeneous contribution to low frequency modes [5].

In the fifth order experiment a medium interacts with two pairs of short pulses at t=0 and Δt_1, and the vibrational coherence is probed by monitoring a scattered light induced by a fifth pulse at $t=\Delta t_1+\Delta t_2$ [5]. According to their calculations, an echo-like siganl is observed at around $\Delta t_1=\Delta t_2$ in the inhomogeneous limit, while for the purely homogeneously broadened case the time-dependence of the fifth order signal is almost identical to that of the third order response when either Δt_1 or Δt_2 is scanned with the other time variable fixed.

We have performed a fifth order optical experiment on liquid CS_2 by non-resonant six-wave mixing at room temperature. A pulse with a duration of 90fs is split into five portions with equal intensities. The scattered light is identified as fifth order signal $I^{(5)}$ by its disappearance when any of the five pulses is blocked and its appearance at the correct phase-matching angle expressed by

$$\mathbf{k_s} = \mathbf{k_p} + (\mathbf{k_2} - \mathbf{k_2}') - (\mathbf{k_1} - \mathbf{k_1}')$$

where $\mathbf{k_s}$ and $\mathbf{k_p}$ are the wave vectors of the fifth order response signal and probe pulse, and $\mathbf{k_i}$ and $\mathbf{k_i}'$ are the wave vectors of the i-th pair of pulses. In order to separate $I^{(5)}$ from scattered lights mainly arising from the third order response, the

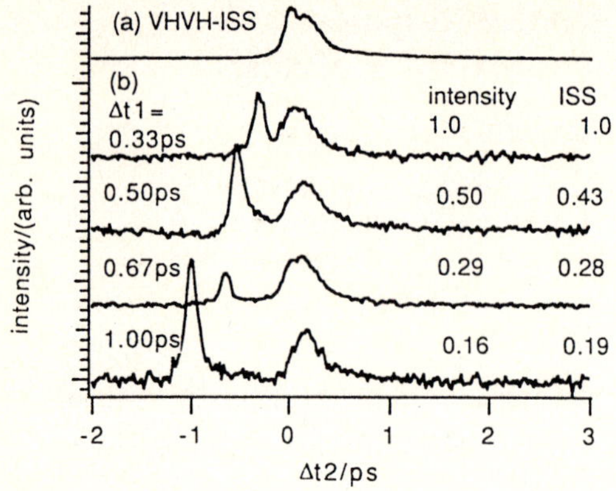

Fig. 1 (a) Impulsive stimulated light scattering signal from CS_2 as a function of delay time of the probe pulse. (b) Fifth order response signal as a function of Δt_2 with different fixed Δt_1.

first and second pairs of pulses are mechanically chopped with different frequencies, and the signal is detected at the frequency of the difference. The polarization condition is chosen so that $(\mathbf{k_s} // \mathbf{k_p} // \mathbf{k_1}' // \mathbf{k_2}') \perp (\mathbf{k_1} // \mathbf{k_2})$.

Figure 1(a) shows the ISS signal from CS_2 taken under the polarization condition of $(\mathbf{k_p} // \mathbf{k_1}') \perp (\mathbf{k_1} // \mathbf{k_s})$ [4]. In Fig. 1(b) the fifth order signals are illustrated as a function of Δt_2 with different fixed Δt_1. Scattered light at $\Delta t_2 = -\Delta t_1$ is mainly due to a third order response. It is notable that the variation of $I^{(5)}$ as a function of Δt_2 is almost identical to that of the ISS signal. In Fig. 1(b) the relative intensity of $I^{(5)}$ at maximum is shown with that of the ISS signal. The change of $I^{(5)}$ as a function of Δt_1 is also quite similar to that of the ISS signal. All the experimental results are consistent with the prediction by Tanimura and Mukamel for the case where the inhomogeneous contribution to the broadening of the low frequency vibrational modes of CS_2 is minor.

References

[1] L.J. Muller, D. Vanden Bout, and M. Berg, *J. Chem. Phys.* **99**, 810 (1993).
[2] R. Inaba, K. Tominaga, M. Tasumi, K.A. Nelson, and K. Yoshihara, *Chem. Phys. Lett.* **211**, 183 (1993).
[3] D. Zimdars, A. Tokmakoff, S. Chen, S.R. Greenfield, M.D. Fayer, T.I. Smith, and H.A. Schwettman, *Phys. Rev. Lett.* **70**, 2718 (1993).
[4] Y.X. Yan, L.T. Cheng, and K.A. Nelson, *Advances in Nonlinear Spectroscopy*, edited by R.J.H. Clark and R.E. Hester, Wiley, New York, 1988, p.299.
[5] Y. Tanimura and S. Mukamel, *J. Chem. Phys.* **99**, 9496 (1993).

Group-Delay Dispersion Measurement of Laser Mirrors by Spectrally Resolved White-Light Interferometry

A.P. Kovács[1], *R. Szipöcs*[2], *K. Osvay*[1], *and Z. Bor*[1]

[1]Department of Optics and Quantum Electronics, JATE University, Dóm tér 9, H-6720 Szeged, Hungary
[2]Optical Coating Laboratory, Research Institute for Solid State Physics, P.O. Box 49, H-1525 Budapest, Hungary

Abstract. Phase dispersion of multilayer structures was measured by evaluating the shape of spectrally dispersed interference fringes over a wide spectral range.

1. Introduction

It is known that the temporal shape of a femtosecond pulse is altered by the frequency dependent group delay of optical elements like mirrors. Specially designed high reflectivity laser mirrors, however, have proved to be a good candidate for controlling both the intra– and extracavity group–delay dispersion (GDD) in femtosecond lasers [1,2]. To construct such a high performance laser, it is essential to know precisely the expected phase shift vs. frequency [3–6]. In this paper we present an experimental technique based on spectrally resolved white–light interferometry (SRWLI) [7] to measure GDD of various multilayer laser mirrors.

2. Description of the Method

The experimental setup was an ordinary Michelson interferometer illuminated by a white–light source (Fig.1). By tilting one of the mirrors in the interferometer, horizontal interference fringes are formed at the exit plane. A vertical stripe cut off by the slit S_2 is dispersed by a transmission grating and subsequently imaged onto a CCD camera. In case of two identical mirrors the spectrally resolved fringes lie exactly in a straight line. Replacing one of the mirrors, however, the relative phase shift due to the two different mirrors can be determined from the slope of the fringes. In our experiment a gold mirror M_G was chosen for reference since it has practically no dispersion [5].

3. Experimental Results

The upper inset in Fig. 1 shows the spectrally dispersed interference pattern if M_D is a gold mirror, while the lower inset displays the same pattern if M_D is a dielectric mirror and the GDD obtained from that. The mirror exhibits a quadratic phase shift vs. frequency. The accuracy of the measurement was enhanced by using

Springer Series in Chemical Physics, Vol. 60 **Ultrafast Phenomena IX**
Editors: P. F. Barbara · W. H. Knox · G. A. Mourou · A. H. Zewail

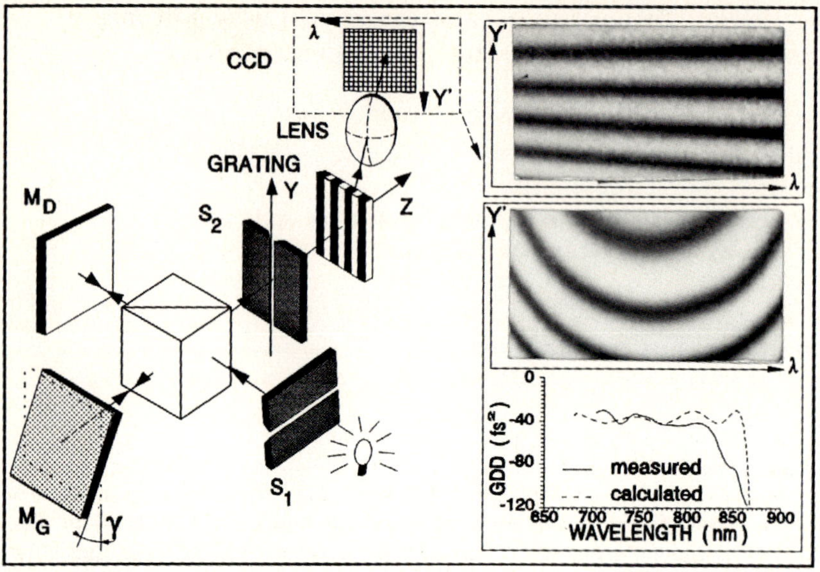

Fig. 1. Spectrally resolved white–light interferometer for GDD measurement of dielectric mirrors (M_D)

four reflections on the dielectric mirrors by inserting an additional mirror identical to M_D and a gold mirror in the object arm (not shown in Fig. 1).

This work has been supported by the U.S.–Hungarian Science and Technology Joint Fund and the OTKA Foundation under grants T–007376 and No. 3055.

4. References

1. J. Kuhl and J. Heppner: IEEE J. Quantum Electron. **QE–22,** 182 (1986)
2. A. Stingl, Ch. Spielmann, F. Krausz and R. Szipőcs: Opt. Lett. **19**, 204 (1994)
3. Z. Bor, K. Osvay, B. Rácz, G. Szabó: Opt. Comm. **78**, 109 (1990)
4. M. Beck, I.A. Wamsley and J.D. Kafka: IEEE J. Quantum Electron. **QE–27**, 2074 (1991)
5. W.H. Knox: Appl. Phys. B **58**, 225 (1994)
6. J. Schwider: Appl. Opt. **31**, 6107 (1992)
7. C. Sainz, J.E. Calatroni and G. Tribillon: Meas. Sci. Technol. **1**, 356 (1990)

Intensity and Phase Measurements of Femtosecond Pulses from a Chirped-Pulse Amplifier

B. Kohler[1], *V.V. Yakovlev*[1], *K.R. Wilson*[1], *J. Squier*[2], *K.W. DeLong*[3], *and R. Trebino*[3]

[1]Department of Chemistry, University of California, San Diego, La Jolla, CA 92093, USA

[2]Center for Ultrafast Optical Science, University of Michigan, Ann Arbor, MI 48109, USA

[3]Combustion Research Facility, Sandia National Laboratories, Livermore, CA 94551, USA

Frequency-resolved optical gating (FROG) is a promising new technique for fully characterizing ultrashort pulses [1]. Knowledge of pulse intensity and phase is important both for understanding ultrashort pulse generation, for interpreting measurements made with ultrashort pulses and for applications such as quantum control which require pulses tailored in phase and amplitude [2]. Here we present FROG measurements of pulses from a chirped pulse amplifier (CPA) which provide insight into dispersive pulse propagation and pulse shaping by frequency-domain filtering in the pulse stretcher.

The Ti:sapphire kHz CPA laser system studied uses 1200 l/mm gratings to stretch and compress initially 70-fs oscillator pulses. FROG traces are recorded in the polarization-gate geometry using a 1/8" thick BK7 glass plate as the nonlinear gate material. The intensity and phase are readily retrieved from each two-dimensional dataset by the generalized projections algorithm.

It is well known that the stretcher and compressor geometries must be carefully designed to compensate for the positive material dispersion of the regenerative amplifier (regen) [3]. A series of FROG traces are recorded as a function of the number of regen cavity roundtrips by adjusting the Pockels cell timing. The frequency-dependent phase, $\phi(\omega)$, obtained from each FROG trace is differentiated twice to obtain the group velocity dispersion (GVD), which is plotted vs. roundtrip number in Figure 1. The slope of the best-fit line is a direct measure of the regen GVD and is equal to 5070 fs^2, in excellent agreement with the value estimated from refractive index data of the cavity components. By differencing the pulse phase from sequential roundtrips, the cavity dispersion can in principle be determined to all orders.

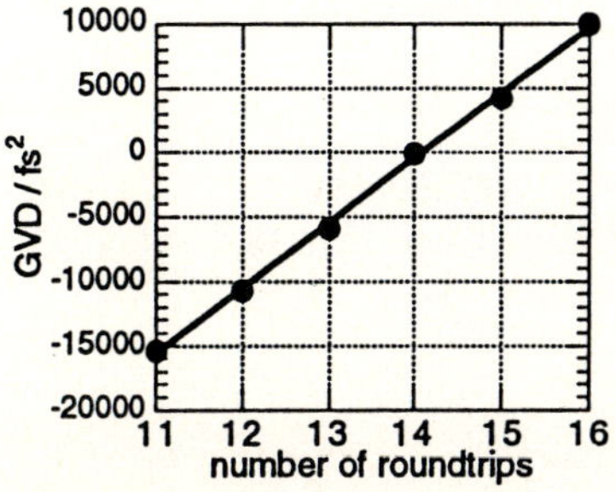

Figure 1. Group velocity dispersion (GVD) of the amplified pulse vs. regen roundtrip number. Each point is determined from the second-derivative of the FROG-retrieved pulse phase. The slope of the best-fit line gives the net regen GVD.

Springer Series in Chemical Physics, Vol. 60 **Ultrafast Phenomena IX**
Editors: P. F. Barbara · W. H. Knox · G. A. Mourou · A. H. Zewail

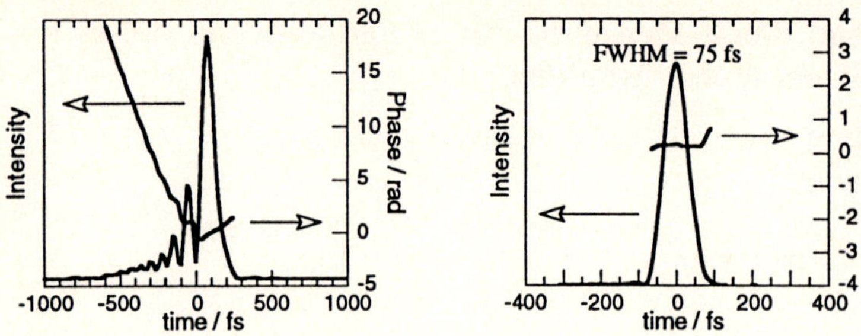

Figure 2. Amplified pulse intensity and phase from misaligned compressor (a) and optimized compressor (b).

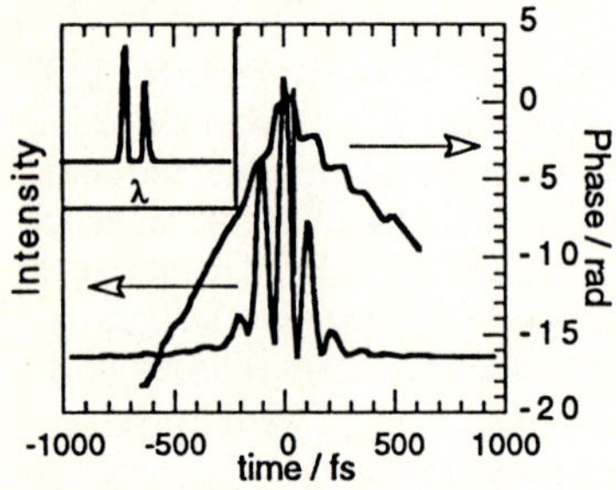

Figure 3. Intensity and phase of an amplified pulse shaped by masking the central portion of the pulse spectrum (inset) with an opaque object in the pulse stretcher.

Figure 2(a) shows the temporal intensity and phase of an amplified pulse which exhibits substantial residual cubic spectral phase, resulting from improper phase compensation in the compressor. Using the measured regen dispersion, the compressor geometry is altered to obtain the pulse intensity and phase shown in Figure 2(b). The frequency-dependent phase is now nearly flat, indicating good phase compensation.

Finally, we demonstrate the use of FROG to characterize a shaped femtosecond pulse. Figure 3 shows the intensity and phase of an amplified pulse which is shaped by amplitude-masking in the pulse stretcher. The shaped spectrum is shown in the inset.

In conclusion, intensity and phase measurements have been used to characterize and optimize ultrashort pulse generation in a chirped pulse amplifier under a variety of conditions.

References

1. R. Trebino and D. J. Kane, J. Opt. Soc. Am. A **10**, 1101 (1993).
2. B. Kohler, J. L. Krause, F. Raksi, C. Rose-Petruck, R. M. Whitnell, K. R. Wilson, V. V. Yakovlev, and Y.-J. Yan, J. Phys. Chem. **97**, 12602 (1993).
3. J. V. Rudd, G. Korn, S. Kane, J. Squier, and G. Mourou, Opt. Lett. **18**, 2044 (1993).

Femtosecond Time Resolution in Scanning Tunneling Microscopy

G. Gerber[1], *F. Sattler*[1], *S. Vogler*[1], *J.Y. Grand*[2], *P. Leiderer*[2], *and R. Möller*[2]

[1]Fakultät für Physik, Universität Freiburg, D-79104 Freiburg, Germany
[2]Fakultät für Physik, Universität Konstanz, D-78434 Konstanz, Germany

Abstract. We have combined femtosecond laser pump-probe techniques with scanning tunneling microscopy (STM) to study ultrafast dynamical phenomena on metal and semiconductor surfaces under ultra high vacuum conditions. For a tantalum surface we observed an ultrafast (τ=130fs) variation of the multiphoton induced photoelectron current, while for a GaAs (110) surface a single exponential decay with τ=11ps describes the measured pump-probe spectrum quite well. This transient spectrum consists of contributions from tunneling electron current and multiphoton induced photoelectron current.

Scanning tunneling microscopy allows to study surfaces with extremely high spatial resolution down to nanometer dimensions. This has changed the field of surface science. However, the time resolution of the STM is limited to the microsecond range due to the tunneling current measurement. Femtosecond laser pump-probe techniques provide a time resolution limited by the temporal pulse duration of the ultrashort laser pulses. However, the spatial resolution is limited to the diffraction limit of the laser beam, which is for visible light of the order of a few micrometer. So, the ultimate solution to study ultrafast dynamical processes on surfaces is to combine these two extremely powerful techniques.

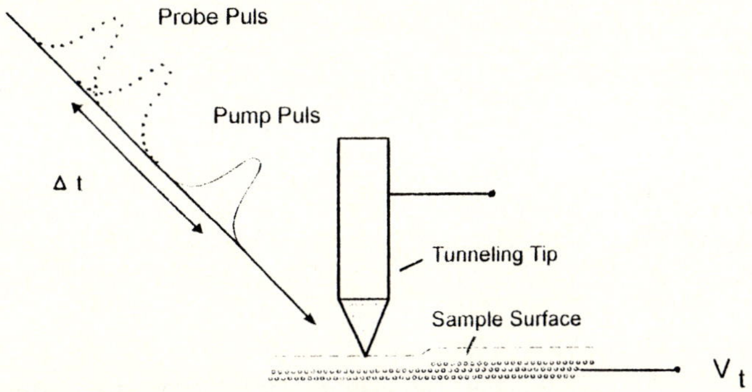

Fig.1 Schematic set-up of femtosecond laser - STM experiment

Springer Series in Chemical Physics, Vol. 60 **Ultrafast Phenomena IX**
Editors: P. F. Barbara · W. H. Knox · G. A. Mourou · A. H. Zewail

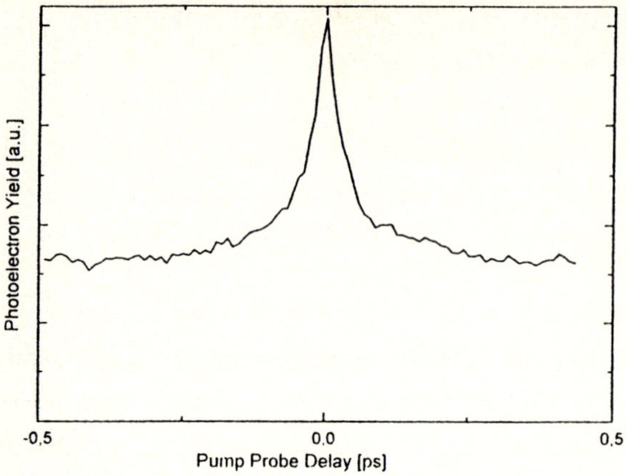

Fig. 2 Transient multiphoton emission of electrons from a tantalum surface.

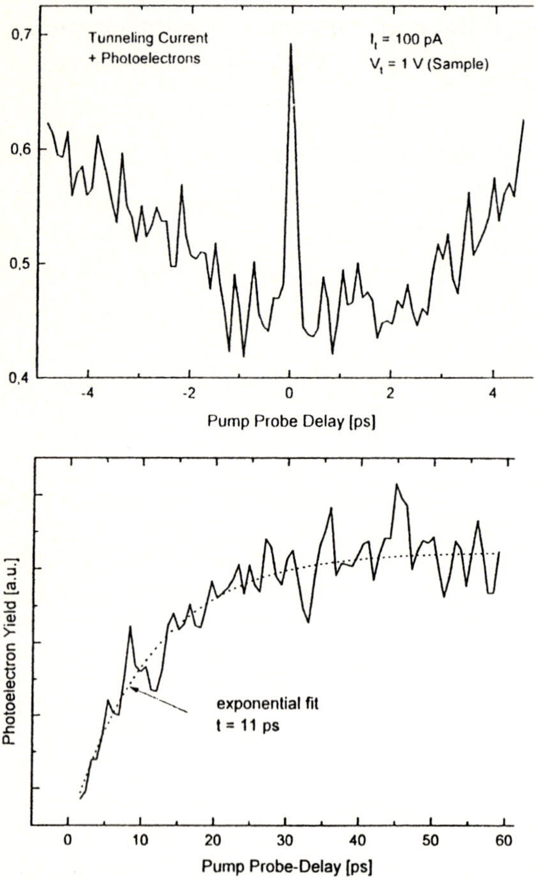

Fig. 3 Transient spectrum from fs-laser interaction with a GaAs (110) surface.

This contribution reports our first results obtained with 40fs light pulses of an amplified home-built Ti:Sapphire laser system directed to the tip-surface gap of a STM. The schematic experimental set-up is shown in Fig.1 The femtosecond time resolution is achieved by scanning the time delay between the identical pump and probe pulses and integrating the photoelectron current. Au and WSe_2 surfaces (on air) are choosen to demonstrate the feasibility of the new experimental technique. The interferometric autocorrelation of the 40fs laser pulse making use of the nonlinear response of the tip-Au surface interaction nicely showed the phase stability of the femtosecond light pulses. The interaction of the fs laser pulse with the Au surface induced an excitation which relaxed on a τ=200fs time scale.

Results of experiments in ultra high vacuum are shown in Fig.2 and Fig.3, where for a tantalum surface an ultrafast (130fs) relaxation is observed while for GaAs (110) a much slower (11ps) process occurs. In both cases the signal is dominated by the multiphoton (n=3-4) induced photoelectron current. Only for GaAs a weaker tunneling current contribution was found. By scanning the tip, surface dynamics can be studied with fs and nm resolution simultaneously.

A Dual-Beam Sub-Picosecond Spectrometer for the 5–11 μm Spectral Region and Its Application to Vibrational Relaxation in the S_1 State

P. Hamm, S. Wiemann, M. Zurek, and W. Zinth

Institut für Medizinische Optik der Universität München,
Barbarastrasse 16, D-80797 München, Germany

Abstract: A dual beam mid-IR spectrometer for the mid-IR with a time resolution of better than 300 fs is developed and to vibrational spectroscopy of an organic dye molecule.

Time resolved pump and probe measurements with probing pulses in the mid-IR spectral region has proven to yield important information on fast reactional and relaxational processes in solid state physics, semiconductor physics, chemical physics, and biophysics. In order to allow new experiments, one requires tunable sub-picosecond pulses in the mid IR and an experimental set-up, which enables the sensitive detection of absorbance changes as small as 0.1 mOD.

The laser system presented here is based on a femtosecond Ti:sapphire laser and regenerative amplifier. A synchronised red shifted second light pulse needed for difference frequency generation is deduced from a small portion of the amplified Ti:sapphire pulse in a tunable travelling wave dye laser (TWDL). The output of the Ti:sapphire laser is mixed with that of the TWDL in a $AgGaS_2$ crystal [1]. By tuning the wavelength of the TWDL and rotating the phase matching angle of the non-linear crystal we are able to obtain femtosecond pulses with between 5 and 11 µm. The path length inside the crystal is only 0.5 mm in order to minimise group velocity dispersion. By this means, mid-IR pulses as short as 300 fs with a pulse energy of more than 10 nJ can be obtained. The spectral bandwidth is independent on the absolute wavelength and is approximately 65 cm^{-1}. At present, a second TWDL is used to generate excitation pulses between 860-1100nm (depending on the laser dye used) with a pulse energy of several micro joules.

Since the bandwidth of the IR pulse is broader than the bandwidth of typical absorption bands of molecules multichannel detection is possible: the probing pulses are dispersed by a grating spectrometer and measured with a 10 element HgCdTe detector array with a spectral resolution between 5 and 10 cm^{-1}. A small part of the probing pulse is split off in front of the sample and is used as reference signal. A measurement time of less than 5 min is required to detect an absorbance change of 0.1mOD.

As a first experiment to demonstrate the ability of the system, the transient IR absorbance changes of the laser dye IR26 (in $C_2H_4Cl_2$)induced by excitation in the S_1 state was measured. Excitation was performed in the short wavelength tail of the S_1 absorbance band at 870 nm. The IR spectrum of the dye shows a intense band at 1530 cm^{-1} well separated from the solvent bands. This band is probably due to the

Springer Series in Chemical Physics, Vol. 60 **Ultrafast Phenomena IX**
Editors: P. F. Barbara · W. H. Knox · G. A. Mourou · A. H. Zewail

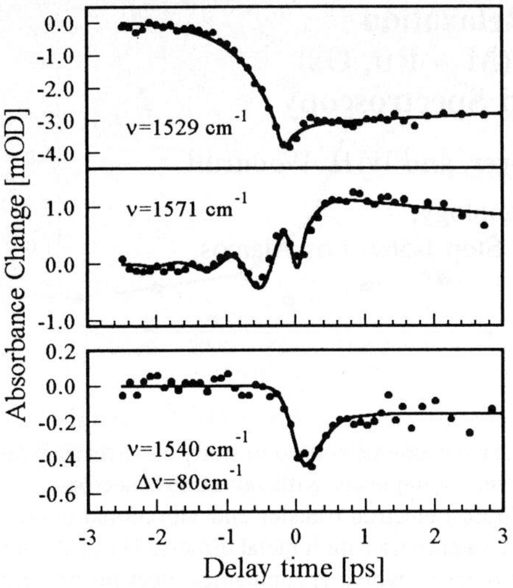

Fig1.: Absorbance changes of dye IR26 induced by subpicosecond light pulses at 870 nm recorded at three probing wavelengths of 1178 cm^{-1}, 1529 cm^{-1} 1571 cm^{-1} (For a detailed discussion see text).

C-C stretching modes of the aromatic system of the molecule and should change significantly upon electronic excitation.

At the center of the absorption band at 1527 cm^{-1} (Fig. 1 top), a negative signal rising more slowly than the experimental time resolution is observed at negative delay times. An additional fast relaxing signal is seen around delay zero. At 1571 cm^{-1} (Fig.1 middle) an oscillating signal with a frequency equal to the detuning with respect to the band center is observed. The behaviour at negative delay times can be explained by the coherent interaction between the polarisation induced by the short probing pulse and the excitation process and occurs whenever a ground state band bleaches with a dephasing time T_2 which is longer than the pulse length [2].

A wavelength integration of the different signal traces over a bandwidth of 50-80 cm^{-1} suppresses the oscillation (see Fig.1 bottom). A fast relaxing signal still remains which is negative above $\approx$1500 cm^{-1} and becomes positive at a frequency below 1500 cm^{-1}. This findings may be explained by a fast thermalisation in the electronic S_1 state: Because of the large excess energy of the pump pulse higher vibrational states are excited. In this non-equilibrium state absorption from hot vibrational levels occurs which is (due to the anharmonicy of the potential surface) at lower energies (<1500 cm^{-1}) than that of the vibrational ground state. After thermalisation (after 200-400 fs) the fast bleaching has disappeared and the induced absorption due to the vibrational ground state in the S_1 remains.

References:

[1] P.Hamm, C.Lauterwasser, W.Zinth, Opt. Lett 18 (1993) 1943
[2] M.Joffre, D.Hulin, A.Migus, A.Antonetti, C.Benoit a la Guillaume, N.Peyghambarian, M.Lindberg and S.W.Koch, Opt. Lett 13 (1988) 276

4. Acknowledgments

This work was performed at Pacific Northwest Laboratory under the auspices of the Division of Chemical Sciences, Office of Basic Energy Sciences, U. S. Department of Energy. Pacific Northwest Laboratory is a multiprogram national laboratory operated for the Department of Energy by Battelle Memorial Institute under Contract No. DE-AC06-76RLO 1830.

5. References

1. H. Graener and G. Seifert, and A. Laubereau, *Phys. Rev. Lett.* **66**, 2092 (1991).
2. Q. Fu, G. Mak, and H. Van Driel, *Optics Lett.* **17**, 1006 (1992); W. Pelouch, P. Powers, and C. Tang, *Optics Lett.* **17**, 1070 (1992).

Part IV

Sources

All-Solid-State Diode-Pumped CW Femtosecond $Cr:LiSrAlF_6$ Lasers

R. Mellish[1], *P.M.W. French*[1], *J.R. Taylor*[1], *P.J. Delfyett*[2], *and L.T. Florez*[3]

[1]Femtosecond Optics Group, Physics Department, Imperial College, London, UK
[2]CREOL, University of Central Florida, Orlando, FL 32826, USA
[3]Bell Communications Research, Red Bank, NJ 07701, USA

Abstract. Transform-limited pulses as short as 220 fs have been generated from an all-solid-state, diode-pumped, true c. w. mode-locked Cr:LiSAF laser.

$Cr^{3+}:LiSrAlF_6$ (Cr:LiSAF) is an attractive alternative to Ti:sapphire, lasing across almost the same spectral range but exhibiting a longer upper-state lifetime and an absorption profile which permits it to be efficiently pumped by laser diodes at 670 nm and by flash-lamps or arc-lamps. It has a greater energy storage capacity and, as a tunable oscillator, its lower lasing threshold and ability to be diode-pumped make it potentially more useful than Ti:sapphire. Kerr Lens Mode-locking of an argon-ion-pumped Cr:LiSAF laser has been demonstrated to yield transform-limited pulses as short as 33 fs for as low as 500 mW pump power [1]. The low pump power requirement demonstrated the potential for diode-pumping and the use of a MQW resonant saturable absorber [2] indicated that an all-solid-state system was feasible. Accordingly the first all-solid-state diode-pumped tunable c.w. and actively mode-locked (Cr:LiSAF) laser was demonstrated with 670 nm InGaAlP laser diodes [3]. These lasers were far from optimised and initially yielded pulses of 100's ps duration. Subsequently we and other workers [4,5] improved on the picosecond results. We now report the first all-solid-state, diode-pumped femtosecond vibronic laser [6].

Initially diode-pumping of Cr:LiSAF was investigated using a simple astigmatically-compensated cavity with an end-pumped rod as described in reference [3]. The two InGaAlP laser diodes used typically provided ~500 mW cw pump power at ~ 670 nm and their output beams were combined using a polarising beamsplitter. Currently the laser threshold is 70 mW and the laser tunes from 799 to 880 nm with up to 135 mW of true c.w. output power.

Active mode-locking was demonstrated using an acousto-optic modulator which yielded full duty-cycle cw mode-locked pulses of ~300 ps duration [3]. We then demonstrated an all-solid-state passively mode-locked laser using RPM with the MQW absorber of reference [2]. Stable, self-starting, cw mode-locked pulse trains at 250 MHz were observed which were as short as ~ 60 ps duration, as measured using a sampling streak camera [5]. When a standard four prism sequence was incorporated into the laser cavity, pulses as short as 600 fs in bursts were observed using an SHG autocorrelator, the individual pulses being separated by the mismatch between the laser and external cavities.

The requirement for high intensities incident at the MQW absorber makes the use of a coupled cavity approach unattractive for mode-locking such a relatively low power laser as the diode-pumped Cr:LiSAF laser. Using the MQW absorber

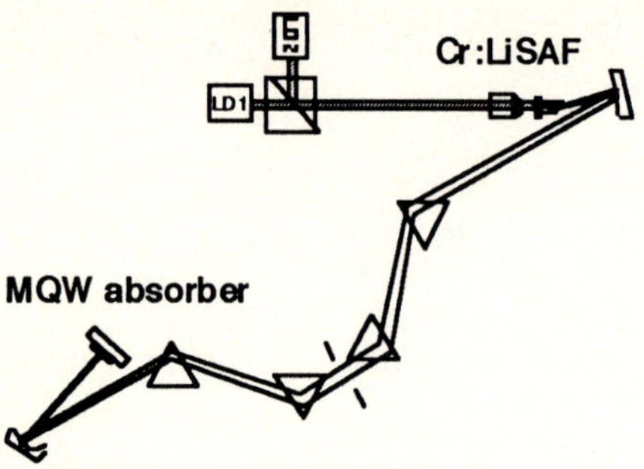

Figure 1. Cavity configuration of Cr:LiSAF laser passively mode-locked with an intracavity MQW absorber

intracavity is much more effective, as was demonstrated for the argon ion pumped system[2]. The cavity shown in figure 1 was therefore constructed. A sequence of four Brewster-angled SF10 glass prisms was used to provide adjustable GVD and a slit was employed at the midpoint to effect spectral control.

Self-starting, c.w. passive mode-locking was obtained at ~846 nm using the MQW absorber intracavity, as shown in figure 1, and transform-limited pulses as short as 220 fs were generated. The time-bandwidth product of these pulses was 0.33. 10 mW of total average output power was obtained for 660 mW of diode pump power.

References

1) N. H. Rizvi, P. M. W. French and J. R. Taylor, *Opt. Lett.* **17**, 1605-1607 (1992)
2) N. H. Rizvi, P. M. W. French, J. R. Taylor, P. J. Delfyett and L. T. Florez *Opt. Lett.* **18**, 983 (1993)
3) P. M. W. French, R. Mellish, J. R. Taylor, P. J. Delfyett and L. T. Florez *Electron. Lett.* **29**, 1263 (1993)
4) R. Mellish, P. M. W. French, J. R. Taylor, P. J. Delfyett and L. T. Florez *Opt. Lett.* **18**, 1934 (1993)
5) F. Balembois, P. Georges and A. Brun, *Opt. Lett.* **18**, 1730 (1993)
6) R. Mellish, P. M. W. French, J. R. Taylor, P. J. Delfyett and L. T. Florez *Electron. Lett.* **30**, 223 (1994)

A Self-Mode-Locked $Ti:Al_2O_3$ Laser Pumped by a Frequency-Doubled Nd:YLF Minilaser

K. Lamb, D.E. Spence, J. Hong, C. Yelland, and W. Sibbett

J.F. Allen Physics Research Laboratories, Department of Physics and Astronomy, University of St. Andrews, St. Andrews, Fife KY16 9SS, Scotland, UK

Abstract. The self-modelocked operation of an all-solid-state Ti:sapphire laser is described. This $Ti:Al_2O_3$ laser was pumped by a frequency-doubled, cw Nd:YLF slab minilaser, and produced femtosecond pulses for pump powers as low as 500mW.

1. Introduction

Recent advances in high power laser diodes operating at 780nm, have led to vigorous research activity in the development of diode-laser pumped, solid-state lasers. To date, pulsed operation of $Ti:Al_2O_3$ lasers has been reported using frequency-doubled, modelocked and Q-switched Nd:YLF [1,2], and frequency-doubled Q-switched Nd:YAG lasers [3]. More recently, commercial frequency-doubled , cw diode-laser pumped Nd:YAG lasers have achieved average powers suitable for pumping a $Ti:Al_2O_3$ laser, which had been specifically designed for low threshold operation at ~200mW [4]. In this paper, we report on the operation of an all-solid-state, self-modelocked $Ti:Al_2O_3$ laser.

2. Experiment and Results

The basic configuration of our regeneratively-initiated, self-modelocked $Ti:Al_2O_3$ laser is similar to the argon-ion-pumped system that has been described elsewhere [5]. In the present case, the gain medium was a 10mm long, Brewster-angled $Ti:Al_2O_3$ rod having a pump absorption coefficient of $1.9cm^{-1}$, and a figure of merit in excess of 250.

The pump source for the $Ti:Al_2O_3$ laser was a diode-laser-pumped, intracavity frequency-doubled, cw Nd:YLF laser. This laser produced a single longitudinal mode output at powers up to 1W at 523nm, in a TEM_{00} beam. Pump powers of approximately 700mW resulted in an average output power of 20mW from the self-modelocked $Ti:Al_2O_3$ laser. For the configuration shown in figure 1, it was observed that the pump power for cw laser oscillation was approximately 400mW, with a threshold for self-modelocking at 500mW.

From the intensity autocorrelation data shown in figure 2, the pulse duration was measured to be 110fs (assuming a $sech^2$ intensity profile). The spectral bandwidth of the pulses was measured to be 9.2nm, as shown by the spectral data in figure 3. These results thus imply a corresponding duration-bandwidth product of ~ 0.46, which is slightly greater than the transform-limit of 0.32.

Editors: P. F. Barbara · W. H. Knox · G. A. Mourou · A. H. Zewail

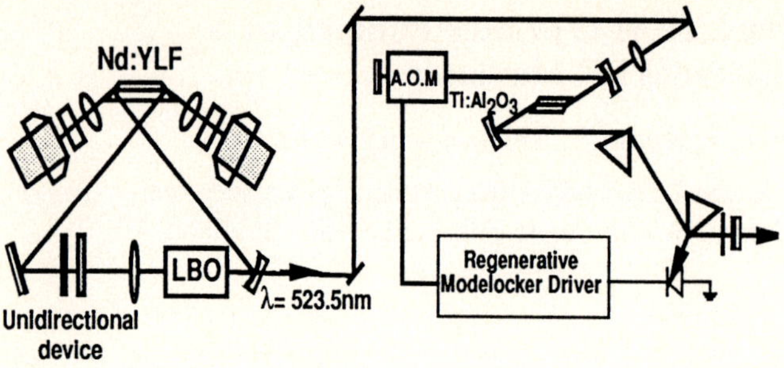

Figure 1: Schematic diagram of the all-solid-state, self-modelocked Ti:Al_2O_3 laser.

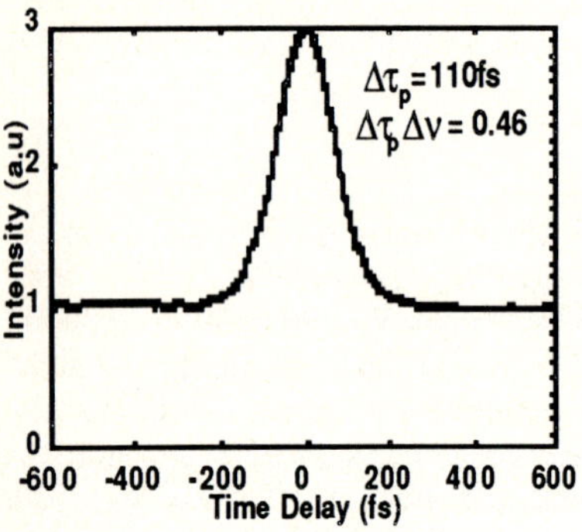

Figure 2: Intensity autocorrelation data.

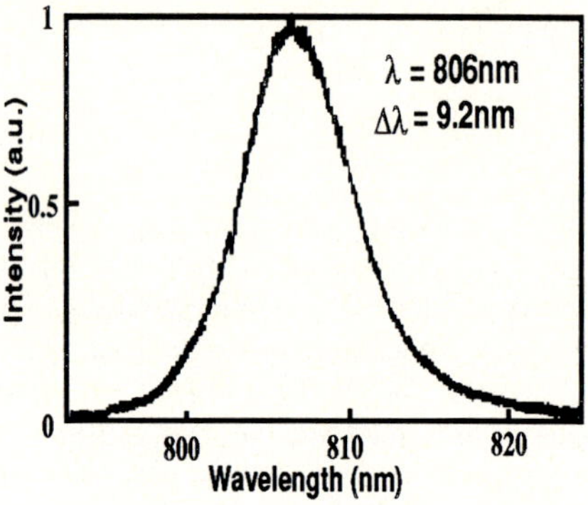

Figure 3: Spectral data.

3. Conclusions

Femtosecond pulses from an-all-solid-state, self-modelocked Ti:Al_2O_3 laser have been generated. At a pump power of 700mW, 110fs pulses with an average power of 20mW were recorded. The threshold for self-modelocked operation was measured to be ~500mW, with a cw threshold for laser oscillation at ~ 400mW.

4. References.

1. G.T Maker and A.I Ferguson. Opt. Lett. **15**, 375 (1990).
2. G.P.A Malcolm and A.I Ferguson. Opt. Commun. **82** 299 (1991).
3. T.R Steele et al. Opt. Lett **16** 399 (1991).
4. J. Harrison et al. Opt. Lett **16** 581 (1991).
5. D.E Spence et al Opt. Lett **16** 1762 (1991).

Sub-100 fs Homogeneously and Inhomogeneously Broadened Nd:Glass Lasers

X.F. Kärtner, D. Kopf, and U. Keller

Institute of Quantum Electronics, Swiss Federal Institute of Technology, ETH Hönggerberg, HPT, CH-8093 Zürich, Switzerland

Abstract: We investigate the effects of intracavity filtering and soliton formation on pulse generation with Nd:glass lasers in different regimes theoretically and experimentally. We show theoretically that even a slow saturable absorber is able to stabilize solitary pulses much shorter than the recovery time of the absorber. We achieve sub-100 fs pulses with quasihomogeneous and inhomogeneous Nd:glass gain media.

Previously, we have introduced a spectral filtering technique and applied it to a passively modelocked Nd:glass laser using an antiresonant Fabry-Perot saturable absorber (A-FPSA) [1,2]. Pulses as short as 130 fs are achieved in a standard dispersion compensated delta - cavity design with a 4 mm Nd:glass plate inserted at Brewster's angle. The Nd:glass laser system is interesting from the point of view of an all solid state laser technology because it can be diode pumped. To clarify the modelocking dynamics in Nd:glass lasers, we performed the following experiments. First we demonstrated regeneratively, actively mode-locked Nd:glass lasers using a phosphate LG760 and silicate LG600 Nd:glass gain media, resulting in stable pulses with 310 fs and 330 fs pulsewidth respectively. Secondly passively modelocked phosphate Nd:glass (LG700) laser with 105 fs and silicate Nd:glass (LG680) laser with 92 fs, see Fig. 1. In the experiments employing the quasihomogenous gain media we have to use the intracavity filter to arrive at these short pulses, whereas in the inhomogeneous case additional intracavity filtering is not required. The passive modelocking experiments used an A-FPSA which is a bitemporal saturable absorber with a fast relaxation time of 150 fs due to intraband thermalization processes and a slow recovery time of 17 ps due to interband recombination processes [2,3]. The different Nd:glass hosts were chosen because the phosphate Nd:glass acts as a quasihomogeneous laser material whereas the silicate Nd:glass is strongly inhomogeneous. Despite the completely different cw-behaviour of the lasers when the homogeneous or the inhomogeneous gain material is used, the modelocking performance is comparable and leads to almost the same pulsewidth in contrast to previously obtained results based on additive pulse mode locking [4].

This demonstrates that the action of the intracavity filter is to flatten the net gain profile, that is the saturated gain minus the frequency dependent filter loss. The flattened net gain allows for shorter pulses in passively as well as actively mode-locked lasers due to reduced losses associated to gain filtering. In the case of inhomogeneous gain media, ideally, the saturated gain would be flat over the inhomogeneous linewidth assuming equally distributed spectroscopic parameters over the full inhomogeneous width. Thus an additional filter does not help to flatten the net gain, which confirms our observations. In all cases actively and passively mode-locked lasers using homogeneously as well as inhomogeneously broadened

Springer Series in Chemical Physics, Vol. 60 **Ultrafast Phenomena IX**
Editors: P. F. Barbara · W. H. Knox · G. A. Mourou · A. H. Zewail

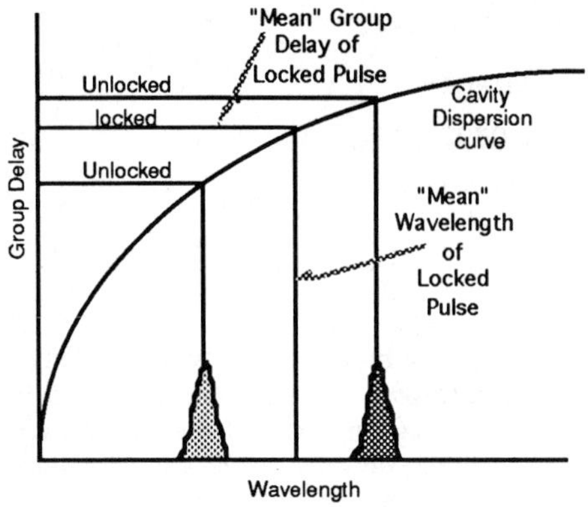

Figure 3. Repetition rate pulling schematic.

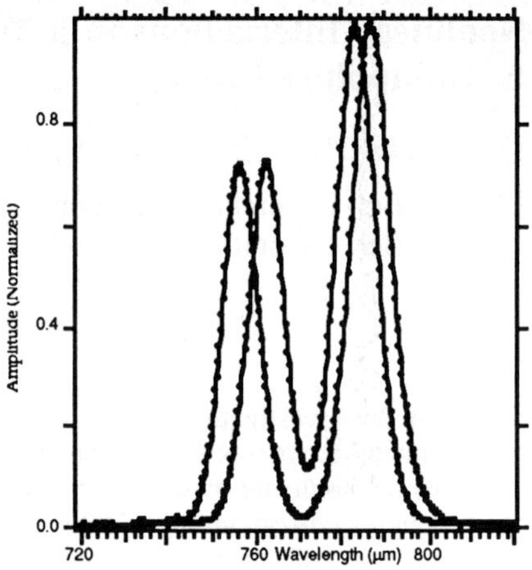

Figure 4. Spectra of both beams at the cavity detuning maxima.

self-focused modelocking results in very strong pulse shaping effects. We then slowly return the length of cavity "A" to its original length. When detuning becomes <3 μm, the cavities again cross lock.

In the second experiment (Fig. 2), the laser is placed in the cross-locked state and then the beam of cavity "B" is physically blocked. When cavity "B" is present, the repetition rate of cavity "A" corresponds to that of the coupled cavities. When cavity "B" is blocked, cavity "A" relaxes to its unperturbed repetition rate. After the block is removed, cavity "A" is frequency-pulled, despite the fact that cavity "B" re-starts in a cw state ("noise"). Even in this state there is clear evidence of oscillation at the cavity "B" round trip time as seen via a photodiode on an oscilloscope. When the table is tapped, the two beams return to the cross-locked state. The time scale in Figs. 1 and 2 is ~1 min.

This phenomena can be understood in terms of the frequency-dependent group delay described by Knox [4]. For the cavity dispersion described by the single generalized curve shown in Fig. 3, the matched condition, corresponds to the group delay being the same "mean" value for both cavities, despite the fact that the spectra are very different. This is the condition resulting in a stable cross correlation. A consequence is the resulting pulse can now also be described by a "mean" wavelength as shown in Fig. 3. The pulses travel together as a *single pulse*, despite very different spectra and a beam overlap in the laser rod which occurs only in the wings of the spatial distribution.

In addition, the wavelengths are pulled, as shown in Fig. 4. The two cross-locked beams are combined at the entrance slit of a spectrometer, and the spectra are recorded at the ±3 μm cavity detuning points. Both wavelengths are shifted, as expected for equal pump intensities. This phenomena is not presently understood and warrants further study

In conclusion, we have investigated cross-modelocking of two independent laser cavities. The frequency-dependent group delay model of mean wavelength and mean group delay can be extended to a pair of cross-locked pulses with disparate spectra and spatial distribution in a nonlinear regenerative system. Further, wavelength pulling is seen to be an additional consequence of the cross-locked condition.

References

1. D.R. Dykaar, and S.B. Darack, Optics Letters, 18 (8), 634 (1993).
2. D.R. Dykaar, W.H. Knox, and S.B. Darack, To appear Optics Letters (1994).
3. W.H. Knox, Optics Letters, 27(1), 514 (1992).
4. W.H. Knox and J.P. Gordon, JOSA B, 10, 2071, (1993)

Ultrashort-Pulse Multi-Terawatt Ti:Sapphire Laser System

C.P.J. Barty, B.E. Lemoff, and C.L. Gordon III

Edward L. Ginzton Laboratory, Stanford University, Stanford, CA 94305, USA

1. Introduction

Chirped pulse amplification (CPA) has become a common technique for circumventing optical damage and nonlinear effects during the amplification of short optical pulses in solid state laser media.[1] Terawatt level pulses with durations ranging from ~1 ps to ~100 fs have been produced by a number of systems. Although much shorter pulse oscillators exist, the implementation of multiterawatt CPA with pulses on the order of 10 fs has not been accomplished. In conventional CPA systems, uncompensatable phase errors in the expansion and recompression process and distortions due to refractive optics produce broadenings which are much greater than 10 fs. In order to alleviate these problems, a novel expander and compressor design has been developed and a lens-free amplification system has been constructed. In addition, gain narrowing during amplification severely limits the final pulse duration. In order to control the pulse spectrum before amplification and thus minimize this problem, spectral masks have also been developed and implemented. A schematic of the laser system is shown in Figure 1.

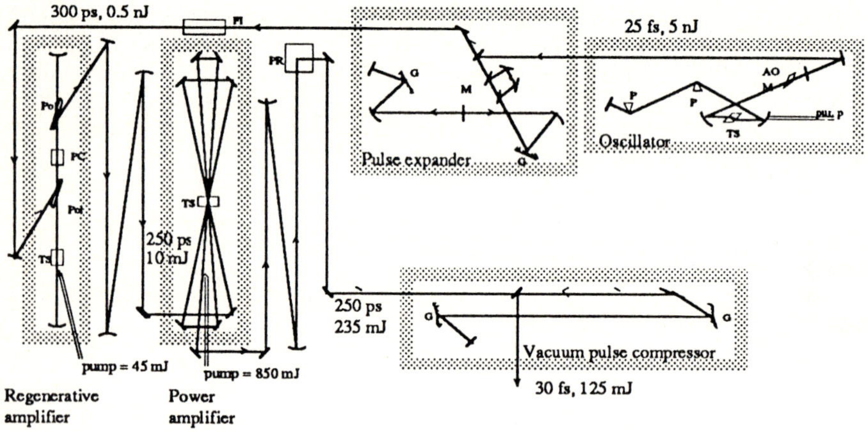

Fig. 1. Schematic of 30-fs multiterawatt laser system.

Springer Series in Chemical Physics, Vol. 60 **Ultrafast Phenomena IX**
Editors: P. F. Barbara · W. H. Knox · G. A. Mourou · A. H. Zewail

2. High fidelity expansion and compression

A 20-fs Ti:sapphire oscillator provides seed pulses for amplification.[2] In order to avoid self focusing in the amplifier medium, optical damage of dielectric coatings and simultaneously reach the saturation fluence of Ti:sapphire (~0.8J/cm^2), it is necessary to temporally stretch the oscillator output to several hundred picoseconds before amplification. By doing so, the peak intensity in the amplifier chain is limited to less than 10 GW/cm^2. To accurately expand and recompress a cylindrical mirror based expander and compressor system is used.[3] This system is capable of temporally stretching a 20-fs pulse by a factor of 15,000 and recompressing it to within 0.1 fs of the original duration. A detailed ray tracing analysis is used to calculate the correct placement, orientation and separation of the expander and compressor elements and the correct amount of optical material necessary to eliminate 2nd, 3rd and 4th order spectral phase distortions.

3. Control of gain narrowing

Because linear phase distortion is controlled in this system, the duration of the amplified pulse is determined by its spectral content and thus by the spectral response of the amplifiers. Gain narrowing severely limits this spectral content. Amplification to 235 mJ is accomplished in a two stage arrangement. The first stage is a regenerative amplifier consisting of one Pockels cell, two polarizers and a 13 mm long Ti:sapphire crystal.[4] The second is a four pass amplifier pumped with 850 mJ of 532-nm radiation. A typical seeded and amplified spectrum corresponds to a pulse of ~45 fs in duration. In order to produce shorter pulses we modify the spectrum of the seed pulse. This is easily accomplished by placing a spatially varying filter of the correct attenuation in the focal plane of the first cylindrical mirror of the pulse expander.[5] By doing so, one may significantly increase the FWHM of the seed pulse spectrum at the expense of reducing the seed pulse energy. Two types of masks are used, a simple circular card and variable transmission gratings etched in chromium films on fused silica substrates. Transform limited pulses with durations as short as 30 fs can be produced in this manner.

Tests of beam quality indicate that the beam is nearly diffraction limited and that focal intensities of > 10^{20} W/cm^2 can be produced with f1 optics. It should also be noted that because the design of this system can accommodate large amounts of optical material in the amplification chain, straightforward scaling to higher energies is possible. Laboratory scale, petawatt (15 J in 15 fs) systems are feasible with currently available optical components.

This work was supported by the U. S. Air Force Office of Scientific Research and the U. S. Army Research Office.

References

1. D. Strickland and G. Mourou, Opt. Comm. 56, 219(1985)
2. B. E. Lemoff and C. P. J. Barty, Opt. Lett., 17, 1367(1992)
3. B. E. Lemoff and C. P. J. Barty, Opt. Lett., 18, 1651(1993)
4. J. D. Kmetec, J. J. Macklin and J. F. Young, Opt. Lett. 16, 1001(1991)
5. A. M. Weiner, J. P. Heritage and E. M. Kirschner, J. Opt. Soc. Am. B, 5, 1563(1988)

Amplification in Ti:Sapphire at the Gain-Narrowing Limit

C.P. Huang, J. Zhou, C. Shi, H.C. Kapteyn, and M.M. Murnane

Department of Physics, Washington State University, Pullman, WA 99164-2814, USA

We have used a multipass Ti:sapphire amplifier system to generate 23 fs pulses at 805 nm with greater than 20 mJ pulse energy. Our system optimizes higher order dispersion compensation using an all-reflective optical design and modest chirped-pulse stretch factor, while still reaching gain saturation in the amplifier.

Design considerations and implementation

There are three main factors that can limit the pulse duration in a laser amplifier using a broadband solid-state media such as Ti:sapphire. These factors are gain narrowing, higher order dispersion, and spectrally limiting optical components. For gain narrowing, our calculations show that, after a gain of 10^7, an infinitely broadband δ-function pulse will spectrally narrow to a bandwidth corresponding to 18 fs.[1] This is an intrinsic limitation and difficult to overcome. However, we can control the other two limitations by proper design of the amplifier.

To minimize total dispersion, and thus higher-order dispersion in the system, we use all reflective optics, and a relatively low CPA stretch factor. Reflective optics reduce the amount of material in the amplifier, and allow broadband, distortion-free operation. However, even after minimizing total dispersion, there will still be residual third-order dispersion which must be compensated. This can be accomplished by using a pair of prisms, which has opposite sign of third-order dispersion to that of the material, stretcher and compressor.[2]

Experiment

Our amplifier uses a chirped-pulse amplification scheme.[3] The input of our amplification system is from a self modelocked Ti:sapphire oscillator that produces transform-limited 11 fs duration pulses, at a center wavelength of 796 nm, and with a FWHM bandwidth of ≈ 62 nm.[4] The initial pulse energy is 5 nJ. The beam passes through the all reflective stretcher which uses a single grating, a single parabolic mirror, and two flat mirrors. The pulses are stretched to approximately 20 ps after the stretcher. The first-stage amplifier uses an 8 mm long, highly-doped (0.23%), Brewster-cut Ti:sapphire rod. It is pumped longitudinally with 90 mJ of 532 nm light from a doubled Q-switched Nd:YAG laser. The amplifier rod is at the focus of four 1-meter radius-of-curvature dielectric mirrors, which form an eight-pass figure "8" amplifier. To suppress ASE, the beam focuses through a four-hole array placed near the crystal, and a saturable absorber is used before the final pass through the amplifier. A pockels' cell and a spatial filter are placed between the first and the

Springer Series in Chemical Physics, Vol. 60 **Ultrafast Phenomena IX**
Editors: P. F. Barbara · W. H. Knox · G. A. Mourou · A. H. Zewail

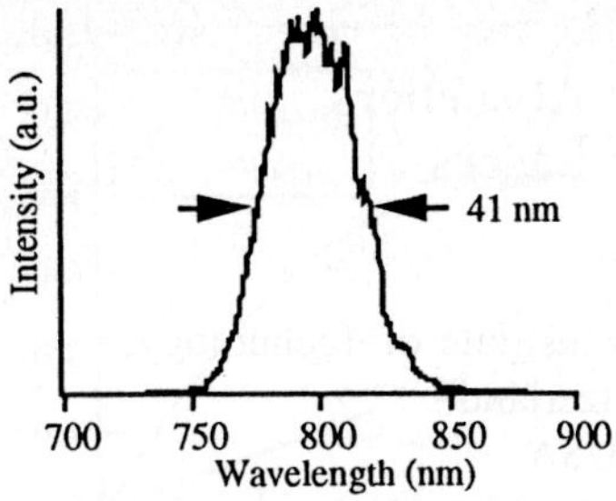

Figure 1.: Typical spectrum of the 50 mJ amplified pulse.

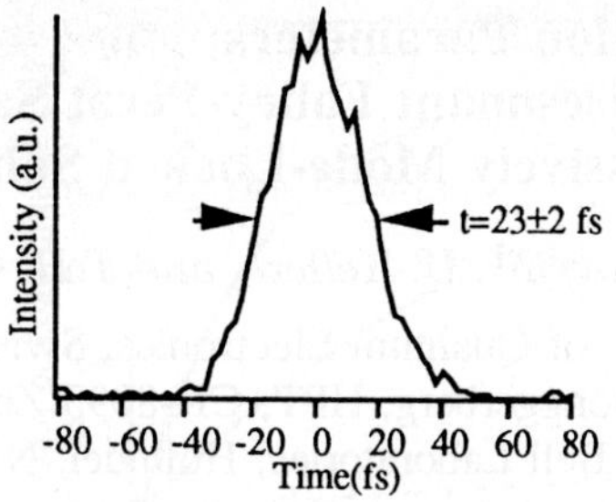

Figure 2.: Typical single shot autocorrelator trace of 50 mJ pulse.

second amplifier to isolate the two stages and eliminate ASE. The second amplifier uses a normal incidence, anti-reflection-coated, 8 mm long crystal. The beam is passed through this amplifier four times without focusing, and is pumped by approximately 400 mJ of green light.

In the first stage amplification, the pulse is amplified to an energy of approximately 2 mJ. We observe strong gain narrowing (to 47 nm bandwidth) in this amplifier, along with a red-shift due to gain saturation. Following the first stage of amplification, we can recompress the output to 21 fs.[1] The beam is then passed four times through the second amplifier stage, to obtain output energies of 50-75 mJ per pulse, with a FWHM bandwidth of 41 nm. A typical spectrum is shown in Figure 1. This bandwidth corresponds to a pulse duration of about 23 fs. Our single shot autocorrelator measurements measure a pulse duration of 23 ± 2 fs, as shown in figure 2. The throughput of compressor is about 36%, resulting in >20 mJ usable output. This corresponds to approximately 1 TW output power in a system which occupies <4' x 12' of table space. The ASE level is <0.3%. The beam profile after the second amplifier is still a very well-behaved Gaussian. This, together with the well behaved spectrum data, indicate that the modest stretch factor does not cause significant nonlinear effects in our system.

In conclusion, we have demonstrated a simple and inexpensive amplifier system which is capable of generating TW level laser pulses with pulse duration of 23 fs near the gain narrowing limit. Our results show that gain narrowing can be the primary limitation in the amplification of ultrashort pulses in broadband solid-state media.

We gratefully acknowledge support from the National Science Foundation, the Washington Technology Centers, and Union Carbide. M. Murnane acknowledges support from a Sloan Foundation Fellowship.

References

1. J.P. Zhou, C. P. Huang, C. Shi, H.C. Kapteyn, M.M. Murnane, Opt. Lett. **19**, 126 (1994).
2. R. L. Fork, C. H. B. Cruz, P. C. Becker, C. V. Shank, Opt. Lett. **12**, 483 (1987).
3. D. Strickland, G. Mourou, Opt. Commun. **56**, 219 (1985).
4. M. T. Asaki, C. P. Huang, D. Garvey, J. Zhou, H. C. Kapteyn, M. M. Murnane, Opt. Lett. **18**, 977 (1993).

Passive Mode-Locking of a Flashlamp-Pumped Ti:Al_2O_3 Laser

B. Bouma[1], *A. Gouveia-Neto*[1], *J. Russell*[2], *R. Sierra*[2], *J.A. Izatt*[2], *U. Keller*[3], *and J.G. Fujimoto*[1]

[1]Department of Electrical Engineering and Computer Science, Research Laboratory of Electronics, Massachusetts Institute of Technology, Cambridge, MA 02139, USA
[2]Candela Laser Corporation, Wayland, MA 01778, USA
[3]Swiss Federal Institute of Technology, ETH Hönggerberg, HPT, CH-8093 Zürich, Switzerland

Nearly all modelocked Ti:Al_2O_3 lasers to date have been cw modelocked systems with relatively low pulse energies. Short pulse energies in the range of microjoules to millijoules can be generated using regenerative and multipass amplifier systems. While these techniques achieve excellent performance, they are relatively costly and complex. In this paper we demonstrate a high performance, all solid-state, modelocked flashlamp pumped Ti:Al_2O_3 laser. The flashlamp pumped Ti:Al_2O_3 laser is a simple, economical, high peak power laser source which can be an attractive alternative technology to conventional oscillator-amplifier approaches. In addition, the development of new techniques for modelocking flashlamp pumped lasers can be generalized to a broad class of new solid state laser materials currently being developed.

Modelocking transient flashlamp pumped systems is extremely challenging because unlike cw lasers, the pulse energy grows rapidly in time. Previous investigations of modelocking Ti:Al_2O_3 have resulted in pulse durations of $\sim$ 100 ps[1,2]. To address the problem of modelocking in flashlamp pumped lasers, we used a combination of active and passive modelocking to achieve optimum pulse shortening during the rapid pulse build-up time. The laser cavity, shown schematically in Figure 1, incorporates a 15 cm long, 8 mm diameter, flashlamp pumped Ti:Al_2O_3 laser rod. The high small signal gain of up to 5 per pass permits a wide range of cavity designs to be implemented. An intracavity telescope was used to achieve the mode specification necessary for stable saturable absorber modelocking and to scale the beam waist for controlled energy extraction and saturable absorption.

Short pulse formation during the rapid pulse build-up time is achieved using a hybrid modelocking technique. Active modelocking rapidly shortens pulses which are long compared to the modulation period while fast saturable absorber modelocking provides a pulse shortening action that increases

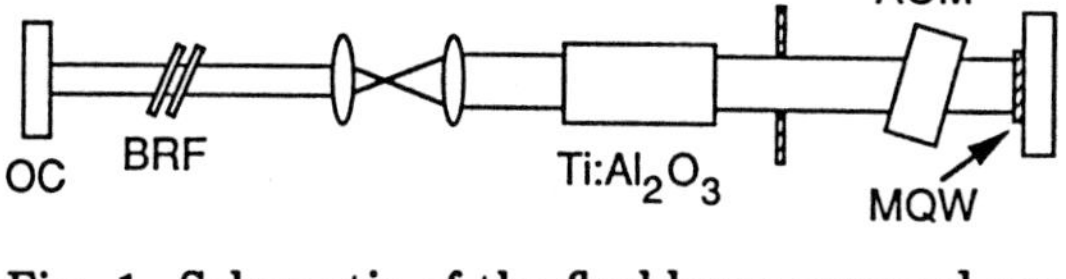

Fig. 1. Schematic of the flashlamp pumped modelocked oscillator. OC: output coupler, BRF: birefringent filter, T: telescope, LH: laser head, A: aperture, AOM: acousto-optic modulator, MQW: multiple quantum well.

Springer Series in Chemical Physics, Vol. 60 **Ultrafast Phenomena IX**
Editors: P. F. Barbara · W. H. Knox · G. A. Mourou · A. H. Zewail

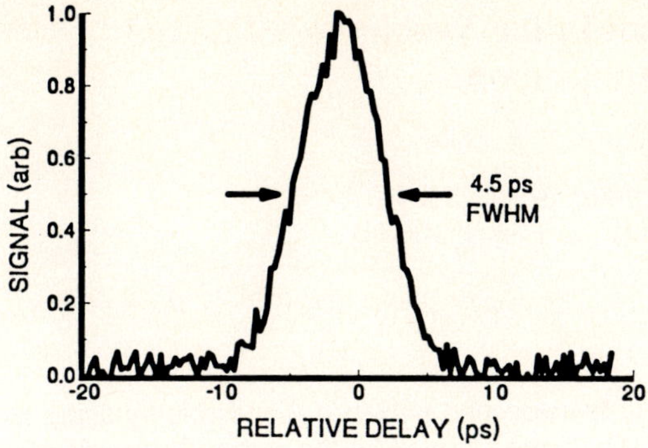

Fig. 2. Slow scan autocorrelation of average modelocked pulse duration in second relaxation spike. The pulse duration is 4.5 ps assuming a $sech^2$ pulse.

inversely as the pulse duration. In this experiment active modulation was provided by a fused silica acousto-optic modulator. When this device was used as the only modelocking element, the minimum pulse duration that could be generated was 100 ps.

To further reduce the pulse duration, we employed an antiresonant semiconductor Fabry-Perot saturable absorber (AFPSA)[3]. A standard noncollinear autocorrelation, shown in Figure 2, indicates a resulting pulse duration of 4.5 ps (assuming a $sech^2$ pulse). The corresponding modelocked pulse energy is $\sim 20\ \mu J$ at a wavelength of 824 nm. Using an intracavity two plate birefringent filter the laser can be tuned from 820 nm to 835 nm.

The pulse energy, duration, and limited tunability demonstrated with this laser are superior to previous flashlamp pumped $Ti{:}Al_2O_3$ systems; however, performance can be significantly improved. Picosecond pulse energies in the millijoule range should be attainable through the use of cavity dumping and by scaling the resonator mode size.

In conclusion, we have demonstrated a new all solid state technique for modelocking flashlamp pumped lasers. Active and passive modelocking used in conjunction with the high gain and high energy storage supplied by flashlamp pumping makes possible the development of a simple, cost effective source of high peak power optical pulses.

A. Gouviea-Neto was visiting from the Departamento de Física, Universidade Federal de Alagoas, Maceió, Brasil. The MQW samples were grown by M. T. Asom of AT&T Bell Laboratories. We gratefully acknowledge helpful scientific discussions with E.P. Ippen and H.A. Haus. This research was supported in part by JSEP grant DAAL03-92-C-0001, AFOSR grant F49620-91-C-0091, and ONR MFEL grant N00014-91-C-0084.

References

[1] S. Oda, Y. Segawa, N. Kodama, Jap. J. Appl. Phys. **28** 1977 (1989).

[2] K. Hamai, K. A. Stankov, H. Jelinkova, I. Prochazka, M. Koselja, *Inst. Phys. Conf. Ser. No. 126*, Sec. I, Paper presented at Int. Symp. Ultrafast Processes in Spectroscopy, Bayreuth, 1991, p. 59.

[3] U. Keller, T. H. Chiu, IEEE J. Quant. Electron. **28**, 1710 (1992).

Tunable Femtosecond Pulses in the Visible by Optical Parametric Amplification

J. Watson, T. Lépine, P. Georges, and A. Brun

Institut d'Optique Théorique et Appliquée, CNRS URA-14,
Bâtiment 503, B.P. 147, F-91403 Orsay Cedex, France

We present the generation of tunable femtosecond pulses in the visible by using an optical parametric amplifier in a BBO crystal pumped by the second harmonic of a Ti:Sapphire laser.

For femtosecond time resolved spectroscopy experiments, it is often interesting to change the wavelength of the pump and the probe. Generally, the production of tunable femtosecond pulses is obtained through spectral continuum generation in order to produce a white light. However, there are some limitations in energy and the spatial profile is altered by self focusing. Optical parametric generation is a nonlinear process which plays an important role in the development of easily and widely tunable coherent sources [1-3].

In this paper we present a femtosecond Optical Parametric Amplifier (OPA) tunable throughout the visible spectrum based on the highly nonlinear BBO crystal. Pulses as short as 350 fs at 560 nm with an energy of 4.5 μJ have been produced in a two-stage OPA pumped by the second harmonic of an amplified titanium sapphire laser working at 780 nm.

An important parameter is the Group Velocity mismatch (GV mismatch) between the pump at 390 nm and the signal in the visible domain as a function of signal wavelength. One has to note that this GV mismatch is always greater than 50 fs per mm of BBO crystal. After two passes in a 6 mm BBO crystal, one would expect this large amount of GV mismatch to severely limit the efficiency of the system.

Our two-stage OPA is presented in figure 1. 150 fs, 2 mJ pulses at 780 nm and at a 20 Hz repetition rate were produced by a Kerr lens modelocked Ti:Sapphire oscillator-regenerative amplifier laser chain. We first frequency doubled the amplified pulses in a 3 mm BBO (type I) crystal and obtained pump pulses at 390 nm with an energy of 150 μJ. Using these blue pulses to pump the OPA allowed us to tune over the entire visible range. We then pumped a 6 mm BBO (type I) crystal to create parametric fluorescence. We separated the 390 nm from the fluorescence and introduced a delay line in one arm to compensate for GV mismatch between the two beams. We then recombined the beams in the same BBO crystal in order to parametrically amplify the on-axis fluorescence. Finally, we amplified the signal beam in a 3 mm BBO (type I) crystal.

With a pump energy of 150 μJ at 390 nm in each arm, we obtained 4.5 μJ pulses around 560 nm with a pulsewidth of 350 fs. Sacrificing pulse length to efficiency, we obtained 10 μJ, 900 fs pulses at 600 nm and, using an 8 mm BBO crystal in the second stage, 29 μJ, 980 fs pulses at 640 nm. The relatively low efficiency of our OPA is due to the high GV mismatch between the pump and the signal compared to results obtained by Banfi et al [2]. In their case, they have a pump at 620 nm and a signal in the near infrared where the GV mismatch is lower (around 20 fs/mm).

Springer Series in Chemical Physics, Vol. 60 **Ultrafast Phenomena IX**
Editors: P. F. Barbara · W. H. Knox · G. A. Mourou · A. H. Zewail

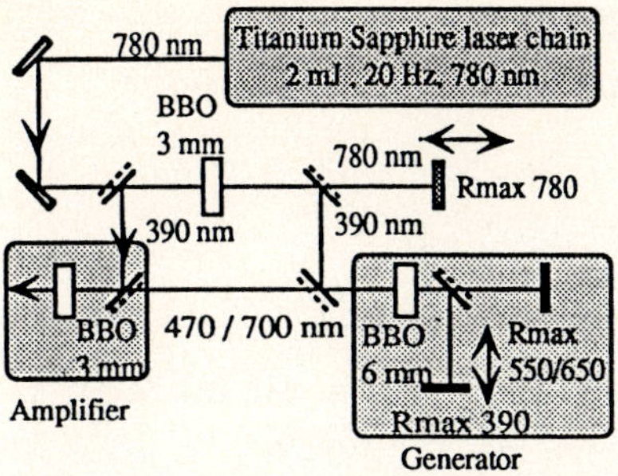

Figure 1. Experimental set-up of the two-stage OPA. The 780 nm pulses pass twice through the BBO doubling crystal to produce two outputs of pump pulses at 390 nm, which are in turn used to pump the two separate stages of the OPA (the double-pass generator and the amplifier).

We also show that to maximize the efficiency of the system, one must precompensate for the Group Velocity Dispersion (GVD) in the crystals by adding negative dispersion to the amplified puises at 780 nm. This is done with the grating compressor following the regenerative amplifier of the Ti:Sapphire laser. In this way, the pump pulse at 780 nm has negative GVD and a duration around 1.2 ps, leading to a signal pulse with the highest energy and the lowest duration.

Our measurements of the wavelength dependence on crystal orientation for the OPA corresponded very well to the theoretical tuning curve and ranged from 470 nm to 700 nm. As regards the GV mismatch, the desire to pump in the blue to extend the tuning range over the visible spectrum has magnified the difficulty of reaching a high efficiency. Precompensating for the GV mismatch by adding a delay line between the passes of the OPA has allowed us to offset this difficulty to some degree.

In conclusion, crystal length of our OPA seems to be a critical parameter : if too short, there is not enough gain, if too long, GV mismatch and GVD can severely limit the efficiency.

References

[1] W. Joosen, H.J. Bakker, L.D. Noordam, H.G. Muller, H.B. van Linden van den Heuvell, "Parametric generation in β-barium borate of intense femtosecond pulses near 800 nm", J.Opt.Soc.Am.B **8**, .2087 (1991).

[2] G.P. Banfi, R. Danielius, A. Piskarskas, P. Di Trapani, P. Fuggi, R. Righini, "Femtosecond traveling-wave parametric generation with lithium triborate", Opt. Lett. 18, 1633 (1993).

[3] J.Y. Huang, J.Y. Zhang, Y.R. Shen, C. Chen, B. Wu, "High-power, widely-tunable, picosecond coherent source from optical parametric amplification in barium borate", Appl.Phys.Lett. **57**, 1961 (1990).

Production of Programable Amplified, Shaped Pulses in Femtosecond Lasers

D. Pinkos[1], *J. Squier*[1], *D. Schumacher*[2], *P. Bucksbaum*[2], *B. Kohler*[3], *V.V. Yakovlev*[3], *and K.R. Wilson*[3]

[1]Center for Ultrafast Optical Science, University of Michigan, 2200 Bonisteel, IST Bldg., Rm. 1006, Ann Arbor, MI 48109-2099, USA
[2]Department of Physics, University of Michigan, Ann Arbor, MI 48109, USA
[3]Department of Chemistry, University of California, San Diego, 9500 Gilman Drive, La Jolla, CA 92093-0339, USA

Abstract. Using liquid crystal modulators to provide independent, progammable amplitude and phase control, we have been able to produce amplified, shaped, femtosecond pulses.

A heretofore unexploited capability of femtosecond chirped pulse amplification systems is the production of shaped pulses. This can be accomplished by phase and amplitude modulation of the pulses prior to amplification by a series of strategically located liquid crystal modulators. The system is suitable for kilohertz amplifiers which produce pulses with peak powers on the order of 10 GW, or 10-Hz amplifiers which can produce TW peak powers. Such a laser source is capable of addressing a variety of challenging scientific problems in a number of fields including chemistry, atomic and plasma physics.

The first demonstration of programmable phase control of femtosecond pulses using liquid crystal arrays was by Weiner et al [1]. More recently, Wefers and Nelson demonstrated independent programmable amplitude and phase control using similar techniques [2]. We have designed two liquid crystal modulators which provide independent amplitude and phase control and can work within the diffractive stretcher system of the laser. To our knowledge, this system has enabled us to produce, for the first time, amplified, shaped femtosecond pulses with a 10-Hz Ti:sapphire amplifier.

In contrast to previous works, our modulators have been designed to work at the conjugates of a stretcher system which provides the stretched seed pulse for a TW laser. In addition, the amplitude and phase control are accomplished in a single pass at the same transform plane, whereas multiple transform planes are required in other configurations [2].

The pulse shaper incorporates all reflective optics to eliminate chromatic aberrations. The input to the pulse shaping apparatus is split, so that the output can be cross-correlated with the original pulse, and thus characterized. Figure 1 shows the experimental and calculated cross-correlation traces. An odd pulse has been created by introducing a phase delay of π radians to only half of the frequency components of the pulse. In this simulation the model did not include diffractive effects or effects due to aberrations. However, we still obtain excellent agreement for these shaped pulses.

In figure 2 the single shot autocorrelation trace of an amplified picket fence is shown. More detailed characterization of these pulses is necessary using

Springer Series in Chemical Physics, Vol. 60 **Ultrafast Phenomena IX**
Editors: P. F. Barbara · W. H. Knox · G. A. Mourou · A. H. Zewail

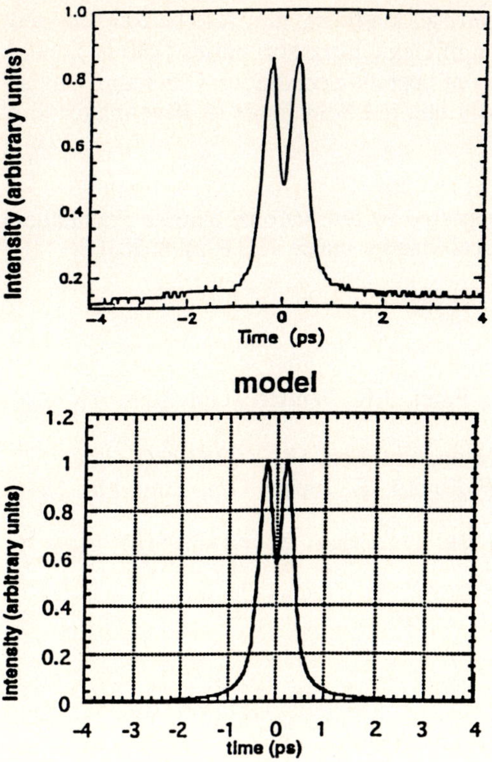

Figure 1. Measured and simulated odd pulse

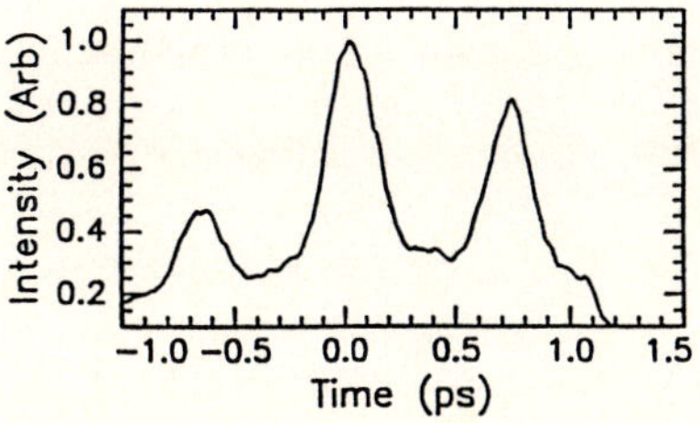

Figure 2. Single shot autocorrelation of an amplified picket fence.

more sophisticated autocorrelation techniques which make it possible to accurately determine the amplitude and phase of not only the amplified pulse, but the seed pulse as well. We are presently using frequency resolved optical gating schemes for this purpose [3,4].

In conclusion, we have demonstrated a programmable pulse shaper for use with amplified pulses. The potential applications of a programmable amplitude and phase modulator system are multi-fold. For instance, it may provide a method of higher order dispersion compensation. Amplitude reshaping effects which are a result of the amplification process may also be compensated. Used in combination, the production of clean, ultrashort pulses (<30 fs) could be made fairly routine under a variety of operating conditions. Other applications include the production of specifically tailored light fields to drive matter to a desired state or target.

This work was partially supported by the National Science Foundation through the Center for Ultrafast Optical Science under STC PHY 8920108.

References

1. A.M. Weiner, D.E.Leaird, J.S. Patel, J.R. Wullert, Opt. Lett. **15**, 326 (1990).
2. Marc M. Wefers and Keith A. Nelson, Opt. Lett. **18**, 2032 (1993).
3. J. Paye, M. Ramaswamy, J. Fujimoto, E. Ippen, Opt. Lett. **18**, 1946 (1993).
4. D.J. Kane, R. Trebino, Opt. Lett. **18**, 823 (1993); R. Trebino, D.J. Kane, J. Opt. Soc. Am. A **10**, 1101 (1993).

All-Solid-State Laser System for the Generation of Tunable Femtosecond Pulses down to 175 nm

F. Seifert, J. Ringling, F. Noack, V. Petrov, and O. Kittelmann

Max-Born-Institut für Nichtlinear Optik und Kurzzeitspektroskopie, D-12474 Berlin, Germany

Abstract. A 1 kHz Ti:sapphire amplifier system is used to produce ≈200 fs pulses in the 175-183 nm spectral region by sum frequency mixing in lithium triborate.

Ultrashort optical pulses tunable in the vacuum ultraviolet (VUV) are of great interest for spectroscopic investigations of various ultrafast processes. The shortest wavelength generated so far by sum frequency mixing in a nonlinear crystal (185 nm) has been achieved using nanosecond pulses [1]. The superior transparency range down to 155 nm and the large nonlinearity of lithium triborate (LBO) are attractive properties for efficient down conversion below 200 nm. However, phase matching in LBO is possible only for two widely separated wavelengths, e.g. in the ultraviolet (UV) λ_{UV}<205 nm and in the infrared (IR) λ_{IR}>1.7 μm, to obtain VUV radiation below 185 nm [2].

To ensure synchronism on a femtosecond time scale we employed a commercial Ti:sapphire laser/regenerative amplifier system (Spectra Physics "Tsunami", Quantronix 4810/20 RGA) whose tunable output pulses (150 fs, 300 μJ) are split into two equal parts to pump an optical parametric generator/amplifier and to generate the 4th harmonic, respectively (Fig. 1). The latter is performed by subsequent sum frequency mixing in three β-barium borate (BBO) crystals. This method produces tunable femtosecond pulses down to 189 nm at microjoule levels [3]. The measured pulse duration of 170 fs and the bandwidth of 0.4 nm correspond to nearly Fourier limited pulses at the 4th harmonic.

Sub-100fs IR pulses are provided by an optical parametric generator/amplifier [4]. A 5 mm long LBO crystal (z-cut, noncritical type II phase matching) is utilized

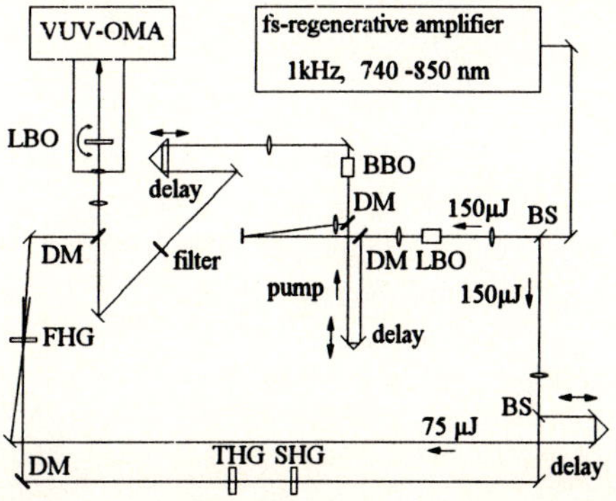

Fig.1: Experimental setup
BS - beam splitter
DM - dichroic mirror
SHG - second harmonic generation
THG - third harmonic generation
FHG - fourth harmonic generation

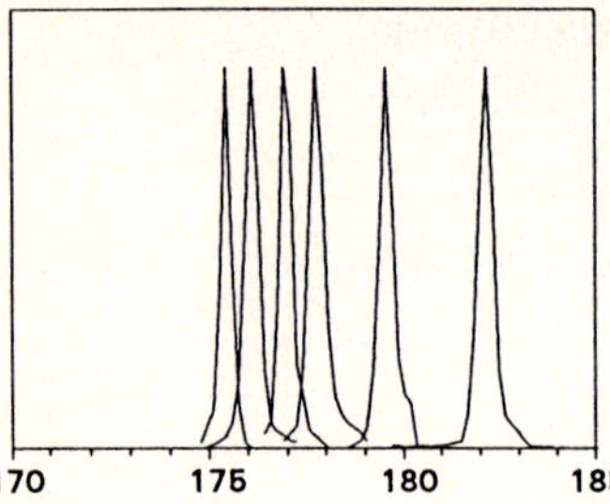

Fig.2: Normalized VUV spectra

to generate spectrally clean seed pulses for further parametric amplification. At room temperature and a pump wavelength of 805 nm we observed a signal wavelength of 1465 nm corresponding to idler pulses at 1800 nm. The amplification of the signal in a 4 mm long BBO crystal ($\theta=20°$ cut, type I phase matching) leads to nearly bandwidth limited 50 fs signal pulses of about 1 μJ energy and sub-100 fs idler pulses at the same energy level [4]. The polarization of both signal and idler pulses is parallel to that of the generated 4th harmonic. The IR beam and the 4th harmonic are combined at a dichroic mirror and focused (by a f=15 cm MgF_2 lens) into a 1mm thick LBO crystal ($\theta=90°$, $\varphi=73°$ cut, type I phase matching) housed in a vacuum chamber to avoid the strong VUV absorption of air. Tuning the Ti:sapphire pump system from 777 to 810 nm we achieved angle tuned ($\varphi=70°...\approx 90°$) phase matched sum frequency mixing of the 4th harmonic (194.2 - 202.4 nm) and the idler pulse at ≈1800 nm. The corresponding VUV spectra in Fig.2 demonstrate tunability between 175.5 nm and 182.2 nm. The spectral width of the VUV pulses at 182.4 nm corresponds to a pulse duration of 130 fs assuming a time-bandwidth product of 0.5. From a detailed analysis of the last nonlinear conversion process taking into account the different group velocities of the three interacting pulses we estimated the pulse width of the VUV signal to be <200 fs.

In conclusion we reported the efficient generation of unprecedented short wavelengths using only solid state materials for both the laser system and the nonlinear optical elements. Phase matching in LBO down to 172.5 nm is expected to be possible by heating the LBO seeder in order to increase the idler wavelength.

References:

1. R. E. Stickel, and F. B. Dunning, Appl. Opt. **17**, 981 (1978)
2. A. Borsutzky, R. Brünger, and R. Wallenstein, Appl. Phys. B **52**, 380 (1991); B. Wu, F. Xie, C. Chen, D. Deng, and Z. Xu, Opt. Commun. **88**, 451 (1992)
3. J. Ringling, O. Kittelmann, F. Noack, G. Korn, and J. Squier, Opt. Lett. **18**, 2053 (1993)
4. V. Petrov, F. Seifert, and F. Noack, Appl. Phys. Lett. **65** (1994), in press

Infrared-Pulse Generation Using a Subpicosecond OPO

J.D. Kafka, M.L. Watts, and J.W. Pieterse

Spectra-Physics Lasers, Inc., 1330 Terra Bella Ave.,
Mountain View, CA 94043, USA

For many ultrafast experiments it is desirable to have synchronized subpicosecond pulses at multiple wavelengths. Recently, synchronously pumped parametric oscillators (SPPO) have been used to satisfy this requirement as well as produce ultrafast pulses at wavelengths beyond 1 μm. In this paper, we demonstrate the synchronism inherent in an SPPO by measuring a 100 fsec cross-correlation between the pump and signal pulses. We then use the synchronization between the signal and idler to generate pulses at longer wavelengths by difference frequency mixing in silver gallium sulfide.

Our pump source consists of a mode-locked Ti:sapphire laser that produces 2 W at 770 nm when pumped by 12 W from an Argon ion laser (Fig. 1). The SPPO contains a temperature tuned 90° phase matched LBO crystal in a linear cavity [1]. This system has generated pulses as short as 40 fsec or average powers as high as 550 mW. We have obtained pulses from 1.1 to 2.4 μm with two sets of optics. For wavelengths shorter than 1.3 μm, we use intracavity prisms for dispersion compensation.

We have measured the cross-correlation between the pump and signal pulses using a BBO crystal. The duration of the pump pulse at 770 nm is 72 fsec after passing through a pair of prisms. The duration of the signal pulse at a wavelength of 1.32 μm is 65 fsec directly from the SPPO. The best cross-correlation is shown in Fig. 2 and is an average of 32 traces over a period of several seconds. The cross-correlation has a FWHM of less than 100 fsec and thus the jitter between the pump and signal pulses is less than 20 fsec. We have also measured the cross-correlation between the signal and idler pulses and typically obtain a FWHM of 180 fsec.

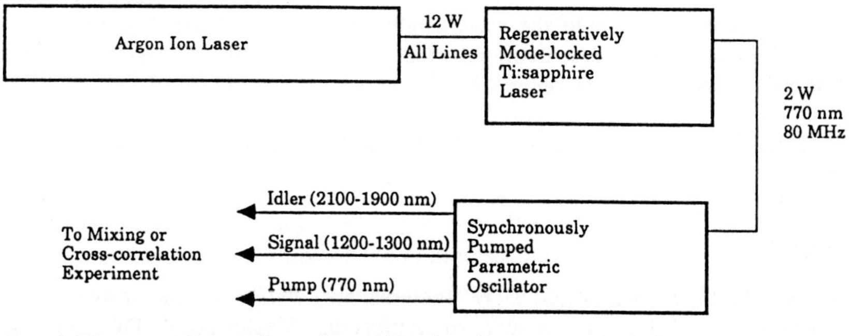

Fig. 1 The experimental configuration for the cross-correlation and for difference frequency generation.

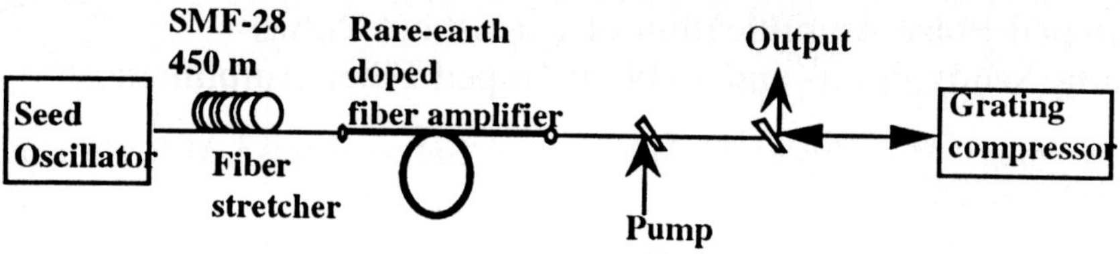

Fig. 1 Schematic of a CPA fiber amplifier system

The experimental set-up for the Er-doped fiber amplifier as reported by Stock et al [6] is similar to that of the Nd-doped fiber amplifier system and is described below. The seed pulses are from an environmentally stable Kerr-type modelocked erbium fiber laser [7]. This source generates band-width limited 250 fs pulses at a repetition rate of 33 MHz with an energy content of up to 100 pJ at a wavelength of 1.567 μm The amplifier consists of a length of 40 m of low-NA erbium doped fiber. The second spectral peak of the Er-doped fiber amplifier is approximately 12 nm shorter than the oscillator center wavelength. Both the oscillator and the amplifier are pumped with a launched power of 200 mW at a wavelength of 980 nm via a variable coupler, where 40 mW are directed to the oscillator and the rest is directed to the amplifier. With a pulse energy of about 30 pJ (average power approximately 900 μW) launched into the amplifier a single-pass gain of 18 dB was achieved, resulting in an output energy of 1.7 nJ (average power approximately 50 mW). After the fiber amplifier a grating pair was employed for pulse recompression. Due to loss on the gratings, after compression the output energy was about 1.4 nJ. The amplified pulses had a typical FWHM pulse width of 440 fs (assuming a $sech^2$ shape) and were approximately bandwidth limited and amplified pulse spectrum of 6 nm [6]. As noted in the Nd-doped amplifier experiment, higher-order dispersion from the stretching stage is uncompensated.

In summary we have demonstrated CPA of ultrashort pulses using both a Nd-doped fiber amplifier and an Er-doped fiber amplifier for the first time. The technique shows the potential of CPA for compact 1.06 μm and 1.55 μm laser systems. With proposed improvements, the CPA technique will be able to reach the micro-Joule saturation energy possible with current rare-earth doped fiber amplifiers in a compact system with sub-100 fs pulses.

The erbium-doped amplifier work was carried out at IMRA, America. MLS, L-MY and GM acknowledge support from the National Science Foundation through the Center for Ultrafast Optical Science under STC PHY 8920108. MEF acknowledges financial support from the Alexander von Humboldt Stiftung. L-MY would like to thank Marcel Bouvier for technical support.

References

[1] Donna Strickland and Gerard Mourou, Opt. Commun. 56 (1985) 219
[2] M. H. Ober, M. Hofer and M. E. Fermann, Opt. Lett. 18 (1993) 367
[3] K. Tamura, E. P. Ippen, H. A. Haus and L. E. Nelson, Opt. Lett. 18 (1993) 1080
[4] G. P. Agrawal, IEEE Phot. Tech. Lett. 2 (1990) 381

[5] W. J. Tomlinson and W.H. Knox, J. Opt. Soc. Am. B 4 (1987) 1404
[6] M. L. Stock et al, Nonlinear Guided Wave Phenomena Topical Meeting, Cambridge (1993) paper PD5
[7] M. E. Fermann, L.-M. Yang, M. L. Stock and M. J. Andrejco, Opt. Lett. 19 (1994),43
[8] C. Rouyer,et al., Opt. Lett. 18 (1993) 214

Femtosecond-Pulse Generation at 1.3 μm from a Praseodymium-Fibre Laser

M.J. Guy, A. Boskovic, D.U. Noske, and J.R. Taylor

Femtosecond Optics Group, Physics Department, Imperial College,
Prince Consort Road, London SW7 2BZ, UK

Abstract. 620 fs pulses have been generated at 1.3 μm by a Praseodymium-doped fluoride fibre laser. Temporal self stabilisation at freqencies up to 440 MHz has been observed.

Praseodymium (Pr^{3+}) doped fluoride fibre has been shown to act as an amplifier providing gain at 1.3 μm, the second low loss window in Silica fibre. Pr^{3+} doped into a fluoride glass host exhibits high gain over a range of ~50 nm and this broad bandwidth has the potential to support femtosecond pulses.

One problem with Pr^{3+} doped single mode fibres is that they tend to be highly normally dispersive at 1.3 μm, making it difficult to exploit the soliton-like pulse compression effects that lead to ultrashort pulses in similar erbium-doped fibre systems. The shortest pulse duration from a Pr^{3+} fibre laser reported until now is 1.6 ps with a time-bandwidth product of 0.85 from a "Figure-of-Eight" laser (F8L) cavity configuration whose overall dispersion was normal[1]. We have built a F8L with anomalous overall dispersion which generates almost transform-limited pulses as short as 620 fs.

The total length of the optimised laser cavity was 166 m, including a 7 m length of Pr^{3+} doped fibre. The dopant concentration was 2000 ppm and the laser was pumped by up to 1 W from a Ti:Al_2O_3 laser tuned to 1.02 μm. The splices between the fluoride and Silica fibres were mechanical and had a measured loss of 0.4 dB each. The anomalously dispersive fibre in the NALM had a minimum dispersion at 1.27 μm and an estimated dispersion of ~ +4.5 ps/nm/km at 1.3 μm.

Figure 1 shows the autocorrelation (with spectrum inset) of 620 fs pulses generated by the laser. The time-bandwidth product was calculated to be ~0.50, but it was noticed that the measured pulse duration changed slightly if the polarisation of the input light to the autocorrelator was rotated. It was found that the laser output could be resolved into two orthogonally polarised components with slightly different pulse durations and spectra, as has recently been observed in a similar erbium-based system[2]. The time-bandwidth product calculated using the appropriate spectral component is 0.32, close to the transform limit for $sech^2$ pulses.

In general this laser exhibited the usual temporal output of F8Ls consisting of bunches of randomly spaced short pulses periodic at the round trip time of the cavity. It also displayed a novel mode of mode-locked operation, however, in which the random stream of pulses would self-organise over a period of a few seconds into a well defined pulse train at harmonics of the round trip frequency, as has recently

Springer Series in Chemical Physics, Vol. 60 **Ultrafast Phenomena IX**
Editors: P. F. Barbara · W. H. Knox · G. A. Mourou · A. H. Zewail

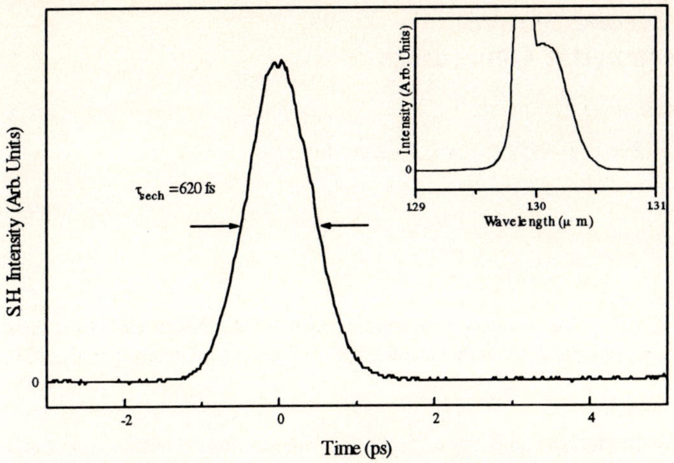

Figure 1. Autocorrelation trace of 620 fs pulses generated from Pr^{3+} laser with spectrum inset.

been observed in a ytterbium-erbium doped fibre ring laser[3]. This was observed both at low harmonics (up to 7th, ~5 MHz) and high harmonics of the cavity, with a maximum repetition rate of 437 MHz, corresponding to the 617th harmonic. These different repetition rates were selected by adjustment of the two polarisation controllers and once established exhibited long term stability. In the case of the low harmonic stabilisation the pulses were generally found to have a square envelope and were ~2 ns long, the exact duration being proportional to the pump power. It was also possible to generate pulses <1 ps in this case. For the higher harmonics, however, the pulses were always in the femtosecond regime.

In addition to the temporal output regimes described above, it was also possible by adjusting the fibre strainers to generate bursts of femtosecond pulses at repetition rates of up to 200 GHz with the number of pulses present in the burst dependent on the pump power. This effect is attributed to spectral beating caused by birefringence in the NALM.

References

[1] T. Sugawa, E. Yoshida, Y. Miyajima and M. Nakazawa, Electron. Lett. **29**, 902 (1993)

[2] M. J. Guy and J. R. Taylor, Electron. Lett. **29**, 2044 (1993)

[3] A. B. Grudinin, D. J. Richardson and D. N. Payne, Electron. Lett. **29**, 1860 (1993)

Tunable Femtosecond Sources and Optical Parametric Generators

F. Salin, F. Estable, and F. Saviot

B.M. Industries, CE 2901, F-91029 Evry Cedex, France

We demonstrate high repetition rate femtosecond sources tunable from 250 nm to 2 μm. These sources are based on Ti:sapphire and use Femtosecond Optical Parametric Generation to achieve broad tunability.

In order to fully exploit the storage capability of Ti:Al_2O_3 we used chirped pulse amplification (CPA). Our B.M.Industries Alpha-1000 source uses a MIRA Ti:sapphire oscillator (Coherent Inc.) producing sub-100 fs pulses tunable from 720 nm to 900 nm. The pulses were stretched up to 300 ps using a all reflective abberation compensated stretcher and then sent to a Ti:sapphire regenerative amplifier. The amplifier was pumped by a B.M. Industries Nd:YLF laser model 621D producing over 15 W of 523 nm pulses at 1 kHz repetition rate. After being amplified the pulses were recompressed to 80 fs. The compressed pulse energy around 800 nm was 900 μJ at 10 W pump power and up to 1.5 mJ (with a somewhat longer duration of 150 fs) at 12 W pump power. Two years ago we introduced a new design based on single grating strecher and compressor mounted on the same rotation table. This set-up allows us to align the compressor and stretcher once and then keep this adjustment regardless of the wavelength. Using this system we produced high power pulses tunable from 740nm to 850 nm. The tuning curve is given in figure 1 for two pump powers and for a single set of optics. In order to extend the tuning range in the UV we frequency doubled the pulses by using a 1 mm uncoated BBO crystal. We achieved up to 48 % efficiency at 400 nm although 40% was typical.. Using a second 1mm BBO crystal cut for third harmonic generation (type I) around 800 nm up to 65 μJ of 267 nm were obtained at 1 kHz repetition rate. Typical stabilities were ±2% for the second harmonic and ±3% for the third harmonic with long term stabilities of ±2.5% simmilar to that of the fundamental.

In order to get real tunability over the visible and the 1-4 μm region we developped optical parametric generators pumped by femtosecond pulses. The beam at the second harmonic was splitted in two parts to pump a parametric generator followed by a parametric amplifier. These two stages uses type I BBO crystals and group velocity compensation is used to obtain the maximum efficiency and shortest output pulses. Group velocity dispersion is a crucial issue when pumping optical parametric amplifiers with femtosecond pulses specially when the pump is in the UV. In BBO type I, the typical walk off due to velocity mismatch between the signal and the pump at 400 nm varies from 50 to 200 fs/mm when the signal tunes from 450 to 700 nm. Nevertheless due to the high gain most of the energy extraction occurs in the last part of the crystals and by a fine adjutment of the delay, pulses as short as the pump can be produced.

The output of the OPG/OPA combination is a collimated beam with a natural divergence of about 3 mrd at 550 nm. We measure the M2 factor of the beam to be 1.2 (i.e. 1.2x diffraction limited) over the whole tuning range. The tuning range extended from 440 nm to 730 nm for the signal and from 2800 nm to 850 nm for the corresponding idler. The energy of the signal has been measured for a total of 250 μJ of 390 nm pump incident on the OPG/OPA. The result is given in figure 2 and shows a relatively constant output at about 30 μJ per pulse. The spectrum of the pulses has been recorded at various point over the tuning range. The bandwith increases when the signal approaches the degeneracy point at 780 nm but stays in the range of 100-250 cm^{-1} over most of

Springer Series in Chemical Physics, Vol. 60 **Ultrafast Phenomena IX**
Editors: P. F. Barbara · W. H. Knox · G. A. Mourou · A. H. Zewail

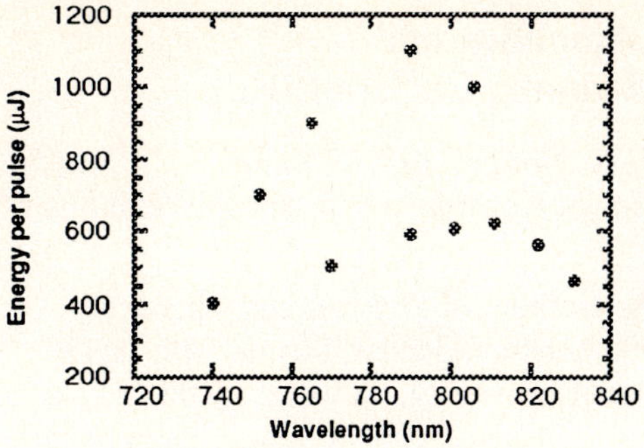

Figure 1. Tuning curve of the amplified pulses. Lower curve pump power=7W; upper curve 12W.

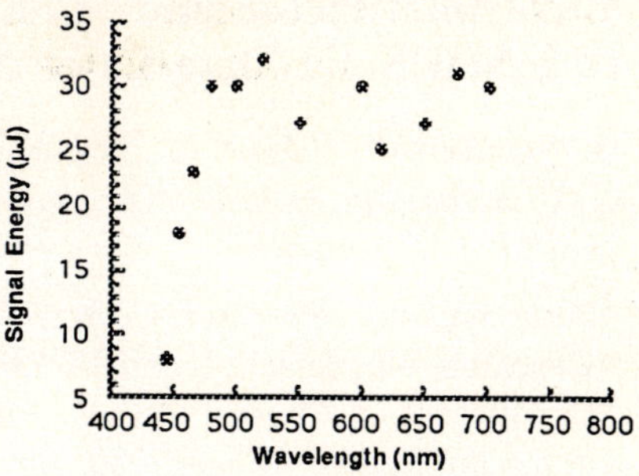

Fig.2: Energie per pulses at the signal wavelength versus wavelength.

the tuning range. We measured the pulse duration at several wavelengths using a scanning autocorrelator and a 100 µm thick KDP crystal. When the delay between the generator and the amplifier is correctly adjusted and if the group velocity dispersion of the pump pulses in the BBO crystals is pre-compensated by using the compressor grating distance, the pulse duration is around 110 fs. The pulse is clean and does not exhibit any wings at least at the measured wavelengths.

In summary we report on the possibility of building all solid-state tunable femtosecond sources. We demonstrated tunability from 250 nm to 2µm with over 100 MW peak power over that range at kilohertz repetition rates. The optical parametric generation proved to be an excellent mean to produce short tunable pulses and the tuning range can be extended by using different crystals and pump wavelengths.

1. F. Salin, J. Squier, G. Mourou, G. Vaillancourt, Opt. Lett. **16**, 1964 (1991).

2. F.Salin and F. Estable,OSA Topical meeting on Advanced Solid State Lasers, Salt Lake city, (1994).

Ultraviolet Picosecond-Pulse Amplification in a New Solid-State-Laser Medium: Ce^{3+}:$LuLiF_4$

N. Sarukura[1], *Z. Liu*[1], *Y. Segawa*[1], *K. Edamatsu*[2], *Y. Suzuki*[2], *T. Itoh*[2], *V.V. Semashko*[3], *A. K. Naumov*[3], *S.L. Korableva*[3], *R.Yu. Abdulsabirov*[3], and *M.A. Dubinskii*[3]

[1]Photodynamics Research Center, The Institute of Physical and Chemical Research, Nagamachi Koeji 19-1399, Aoba-ku, Sendai, Miyagi 980, Japan
[2]Department of Applied Physcis, Tohoku University, Aoba, Aramaki, Aoba-ku, Sendai, Miyagi 980, Japan
[3]Kazan State University, Lenin Street 18, 420008 Kazan, Russia

Abstract. Ce^{3+}:$LuLiF_4$ is shown to have a broad gain spectrum, which is attractive for short pulse applications. Sufficient small signal gain over 6 dB/cm was observed. Furthermore in a confocal 4-pass amplifier configuration, the gain can be enhanced to 20-dB for a cw He-Cd laser light and 17-dB for picosecond pulses.

Recent progress in tunable solid-state-laser materials in the infrared region has enabled development of completely new ultrashort-pulse laser sources. Such innovations are also expected in the ultraviolet region, as several Ce^{3+}-doped tunable ultraviolet laser crystals have been developed including $YLiF_4$,[1] LaF_3,[2] $LiCaAlF_6$,[3] $LiSrAlF_6$,[4] and $LuLiF_4$.[5] In spite of their vast potentiality as new categories of laser media for short-pulse applications, there has been little work on these crystals. In this paper, we will describe the some spectroscopic properties of Ce^{3+}:$LuLiF_4$ and its application to ultraviolet light amplification including picosecond pulses. The fluorescence spectrum (300-345 nm) has two peaks at around 308 nm and 327 nm (Fig. 1). The absorption and excitation spectra matches well with the KrF-excimer-laser wavelength (248 nm). The fluorescence lifetime was 40 2 nsec under a KrF excimer laser excitation. The small signal gain, pumping power dependence, saturation fluence, and gain curve were measured for 1-cm Ce^{3+}:$LuLiF_4$ crystal which was pumped from the side window on the crystal by a 10-nsec KrF laser. Small-signal gain and saturation fluence were evaluated using the second harmonic (325 nm) of a nanosecond DCM-dye laser as a probe. The relative timing of the probe dye laser and the pumping excimer laser was controlled by a synchronizer to obtain the largest possible gain. The single-pass gain in the small-signal region was over 6-dB/cm (4.3 times) with 0.5-J/cm^2 excitation. The saturation fluence assuming Frantz-Nodvick relation was measured to be ~50 mJ/cm^2, which is much higher than that of organic dyes (~ mJ/cm^2). These results indicate its high potentiality as a power amplification medium. The gain spectrum was measured using the second harmonic of a tunable mode-locked laser as a probe, and a KrF-laser as a pumping source. The gain spectrum from 323 nm to 335 nm was measured (Fig. 2), and it is broad enough to amplify tunable femtosecond pulses. To further demonstrate large small-signal gain, the Ce^{3+}:$LuLiF_4$ amplifier was designed with a side-pumped confocal 4-pass configuration similar to the coaxially pumped confocal multi-pass configuration of Ti:sapphire amplifier.[6] The amplifier consisted of a gain medium located at the beam waist of a confocal lens pair (15-cm focal-length) and two turning prisms for small displacement of each pass (Fig. 3). This configuration allowed the passes to overlap well in a small pumped region. A side pumping scheme is necessary because of rather high

Springer Series in Chemical Physics, Vol. 60 **Ultrafast Phenomena IX**
Editors: P. F. Barbara · W. H. Knox · G. A. Mourou · A. H. Zewail

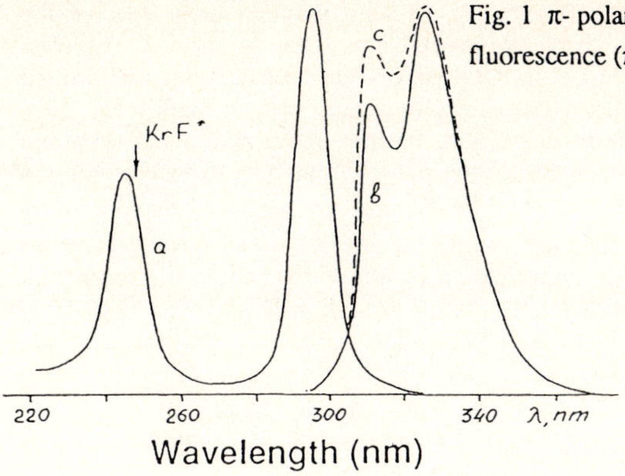

Fig. 1 π- polarized absorption (a) and normalized fluorescence (π- polarization, b and σ- polarization, c)

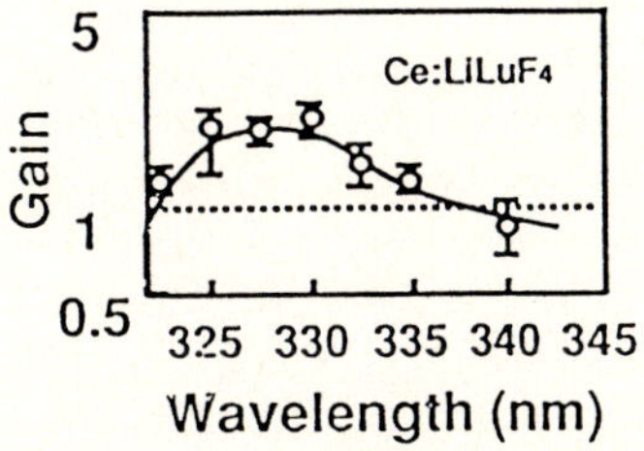

Fig. 2 Gain spectrum under the pumping of KrF laser

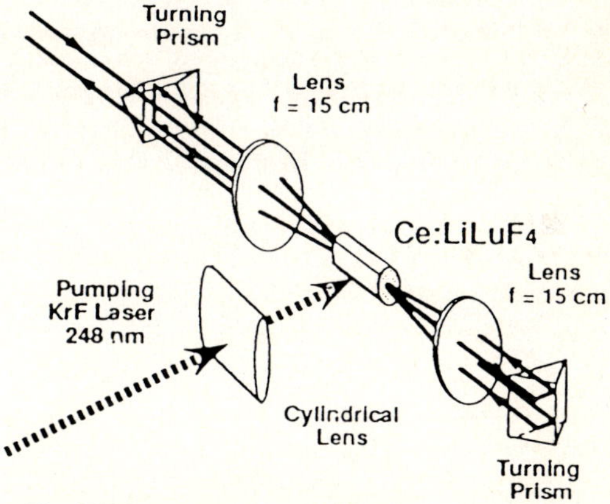

Fig. 3 The side-pumped confocal 4-pass Ce^{3+}:$LuLiF_4$ amplifier.

absorption (~3 cm^{-1}) at the pumping wavelength. The 1-cm crystal was cut cylindrically, and a flat window was polished on the side. ~10-mJ, 10-nsec KrF-laser pulses were weakly focused by a 10-cm focal-length cylindrical lens. The pumping laser operated at a 3-Hz repetition rate to avoid the possible color-center formation effects.[7] The 4-pass gain for the cw He-Cd laser was ~20 dB for the transmitted light. This small-signal gain is sufficiently high for the practical use. Using the above laser material and amplifier configuration, an all-solid-state ultraviolet picosecond laser system was designed. The seed picosecond ultraviolet pulses at 325 nm were generated by the sum-frequency of 835-nm picosecond pulses from a Ti:sapphire amplifier and the second harmonic of a amplified typically up to 100-μJ level by a double-side-pumped confocal 4-pass Ti:sapphire amplifier.[6] For efficient sum-frequency generation, the peak position of the 532-nm, 7-nsec pulses was adjusted to the largest pulse among the amplified picosecond pulse train by an optical delay (40-nsec). In this way, a single ~1-μJ, 325-nm seed pulse was generated at a 10-Hz repetition rate using a 5-mm thick BBO crystal cut at 35 degrees. The duration of the ultraviolet pulses was evaluated to be less than 3-psec using a streak camera. The confocal 4-pass Ce^{3+}:$LuLiF_4$ amplifier as described before was pumped by 100-mJ, KrF-laser pulses at a 0.5-Hz repetition rate synchronized to 10-Hz ultraviolet seed pulses with 0.5 J/cm^2 fluence. 17-dB gain was demonstrated for the transmitted 325-nm picosecond pulses. Using the streak camera, the duration of amplified ultraviolet pulses was evaluated to be less than 2.9 psec, but the measurement was limited by the resolution of the apparatus. This result shows that there was no significant pulse width broadening in the amplifier due to dispersion. Further optimization will improve the gain factor, and the amplification of much shorter pulse can also be expected with the full use of its bandwidth. In conclusion, $Ce^{3+}:LuLiF_4$ was proven to be an ultraviolet amplification medium with broad bandwidth which is attractive for ultrashort pulse applications. Using this material, direct amplification of ultraviolet pulses in solid-state media was demonstrated for the first time.

1. D.J. Ehrlich, P.F. Moulton, and R. M Osgood,Jr., Opt. Lett. 4, 184 (1979).
2. D. J. Ehrlich, P. F. Moulton, and R. M Osgood, Jr., Opt. Lett. 5, 339 (1980).
3. M.A.Dubinskii,V.V.Semashko,A.K.Naumov,R.Y.Abdulsabirov, and S.L.Korableva, J.Mod.Opt.,40,1, (1993).;
4. J.F. Pinto, G.H. Rosenblatt, L. Esterowitz, and G.J. Quarles, LEOS '94 , Paper PDI. 4.
5. M.A.Dubinskii, R.Y.Abdulsabirov, S.L.Korableva, A.K.Naumov, and V.V.Semashko, IQEC '92, Paper FrL2.
6. N. Sarukura and Y. Ishida, J. Quant. Electron. 28, 2134 (1992).
7. Ki-Soo Lim and D. S. Hamilton, J. Opt. Soc. Am. B7, 1401 (1989).

Soliton Compression in a Nonlinear Amplifying-Loop Mirror

S. Gray[1], *A.B. Grudinin*[1], *W.H. Loh*[1], *D.N. Payne*[1], *and V.V. Afanasjev*[2]

[1]Optoelectronics Research Centre, Southampton University, Southampton SO17 1BJ, UK

[2]General Physics Institute, Russian Academy of Sciences, 38 Vavilov Street, 117942 Moscow, Russia

We have experimentally demonstrated soliton compression in a NALM with a low fraction of non-soliton component and switching efficiency of about 60%. The principle of operation of an all fibre tunable source of femtosecond pulses has been shown.

1. Introduction

The Nonlinear Amplifying Loop Mirror [1] posseses both compression and pulse shaping characteristics [2-4]. In this paper we study soliton compression in a NALM and demonstrate the principle of operation of an all fibre tunable source of femtosecond pulses.

The soliton source was a passive harmonic modelocked fibre soliton laser generating 2.5ps pulses (fwhm) at a repetition rate of 50 MHz [5]. The laser output was connected to the NALM through an isolator to prevent feedback into the laser cavity. The NALM was formed between the output ports of a 50/50 coupler and comprised a 3 metre long Yb^{3+}/Er^{3+} codoped fibre amplifier and 50 metres of standard telecom fibre with group velocity dispersion D=17 ps/(nm.km).

2. Compression and Switching Characteristics of the NALM

To characterise the NALM transmission in the soliton regime we studied the spectral and temporal characteristics of the output pulses and the switching efficiency for different gain levels. The switching efficiency is given by P_{out}/GP_{in} where P_{in} and P_{out} are the powers at the NALM input and output and G is the amplifier gain.

Fig. 1 shows the gain dependence of the compression ratio and switching efficiency. The maximum switching efficiency occurred for 13.5 dB gain and maximum compression occurred at 15 dB gain.

The experimental results indicate that there is an optimal gain, G_{opt}, corresponding to the maximum switching efficiency where a clean output pulse is observed. For gain exceeding G_{opt} spectra and autocorrelations of multisoliton pulses are observed at the NALM output. For $G<G_{opt}$ the time-bandwidth product of the output pulses was close to 0.31 while the pulsewidth varied by over a factor of two (Fig. 1).

Editors: P. F. Barbara · W. H. Knox · G. A. Mourou · A. H. Zewail

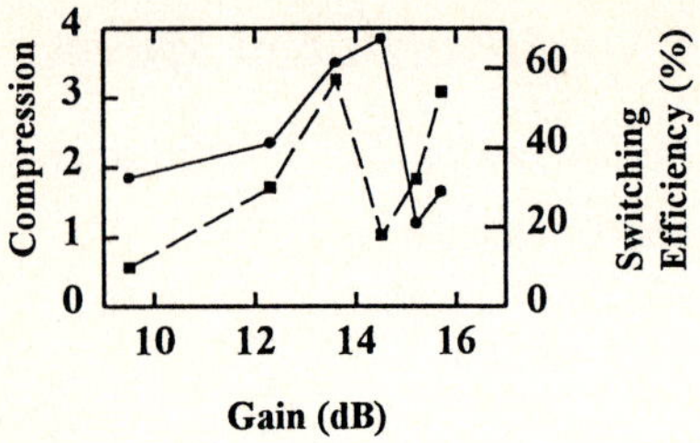

Fig. 1. Gain dependance of compression (solid line) and switching efficiency (dashed line)

3. Tunable All Fibre Source of Femtosecond Pulses

The gain dependence of the pulsewidth gives rise to the possibility of developing a tunable source of femtosecond pulses based on the soliton self frequency shift. Due to the effect of Raman gain the central wavelength of the soliton, λ_0, experiences a stokes frequency shift given by $d\lambda_0/dz(\mathrm{nm/m}) \simeq 2.3.10^{-5} D/\tau^4$ where D is the dispersion (ps/nm.km) and τ is the pulsewidth (ps) [6].

By using an auxiliary fibre at the NALM output the pulsewidth variation can be translated into a controllable wavelength shift. To demonstrate the operation of such a source we shortened the laser cavity which decreased the pulsewidth to 1.3 ps and reduced the length of the undoped fibre in the NALM to 3 metres. In this configuration by setting the NALM gain to values up to 15 dB we observed reduction of the pulsewidth down to 250 fs while the time-bandwidth product changed from 0.32 to 0.28.

For 250 fs solitons travelling in a lossless fibre with dispersion D=17ps/nm.km we expect the self frequency shift to be $\delta\lambda$=10nm/100m. Fig. 2 (a and b) shows the spectra of the pulses after propagation along 300 metres of the auxillary fibre for NALM gains of 14.5 dB and 15 dB respectively. A small gain change shifts the central pulse wavelength from 1544 nm to 1549 nm for 14.5 dB gain and to 1574 nm for 15 dB gain giving a tuning range of 30 nm.

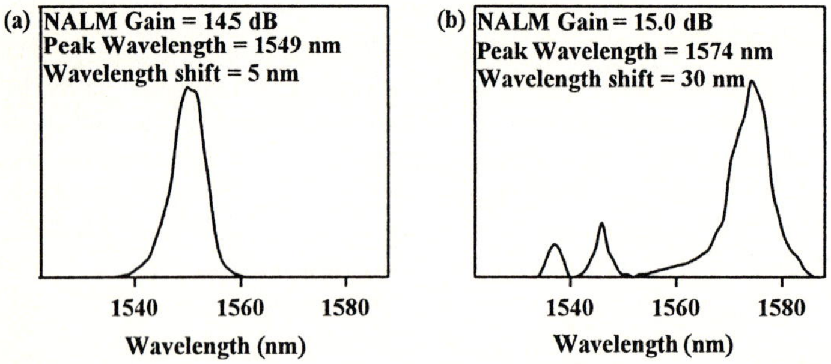

Fig. 2. Spectra of pulses after propagation through 300 metres of standard telecom fibre.

4. Conclusion

In conclusion, we have experimentally studied soliton propagation through a NALM and demonstrated soliton compression with only a 10% fraction of non soliton component and a switching efficiency of 60%. The pulsewidth dependence on gain allows the development of an all fibre tunable source of femtosecond pulses and the principle of operation of such a source has been shown.

References

1. Fermann, M.N., Haberl, F., Hofer, M. and Hochreiter, H. Opt. Lett., 1990, **15**, pp 752-754.
2. Blow, K., Doran, N.J. and Nayar, B.K., Opt. Lett., 1989, **14**, pp 754-756.
3. Islam, M.N., Sunderman, E.R., Stolen, R.H., Pliebel, W., and Simpson, J.R., Opt. Lett., 1989, **14**, pp 811-813.
4. Smith, K., Greer, E.L., Doran, N.J., Bird, D.M. and Cameron, K.H., Opt. Lett., 1992, **17**, pp 408-410.
5. Grudinin, A.B., Richardson, D.J. and Payne, D.N., Electron. Lett., 1993, **29**, pp 1860-1861.
6. Gordon, J.P., Opt. Lett., 1986, **11**, pp 662-664.

Femtosecond Cr^{4+}:YAG Laser

H. Nathel[1], *A. Sennaroglu*[2], *and C.R. Pollock*[2]

[1]Lawrence Livermore National Laboratory, P.O. Box 808, L-524, Livermore, CA 94551, USA

[2]School of Electrical Engineering, Cornell University, Itahca, NY 14853, USA

1. Introduction

Recent advances in mode-locked solid-state lasers have resulted in pulses as short as 8.5 fsec in the 800 nm region from Ti:sapphire oscillators [1]. Using similar techniques, sub-50 fsec pulses [2-4] have been generated in another solid-state laser system, Cr:forsterite, which operates in the 1.2 to 1.3 μm range. We report here results from both a regeneratively-initiated and self-initiated, mode-locked Cr:YAG laser [5] which is tunable from 1.51 to 1.53 μm. One hundred and twenty femtsosecond, nearly transform-limited pulses have been generated with peak output powers of 45 kW. The stable, high peak power pulses and room temperature operation of this laser make it a very suitable alternative to the cumbersome, cryogenic mode-locked NaCl laser [6] commonly used in both narrow bandgap semiconductor and optical communications research.

Lasing action in chromium-doped yttrium aluminum garnet (Cr^{4+}:YAG), first reported by Shestakov [7], occurs over the wavelength range from 1.34 to 1.60 μm. Due to the broad tunability, Cr:YAG is a good candidate for femtosecond pulse generation. Active and moving mirror mode-locking of a Cr:YAG laser have resulted in 26 psec and 3 psec pulses at 1.44 μm [8,9], respectively. However citing poor crystal quality resulting in low intracavity power, Kerr lens mode-locking was unachievable and consequently femtosecond pulse generation was not observed.

2. Experimental Results

We have constructed a folded, astigmatically compensated z-cavity laser cavity which has been detailed elsewhere [5]. With no intracavity elements present, an absorbed pump power slope efficiency of 15% with threshold of 1.04 W was measured with the output of the laser centered at 1.45 μm. Continuous wave output power of 840 mW was obtained. With the inclusion of intracavity elements (A.O.M. and the prism pair), the slope efficiency dropped to 10%, and the output laser shifted to 1.49 μm, mainly due to wavelength dependent absorption and scattering from surface imperfections in the intracavity elements. The threshold absorbed pump power increased to 1.82 W.

To initiate mode-locking, 4% of the output beam was picked off via a beamsplitter and sent to an InGaAs photo diode which detected the cavity mode beating signal. The signal was frequency-divided and amplified with adjustable phase shift to drive the A.O.M. that had 0.4% modulation depth. Optimization of the cavity to produce self focusing and induce Kerr nonlinearities resulted in a femtosecond pulse train. Figures 1(a) and 1(b) show the background-free intensity

Springer Series in Chemical Physics, Vol. 60 **Ultrafast Phenomena IX**
Editors: P. F. Barbara · W. H. Knox · G. A. Mourou · A. H. Zewail

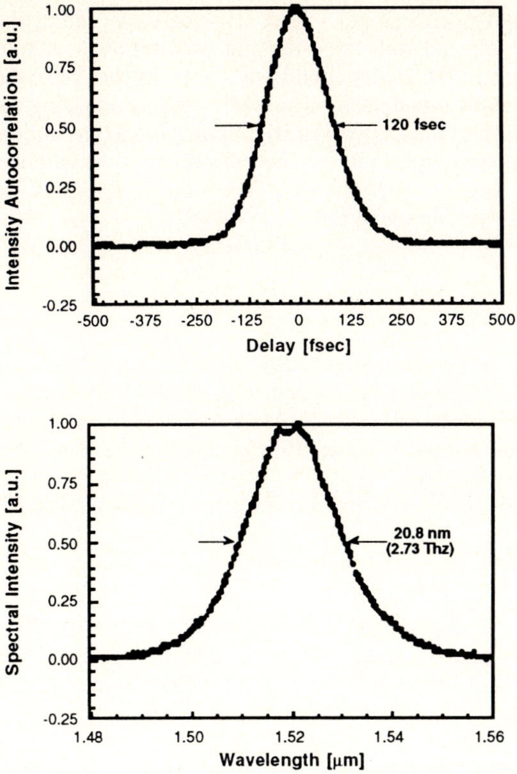

Figure 1. (a) The background-free autocorrelation, and (b) spectrum of mode-locked pulses.

autocorrelation and the spectrum, respectively, of the mode-locked pulses. Assuming a $sech^2$ intensity profile the pulse width was measured to be 120 fsec. A simultaneous measurement of 20.8 nm spectral bandwidth yielded a time-bandwidth of 0.327 verifying that the pulses were nearly transform-limited. The onset of mode-locking resulted in the peak wavelength to shift to 1.52 nm and a reduction in output power from 460 mW to 440 mW. By placing an aperture between the prism and the high reflector, the output wavelength could be tuned over the range from 1.51 to 1.53 µm. Operation to longer wavelengths was inhibited by the optics coatings. At shorter wavelengths, tuning resulted in the collapse of the broad spectral bandwidth resulting in approximately 20 psec pulses. Losses introduced by atmospheric water absorption together with excited-state absorption (ESA) in the Cr:YAG crystal are the dominant factors inhibiting femtosecond pulse formation at shorter wavelengths (below 1.48 µm). The role of ESA as a net negative saturable absorber quenching self-mode-locked operation has been studied and suggested previously [8,10].

With the A.O.M. turned off, self-sustained operation in excess of one hour was observed. The measured peak power and energy per pulse were 45 kW and 5.4 nJ, respectively. A threshold output power of approximately 100 mW was required to observe the onset of femtosecond pulse formation (self or Kerr lens mode-locking) corresponding to 1 MW intracavity peak power. In a separate experiment, the A.O.M. was removed from the laser and self-mode-locking was initiated without applying any periodic perturbation to the cavity round-trip gain. In this case, the mode beating signal detected by the InGaAs detector was maximized by adjusting the curved mirrors around the laser rod and the cavity was slightly misaligned to initiate the femtosecond pulse train. Alternatively, one of the prisms could be slightly translated to initiate the pulse train. The output pulse characteristics were indistinguishable from the regeneratively-initiated case.

3. Conclusions

As mentioned previously, the femtosecond laser source described here should serve as a fascile replacement to the cumbersome NaCl color center laser. Additionally, the femtosecond Cr:YAG laser should lend itself beautifully to an all solid-state system, either being pumped by a diode pumped Nd:YAG or Nd:YLF laser or alternatively a high power diode laser operating at 980 nm.

4. Acknowledgments

This work was supported by the U. S. Department of Energy under contract W-7405-Eng-48, the National Science Foundation under grant ECS-9111838, the Joint Services Electronics Program, and the Materials Science Center at Cornell University.

Added note in proof: The Imperial College group has obtained 70 fsec pulses with 120 mW of output power and has also identified atmospheric water absorption as an inhibiting factor in femtosecond operation of Cr:YAG below 1.48 μm.

5. References

1. M. M. Murnane, H. C. Kapteyn, J. Zhou, C. P. Huang, D. Garvey, and M. T. Asaki, *Tech. Dig. OSA Ann. Meet.*, Toronto, Canada, paper WD1, page 103 (1993).
2. A. Sennaroglu, C. R. Pollock, and H. Nathel, *Opt. Lett. 18(10)*, 826 (1993).
3. A. Seas, V. Petricevic, and R. R. Alfano, *Opt. Lett. 18(11)*, 891 (1993).
4. V. Yanovsky, Y. Pang, F. Wise, and B. I. Minkov, *Opt. Lett. 18(18)*, 1541 (1993).
5. A. Sennaroglu, C. R. Pollock, and H. Nathel, *Opt. Lett. 19(6)*, 390 (1994).
6. C. P. Yakymyshyn, J. F. Pinto, and C. R. Pollock, *Opt. Lett. 14,* 621 (1989).
7. A. V. Shestakov, N. I. Borodin, V. A. Zhitnyuk, A. G. Ohrimtchyuk, and V. P. Gapontsev, *Digest of CLEO*, Baltimore, Md., paper CThS82, (1993).
8. P. M. W. French, N. H. Rizvi, J. R. Taylor, A. V. Shestakov, *Opt. Lett. 18(1)*, 39 (1993).
9. P. J. Conlon, Y. P. Tong, P. M. W. French, and J. R. Taylor, *Tech. Dig. OSA Top. Meet. Adv. ASSL*, Salt Lake City, Utah, poster AMC2, page 15 (1994).
10. A. Sennaroglu, C. R. Pollock, and H. Nathel, submitterd to *JOSA B*.

Sub-40 fs Pulse Generation in a KTP-Based Parametric Oscillator

D.T. Reid, J.M. Dudley, M. Ebrahimzadeh, and W. Sibbett

J.F. Allen Research Laboratories, Department of Physics and Astronomy, University of St. Andrews, North Haugh, St. Andrews, Fife KY16 9SS, Scotland, UK

Abstract. Under near-zero dispersion conditions sub-40 fs duration transform-limited pulses are generated by a Ti:Al_2O_3-pumped optical parametric oscillator. With net positive dispersion the output signal pulses exhibit behaviour characteristic of soliton propagation

1. Introduction

The demand for femtosecond sources at new wavelengths in the near to mid-infrared together with the emergence of the self-mode-locked Ti:sapphire laser as a convenient pump source has led recently to the development of femtosecond optical parametric oscillators. We report here the operation of a synchronously pumped femtosecond optical parametric oscillator (OPO) based on a 1.5 mm thick crystal of KTP cut for non-critical phase-matching. Pumped by 85 fs pulses from a self-modelocked Ti:sapphire laser the OPO is tunable over the bandwidth of the mirrors from 1.1 - 1.3 µm and pump tuning over 100 nm is demonstrated. Our intracavity dispersion-compensated OPO has been operated in both three-mirror and four-mirror standing-wave geometries using single-pass pumping and exploiting a variety of mode-matching configurations. In addition, the observation of soliton generation in the femtosecond OPO is described.

2. Results and discussion

A typical spectrum and interferometric autocorrelation of the signal pulses produced in the three-mirror configuration are presented in Figure 1 and correspond to bandwidth-limited 39 fs pulses with intracavity peak powers approaching 1 MW [1]. The shortest pulses are observed when operating at near-zero intracavity GVD.
A comparison of both three-mirror and four-mirror standing wave cavity configurations has been made. We have demonstrated power conversion efficiencies from the pump to the signal of up to 21%. Optimum operation was achieved when both the pump and signal modes were focused in the crystal to beam waists of radius < 20µm. Measured thresholds of the non-dispersion-compensated oscillators were as low as 230 mW while the insertion of dispersion compensation using a pair of intracavity SF14 prisms increased the threshold to around 300 mW. With dispersion compensation, pulse durations as short as 38 fs were measured for the three-mirror oscillator whereas the four-mirror design generated pulses as short as 56 fs. We attribute the longer pulse duration of the 4-mirror cavity to higher third-order

Editors: P. F. Barbara · W. H. Knox · G. A. Mourou · A. H. Zewail

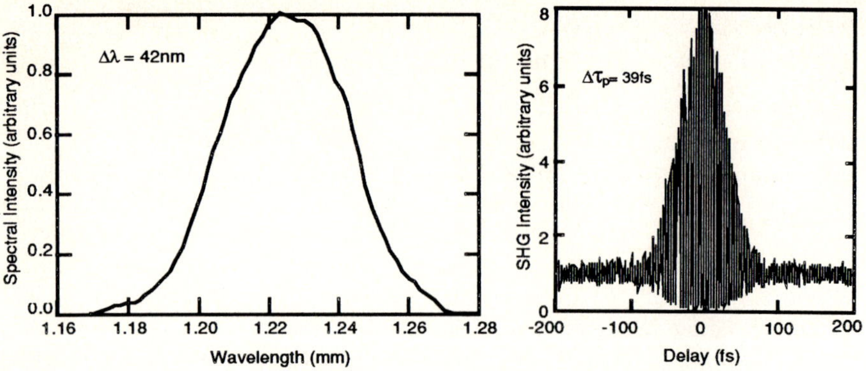

Figure 1 Spectrum and interferometric autocorrelation of the signal pulse

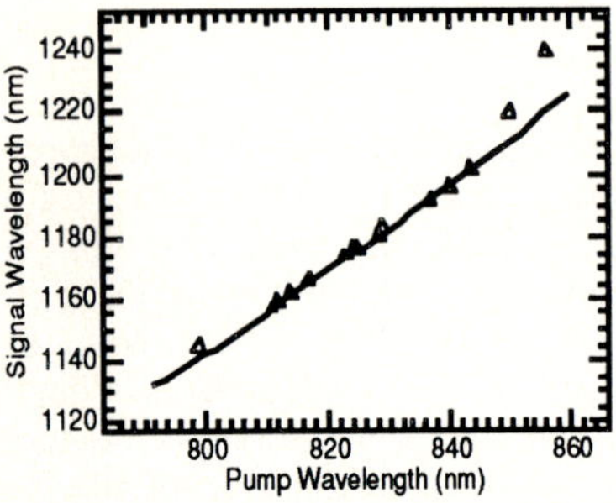

Figure 2. Experimental pump tuning data and predicted theoretical curve

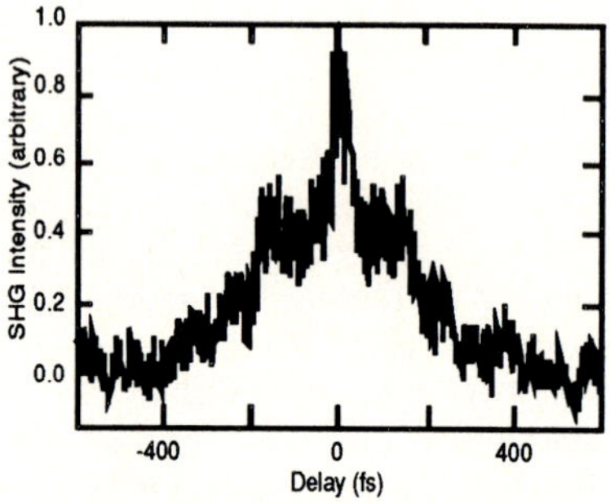

Figure 3. Autocorrelation profile characteristic of N = 3soliton

dispersion associated with two additional mirror reflections. Each of the OPO cavity configurations used provided extensive tuning achievable either by changing the crystal orientation or by altering the wavelength of the pump laser. Results of pump laser tuning are presented in Figure 2 together with the predicted tuning curve derived from Sellmeier data for KTP [2]. Under appropriate conditions of net positive intracavity dispersion (~300 - 700 fs^2) the output of the OPO has features characteristic of soliton propagation. A periodic intensity modulation of frequency between 0.5 MHz and 2.2 MHz is observed and corresponding pulse duration measurements are consistent with the soliton theory reported by Salin *et al* [3]. This behaviour is accompanied by a shift in the signal spectrum towards longer wavelengths as in agreement with theoretical predictions [4]. Under these circumstances strong components are produced at $\lambda > 1300$ nm without requiring to tune the pump laser to longer wavelengths. The autocorrelation profile of the output pulses presented in Figure 3 exhibits a triple-peaked structure characteristic of third order soliton propagation [5]. The noise present on the autocorrelation profile is typical of the OPO when operated in this configuration and the monitored real-time autocorrelation of the pulse showed rapid fluctuations.

REFERENCES

1 J. M. Dudley, D. T. Reid, M. Ebrahimzadeh and W. Sibbett, Opt. Comm. **104**, 419 (1994)
2 J. D. Bierlein and H. Vanherzeele, J. Opt. Soc. Am. B **6**, 622 (1989)
3 F. Salin, P.Grangier, G. Roger and A. Brun, Phys. Rev. Lett. **56**, 1132 (1986)
4 P. K. A. Wai, C. R. Menyuk, H. H. Chen and Y. C. Lee, Opt. Lett. **12**, 628 (1987)
5 L. F. Mollenauer, R. H. Stolen and J. P. Gordon, Phys. Rev. Lett. **45**, 1095 (1980)

Use of Monolithic Tunable Laser Diodes for Chirped-Pulse Amplification in Fiber Amplifiers

A. Galvanauskas, M.E. Fermann, and D. Harter

IMRA America Inc., 1044 Woodridge Avenue, Ann Arbor, MI 48105, USA

Abstract. Three-section distributed-Bragg-reflection diode lasers were studied as pulse sources for chirped pulse amplification in a fiber amplifier. It was shown that ultrashort microjoule pulses can be obtained at a variable repetition rate.

We have studied the use of tunable three-section distributed-Bragg-reflector (DBR) diode lasers as pulse sources for chirped pulse amplification (CPA) in a fiber amplifier. The broad continuous-tuning range of the laser allows the generation of nanosecond linearly chirped pulses, which are amplified in a fiber amplifier up to microjoule energies and then compressed near to transform-limited duration. The energy of these ultrashort pulses is more than two orders of magnitude higher than any previously reported energies from a compact diode laser and fiber source.

The key of chirped pulse amplification is that long pulses with sufficiently low peak powers are amplified to avoid the distortions caused by nonlinear effects in the amplifying medium [1]. Long positively chirped pulses can be obtained by fast tuning of the emission wavelength of a DBR laser [2]. This structure contains a gain section, a phase-control section and a Bragg reflector section. The wavelength can be shifted continuously by introducing simultaneously changes in the refractive indices of the Bragg and the phase-control sections. An increase in free-carrier concentration lowers the refractive index and shortens the wavelength. By applying current pulses with sufficiently short rise times and properly chosen magnitudes to the tuning sections linear wavelength-shifts can be attained [3].

We studied the generation of linearly chirped pulses for CPA experimentally and theoretically. Numerical investigations of the carrier dynamics in the tuning sections were based on the carrier rate equation. We analyzed linearity, bandwidth and the duration of the linearly chirped pulses, which are important for undistorted amplification and compression using CPA. Wavelength evolution of the fast-tuned DBR laser diodes was characterized experimentally using a monochromator and an ultrafast streak camera. In Fig. 1 spectrally resolved streak-camera image of a wavelength-tuned pulse from a DBR laser is shown.

The spectral extent of the chirp increases with the magnitude of the tuning pulses. The highest chirp experimentally obtained was 12.5 nm. However, the shortest compressed pulses after a diffraction grating compressor (Fig. 2) were 1.8 - 2 ps at 4 - 6 nm of chirp bandwidth (~ 2 times the transform limit). At higher chirp bandwidths compressed pulses were longer. For attaining transform limited pulses after the linear compression with a grating pair deviation from the linearity over the bandwidth Δf in the initial chirp should be $\Delta t \sim 1/\Delta f$. *E.g.* for attaining 10 nm bandwidth transform limited pulses this deviation over the entire bandwith should be

Springer Series in Chemical Physics, Vol. 60 **Ultrafast Phenomena IX**
Editors: P. F. Barbara · W. H. Knox · G. A. Mourou · A. H. Zewail

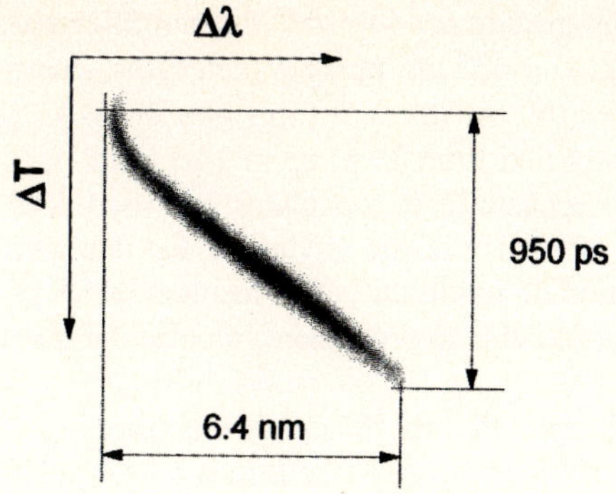

Fig. 1 Streak camera image of a linearly chirped pulse from a DBR diode laser

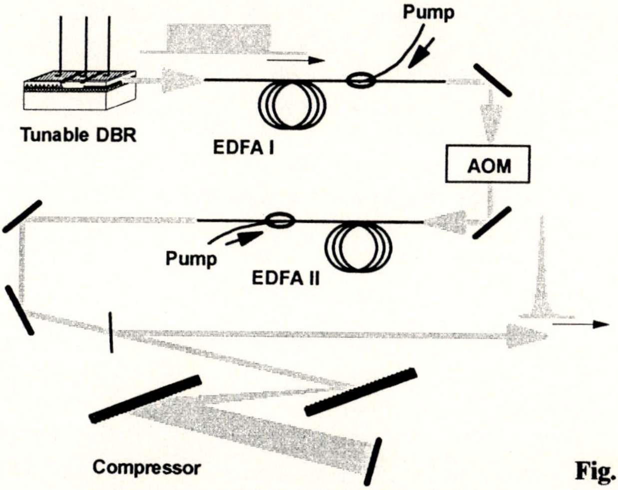

Fig. 2 Experimental set-up

less than 500 fs. Experimentally we measured that for 6 nm bandwidth it was $\Delta t \sim 2$ ps. For higher chirp bandwidth deviation from linearity was reaching several picoseconds. There are no principal obstacles to achieve higher chirp linearity by increasing the tuning accuracy by about an order of magnitude.

Our studies revealed that this fast-tuning technique has a number of properties that makes it uniquely attractive for attaining high-energy ultrashort pulses. The chirped pulses from a DBR diode laser have sufficient linearity and bandwidth for potentially attaining femtosecond transform-limited compressed pulses. The tuning range of these InGaAsP/InP devices is typically within the gain band of erbium-doped fiber amplifiers (EDFA). The duration (as well as the bandwidth) of the chirped pulses is electrically controllable from about 200 ps to 2 ns. Hence even at μJ energies in the amplifier, the peak power of the pulses can be chosen to be sufficiently low to avoid nonlinear interaction.

The experimental set-up for chirped pulse amplification in a fiber amplifier is shown in Fig. 2. It consists of a DBR laser diode, cascaded EDFA with an optical gate between the two stages and a diffraction-grating compressor. The amplifier was pumped with two diode lasers with 1.48 μm and 980 nm pump wavelengths. Pump powers were respectively 70 mW and 900 mW. At the repetition rate of 10 kHz pulses from the tunable diode laser were amplified from 25 pJ up to 2 μJ (total gain of 49 dB). As an optical gate a fast digital acousto-optic modulator was used. The gate width was 150 ns and the diffraction efficiency of the modulator for this gate width was 70 %. Preliminary, the compression of amplified pulses resulted in 1.8 ps pulses with ~ 1 μJ of remaining energy. Considerable improvement in pulse duration and remaining energy is expected.

We believe that this technique will open the way for a new compact high-power ultrafast technology. The laser diode can be operated both in a single-pulse mode and with variable repetition rates up to tens of MHz. It is monolithic and does not exhibit the stability problems that are usual in modelocked sources. Therefore, such highly reliable semiconductor-lasers and fiber source can boost many practical applications of ultrafast techniques.

References

1. D. Strickland and G. Mourou, Opt. Commun. **56**, 219 (1985)
2. A. Galvanauskas, A. Krotkus, J. A. Tellefsen, M. Öberg, and B. Broberg, Appl. Phys. Lett. **60**, 145 (1992)
3. A. Galvanauskas, P. Blixt, and J. A. Tellefsen, Appl. Phys. Lett. **63**, 1742 (1993)

Coherent Pulse Shaping in a Self-Mode-Locked Ti:Sapphire Laser

J.D. Harvey[1], *J.M. Dudley*[1], *P.F. Curley*[2], *C. Spielmann*[2], *and F. Krausz*[2]

[1]Physics Department, University of Auckland, Private Bag 92019, Auckland, New Zealand

[2]Abteilung Quantenelektronik und Lasertechnik, Technische Universität Wien, Gusshausstrasse 27, A-1040 Wien, Austria

Spectral and temporal measurements of ≈ 10 fs pulses from a self-modelocked Ti:sapphire laser were found to exhibit characteristics of coherent pulse propagation.

Recent experiments optimising the operation of a self-modelocked Ti:sapphire laser have resulted in the generation of exceptionally broad modelocked spectra with FWHM of up to 150 *nm* and pulse durations ≈ 10 *fs* [1]. In addition these pulses have been associated with the appearance of a prominent double-peaked structure in the optical spectrum and a modulation in the interferometric autocorrelation. The measured time-bandwidth products of > 0.6 represent a striking deviation from the expected near-sech^2 modelocked pulses. Here we consider the possibility that the observed features occur because of a coherent interaction between the circulating pulse and the Ti:sapphire gain medium.

Existing models of pulse evolution in passively modelocked lasers [2,3] use a rate-equation approach to describe the interaction between the circulating pulse and the laser gain medium which is justified if the pulse bandwidth is only a small fraction of the available transition bandwidth. Studies in a high-Q configuration of the modelocked argon laser, however, have shown that as the pulse bandwidth approaches the transition bandwidth, the coherent interaction between the circulating pulse and the laser gain medium becomes important, and can lead to an oscillatory ringing structure on the pulse envelope as a result of the coherent Rabi oscillation of the atomic polarisation in the gain medium [4].

The coherent interaction in the argon laser is a strong-field interaction in the limit where the interaction Rabi frequency Ω_R is comparable to the argon transition linewidth. This condition for strong coherent coupling is not met in current modelocked solid-state lasers because their broad bandwidths imply that Ω_R is much less than the homogeneous linewidth. Nonetheless, if the modelocked laser is operated so that centre wavelength of the modelocked pulse spectrum is detuned from the centre of the laser transition, then the induced coherent polarisation contains an oscillatory component at the detuning frequency Δ which can cause a perturbation on the trailing edge of a pulse, if the pulse duration is comparable to the dephasing time. The results of numerical simulations based on the Maxwell-Bloch equations showing the effect of this coherent interaction are shown in Figure 1 where an initially smooth but detuned gaussian pulse is seen to develop weak coherent ringing on the trailing edge after making repeated passes through a gain medium.

Analysis of the experimental results in [1] based on a model assuming coherent ringing has been carried out [5] and gives results in good agreement with experiment. While existing models of optimised femtosecond oscillators include

Editors: P. F. Barbara · W. H. Knox · G. A. Mourou · A. H. Zewail

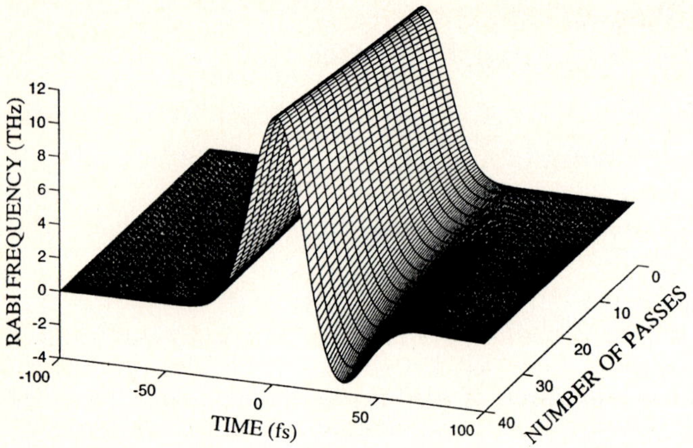

Figure 1. Numerical Simulations showing how an incident 10 fs pulse detuned 30 THz from line centre acquires an oscillatory perturbation on its trailing edge after 40 passes through a Ti:sapphire medium with gain of 3% and approximately equal single-pass loss.

the effects of self-phase modulation and dispersive perturbations, these investigation indicate that coherent effects will need to be included in any detailed quantitative analysis.

References

1. Ch. Spielmann, P. F. Curley, T. Brabec, and F. Krausz, "Ultrabroad-band femtosecond lasers," IEEE J. Quantum Electron. (in press)
2. O. E. Martinez, R. L. Fork, and J. P. Gordon, J. Opt. Soc. Am. B **2**, 753 (1985)
3 H. A. Haus, J. G. Fujimoto, and E. P. Ippen, IEEE J. Quantum Electron. **QE-28**, 2086 (1992)
4 J. D. Harvey, R. Leonhardt, P. D. Drummond, and S. J. Carter, Phys. Rev. A **40**, 4789 (1989)
5. J. D. Harvey, J. M. Dudley, P. F. Curley, C. Spielmann and F. Krausz, "Coherent Effects in a Self-Modelocked Ti:Sapphire Laser", Optics Letters, *in press* (1994)

Intracavity Doubling in Ti:Sapphire

M.T. Asaki, S. Backus, C. Baldwin, C. Shi, M.M. Murnane, and H.C. Kapteyn

Department of Physics, Washington State University, Pullman, WA 99164-2814, USA

We have demonstrated the use of intracavity frequency doubling in a self-modelocked Ti:Sapphire laser to generate pulses as short as 14 fs in the blue region of the spectrum. BBO, LBO and KDP were used as the doubling crystals in a standard Ti:Sapphire laser cavity. LBO gave an 18 nm bandwidth with 30 mW of average power in each of two arms. Pulse durations as short as 14 fs have been measured in this configuration. BBO gave an 11 nm bandwidth with 60 mW of average power in each of two beams. Pulse durations as short as 17 fs have been measured for this case. Some tunability of the fundamental, and hence the second harmonic, was possible.

Introduction

Recently there has been rapid progress in the development of stable, high-average-power sources of optical pulses as short as 10 fs in the near infrared.[1] Frequency doubling can extend the wavelength range of these ultrashort-pulse sources into the blue.[2-3] For our work, we have chosen intracavity frequency doubling instead of extracavity doubling because the intracavity set-up is both simpler and more efficient, due to the high intracavity power.

Initial Results

Our initial setup was based on a self-mode-locked Ti:Sapphire laser, which uses a 4.5-mm-long Ti:Sapphire crystal and can generate pulses as short as 10 fs around 800 nm.[1] To generate the ultrashort blue light, we incorporated a second-harmonic-generation crystal between two 10-cm radius-of-curvature mirrors, at the end of the laser opposite the prism pair. The mirrors of this second focus are broadband high reflectors in the red, and are partially transmitting (50-75%) in the blue. We used a normal incidence, 100 micron thick LBO crystal for the frequency doubling. The output blue bandwidth was 19.2 nm FWHM centered at 416 nm, with an average power of 25-30 mW in each of two beams. After recompression, the pulse duration was measured to be as short as 14 fs, assuming a sech^2 pulse shape.[4]

Current Setup

Our most recent setup is also based on the Ti:Sapphire laser system described above. However, the focusing cavity for the doubling crystal consists of a pair of lenses placed between the Ti:Sapphire rod and the prism pair, as shown in Figure 1.

Springer Series in Chemical Physics, Vol. 60 **Ultrafast Phenomena IX**
Editors: P. F. Barbara · W. H. Knox · G. A. Mourou · A. H. Zewail

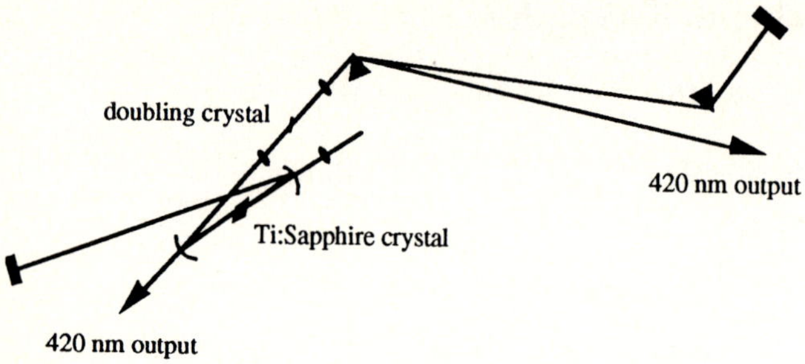

Figure 1: Laser setup for intracavity frequency doubling.

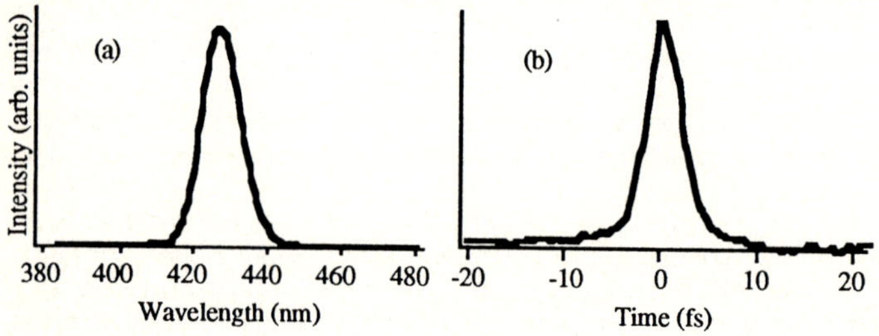

Figure 2: (a) The second harmonic spectrum generated with a BBO doublind crystal; (b) autocorrelation trace of a 17 fs FWHM blue pulse.

The blue light beams are coupled out of the laser through one of the Ti:Sapphire cavity mirrors and through the first prism. In this setup we used a 100μ brewster-cut BBO crystal. The output blue bandwidth is 11 nm FWHM, and we obtain 60mW of average power in each arm. The fundamental has some tunability, allowing the center wavelength of the second harmonic to range from 419 nm to 460 nm.
A typical spectrum of the second harmonic is shown in Fig. 2(a). After recompression of the blue output, pulse durations of 17-21 fs can be measured. Fig. 2(b) shows an experimental pulsewidth measurement with a FWHM of 17 fs.

We gratefully acknowledge support from the Murdock Trust, National Science Foundation, and the Washington Technology Centers. M. Murnane acknowledges support from a Sloan Foundation Fellowship. M. Asaki acknowledges support from a National Science Foundation Fellowship.

References.

1. M.T. Asaki, C.P. Huang, D. Garvey, J. Zhou, H.C. Kapteyn, M.M. Murnane, Opt. Lett. **18**, 977 (1993).
2. V. Petrov, D. Georgiev, and U. Stamm, Appl. Phys. Lett. **60**, 1550 (1992)
3. R. J Ellingson, and C. L. Tang, Appl. Phys. Lett. **17**, 343 (1992)
4. S. Backus, M.T. Asaki, C. Shi, H.C. Kapteyn, and M.M. Murnane, Opt. Lett. **19**, 399 (1994).

Optical Parametric Amplification of the White-Light-Continuum Components at 250 kHz from a Ti:Sapphire Regenerative Amplifier

M.K. Reed, M.K. Steiner-Shepard, and D.K. Negus

Coherent Laser Group, 5100 Patrick Henry Drive, Santa Clara, CA 95054, USA

Abstract. A Ti:sapphire mode-locked regenerative amplifier system pumped with a single argon ion laser is capable of producing 4μJ, 170fs pulses of 800nm wavelength at 250kHz repetition rate. Part of this output is used to generate a high-beam-quality white-light continuum while the rest is doubled to 400nm and used to parametrically amplify visible components of this continuum to more than 100nJ. Synchronized with the visible signal pulse tunable from 450 to 700nm is an infrared idler pulse tunable from 2.4 to 0.9μm.

The output of our Ti:sapphire mode-locked regenerative amplifier system will generate broadband white-light when focused into transparent materials. For energies less than 2μJ we find that focusing the beam into sapphire leads to whole-beam self-focusing characterized by a collapse into an exceptionally stable single filament. The high intensities in this filament allow self-phase-modulation to create a white-light continuum extending across the entire visible spectrum The threshold for this single filament continuum generation is less than 0.5μJ but for input energies >2μJ the collapse appears to involve multiple unstable filament formation attendant with much larger spatial and frequency fluctuations on the continuum beam. The single filament collapse in the 1 to 2μJ regime leads to a gaussian output beam at all wavelengths and therefore appears to be optimum for high beam quality continuum generation.

A stable white-light continuum is generated with a fraction of the available compressed 4μJ energy from the regenerative amplifier. The remaining energy is able to be frequency doubled and used to pump an optical parametric amplifier (OPA) to enhance a component in this spectrum. A phase matching analysis of a BBO type I optical parametric amplifier (OPA) pumped at 400nm shows that rotating the crystal propagation angle, theta, from 29 to 21 degrees allows signal and idler tuning across the entire visible and near infrared spectrum. The layout for the continuum generation and OPA is shown in Figure 1. The input 800nm beam is split with a 75% reflector. The 1 μJ portion is focused into a sapphire crystal to generate white-light and the output is collimated with an achromat. The 3μJ portion of the 800nm beam is frequency-doubled in a 1mm thick BBO crystal to give 1μJ of energy at 400nm. This beam is recollimated and recombined on a dichroic 400nm reflector with the white-light beam after it has followed a path-length-matching optical delay line. The two colinear and cotemporal pulses are focused into a 1 mm thick BBO crystal of the same cut as the BBO SHG crystal. It is possible to significantly amplify a component of the whitelight at wavelengths from 450nm to 700nm by rotating the BBO crystal and reoptimizing the optical delay to compensate for chirp in the continuum. Further amplification is then possible by splitting and recombining the white-light and 400 nm pulses for a second pass through the BBO OPA crystal. The whitelight output is recollimated and separated from the 400nm pump with the dichroic reflector to allow measurement of the signal and idler energies, bandwidths and pulsewidths. The results are shown in Figures 2 and 3. For the OPA signal around 580nm the power gained in amplification is 35mW, or 140 nJ per pulse. The total gain from the initial continuum level is over 400 times or e^6. The signal power curve is fairly flat for tuning across the visible from 500 to 700 nm and falls off with wavelength for the idler as expected. The autocorrelation of the signal output was measured using thin KDP crystals. Pulsewidths, calculated assuming a gaussian pulseshape as 0.707 times the autocorrelation width, are near 150 fs for all wavelengths. The output has a time-

Editors: P. F. Barbara · W. H. Knox · G. A. Mourou · A. H. Zewail

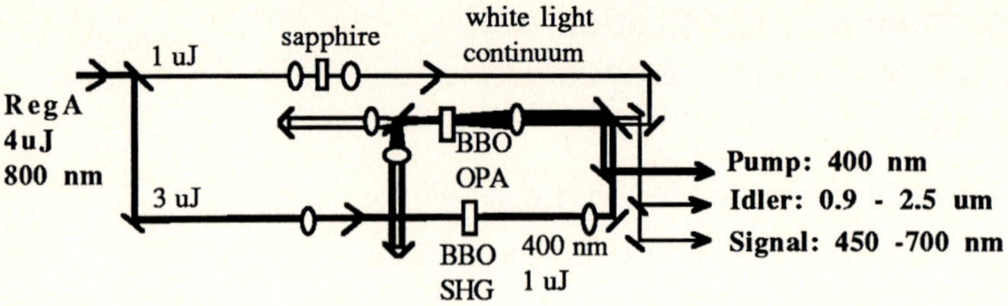

Figure 1: White-light generation and parametric amplification layout.

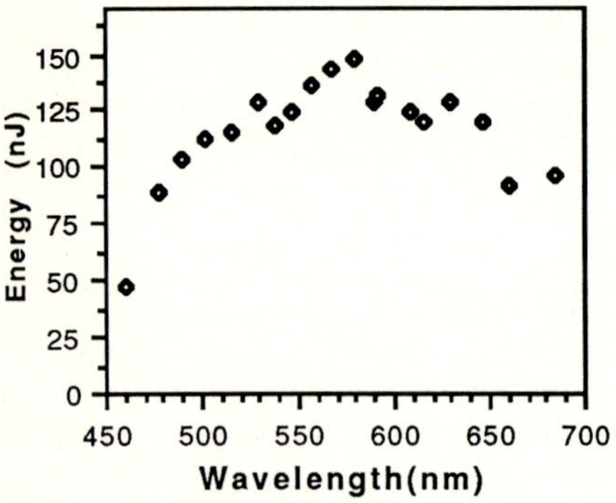

Figure 2: OPA signal energy versus wavelength.

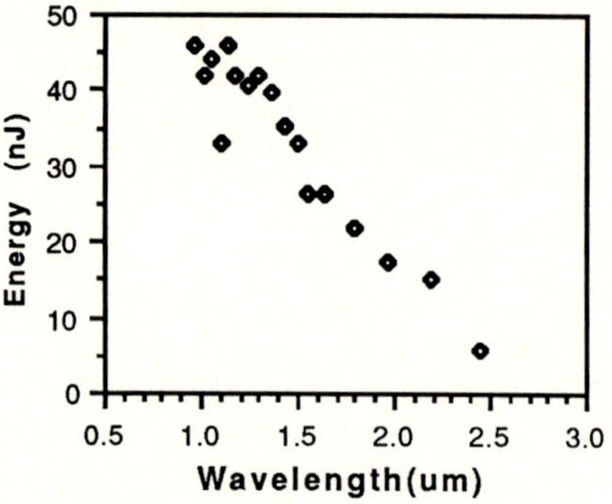

Figure 3: OPA idler energy versus wavelength.

bandwidth product of about 2 times the minimum from 450 to 600nm and then diverges to a larger product as the signal wavelength approachs the 800nm degeneracy. The idler time-bandwidth product, which we could measure for wavelengths <1.18 microns, was lower than the corresponding signal. The amplified beams have symmetric Gaussian shapes and the output pulses show approximately 5% peak-to-peak noise.

In conclusion, we have demonstrated that the output from an ion laser pumped Ti:sapphire modelocked regenerative amplifier system, the *Mira RegA 9000*, is able to be efficiently converted in an optical parametric amplifier to the visible and infrared. The quality of the output and the ease of tuning from 450nm to 700nm and 0.9 to 2.4 microns make this system an excellent source of ultrafast pulses. In addition, the signal and idler can both be efficiently converted to second harmonic with a singlepass through BBO or LBO, thereby affording complete spectral coverage from 230 nm to 2.5 microns.

References

1. T.B.Norris, "Femtosecond pulse amplification at 250 kHz with a Ti:sapphire regenerative amplifier and application to continuum generation", Opt. Lett. 17, p 1009 July 15 1992.

40 GHz Soliton-Train Generation Through Multisoliton Pulse Propagation in a Dispersion-Varying Optical-Fiber Circuit

D.J. Richardson[1], *A.V. Shipulin*[2], *E.M. Dianov*[2], *and D.N. Payne*[2]

[1]Optoelectronics Research Centre, Southampton University, Southampton SO9 5NH, UK

[2]General Physics Institute, 38 Vavilova Street, Box 117333, Moscow, Russia

Abstract. We experimentally demonstrate 40 GHz soliton train generation using a novel optical beat-signal to soliton train conversion technique based on multisoliton compression effects during propagation in a dispersion varying optical fiber.

Recent advances in telecommunications have led to a number of impressive high speed data transmission results at rates as high as 100 GBit/s. The development of novel high-frequency pulse generation schemes is thus of great practical relevance. One particular class of technique that has been the subject of considerable recent interest is based upon the transformation of a beat-signal into a train of solitons as a result of nonlinear propagation in Dispersion Decreasing Fiber (DDF) [1]. Unfortunately, the basic technique is not applicable for pulse repetition rates below ~50 GHz and is therefore of little interest for current telecommunication applications. In this paper we experimentally demonstrate that a simple combination of spectral enrichment in dispersion-shifted fiber (DSF) and multisoliton compression in standard fiber prior to propagation in a DDF, can be used to extend the basic technique to lower (<40 GHz) repetition frequencies, such rates are within the range of telecommunication interest.

The experimental configuration is illustrated in Fig.1. The outputs from two, pig-tailed, single frequency DFB lasers were combined to create a beat-signal and then amplified in a, Er^{3+}/Yb^{3+} doped fiber amplifier. The amplified beat-signal was passed through 1km of DSF (D=0.5 ps/nm.km) where four-wave mixing leads to spectral sideband generation. The resulting signal is then passed through 1km of standard fiber (D=16 ps/nm.km) where it undergoes multisoliton compression by a factor of ≈2 before entering the 3.5 km DDF. The dispersion varied along the DDF length in a hyperbolic fashion from D_{inp}=11 ps/nm/km to D_{out}=0.8 ps/nm/km at the output.

The best quality pulses were obtained at a wavelength separation of 0.318 nm (R=40 GHz repetition rate) and an input beat-signal power of 135mW. An autocorrelation trace and optical spectrum of the trains so obtained are shown in Fig.2 where we see that we have obtained a 40 GHz train of 3.4 ps pulses (assuming a $sech^2$ form) and corresponding to a mark-space ratio (MSR) of ~7.3:1. The measured autocorrelation trace shows a difference in the wings of the pulses from that expected for a train of solitons (dashed line) with an MSR of

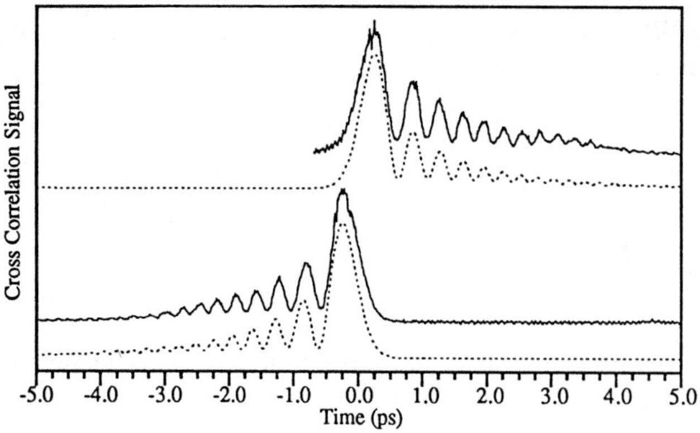

Figure 2. Experimental (solid lines) and calculated (dashed lines) cross correlations for a cubic rf phase sweep. Top: negative cubic phase sweep; bottom: positive cubic phase sweep.

100MHz) are readily available; phase and amplitude modulated rf pulses are easily characterized as well. The rapid update rate makes the technique suitable for incorporating feedback from experimental observables, which appears to be a promising approach for generating useful shaped waveforms. [1]

The most important advantages are the available resolution with existing AOMs (up to 2000 resolvable frequency elements using commercial components only) and the continuous nature of the transmission mask. This is particularly useful if complex phase modulation is required (for example, for dispersion compensation to shorten amplified pulses to the transform limit). Pixel gaps can cause artifacts and limit the overall phase shift possible; as phase jumps between pixels become larger cross talk and calibration errors can cause distortions [4].

Figure 2 shows cross-correlations corresponding to a cubic spectral phase sweep imposed on the output of a Tsunami Ti:sapphire laser, corresponding to a quadratic frequency sweep on the r.f. pulse; higher order phase sweeps can be easily generated. This shows both greater total phase shifting capability and also greater accuracy of the phase sweep than in [4]. These results were obtained with a 4mm aperture, 50MHz bandwidth TeO_2 modulator; we are now using a 4cm aperture, 100MHz bandwidth modulator which should give dramatically increased resolution.

In summary, acousto-optic modulation inside a zero dispersion line provides a simple, extremely flexible, and broadly applicable technique for fs pulse shaping with both phase and amplitude modulation, and we expect this approach will tremendously simplify future developments of advanced laser systems and quantum molecular control. We gratefully acknowledge the financial support of the National Science Foundation and The Office of Naval Research.

References

[1]For a review see W. S. Warren, H. Rabitz and M. Dahleh, Science 259, 1581 (1993)
[2]C. W. Hillegas, J.X. Tull, D. Goswami, D. Strickland and W.S. Warren, Opt. Lett. 19, 10 (1994)
[3]A. M. Weiner, J. P. Heritage and R. N. Thurston, Opt. Lett. 10, 609(1985)
[4]A.M. Weiner, D.E. Leaird, J.S. Patel and J.R. Wullert,IEEE J.Quan.El. 28, 908(1992)
[5] M. Wefers and K. Nelson, Science 262, 1382 (1993); Opt. Lett. 18, 2032 (1993)

Part V

High-Intensity and Nonlinear Effects

Climbing a Ladder System by Frequency-Chirped Laser Pulses

L.D. Noordam[1], *P. Balling*[1], *B. Broers*[1], *J.H. Hoogenraad*[1], *D.J. Maas*[1], *P.W. van Amersfoort*[2], *H.B. van Linden van den Heuvell*[1], *A.F.G. van der Meer*[2], *and R.B. Vrijen*[1]

[1]FOM-Institute AMOLF, 1098 SJ Amsterdam, The Netherlands
[2]FOM-Institute Rijnhuizen, 3430 BE Nieuwegein, The Netherlands

Introduction

Excitation of vibrations of chemical bonds in molecules is of considerable interest for controlling chemical reactions. Ever since the advent of the laser, chemists have been dreaming about the possibility to use laser light to selectively excite molecular bonds all the way up to the dissociation limit. Direct single photon excitation to a highly excited vibrational state is forbidden due to the very small overlap between the initial and final state. This necessitates a step-wise excitation. So far two complications have frustrated this quest for bond-selective chemistry. Firstly, the anharmonicity of a chemical bond makes it difficult to attain a high level of vibrational excitation: although the ground state can be depleted in favour of the first excited state upon irradiation with a precisely tuned laser, further excitation steps will require slightly different frequencies. Secondly, the time scale on which the vibrational energy in a given bond delocalizes by redistributing itself through the molecule is very short (picosecond range).

Very short laser pulses present a solution to the problem. Intrinsically they have a large enough bandwidth to overcome the variation in spacing of the vibrational ladder due to anharmonicity. Furthermore, their pulse durations are smaller than the time scales for intra-molecular energy flow. It has been suggested [1] that the vibrational excitation can be enhanced by applying short pulses whose frequency is swept ('chirped') in order to follow the change in size of the vibrational quanta. Calculations have indicated that in this way 100% of the population can be transferred from the ground vibrational state to a highly excited vibrational state, without leaving population in intermediate states. In principle the final state can be as high as the vibrational continuum, resulting in bond-selective dissociation into two fragments.

In this contribution we experimentally demonstrate 100% population transfer in an atomic model system [2,3]. In a three-step ladder in rubidium we achieved full population transfer from the electronic ground state (5s) to the final state 5d. As a next step towards molecular dissociation we performed experiments on multiphoton ionization of Rydberg atoms with a FIR free electron laser, and observed four-photon ionization in which all intermediate steps are resonant.

Experimental demonstration of population sweeping in a ladder system

Recently we have demonstrated [2] in an atomic model system with three levels that 100% population transfer can be obtained from the initial state to the final state by sweeping the

Springer Series in Chemical Physics, Vol. 60 **Ultrafast Phenomena IX**
Editors: P. F. Barbara · W. H. Knox · G. A. Mourou · A. H. Zewail

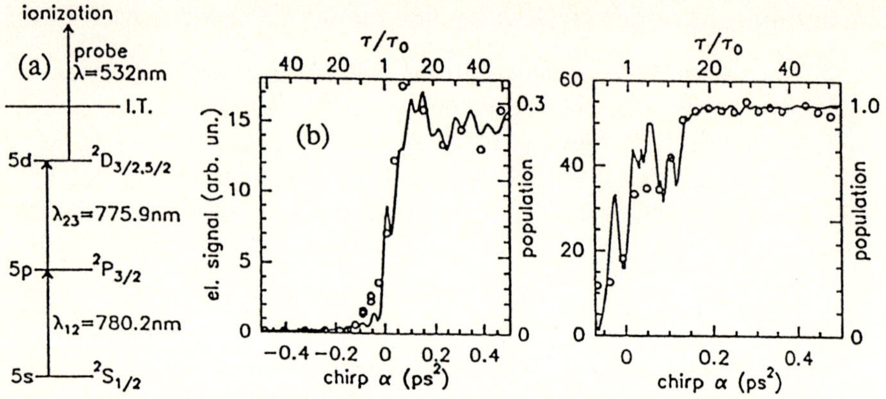

Fig. 1 (a) Energy levels of the three-state ladder in rubidium (5s,5p,5d). The population of the upper state is probed afterwards by photoionization by a second laser. (b) Measured enhancement of the upper state population for the chirped pulses; blue first ($\alpha<0$) results in almost no population of the upper state; red first ($\alpha>0$) enhances the population dramatically compared to no chirp ($\alpha=0$). At high fluence (500 μJ/cm^2) all population is transferred to the final state [c].

frequency. Instead of *vibrational* levels in a molecule, in the atomic model system three *electronic* levels are studied. Rubidium atoms in the electronic ground state (5s) are excited via the intermediate state (5p) to the final electronic state (5d). Since the transition frequencies for these two steps are not identical (see figure 1) we have a quantum system that resembles three vibrations in an anharmonic molecular potential. With our laser system we produce sub-picosecond pulses with sufficient bandwidth to include both steps, and by means of a grating device we are able to vary the chirp the pulses continuously. In the experiment, after the short pulse we have probed the population of the upper state (5d) with another laser. As can be seen in fig. 1 indeed 100% population is transferred.

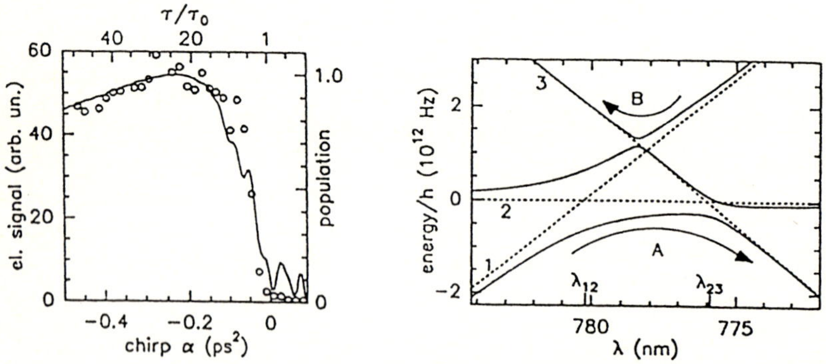

Fig. 2 (a) As figure 1c, at high intensities the counterintuitive sweep blue first (775.9 nm, $\alpha<0$) results in 100% population of the upper state. (b) Corresponding dressed level scheme of the states 5s (1), 5p (2), and 5d (3), path A corresponds to population transfer by two sequential adiabatic passages by a intuitively chirped pulse ($\alpha>0$); path B corresponds to the counterintuitive frequency sweep.

A surprise

At somewhat higher laser intensities we found (see figure 2) that the counterintuitive frequency sweep (applying the frequency for the second step first) gave 100% population transfer. The population transfer by this, at first glance, surprising frequency sweep is confirmed by solving the Schrödinger equation for this system numerically. In fact, in the counterintuitive sweep the population of the intermediate state is almost avoided and is thus well suited for systems with a short lived intermediate state.

Dressed level model

The physics of the process is best explained by a dressed level model (see fig. 3a) at the intermediate level |2>. We dress the lower state |1> by adding one photon |1+ω> and the upper state |3> by subtracting one photon |3-ω>. In figure 3a the dressed levels are drawn as a function of the laser frequency ω. The level crossings are avoided due to the coupling of states |1> - |2>, and |2> - |3>, no direct coupling between |1> - |3> is possible, only by mixing of state |2> character in one of the dressed levels.

In the case of an intuitive sweep we start in level |1> in the lower left corner of fig. 3a (path A). If now the adjustment of the laser frequency is sufficiently slow (on the scale of the Rabi frequency), the entire population will remain in this dressed state. At the end of the sweeping pulse the dressed state is a pure |3> state: the population is transferred in an adiabatic way. This mechanism is, in contrast to Rabi pumping, insensitive to the precise intensity and duration of the laser pulse.

In the case of an counter-intuitive sweep we start again in level |1> but in the upper right corner of fig. 2b (path B). If the laser intensity is sufficiently high to open the |1> - |3> crossing (by mixing state |2> character via the other crossings) again the population can be

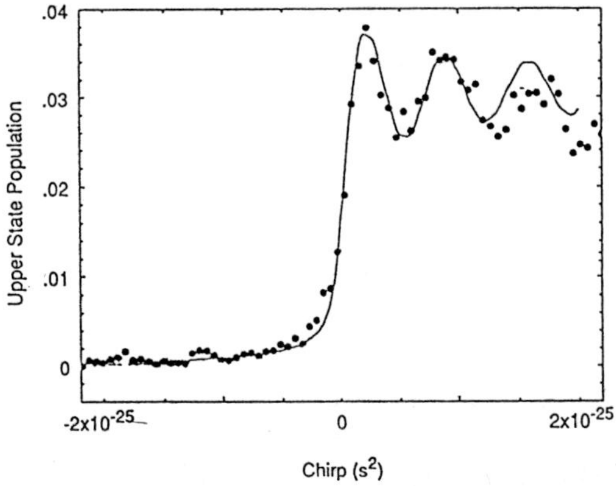

Fig. 3 Measured enhancement of the upper state population at moderate intensities. The wiggles are due to interference in the upper-state amplitude of direct 2-photon excitation and sequential 1+1-photon excitation.

Coherence Properties of High-Order Harmonic Radiation

T. Ditmire[1], K.S. Budil[1], J.K. Crane[1], H. Nguyen[1], M.D. Perry[1], P. Salieres[2], and A. L'Huillier[2]

[1]Lawrence Livermore National Laboratory, P.O. Box 808, L-443, Livermore, CA 94550, USA

[2]Service des Photons, Atomes et Molecules, Centre d'Etudes de Saclay, CEA, F-91191 Gif-sur-Yvette, France

Abstract. The results of a series of experiments to characterize the coherence properties of xuv radiation produced by high-order harmonic generation in helium, neon and argon are reported and compared to predictions from an effective order model. The harmonics exhibit smooth, near gaussian spatial profiles, and have a divergence that is approximately constant (< 12 mrad) in the plateau region and decreases (≈ 4 mrad) in the cutoff for f/17 focusing. For a bandwidth limited , 140 fsec incident pulse, we measure a harmonic line width of $\Delta\lambda/\lambda \approx 2 \times 10^{-3}$ at 30.3 nm. By reducing the spectral width of the driving pulse, harmonics with $\Delta\lambda/\lambda \approx 2 \times 10^{-4}$ can be produced. Absolute conversion efficiency as high as 10^{-7} for harmonic radiation as short as 20 nm has been achieved by using 400 fsec, 526 nm pulses from an Nd:Glass laser.

1. Introduction

The understanding of high order harmonic generation in rare gases by a very intense, short pulse laser has progressed significantly in the past few years. The detailed characterization of the generated harmonic fields, however, is only now being addressed. In particular, the spatial and temporal coherence of the high order harmonics in addition to the absolute conversion efficiency must be optimized for use in applications. Spatial profile measurements of high order harmonics produced by a 1 ps laser have been recently reported by Tisch et. al.[1], and L' Huillier et. al. reported measurements of the spatial profiles of the 13th harmonic of a 40 ps Nd:YAG laser.[2] Some work has also been reported on the spectral characteristics of the harmonics. In particular, the spectral blue shift of the harmonics due to the ionization and plasma formation have been reported by two groups.[3,4] In this paper, we report on measurements of the absolute conversion efficiency, spatial distributions and the spectral line widths of high order harmonics generated in helium, neon and argon and compare these results to predictions from an effective order model. Measurements of coherent xuv radiation at 20 nm with a peak brightness on the order of 10^{24} photons/(mm^2-mRad-sec-0.1% bandwidth) are presented.

2. Spatial Distribution Measurements

The harmonic spatial distribution experiments were conducted by generating harmonics of a short pulse $Cr:LiSrAlF_6$ (LiSAF) laser[5] operating at 825 nm and capable of producing pulse energies in excess of 200 mJ. Special care has been taken to characterize the far-field focal profile of the laser which is approximately two-times

Springer Series in Chemical Physics, Vol. 60 **Ultrafast Phenomena IX**
Editors: P. F. Barbara · W. H. Knox · G. A. Mourou · A. H. Zewail

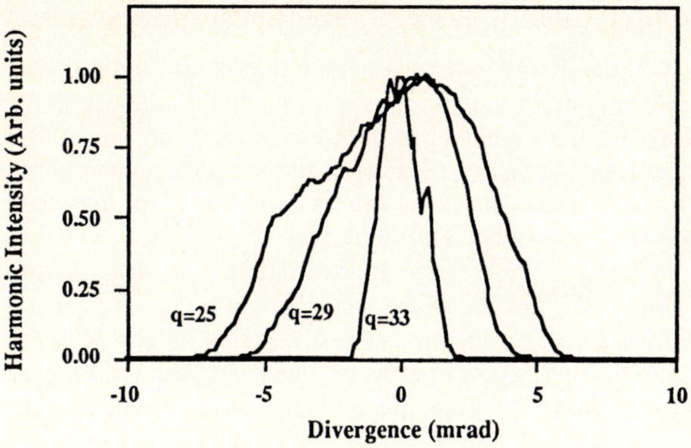

Figure 1: Integrated line outs of selected harmonics (25th, 29th and 33rd) produced in argon at a laser intensity of 4 x 10^{14} W/cm^2.

diffraction limited with a 70 μm focal spot (1/e^2) using an f/17 focusing lens. For the spatial profile measurements the laser was focused at f/17 into a gas plume provided by a pulsed gas jet. The estimated atomic density for these experiments could be varied between 10^{17} atoms/cm^3 and 10^{18} atoms/cm^3. The measured divergence of the focused, 825 nm beam was 62 mrad (1/e^2). The harmonics produced were collected by a flat field spectrometer consisting of a flat, variable line spaced gold grating at grazing incidence and a single, flat, carbon folding mirror. A CsI coated dual micro-channel plate (MCP) detector was located at the image plane of the spectrometer, 59 cm from the source of the harmonics. A phosphorus screen coupled to the MCP detector was read by an 8-bit CCD camera yielding simultaneous spectral and spatial information on each laser shot.

The experiments were conducted by averaging 5 laser shots in a 10% energy window. For intensities below the ionization saturation intensity, the images are extremely reproducible on a shot to shot basis. In figure 1, the line outs of three harmonic profiles generated in argon at an intensity of 4.0 x 10^{14} W/cm^2 are shown. The 25th and the 29th harmonics represent harmonics in the plateau region of the spectrum and the 33rd harmonic is the highest order harmonic detectable in the cutoff with the MCP detector. The harmonic distributions exhibit near gaussian profiles with little or no substructure. The divergence of the harmonic radiation decreases slightly with increasing order in the plateau region and rapidly in the cutoff. The divergence for the harmonics in argon range from 12 mrad for the orders from the 17th to the 25th down to 3 mrad for the 33rd harmonic.

Harmonics produced in neon show a similar behavior as those produced in argon. Like harmonics produced in argon, the distributions are smooth and exhibit a divergence that decreases from around 13 mrad for the lower orders down to 6 mrad for harmonics at the cutoff. In general, we find that the harmonics exhibit a divergence that is less then that predicted by lowest order perturbation theory.[6]

This can be explained by considering the fact that the harmonic yields typically scale with the laser intensity with an order that is typically much less than the actual process order[4]. Under our experimenta conditions, the confocal parameter, b, is much longer than the length of the generating medium (gas jet), L. In this limit (b>>L), perturbation theory predicts that a harmonic with order q will have a divergence that is $1/\sqrt{q}$ times the laser divergence. If, however, the harmonic varies with an effective nonlinear order, p, then the divergence of the harmonic is $\sqrt{p}/q$ times the laser divergence. Since p is typically much smaller than q for the harmonics out in the plateau (p ≈ 5-8) we would expect the higher order harmonics to exhibit a smaller divergence than predicted by perturbation theory. This prediction is consistent with our measurements throughout the rare gases provided the laser intensity is less than approximately twice the saturation intensity for ionization, I_{sat}.

When the peak laser intensity is greater than approximately twice I_{sat}, plasma formation occurs early in the pulse and refraction of the incident light produces structure on the harmonic spatial profiles and rapidly increases the divergence (along with the previously reported spectral blue shifting). Similar observations have been reported by Tisch, et al.[1] A detailed model accounting for these effects is currently in development.

3. Spectral Profile Measurements

The spectral distributions of selected high order harmonics produced by the Cr:LiSAF laser were conducted using a reimaging monochrometer, (described in ref 4). The monochrometer had an estimated resolution of .03 nm, and the spectral width of the laser for these experiments was 8.5 nm, $\Delta\lambda/\lambda = 1 \times 10^{-2}$. Profiles of the 27th harmonic in argon and neon produced at an incident intensity, near I_{sat} for each atom are shown in figures 2a and 2b respectively. Accounting for the resolution of the spectrometer we find that the fractional bandwidth of the harmonics was $\Delta\lambda/\lambda \approx 2 \times 10^{-3}$ in argon and was $\Delta\lambda/\lambda \approx 5 \times 10^{-3}$ in neon. This range of spectral line widths is comparable to the line widths predicted by perturbation theory which predicts a fractional bandwidth of $\Delta\lambda/\lambda \approx 2 \times 10^{-3}$ for the 27 th harmonic.

As the intensity increases above the saturation intensity, the bandwidth of the harmonic also increases, similar to the increase in divergence. This can be explained

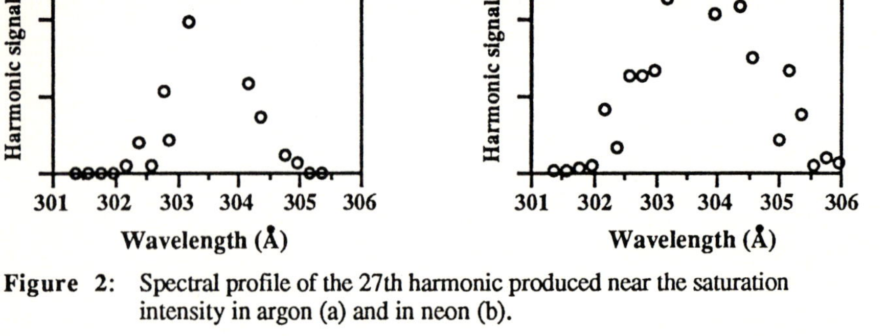

Figure 2: Spectral profile of the 27th harmonic produced near the saturation intensity in argon (a) and in neon (b).

self-phase modulation of the laser pulse created by the time-dependent refractive index experienced by the leading edge of the pulse resulting from ionization of the gas.[4] At an intensity of 3×10^{15} W/cm^2, the bandwidth of the 27th harmonic produced in neon increases to $\Delta\lambda/\lambda \approx 8 \times 10^{-3}$, well above the transform limited prediction of either perturbation theory or the effective order model which predicts that the normalized spectral width of the harmonic should exhibit a $\sqrt{p}/q$ dependence on the fractional bandwidth of the laser, $(\Delta\lambda/\lambda)_q = \sqrt{p}/q\ (\Delta\lambda/\lambda)_{laser}$.

4. Conversion Efficiency Measurements

The absolute energy yields of harmonics produced in helium and neon by a frequency doubled Nd:Glass laser were studied under conditions identical to those in which the spatial and spectral distributions were studied. The harmonic yields were measured using an absolutely calibrated X-ray CCD camera which can accurately determine the yield of incident photons over the spectral range 2 to 40 nm. The absolute yields of harmonics produced in helium with three different focal ratios are shown in figure 3 for a peak intensity of 6×10^{15} W/cm^2. The harmonic yield increases linearly with the increasing confocal parameter, b obtained with weaker focusing. Energies in excess of 10 nJ can be generated for harmonics in the plateau. We have generated individual harmonics with an energy as high as 58 nJ at 25 nm with 500 mJ of green light. An optimum conversion efficiency of 2×10^{-7} was obtained at 25 nm (40 nJ produced) decreasing to 1×10^{-7} at 20 nm (25th harmonic) with 200 mJ incident at 526 nm and f/70 focusing into helium. Measured harmonic yields in the plateau in neon are very close to those values plotted in figure 3 for helium with the exception that the harmonics exhibit a cut-off at the 21st harmonic.

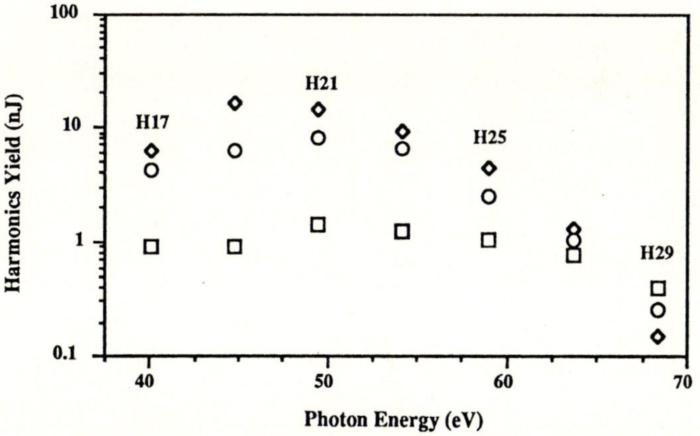

Figure 3: Absolute energy yield of harmonics produced in helium by 526 nm laser light at a peak intensity of 6×10^{15} W/cm^2 for three focal ratios. Squares (□) are for f/25 focusing, circles (○) for f/50 focusing, and diamonds (◇) are for f/70.

5. Conclusion

Measurements of the angular distributions, spectral line width and absolute conversion efficiency of high order harmonic radiation indicate a source peak brightness on the order of 10^{24} photons/(mm^2-mrad-sec-0.1% bandwidth) at 20 nm produced by table-top subpicosecond lasers. We find that the angular distributions are smooth, near Gaussian distributions, without structure when the incident intensity is less than approximately $2xI_{sat}$. The harmonics exhibit a fractional line width of $2x10^{-3}$ for a 140 fsec laser ($\Delta\lambda/\lambda_{laser}=1x10^{-2}$). Measurements of the absolute harmonic conversion efficiency of a 600 fs, 526 nm laser pulse in neon and helium under similar conditions show that individual harmonics can be produced in excess of 20 nJ can be generated at 20 nm with a conversion efficiency $>10^{-7}$.

This work was performed under the auspices of the U. S. Department of Energy by Lawrence Livermore National Laboratory under contract W-7405- Eng-48.

References

1 J. W. G. Tisch, et al, Phys. Rev. A. **49**, R28 (1994).
2 A. L'Huillier, and Ph. Balcou, Las. Phys. **3**, 654 (1993).
3 J. J. Macklin, J. D. Kmetec, and C. L. Gordon, Phys. Rev. Lett. **70**, 766 (1993).
4 C-G. Wahlstrom, et al, Phys. Rev. A. **48**, 4709 (1993).
5 T. Ditmire, M. D. Perry, Opt. Lett. **18**, 426 (1993).
6 P. Saliere, A. L'Hullier, T. Ditmire, K.S. Budil and M.D. Perry, J. Phys. B, (1994).

Use of Self-Trapped Filaments in Air to Trigger Lightning

X.M. Zhao[1], *J.-C. Diels*[1], *A. Braun*[2], *X. Liu*[2], *D. Du*[2], *G. Korn*[2], *G. Mourou*[2], *and J.M. Elizondo*[3]

[1]Department of Physics and Astronomy, University of New Mexico, Albuquerque, NM 87131, USA
[2]The Center for Ultrafast Optical Science, University of Michigan, Ann Arbor, MI 48109, USA
[3]Tetra Corporation, 3701 Hawkings Street, Albuquerque, NM 87109, USA

Abstract. High energy (millijoule) femtosecond laser pulses are trapped into filaments of tens of micrometer diameter. These filaments are used to trigger a high voltage lightning discharge across a wide gap.

The possibility of triggering lightning with laser pulses has been investigated. Theoretical analyses show that a continuous ionized path of several tens of meters is sufficient to initiate a lightning discharge at 1/5 or less of the natural lightning threshold. In recent experiments, low energy ($\approx 200\ \mu$J) femtosecond ultra-violet laser pulses are focused to generate a column of week ionization between two electrodes spaced by 25 cm. Discharges were induced with this technique at half the self breakdown field of air [1]. Self-trapped filaments [2] create a continuously ionized path over very long distances, which provide the main requirement for "seeding" and guiding a discharge over long distances with minimal laser energy.

The lase system consists of a Ti:sapphire oscillator combined with a Ti:sapphire amplifier using the technique of chirped pulse amplification produces 200 fs pulses at 780 nm with an energy of up to 30 mJ. When the peak power of a pulse exceeds a critical power, the laser beam starts self-focusing into a small transverse scale filament. After propagating a distance of about 10 meters in free space, a 1.2 cm diameter beam is self-trapped into a filament of 50 to 100 micrometers. Such a filament was observed to propagate over a distance of 20 meters (the maximum space available).

In a retarded time frame of reference, the evolution with propagation distance z of the laser field $\tilde{\mathcal{E}}(r,t)$, within the paraxial slowly varying envelope approximations, is given by:

$$\frac{\partial}{\partial z}\tilde{\mathcal{E}} = -\frac{i}{2k}\left(\frac{\partial^2}{\partial r^2} + \frac{1}{r}\frac{\partial}{\partial r}\right)\tilde{\mathcal{E}} - i\frac{\omega}{2c}\left[n_2|\tilde{\mathcal{E}}|^2 + n_e\right]\tilde{\mathcal{E}} - \frac{ik''}{2}\frac{\partial^2}{\partial t^2}\tilde{\mathcal{E}}, \qquad (1)$$

where k and k'' are the wave vector and its second derivative with respect to frequency, $n_2 = 5 \times 10^{-19}$ cm^2/W is the (Kerr) coefficient of nonlinear index of air, n_e is the (negative) index of the photoelectrons, and has a higher order field dependence as the Kerr nonlinearity. The continuous confinement of the energy inside the filament is attributed to a balance between the self-focusing n_2 term,

Optical Breakdown in Wide-Band-Gap Materials as a Function of Pulse Duration

D. Du, X. Liu, G. Korn, J. Squier, and G. Mourou

Center for Ultrafast Optical Science, The University of Michigan,
2200 Bonisteel Blvd., IST Bldg., Rm. 1006, Ann Arbor, MI 48109-2099, USA

Abstract. We report the first observations of dramatic increase in laser-induced breakdown threshold in SiO_2 and MgF_2 as the pulse width decreases below ten picoseconds.

The phenomena of optical breakdown of normally transparent optical materials such as fused silica has been studied for three decades[1]-[3]. Through the nonlinear photon-matter interaction (electron avalanche process or multiphoton ionization), a plasma is generated at the focal spot and laser energy is deposited to material, which heats the material that can lead to the irreversible damage to the material. The breakdown threshold is pulse-width dependent, and an empirical scaling law of breakdown threshold exists: $F_{th} \sim \tau^{1/2}$, for pulse duration greater than 10 ps.

We systematically studied the pulse-width dependence of laser-induced optical breakdown threshold in SiO_2 and MgF_2, with laser pulse width from 150 fs to 7 ns, using a chirped-pulse amplication (CPA) laser system[4]. A Ti:sapphire-based CPA laser system was used in the experiments. The sample thickness of SiO_2 is $150 \mu m$, and $500 \mu m$ for MgF_2. Fig. 1 shows the experimental result. The empirical scaling law of breakdown threshold as a function of pulse width failed when pulse width became less than 10 ps. We observed, for the first time, that the fluence breakdown threshold increased when pulse width decreased below ten picoseconds in SiO_2 and MgF_2. This surprising result can be explained on the basis of avalanche ionization theory. The impact ionization coefficient α is given by[5]

$$\alpha(E) = \frac{eE}{U_i} \exp\left(-\frac{E_i}{E(1+E/E_p)+E_{kT}}\right),$$

where E_{kT}, E_p, and E_i are threshold fields for electrons to overcome the decelerating effects of thermal, phonon, and ionization scattering, respectively. Breakdown is reached when $\alpha v_d \tau \sim 30$, where v_d is the electron drifting speed. When $E > \sqrt{E_p E_i}$ is satisfied, $\alpha \sim E$, the breakdown threshold $F_{\text{th}} \propto E^2 \tau_p \sim 1/\tau_p$, i.e., F_{th} will increase for ultrashort laser pulses.

Calculation of the impact ionization coefficient α gives good agreement between theory and experiment as shown in Fig. 2. It leads to the conclusion that avalanche ionization is the dominant mechanism in laser-induced breakdown in wide-band-gap materials down to 150 fs.

In conclusion, we have observed the increase of breakdown threshold when the laser pulse width becomes less than a few picosecond. The scaling of break-

Springer Series in Chemical Physics, Vol. 60 **Ultrafast Phenomena IX**
Editors: P. F. Barbara · W. H. Knox · G. A. Mourou · A. H. Zewail

Laser–Solid

Z. Jiang[1],*, J.(
C.Y. Chien[2], S

[1]INRS-Energie
Varennes, Qué
[2]Center for Ultr
2200 Bonisteel
[3]CEA Limeil, V
*On leave from
Shanghai, Peo
**On leave from

We discuss the phy
matter at very high
of the critical elem
state will offer a
interest for astroph

1. General Co

During the last fift
for, among other as
interaction of a nan
a very inhomogene
length, large comp
mainly absorbed i
conduction to the s

Recent breakthro
scale laser system
W/cm^2 and plans f
power and irradian
Amplification (CPA
[2].

The applications
features. The first i
during the heating
gradient scale lengt
place at near solid
resulting extremely
to modify the pl
ponderomotive forc
for densities well a
induce very large e

Springer Series in Chemi
Editors: P. F. Barbara · W

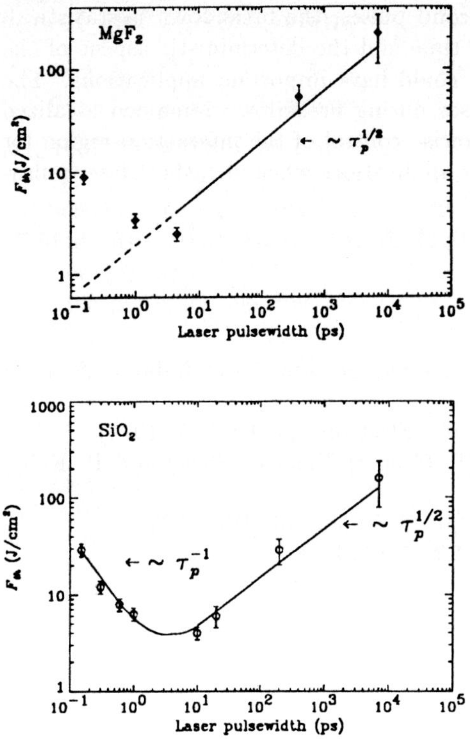

Figure 1: Breakdown threshold as a function of pulse-width

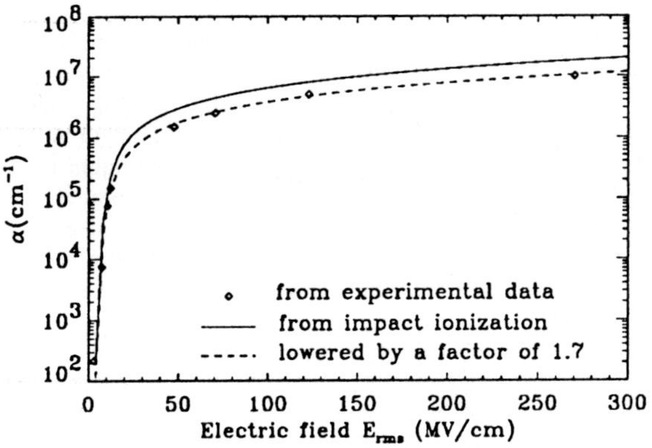

Figure 2: Impact ionization rate per unit distance α of SiO_2 determined from experiment and theory as a function of electric field

down threshold changes from $F_{\text{th}} \sim \tau_p^{1/2}$ to $F_{\text{th}} \sim 1/\tau_p$ at short pulse limit. Optical breakdown remains an electron avalanche dominant process for pulse width down to 150fs. Photoionization is responsible for the initial electron generation. Due to the assistance of photoionization, breakdown with short pulses

is determinis
The short la
breakdown v
plasma gene
at threshold,
material pro
are used.

Referen

[1] Proceed
in Optic
[2] N. Bloe
[3] S. C. J
Opt. En
[4] D. Stric
[5] K. K. T

2. What Kind of New Physics?

Many works on sub-picosecond laser solid matter interaction have concentrated up to now on generating short pulses of x-rays with sub-picosecond laser pulses [5] and the related transient and non-local atomic physics has been explored [6]. In addition new phenomena, such as the generation of far infrared radiation [7], ion collimation [8] and strongly coupled Raman instability [9], have been observed.

Here we introduce and discuss some elements related to the challenging problem of producing 1D, hot, uniform solid density plasmas. The aim of these experiments is to use judiciously the radiation pressure of a very high contrast pulse to balance the dense slab thermal pressure during laser heating. Carefull optimization is required. On one hand we must overcome the thermal and hydrodynamic mechanisms tending to spread the plasma when irradiance is too low. On the other hand we must determine the limit above which the irradiance is so high that we enter a regime of 2D/3D profile modifications by radiation pressure (hole boring) [3].

Results are very promising for the achievement of isochoric heating (the term used to indicate heating at constant density) of matter at near solid density, up to extreme temperatures. Production of matter in such a state offers a unique way to address experimentally in the laboratory some problems of great astrophysical interest, namely the opacity of dense matter and the physics of hot strongly correlated plasmas [10]. Furthermore this is also of importance for atomic physics and will give us the possibility to study among other aspects i) the ionization dynamics in hot dense plasmas with many electron atoms [11], ii) the physics of quasi-molecular states [12] and iii) the statistical physics of hot strongly correlated plasmas.

3. Critical Elements for Solid Density Plasmas

The ultimate plasma we need to do clean experiments is ideally hot (electron temperature in the keV range) still (ions at rest) and at solid density (6×10^{23} cm^{-3} for Li-like solid Al for instance). Producing such a tailored plasma is a complex art which requires i) a strict control of several crucial laser pulse parameters, ii) detailed target design and iii) an excellent experimental characterization of the plasma.

For the particular problem of forming a hot solid density nearly uniform plasma the key laser pulse issues are i) a high contrast ratio to avoid any expansion of the matter before the main pulse arrival and ii) an optimally high intensity (10^{17}-10^{19} W/cm^2).

While high intensities for short pulse duration are features obtained with the technique of chirped pulse amplification (CPA), an undesirable by-product of this technique is the production, at the wavelength of 1 µm, of a long prepulse at very low relative intensity (typically 10^{-6} of the main pulse), for a contrast of 10^6. This contrast is not adequate at the intensities of interest here, where a contrast of something like 10^{10} is required to avoid excessive energy deposition at lower density in the preplasma created by the long prepulse. The contrast has been improved in our experiments by using the second harmonic generation (i.e. conversion to green light) in non-linear crystals which will not double the weak prepulse. By using appropriate crystals we have been able to generate a maximum energy of 900 mJ in 300 fs green pulses, giving three terawatts on target with no prepulse, and intensities up to 4×10^{18} W/cm^2 corresponding to $I\lambda^2 = 10^{18}$ W µm^2/cm^2 (relativistic effects threshold). With such pulses, at 10^{18} W/cm^2, we have observed that the Li-like emission (at 1.5 keV), for an Al target, was radiated at an averaged (time integrated) density of 4×10^{23} cm^{-3}, which indicates the production of a hot, near solid density plasma. Our measurements show that there is an optimum condition to produce emission at solid density, for a given isoelectronic sequence.

Thermal electron conduction (including the non local transport which arises in steep-gradient, short scalelength plasmas) is the important loss mechanism during the laser heating of solid density targets. Our study with multilayer targets at 3×10^{18} W/cm^2 (Al layer on a Si substrate) show that around 2000 Å of solid material is heated by the thermal wave and that hot electron penetration is typically a few micrometers in Aluminum. The problems of energy spread and possible inhomogeneity should be limited by the use of targets in the form of thin films (thicknesses fo a few skin depths which also enhance the suprathermal electron energy deposition in a limited mass). Ultimately a double-sided irradiation might well be appropriate to confine the hot plasma by the effects of the radiation pressure on both sides of the target.

4. Application to Ultrafast X-Ray Sources

Solid density plasmas have been proposed as bright sources of ultrafast X-ray pulses. In our experiments with green 300 fs pulse, and no prepulse, we observed an increase of the X-ray yield, and of line emission intensities, with the laser intensity. A power law dependence of $I_L^{1.5}$, where I_L is the laser intensity, is observed for the Al He_α resonance line (7.76 Å) for 10^{17} W/cm^2 $\leq I_L \leq 10^{18}$ W/cm^2. The conversion efficiency in the small bandwidth Al He_α line emission is around 2 x 10^{-3}% in 2π at 10^{18} W/cm^2 in our experimental conditions. The Figure 1 indicates the photon number measured in 2π srd as a function of the wavelength for various kind of emissions, at 3 x 10^{18} W/cm^2 (600 mJ in the 300 fs green pulse). Around 10^{10} ph are emitted in cold K_α radiation (produced by hot electrons penetrating the solid target). Up to 2 x 10^{11} ph are obtained with the Ni-like unresolved transition array (UTA) from Ta targets (with 3% bandwidth). Our results also indicate that the window of optimum conditions to obtain simultaneously the shortest and brightest radiation pulse is relatively narrow for a given wavelength. For keV X-rays, lower intensity (less than 10^{17} W/cm^2) will produce short pulse but low brightness because the temperature and the ionization are too low. As the laser intensity is increased the conversion efficiency increases but above some laser intensity (typically 10^{18} W/cm^2 for Al keV radiation) the plasma is overheated and emission duration starts to dramatically increase because of recombination processes and plasma expansion.

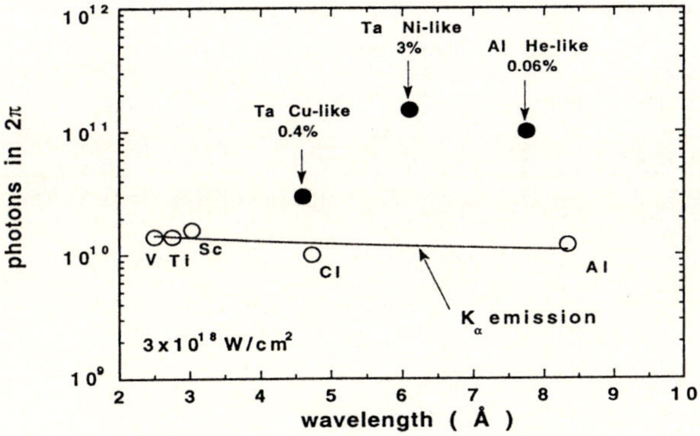

Fig. 1: Photon number produced in 2π srd with a 300 fs green (0.5 µm) laser pulse at 3 x 10^{18} W/cm^2 (no prepulse) as a function of wavelength. Open circles represent K_α emission and solid dots represent UTA emission from Ta in 0.4% and 3% bandwidth and Al He_α resonance emission.

5. Perspectives

Access to new regimes of laser matter interaction has been made possible by the recent advent of compact high intensity lasers for laboratories of relatively modest size. Production of matter heated, at constant density, to extreme temperature (in the keV range) offers a unique way to address experimentally in our laboratory some basic plasma physics problems and some of considerable interest for atomic physics and astrophysics. Furthermore the prospects for the generation of a bright, ultrafast X-ray source are very good and by choosing the appropriate emitting target, laser intensity and pulse shape the conversion efficiency as well as the X-ray duration may be optimized and we can plan to use these source, among other aspects, for molecular dynamics problems [13].

6. Acknowledgement

The authors would like to thank T.W. Johnston, R. More, H. Pépin, D. Umstadter and R. Walling for many helpful discussions. This work is supported in part by NSERC and Ministère de l'Éducation du Québec.

References

[1] P. Maine, D. Strickland, P. Bado, M. Pessot and G. Mourou, IEEE J. Quantum Elec. 24, 398 (1988).

[2] G. Mourou and D. Umstadter, Phys. Fluids B4, 2315 (1992); J.C. Kieffer, P. Audebert, M. Chaker, J.P. Matte, H. Pépin, P. Maine, D. Meyerhofer, J. Delettrez, D. Strickland, P. Bado and G. Mourou, Phys. Rev. Lett. 62, 760 (1989); M. Murnane, H.C. Kapteyn and R.W. Falcone, Phys. Rev. Lett. 62, 155 (1989); H.M. Milchberg, R.R. Freeman, S.C. Davey and R.M. More, Phys. Rev. Lett. 61, 2304 (1989).

[3] X. Liu and D. Umstadter, Phys. Rev. Lett. 69, 1935 (1992); S.C. Wilks, W.L. Kruer, M. Tabak and A.B. Langdon, Phys. Rev. Lett. 69, 1383 (1992).

[4] T.W. Johnston, Y. Beaudoin, M. Chaker, C.Y. Côté, J.C. Kieffer, J.P. Matte, H. Pépin, C.Y. Chien, S. Coe and G. Mourou, SPIE 2041, 206 (1993).

[5] B. Soom, H. Chen, Y. Fisher and D.D. Meyerhofer, J. Appl. PHys. 74, 5372 (1993); J.C. Kieffer, M. Chaker, J.P. Matte, H. Pépin, C.Y. Côté, Y. Beaudoin, T.W. Johnston, C.Y. Chien, S. Coe, G. Mourou and O. Peyrusse, Phys. Fluids B5, 2676 (1993); G.A. Kyrala, R.D. Fulton, E.K. Waklin, L.A. Jones, G.T. Shappert, J.A. Cobble and A.J. Taylor, Appl. Phys. Lett. 60, 2195 (1992); P. Audebert, Europhys. Lett. 19, 189 (1992); D.G. Stearns, O.L. Landen, E.M. Campbell and J.H. Scofield, Phys. Rev. A 37, 1684 (1988); J.D. Kmetec, C.L. Gordon, J.J. Macklin, B.E. Lemott, G.S. Brown and S.E. Harris, Phys. Rev. Lett. 68, 1527 (1992).

[6] J.P. Matte, J.C. Kieffer, S. Ethier, M. Chaker and O. Peyrusse, Phys. Rev. Lett. 72, 1208 (1994).

[7] H. Hamster, A. Sullivan, S. Gordon, W. White and R.W. Falcone, Phys. Rev. Lett. 71, 2725 (1993).

[8] A.R. Bell, F.N. Beg, Z. Chang, A.E. Danger, C.N. Danson, C.B. Edwards, A.P. Fews, M.H.R. Hutchinson, S. Luan, P. Lee, P.A. Norreys, R.A. Smith, P.F. Taday and F. Zhou, Phys. Rev. E 48, 2087 (1993).

[9] C.B. Darrow, C. Coverdale, M.D. Perry, W.B. Mori, C. Clayton, K. Marsh and C. Joshi, Phys. Rev. Lett. 69, 442 (1992).

[10] F.J. Rogers and C.A. Iglesias, Science 263, 50 (1994).

[11] C. Bauche-Arnoult, J. Bauche, E. Luc-Koenig, J.F. Wgart, R.M. More, C. Chenais-Popovics, J.C. Gauthier, J.P. Geindre and N. Tragin, Phys. Rev. A 39, 1053 (1989).

[12] E. Leboucher-Dalimier, A. Poquérusse and P. Angelo, Phys. Rev. E 47, 1467 (1993).

[13] J.P. Bergsma, M.H. Coladonato, P.M. Edelsten, J.D. Kahn, K.R. Wilson, D.R. Fredkin, J. Chem. Phys. 84, 615 (1986).

Electron Energies in Field-Ionized Laser Plasmas

T.E. Glover[1], *T.D. Donnelly*[1], *E.A. Lipman*[1], *A. Sullivan*[1], *J.C. Crane*[2], *M.D. Perry*[2], *and R.W. Falcone*[2]

[1]Department of Physics, University of California at Berkeley, Berkeley, CA 94720, USA

[2]Lawrence Livermore National Laboratoriy, University of California, Livermore, CA 94550,USA

Abstract. We present the first sub-picosecond electron energy measurements in plasmas of interest to recently proposed recombination x-ray laser schemes.

Recently, several authors have proposed that x-ray lasers pumped by electron-ion recombination are feasible if cold electrons can be produced from the interaction of a high-intensity laser pulse with a gas sample [1]. Burnett and Corkum [2] have used a model based on tunneling ionization to predict that cold electrons can be produced from such an interaction. Initial experimental tests of such a prediction yielded conflicting results [2]. Measurements have not previously been performed simultaneously at high density and on sub-picosecond time-scales relevant to transient recombination lasers [1].

In this work we use Thomson scattering and radiative recombination to measure electron energies at gas densities and time-scales relevant to transient recombination lasers. Our Thomson scattering experiments represent the first sub-picosecond temperature measurements of plasmas ionized by high-intensity optical ionization. Our recombination experiments constitute the first measurements of electron energy *distributions* in plasmas of interest to recent recombination x-ray laser schemes [1].

Thomson scattering experiments were performed in He. An initial (125 fs, $2x10^{17}$ W/cm^2) laser pulse at 800 nm ionized the He gas and a second sub-picosecond pulse at 400 nm probed the preformed plasma. Thomson spectra were taken at pressures ranging from 3 - 75 torr. A representative spectrum taken at 50 torr is shown in Fig. 1. The data has been fit to a theoretical Thomson spectrum (for He^{2+}) convolved with the response of the spectrometer and the probe laser spectrum. We obtain an electron temperature of 40 eV. While our fit to the short wavelength side of the data in Fig. 2 is good, an asymmetry results from stimulated Raman scattering (SRS) at long wavelengths. Our measured electron temperature (as given by the short wavelength electron feature) is unaffected by the presence of SRS. We experimentally determined that plasma cooling was insignificant by varying the pump - probe delay. Additionally, the measured electron temperature did not change over a 3 to 75 torr pressure variation; this is consistent with calculations that show

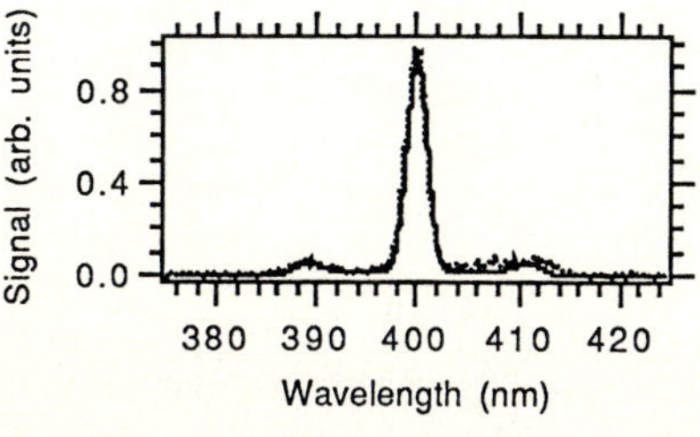

Figure 1. Experimental (dashed line) Thomson scattered spectrum for 50 torr of doubly ionized helium. The solid line is a theoretical spectrum for a pressure of 50 torr, an electron temperature of 40 eV and an ion temperature of 3.6 eV.

Springer Series in Chemical Physics, Vol. 60 **Ultrafast Phenomena IX**
Editors: P. F. Barbara · W. H. Knox · G. A. Mourou · A. H. Zewail

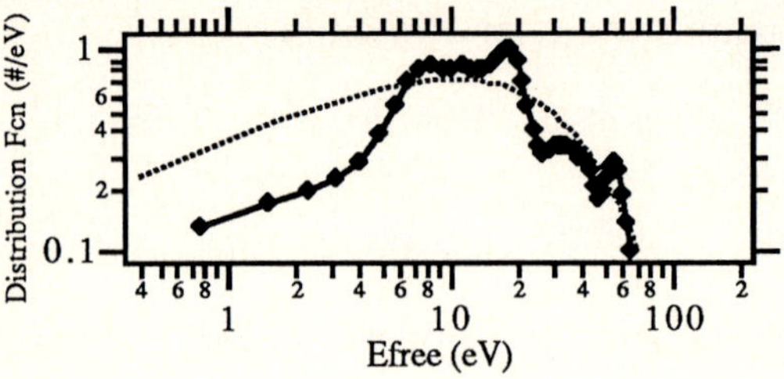

Figure 2. Measured initial distribution of electrons from doubly-ionized He. Also shown is a Maxwellian distribution for KT = 20 eV.

that electron heating is dominated by tunneling ionization. The predicted temperature is 40 eV, in good agreement with measurement. The central scattering feature, called the "ion feature," at 400 nm in Fig. 1 provides information on the ion temperature. Our data indicates that the ion temperature ranges from approximately 0.5 eV at 3 torr, to approximately 5 eV at 75 torr.

We also measured radiative recombination spectra in He^{2+} using 616 nm, linearly and circularly polarized, 160 fs pulses focused to $2x10^{16}$ W/cm^2. XUV fluorescence was spectrally and temporally resolved on picosecond timescales. The initial distribution of electron energies obtained from the He^{2+} plasma with linearly polarized pulses is shown in Fig. 2. The data has been best fit to a Maxwellian distribution of 20 eV. The average energy in the measured distribution is 30 eV. The average energy expected from the tunneling model is 32 eV, implying good agreement with the tunneling model. The measured distribution, however, exhibits a deficit of low energy electrons in comparison to both tunneling and Maxwellian distributions. This result is in contrast to earlier measurements in He [2] performed at low gas density which show good agreement between the measured distribution and a Maxwellian distribution. At late times, our measured distribution cools and thermalizes so that the agreement between the measured distribution and a Maxwellian distribution is good at late times . The discrepancy between our initial He distribution measured at high density and the He distribution measured at low density [2] may be due to the importance (at high density) of collisional ionization by laser driven electrons.

In summary, Thomson scattering and radiative recombination were used to measure electron energies in plasmas produced by high intensity optical ionization. The average electron energies are in good agreement with predictions based on the tunneling ionization model [2]. The measured distribution functions, however, show a deficit of low energy electrons in comparison to present theory.

References

1.Amendt,*et al.*,Phys.Rev.Lett.,**66**(1991);Burnett et *al.*, IEEE J.Q.E.,**26**(1990).

2.Burnett *et al.*, J. Opt. Soc. Am. B,**6**(1989);Corkum, et *al., Phys.* Rev. Lett., 62(1989);U. Mohideen, *et al., Phys.* Rev. Lett., 71(1993).

Mode Properties of a High Intensity Plasma Waveguide

H.M. Milchberg, C.G. Durfee III, and J. Lynch

Institute for Physical Science and Technology, University of Maryland, College Park, MD 20742, USA

Recently, we demonstrated a technique for the optical guiding of intense laser pulses, in which a plasma refractive index channel was produced through the hydrodynamic evolution of a laser breakdown spark in an ambient gas[1]. Here, we present measurements showing that our plasma light pipe can support distinct electromagnetic mode propagation at intensities up to 10^{14} W/cm^2. In principle, guided intensities exceeding 10^{18} W/cm^2 should be possible.

A condition on the modes that will be supported by a plasma refractive index channel can be estimated by assuming an index variation $n(r)=1-1/2\ N_e(r)/N_{cr}$ (for $N_e/N_{cr} << 1$). It can be shown[2] that a linear waveguide analysis is applicable in a plasma waveguide for intensities considered here ($\leq 10^{14}$ W/cm^2). For guides with a wide range of curvature and $n(r)=n_{min}$ (or $N_e(r)= N_e(r_m)=N_e^{max}$) for $r>r_m$, approximate *cutoff* and *guiding* conditions, respectively, can be obtained[2]:

$$N_e^{max} - N_e^{min} = \Delta N_e(r_m) \geq (2p+m+1)^2/\pi r_e r_m^2 \ ; \quad \Delta N_e(w_{ch}) = (2p+m+1)^2/\pi r_e w_{ch}^2 \quad (1)$$

Here r_e is the classical electron radius, $N_e^{min}=N_e(0)$, $\Delta N_e(w_{ch}) = N_e(w_{ch})-N_e(0)$, $p\geq 0$ and $m\geq 0$ are radial and azimuthal mode indices, and w_{ch} is the 1/e field radius of the p=0, m=0 mode. From Eqn. (1), lowest order or p=0, m=0 (Gaussian) propagation of a spot size $w_0=w_{ch}=10\mu m$ requires $\Delta N_e \approx 10^{18}$ cm^{-3}. Guiding of higher order modes requires a greater electron density difference and/or channel radius r_m. As seen from Fig. 1 of ref. 1, , the electron density decreases beyond the shock for $r>r_m$, so that it is to be expected that a mode below but near the cutoff of Eqn. (1) would tunnel or "leak" through the electron density peak of the shock. We present evidence of this "leaky mode" effect in another paper[2].

The laser configuration and experimental setup for channeling is described in ref. 1. The channel widening with delay is verified experimentally in Fig. 1, for a gas

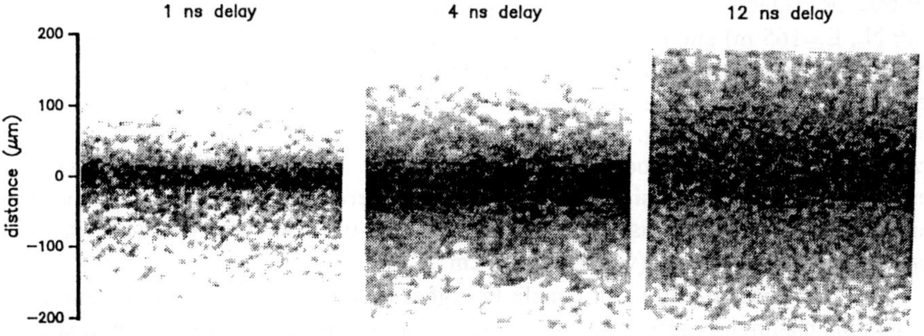

***Fig. 1.** Laser scattering image of channel centre region for delays 1, 4, and 12 ns. (30 torr Xe/95 torr N_2, E_1=100 mJ, E_2=15 mJ)*

Springer Series in Chemical Physics, Vol. 60 **Ultrafast Phenomena IX**
Editors: P. F. Barbara · W. H. Knox · G. A. Mourou · A. H. Zewail

Self-Channeling of High Peak-Power Femtosecond Laser Pulses in Air

A. Braun, G. Korn, X. Liu, D. Du, J. Squier, and G. Mourou

Center for Ultrafast Optical Science, University of Michigan,
2200 Bonisteel Blvd., IST Bldg., Rm. 1006, Ann Arbor, MI 48109-2099, USA

The peak power of ultrashort laser pulses (100fs) containing 1 mJ of energy is greater than the critical power for self-focusing in air. For the first time, it was observed that these pulses self-channel into a filament with intensities of 5×10^{13} W/cm^2 and propagate through distances greater than 30 meters. We attribute this stabilization to an interplay between self-focusing and the combined effects of diffraction and self-defocusing from the plasma created by the intense laser pulse. The study of long interaction length, intense laser pulses [1] have interesting applications including laser induced lightning [2], laser pumped x-ray sources [3], and laser-plasma based accelerators [4]. Furthermore, these measurements infer that the pulse is spatially and temporally trapped into a 3-dimensional light bullet [5].

The propagation studies were carried out with a Ti:sapphire CPA laser system which delivers up to 50 mJ per pulse at a reptition rate of 10 Hz and pulsewidths between 150 - 200 fs. The spectrum is centered at 780 nm. At low energy the beam was collimated down the hallway. When the energy exceeded 2 mJ per pulse, whole beam self-focusing occured. As the energy exceeded 5 mJ, a small scale filament began to form. By increasing the energy, the filament formation occurred earlier. The diameter of the filament was measured by imaging its spatial profile off a diffuser at grazing incidence. This technique avoids damage to the diffuser due to the high intensity in the filament. A lineout obtained with a 2D CCD detector after a propagation distance of 30 m and an energy of 20 mJ per pulse after the compressor gives a diameter of 80 μm FWHM (Figure 1).

The energy in the filament was also measured at different locations along the hallway. The normalized energy per filament is roughly constant at about 0.75 mJ. An intensity of about 5×10^{13} W/cm^2 in the filament is calculated using the measured values of the energy and diameter, and assumming no pulsewidth change during the propagation in the channel. The spectrum of the pulse was also measured after the formation of the trapped filament. This spectrum is broadened and modulated with respect to the spectrum directly out of the compressor, yet was measured to be relatively constant as the trapped pulse propagated. Since the intensity of the pulse is clamped due to the defocusing from the generated plasma, and the energy, diameter, and spectrum are measured to be a fairly constant during propagation in the channel, we can infer that the pulse duration is also fairly constant in the channel. Also, since the relative magnitude of both the positive group velocity dispersion from the air and the negative group velocity dispersion from the plasma are roughly the same, any change in the pulsewidth once the pulse is in the channel will be small.

As is well known, the trapped state of self-focusing is not stable if n_2 alone is the mechanism [6]. In our case, the negative lens effect of the plasma produced by the

Springer Series in Chemical Physics, Vol. 60 **Ultrafast Phenomena IX**
Editors: P. F. Barbara · W. H. Knox · G. A. Mourou · A. H. Zewail

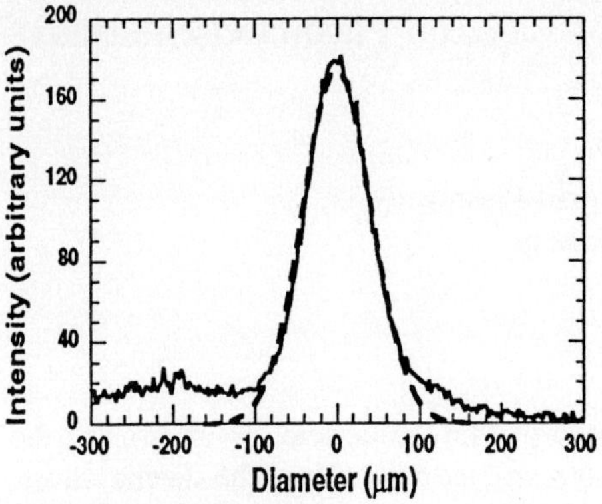

Figure 1: Filament diameter with a Gaussian fit shows a FWHM of 80 μm.

intensity-dependent ionization can counter-balance the self-focusing [7]. Using the value of n_2 for air to be 5 x 10^{-19} cm^2/W, the index change due to n_2I is 2.5 x 10^{-5}. An ionization of 1 x 10^{17} cm^{-3}, or 3.7x 10^{-3} of atmospheric air concentration, will balance the self-focusing at λ=780 nm. As a significant portion of air starts to be ionized at this intensity, it is reasonable to believe that ionization can stabilize self-focusing, making filament channeling possible. The propagation of the trapped filament also resulted in white light generation and stimulated Raman scattering. These losses, along with the plasma dependent loss, will eventually cause the channeled filament to disappear. A computer simulation solving the cylindrically symmetric nonlinear Schroedinger equation for the electric field including self-focusing, self-defocusing from an ionization dependent refractive index change, and an intensity dependent loss confirms the measured parameters for the field in the channel and makes the physical picture more reasonable. The simulation results show that a lower ionization density is needed to stablize self-focusing because of the inclusion of diffractive effects. Furthermore, we are studying the role the background energy has on the propagation of the trapped pulse.

This work was supported by the National Science Foundation through the Center for Ultrafast Optical Science under STC PHY 8920108.

References

[1] C.G. Durfee and H.M. Milchberg, Phys. Rev. Letters **71**, 2409 (1993).
[2] X.M. Zhao, C.Y. Yeh, J.-C. Diels, and C.Y. Wang, *Ultrafast Phenomena VIII*, eds J.-L Martin, A. Migus, G.A. Mourou, A.H. Zewail (Springer-Verlag, Berlin, 1993) p. 264.
[3] S.C. Rae and K. Burnett, Phys. Rev. A **46**, 2077 (1992).
[4] T. Tajima and J. M. Dawson, Phys. Rev. Lett. **43**, 267 (1979); P. Sprangle, E. Esarey, A. Ting, and G. Joyce, Appl. Phys. Lett. **53**, 2146 (1988).
[5] Y. Silberberg, Opt. Lett. **15**, 22 (1990).
[6] Y.R. Shen, *The Principles of Nonlinear Optics*, (Wiley-Interscience, 1984), pp. 303-324.
[7] X. Liu and D. Umstadter, OSA Proc. on Shortwavelength V, **17**, (1993).

Observation of the Dynamics of Electron-Plasma Oscillations in Femtosecond Laser-Produced Plasmas

D. von der Linde and H. Schüler

Fachbereich Physik, Institut für Laser- und Plasmaphysik,
Universität Essen, D-45117 Essen, Germany

Abstract: Electron density oscillations in a plasma are detected by measuring the second harmonic. Pump-probe experiments reveal the dynamics of the plasma waves.

One of the mechanisms of second harmonic generation (SH) in plasmas is coherent scattering of the fundamental light wave from the electron density fluctuations associated with plasma waves [1]. Using SH polarization selection rules [2] and measuring properly selected SH components, one can directly observe electron plasma waves. We report here SH experiments in which the ultrafast dynamics of electron plasma waves have been directly measured for the first time.

An intense pump pulse produced a microplasma on the surface of a glass substrate, and the plasma was interrogated by a weak probe pulse. The laser pulses were obtained from an amplified CPM dye laser (λ=620 nm). The duration, focal diameter and peak intensity of the *pump* beam on the target surface were, respectively, 120 fs, 7 µm, and 6×10^{15} W/cm^2. The peak intensity of the *probe* pulse was ten times less than the pump, but otherwise pump and probe were the same.

The p-polarized pump and the s-polarized probe beam were incident on the target surface at θ_{pump}=37° and θ_{probe}=50°, respectively. A diaphragm positioned approximately in the direction of the bisector of the reflected beams blocked the latter and passed the SH due to sum frequency mixing of pump and probe. A polarization analyser placed in the SH beam after the diaphragm suppressed scattered light from the strong pump beam and ensured that only s-polarized SH reached the detector.

The induced SH current density in a plasma can be written [2]:

$$\mathbf{J}_{2\omega}^{(2)} = \frac{ie}{4\pi m\omega}\left[\frac{\omega_p^2}{4\omega^2}\nabla \mathbf{E}_\omega^2 + \mathbf{E}_\omega \nabla\cdot\mathbf{E}_\omega\right] \tag{1}$$

$\mathbf{E}_\omega$ is the electric field vector of the wave at the frequency ω, ω_p is the local plasma frequency, and the other symbols have their usual meaning. The second term in (1) is directly proportional to the first order electron density perturbation $n_\omega^{(1)}$ associated with the electron plasma wave driven by the fundamental field, $n_\omega^{(1)} \propto \nabla\cdot\mathbf{E}_\omega$. For a p-polarized pump field $\mathbf{E}_{pump}$ and a s-polarized probe $\mathbf{E}_{probe}$ the component of the SH current density corresponding to s-polarization is (α is some constant):

$$\mathbf{J}_{2\omega\perp}^{(2)} = \alpha\,\mathbf{E}_{probe}\nabla\cdot\mathbf{E}_{pump} \tag{2}$$

The angle $\theta_{2\omega}$ of the SH generated by this current is approximately given by the bisector of the reflected beams. It follows that for the chosen experimental configuration the

Springer Series in Chemical Physics, Vol. 60 **Ultrafast Phenomena IX**
Editors: P. F. Barbara · W. H. Knox · G. A. Mourou · A. H. Zewail

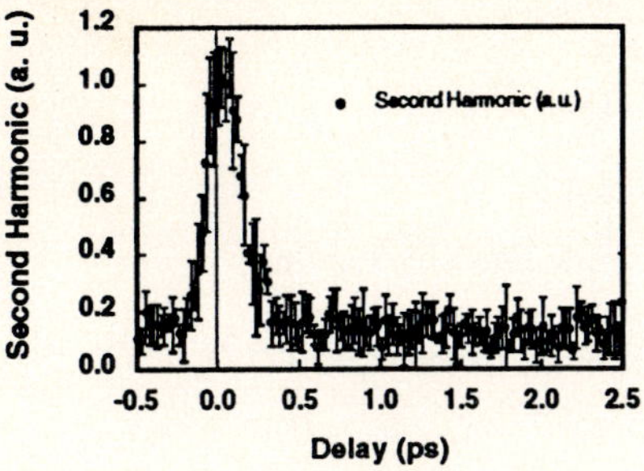

FIG. 1: Measured s-polarized SH of the probe pulse as a function of delay time.

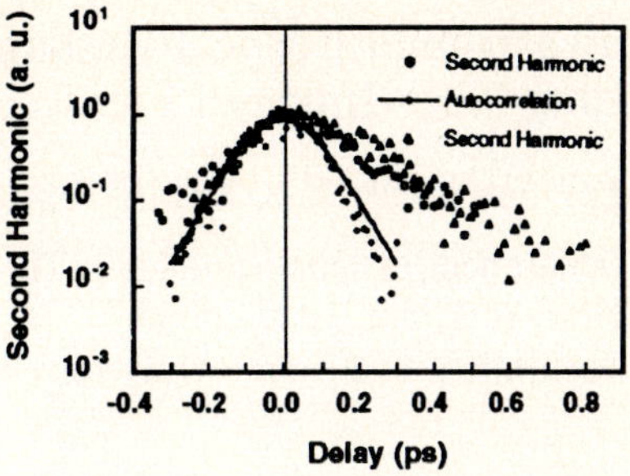

FIG. 2: SH versus delay time and autocorrelation function of the laser pulses. Two different runs for the SH are shown.

selected SH component is proportional to the squared *amplitude of the electron plasma wave* generated by the pump pulse.

An example of the observed time dependence of the s-polarized SH is shown in Fig. 1. It can be seen that the s-SH is concentrated in a sharp peak of about 200 fs, which appears to be slightly asymmetric and shifted by approximately +100 fs with respect to zero delay time.

In Fig. 2 the s-polarized SH is compared with the measured intensity autocorrelation of the laser pulses. The autocorrelation data can be represented, to a good approximation, by a squared hyperbolic secant function (solid line in Fig. 2). The comparison shows that the decrease of the SH for positive delay times is significantly slower than that of the autocorrelation. The decay of the SH signal can be fitted by an exponential function with a time constant of (160 ± 10) fs.

The distinct difference between the time dependence of the SH from the plasma and the autocorrelation of the laser pulses provides strong evidence of a two-step SH generation process that can be described as follows. Firstly, a thin surface layer of the material is ionized by the pump pulse and a rapidly expanding plasma is formed. The p-polarized pump pulse couples with the longitudinal modes of the inhomogeneous plasma and drives coherent electron density oscillations at the frequency of the light. Secondly, the probe pulse is scattered from these charge density fluctuations. This process gives rise to anti-Stokes scattered light at the second harmonic frequency. Thus, in the pump-probe experiment measuring this SH signal, the rise and decay of the electron density oscillations is mapped out.

References

1. N. G. Basov, Y. A. Zakharenkov, N. N. Zorev, A. A. Rupasov, G. V. Sklizkov, A. A. Rupasov, and A. S. Shikanov, *Heating and Compression of Thermonuclear Targets by Laser Beam*, Cambridge University Press (Cambridge, 1986).
2. D. von der Linde, in: *Laser Interaction with Atoms, Solids and Plasmas*, NATO Advanced Study Institute, Cargèse, Corsica, 1992.

Intrastromal Corneal Reshaping Using Near-Infrared Picosecond Laser Pulses

M.G. Speaker[1], *M.S. Habib*[1], *and T. Juhasz*[2]

[1]Department of Ophthalmology, The New York Eye and Ear Inf., New York, NY 10003, USA

[2]Department of Physics, University of California, Irvine, CA 92717, USA

Abstract. Picosecond intrastromal photorefractive keratectomy is utilized for refractive corneal surgery. Basic principles and animal experiments are discussed.

Picosecond lasers have great potential for corneal intrastromal surgery due to their well localized surgical effect. Damage mechanism of picosecond lasers begins with optical breakdown in the focus of the laser beam. The plasma creation is followed by the generation of the shock wave and cavitation bubble. Tissue removal is due to the effect of the laser plasma [1,2]. Using multiple picosecond pulses it is possible evaporate a disk-shaped volume inside the cornea by delivering the pulses along a spiral pattern. The collapse of the created intrastromal cavity results in a thinning of the central cornea and, therefore, a myopic refractive correction. The method is referred as intrastromal photorefractive keratectomy (ISPRK). ISPRK is attractive because it obviates the need for removal of the corneal epithelium, therefore it has the potential for avoiding wound healing responses.

The efficiency of the intrastromal tissue removal is strongly influenced by the development of the cavitation bubble. If the cavitation generated by the first laser pulse is greater than the actual plasma size, it may interact with the surgical effect of the next pulse. Therefore, the size of the cavitation at the moment of arrival of the next pulse should approach that of the plasma ball. To find the best laser parameters for intrastromal tissue removal, we have investigated the time and the energy dependence of the cavitation bubble diameter in bovine corneas with time resolved photography. Fig. 1. displays the development of the cavitation generated by 20 ps pulses with 270 μJ energy when the laser is focused to a 20 μm spot size. As is seen in the figure the cavitation undergoes a transient oscillation, then reaches a final size after approximately 500 μs. The bubble then slowly decreases and

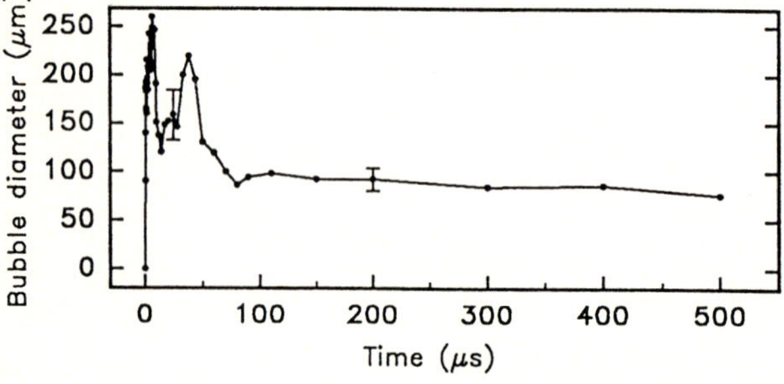

Fig.1.

Springer Series in Chemical Physics, Vol. 60 **Ultrafast Phenomena IX**
Editors: P. F. Barbara · W. H. Knox · G. A. Mourou · A. H. Zewail

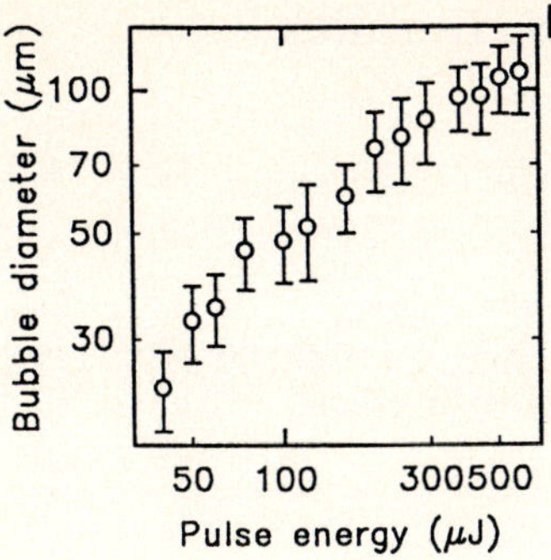

Fig. 2.

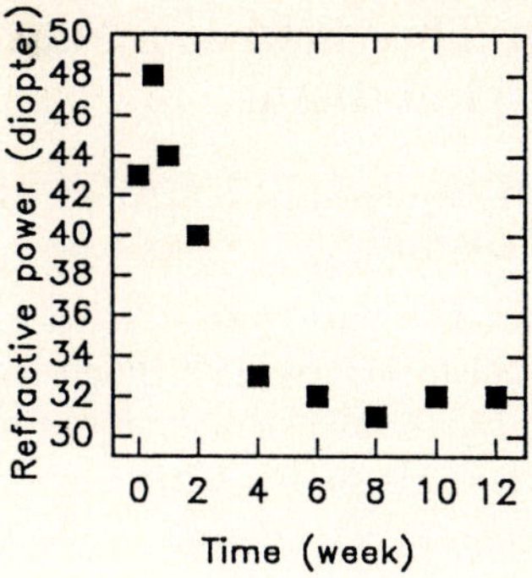

Fig. 3.

completely disappears after 5 to 30 minutes. Fig. 2. shows the dependence of the bubble diameter on the laser pulse energy. These experimental results indicate that intrastromal tissue removal should be performed with pulse energies close to the threshold of the breakdown when only small cavitation bubbles are generated.

In the animal experiment output pulses from a regeneratively amplified Nd:YLF laser (ISL, Model 4000) were used to perform intrastromal tissue removal. The pulse duration is 40 ps and the focus spot diameter is 10 μm. Pulse energies between 20 and 40 μJ were chosen which are close to the threshold. Spiral patterns were placed in the upper third of the central cornea of 12 cats with 6 mm in diameter. The spot separation in the pattern and the layer separation between patterns were chosen to be 15 μm according to the measurement of the cavitation bubble diameter. The effects of 3 and 5 layer treatments were investigated. Post-operative corneal topography, ultrasonic biomicroscopy and ultrasonic pachymetry were performed at scheduled intervals. The follow up ranged from two weeks to 6 months. Corneal thickness was increased an average of 50 μm at 24 hrs. The corneas were thinned over a period of 2 weeks and central flattening was apparent at 3 weeks. Central topographic flattening up to 11 diopters was achieved which was associated with a central thinning of 120 μm. The average refractive power of the eyes at the center of the cornea as a function of time, as it is measured with corneal topography, is displayed in Fig. 3. The central topographic change of the corneas was stable as far as 6 months and no evidence of regression was observed.

1. M.H. Niemz, et al., Lasers and Light in Ophthalmol., 5, 145 (1993).
2. A. Vogel, et al., IEEE J. of Quant. Electron. 26, 2240 (1990).

Damage Threshold as a Function of Pulse Duration in Biological Tissue

D. Du[1], *J. Squier*[1], *R. Kurtz*[2], *V. Elner*[2], *X. Liu*[1], *G. Güttmann*[1;2], *and G. Mourou*[1]

[1]Center for Ultrafast Optical Science, University of Michigan, 2200 Bonisteel Bldv., IST Bldg., Rm. 1006, Ann Arbor, MI 48109-2099, USA
[2]Kellogg Eye Center, University of Michigan, 1000 Wall Street, Ann Arbor, MI 48105, USA

Abstract. Damage thresholds for corneal tissue are measured as a function of pulse duration, revealing strong departure from previous findings.

We measured damage thresholds for corneal tissue over a range of pulse durations (from 150 fs - 7 ns) that could not be examined previously with a single laser source, using a Ti:sapphire chirped-pulse amplification (CPA) laser [1]. Our results depart from previous findings [2,3], which had to use multiple laser sources to investigate a similar pulsewidth range.

For ultrashort pulse durations (<10 ps), the primary mechanisms for tissue damage have not been described. For certain applications, such as ophthalmic surgery, collateral damage must be minimized. By understanding the mechanisms and defining the breakdown thresholds for ultrashort pulse durations, we hope to reduce collateral damage for these precision applications.

Laser-induced optical breakdown consists of three general steps: free electron generation and multiplication, plasma heating and material deformation or breakdown. Avalanche ionization and multiphoton ionization are the two processes responsible for the breakdown. The laser-induced breakdown threshold in dielectric material depends on the pulsewidth of the laser pulses. An empirical scaling law of the fluence breakdown threshold as a function of the pulsewidth is given by $F_{th} \propto \sqrt{\tau_p}$, or alternatively, the intensity breakdown threshold, $I_{th} = F_{th}/\tau_p$. Although this scaling law applies in the pulsewidth regime from nanosecond to tens of picoseconds, recently, we found that the breakdown threshold does not follow the scaling law when laser pulses are shorter than ten picoseconds for SiO_2 [4]. Therefore, a systematic investigation of the breakdown threshold of cornea tissue as a function of the laser pulsewidth should enable us to identify the best operating parameters.

We performed a series of experiments to determine the breakdown threshold of cornea as a function of laser pulsewidth between 150 fs - 7 ns, using a CPA laser system . One advantage of this laser system is the laser pulsewidth can be varied while all other experimental parameters (spot size, wavelength, energy, etc.) remain unchanged. The laser was focused to a spot size (FWHM) of 26 μm. The plasma emission was recorded as a function of pulse energy in order to determine the tissue damage threshold. Histology was also performed to assess tissue damage.

Breakdown thresholds calculated from plasma emission data revealed deviations from the scaling law, $F_{th} \propto \sqrt{\tau_p}$. As shown in Figure 1, the scaling law of the fluence threshold is true to about 10 ps, and fails when the pulse shortens to less than a few picoseconds. The standard deviation of breakdown threshold measurements

Springer Series in Chemical Physics, Vol. 60 **Ultrafast Phenomena IX**
Editors: P. F. Barbara · W. H. Knox · G. A. Mourou · A. H. Zewail

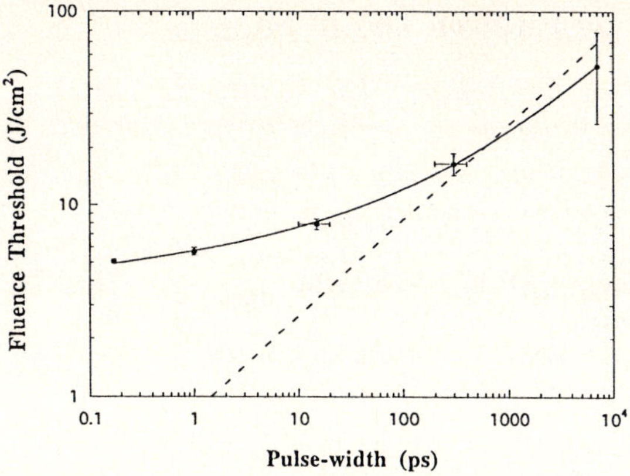

Figure 1. Damage threshold of cornea as a function of pulse duration.

decreased with shorter pulses. Histological analysis reveals more localized ablation with pulses less than 10 ps.

The breakdown threshold for ultrashort pulses (< 10 ps) are less than longer pulses and have smaller standard deviations. The reduced collateral damage caused by ultrashort laser pulses show that these systems may have a future in ophthalmic and other precision surgical procedures.

This work was partially supported by the National Science Foundation under STC PHY 8920108 and the Midwest Eye Bank and Transplantation Centers.

References

[1] J. Squier, F. Salin, G. Mourou, Opt. Lett. **16**, 324

[2] D. Stern, R. Schoenkin, C. Puliafito, E. Dobi, R. Birngraber, J. Fujimoto, Arch. Opthalmol. **107**, 587 (1987)

[3] R. Birngruber, C. Puliafito, A. Gawande, W. Lin, R. Schoenlein, J. Fujimoto, IEEE J. Quantum Electron. **23**, 1836 (1987)

[4] D. Du, X. Liu, G. Korn, J. Squier, G. Mourou, to be published in App. Phys. Lett..

Ultrafast Observation of an Electron Gas in SiO_2

P. Audebert[1], *Ph. Daguzan*[2], *A. Dos Santos*[3], *J.C. Gauthier*[1], *J.P. Geindre*[1], *S. Guizard*[2], *G. Hamoniaux*[3], *K. Krastev*[2], *P. Martin*[2], *G. Petite*[2], *and A. Antonetti*[3]

[1]Laboratoire pour l'Utilisation des Lasers Intenses, Ecole Polytechnique, F-91128 Palaiseau, France

[2]Commissariat à l'Energie Atomique, DRECAM/SRSIM, F-91191 Gif-sur-Yvette, France

[3]Laboratoire d'Optique Appliquée, ENSTA, Batterie de l'Yvette, F-91120 Palaiseau, France

Abstract: Interferences in the frequency domain with sub-picosecond resolution are used to observe the ultra-fast excitation and relaxation of a photoexcited electron gas in $Si0_2$.

When an ultra-short and intense laser field propagates inside a dielectric medium, it induces a strong polarization field and, if the intensity is high enough, it produces a high density of electrons and holes. In this paper, we report on an experiment based on a new phase-sensitive technique [1,2] which allows the direct visualization of the polarization and the excitation density with a resolution of 120 femtosecond in time and 5 μm in space. To our knowledge, the very first relaxation events, concerning the *photogenerated carriers in the conduction band*, remain an unexplored domain in wide band-gap insulators.

The principle of this experiment is the following: the target is probed with two collinear femtosecond pulses separated by a fixed time delay (1 ps in this experiment). These two pulses are then analyzed in a spectrometer. The area of the sample of interest is imaged on a slit of variable aperture at the entrance of the spectrometer. The power spectrum of such a sequence is a fringe system with a period inversely proportional to the pulse separation. This interference pattern is visualized on a CCD camera: one axis represents the wavelength, while the radial spatial dimension is represented on the perpendicular axis. If a high intensity pump pulse (ranging from 10^{12} W/cm^2 to a few 10^{13} W/cm^2 in this experiment) is focused on the sample at some delay between the two probe pulses, the second probe pulse will experience a disturbed dielectric constant. This leads to a local distortion of the interference pattern. The induced phase shift can thus be recorded and analyzed as a function of delay and space. The absorption of the probe pulse can simultaneously be obtained by measuring the fringe contrast directly on the interference pattern. For this experiment we use as a pump a colliding-pulse mode-locked (CPM) dye laser which emits 70 fs pulses at 620 nm amplified to energies up to 3 mJ. Probe pulses are produced with a part of this beam by continuum generation in a water cell and frequencies around 560 nm are selected and amplified. Their time duration is 60 fs. The samples are 100 μm thick high purity amorphous (a-SiO_2) and crystalline (α-SiO_2) silicon-dioxide.

In figure 1, the phase shift between probe pulses is shown as a function of space and time for a-SiO_2. The pump intensity is set at 2.7×10^{13} W/cm^2. The spot size is 25 μm (FWHM). We observe a very rapid evolution of the phase shift as a

Springer Series in Chemical Physics, Vol. 60 **Ultrafast Phenomena IX**
Editors: P. F. Barbara · W. H. Knox · G. A. Mourou · A. H. Zewail

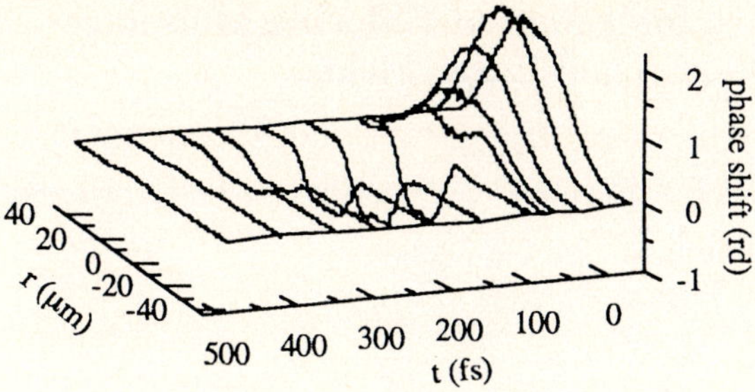

Fig1 : phase shift (in radian) as a function of time (fs) and space (µm).

function of the time delay between the pump pulse and the second probe pulse. It is positive for time delay close to zero, becomes suddenly negative and finally goes back to a value close to zero. Concerning the space dependence, the positive phase shift occurs in a broader region, the half width of the signal is larger than in the case of negative phase shift. This result is interpreted as follows: the positive phase-shift is due to an induced phase-modulation and occurs as far as the pump pulse and the second probe pulse temporally overlap. Note that the temporal resolution (120 fs) can be directly observed in figure 1. On the other hand, a free-electron population, produced by multiphoton absorption, induces a negative phase shift which becomes visible even before the pump pulse has ended. Let us note that a decrease of the phase shift implies a decrease of the free-electron density. More precisely, electrons must be trapped into levels deeper than the photon energy, otherwise they would be seen as free by the laser field. The trapping of the gas is extremely fast and its time dependence is well fitted by an exponential law with a time decay of 150 fs. As a consequence, the gas has "disappeared" before expanding in a measurable proportion. We find that this mean time does not depend on the pump intensities. The similarity of the results between α-quartz and silica proves that these traps have an intrinsic character because the density of extrinsic defects in amorphous silica is known to be greater than in the crystallized form. We attribute the rapid variation of the excitation density to the formation of E' centers whose kinetic of formation has been recently measured [3].

[1] E. Tokunaga, A. Terasaki and T. Kobayashi, Opt.Lett. **18**, 370 (1992).
[2] J. P. Geindre, P. Audebert, F. Falliès, A. Rousse, J. C. Gauthier, A. Antonetti, J. P. Chambaret, G. Grillon, A. Mysyrowicz, to be published.
[3] P. N. Saeta, B. I. Greene, Phys. Rev. Lett. **70**, 3588 (1993).

Multielectron Dissociative Ionization of Chlorine Molecules in Intense Femto- and Picosecond Laser Fields: A Puzzling Result

M. Schmidt, C. Cornaggia, P. D'Oliveira, P. Meynadier, and D. Normand

CE, DSM/DRECAM/SPAM, Bâtiment 524, CE Saclay,
F-91191 Gif-sur-Yvette Cedex, France

Abstract: The multielectron dissociative ionization of the chlorine molecule shows identical near Coulomb explosion with femto- and picosecond pulses of same intensity.

During the last few years the interaction of diatomic molecules with very intense pico- and femtosecond laser fields has become a field of intense research [1 - 5]. One essential issue of the multielectron dissociative ionization (MEDI) studies concerns the interaction dynamics, that is the competition between the removal of the electrons and the dissociation of the molecular ions [2]. The interaction of ultrashort laser pulses with heavy molecules has been suggested to avoid this competition since ionization would occur on a time scale where nuclear motion is essentially frozen and the Coulomb repulsion would follow in field-free conditions. Indeed, experiments performed on I_2 with 90 fs pulses [3] yield energy releases E_{exp} which are closer ($\approx$ 75%) to those predicted by the simple Coulomb model (E_{Cb}) than in all experiments with lighter molecules ($E_{exp}/E_{Cb} \approx 50\%$) [1, 2, 4]. In order to determine the respective role of the specific molecular electronic structure and of the ultrashort pulse excitation scheme, we have investigated the MEDI of Cl_2 with femto- and picosecond pulses.

In this experiment we used a commercial Ti:sapphire system of the new DRECAM laser facility delivering 150 fs pulses at 790 nm and a picosecond synchro-pumped dye laser system delivering 2.5 ps pulses at 616 nm [4]. Beam focalisation allowed to reach laser intensities of at least 5×10^{15} W/cm^2. The fragment-energy measurement is based on the forward-backward splitting of the ionic fragments in a time-of-flight (TOF) mass-spectrometer [2, 4].

Figure 1 shows the comparison of the Cl^{2+} and Cl^{3+} MEDI spectra, as a part of typical TOF mass-spectra of Cl_2, obtained after interaction with femtosecond (a) or picosecond (b) laser pulses at the same intensities of 5×10^{15} W/cm^2. As a matter of fact, the two spectra are almost identical, indicating that the kinetic energy releases are independent of the pulse duration. The identification of the different fragmentation channels has been achieved using high resolution TOF spectra and covariance mapping [1, 5]. The measured respective kinetic energy releases E_{exp} for the different fragmentation channels of molecular ions up to Cl_2^{8+} have been compared to the values E_{Cb} expected from pure Coulqmb repulsion assuming the neutral ground-state internuclear distance ($R_o = 2.0$ A). Interestingly, a nearly constant energy ratio E_{exp}/E_{Cb} of about 70 % is found for all fragmentation channels.

Springer Series in Chemical Physics, Vol. 60 **Ultrafast Phenomena IX**
Editors: P. F. Barbara · W. H. Knox · G. A. Mourou · A. H. Zewail

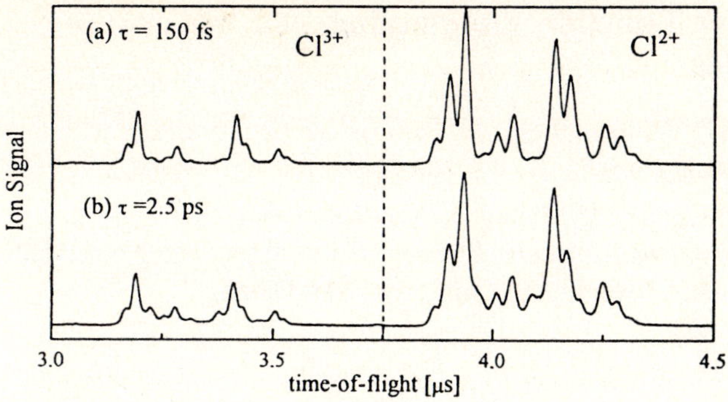

Figure 1: Part of the TOF ion mass-spectra of Cl_2 showing Cl^{2+} and Cl^{3+} fragment ions (conditions see text). The laser polarization is parallel to the detection axis.

The present result is the first example where near Coulomb energies are found in the long-pulse regime and raises the question of the MEDI interaction dynamics. The constant energy ratios E_{exp}/E_{Cb} observed in this experiment suggest, that the molecular ions follow a simple Coulomb explosion at $R \approx 1.4 \times R_o$ indicating that the internuclear distance of the ions essentially remains frozen during the interaction (150 fs). The short ground-state vibrational period of the molecule (60 fs), however, indicates that the molecular ions should already start to dissociate in the leading edge of the laser pulse. This result demonstrates that the electronic structure of the molecule must play an important role in the MEDI process [6], either through screening effects or through stabilization effects [7, 8].

References

[1] Normand and C. Cornaggia 1993 *Laser Physics* **3, 3** 664 and references therein.

[2] C. Cornaggia, J. Lavancier, D. Normand, J. Morellec, P. Agostini, J.P. Chambaret and A. Antonetti 1991 *Phys. Rev. A* **44, 7** 4499.

[3] D.T. Strickland, Y. Beaudoin, P. Dietrich and P.B. Corkum 1992 *Phys. Rev. Lett.* **68, 18** 2755.

[4] C. Cornaggia, J. Lavancier, D. Normand, J. Morellec and H.X. Liu 1990 *Phys. Rev. A* **42, 9** 5464.

[5] L.J. Frasinski, K. Codling and P.A. Hatherly 1989a *Science* **246** 1029.

[6] C. Cornaggia, D. Normand and J. Morellec 1992 *J. Phys. B: At. Mol. Opt. Phys.* **25** 415

[7] A. Giusti-Suzor and F.H. Mies 1992 *Phys. Rev. Lett.* **68, 26** 3869.

[8] A.Zavriyev, P.H. Bucksbaum, J. Squier and F. Saline 1993 *Phys. Rev. Lett* **70, 8** 1077.

Programmable Wave-Function Engineering in Cs Rydberg States

P.H. Bucksbaum[1], *D.W. Schumacher*[1], *D. Pinkos*[1], *J.H. Hoogenraad*[2], *J.L. Krause*[3], *and K.R. Wilson*[3]

[1]Physics Department, University of Michigan, Ann Arbor, MI 48109, USA
[2]FOM-Institute for Atomic and Molecular Physics, Amsterdam, The Netherlands
[3]Department of Chemistry, University of California, San Diego, La Jolla, CA 92093, USA

Abstract. We use amplified programmable shaped optical pulses to engineer Rydberg wavefunctions. The method employs a computer-controlled liquid crystal pulse shaper[1]. The wavefunctions, which are nonstationary states, are monitored by observing temporal interference via the optical Ramsey method[2,3].

A non-stationary atomic state -- i.e. a wavepacket -- can be the starting point for a quantum control experiment. In principle, any arbitrary set of initial conditions for a wavepacket superposition can be obtained by manipulating the light that creates the excited state. We use this idea to control wavefunction formation by controlling the degree to which various states are excited.

We start with 110fs 785 nm pulses containing about 10 nm of bandwidth, from a Ti:Sapphire Kerr lens modelocked laser. We send the light through a zero-dispersion grating-pulse expander. In the middle of the expander, where the light is focused, but dispersed in the horizontal plane, we insert two liquid crystal displays (LCD)[1], each with 128 independent pixels that each produce independent computer-controlled phase delays or polarization shifts. With an external polarizer, the two displays provide both phase and amplitude control over each color in the pulse. The pulses undergo chirped-pulse amplification to 1 mJ. They are characterized using a FROG[4] as well as other more traditional techniques.

The shaped light is split into two identical spatially overlapped pulses with a variable time separation between them using a Michelson interferometer. Each pulse in the pair excites Cesium from a pre-excited 7s state to a sculpted wave function or "wavepacket", i.e. a coherent superposition of np states centered around n=27. The two wavepackets are phase-coherent with respect to each other. The change in their interference with time delay creates an interferometric autocorrelation of the complex scalar wavefunction. We observe this via ramped-field ionization of the final Rydberg populations. In the weak field limit this autocorrelation reveals the spectral amplitude function of the pulse, much as a Michelson autocorrelation of a laser pulse. In strong fields, phase information is also obtained.

In the following example, obtained in the weak field regime, we show how pulse shaping can change the return, collapse, and revival[5] of a Rydberg wavepacket:

Springer Series in Chemical Physics, Vol. 60 **Ultrafast Phenomena IX**
Editors: P. F. Barbara · W. H. Knox · G. A. Mourou · A. H. Zewail

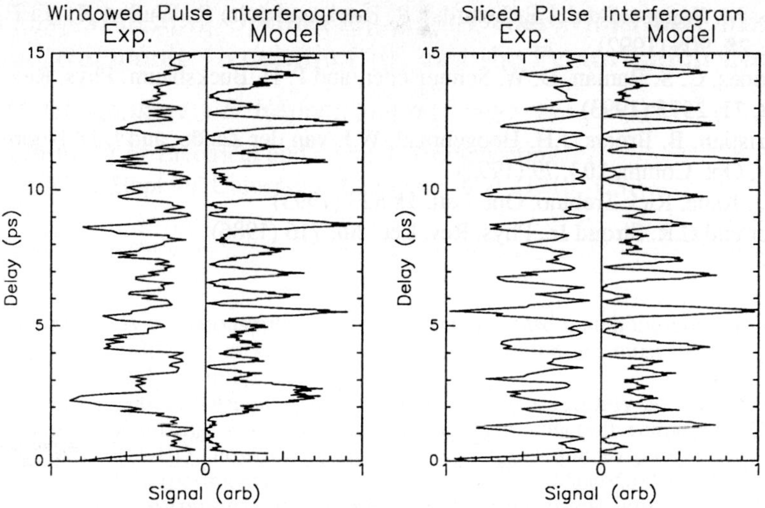

Figure 1 Left: Wavepacket autocorrelation for a "windowed" spectrum in which only Rydberg states 26p through 30p are present, shown mirrored with a calculation for comparison. Right: "Sliced" spectrum in which only 26p, 27p, and 29p, are excited. The wavepacket return time is halved.

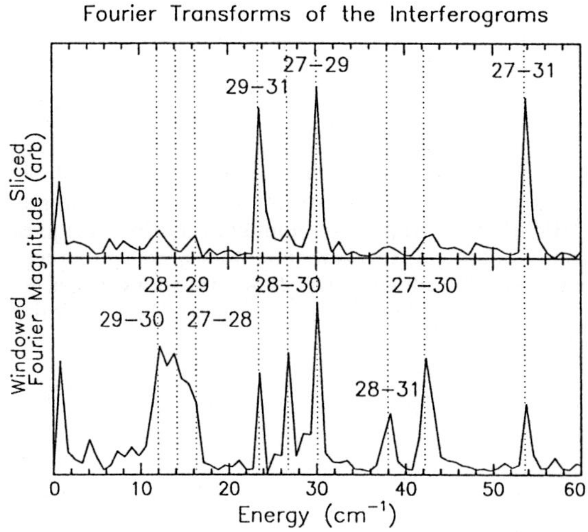

Figure 2. Fourier transforms of the above interferograms. Peaks correspond to the energy differences present in the system. The peak heights are direct measures of the interaction strengths of the various transitions. Various energy differences are labeled.

same for both weak and strong fields. In contrast to Ref. [2], however, we do not presume any particular expression for the amplitude p; for our conclusions to be valid, it is enough if the amplitude p is a positive, monotonic, rapidly increasing function of E_j and, therefore, a rapidly decreasing function of r,u. Consequently, we will neglect contributions to the generated field from the points of a medium that are sufficiently far from the focal point. Then the electric field $\mathcal{E}(r',z')$ generated in a homogeneous medium and observed at a point (r',z') in the far-field area $u'=2z'/b \gg 1$, $u' \gg u$, can be written as

$$\mathcal{E}(r',u',M) \sim \int cos(Mu)\,[F_1 + i F_2]\,du, \quad F_1 \approx \int dr\, f, \quad F_2 \approx \int dr\, 2kr^2 f/bu',$$
$$M = b\Delta k/2 + (m-l), \quad f = rJ_0(2kr'r/bu')\,p(r,u)/u', \tag{3}$$

where J_0 is the ordinary Bessel function of the zeroth order, $\Delta k = k - k'$, $k' = mk_1 - lk_2$, and k is the wavevector of the generated field. For assumed p, $F_{1,2}(u) > 0$ at any u, and since $cos(0) \geq cos(Mu)$ for any u and $M \neq 0$, both the real and the imaginary part of $\mathcal{E}$, and, therefore, the generated field amplitude $|\mathcal{E}|$, are maximal if $M=0$, which corresponds to Eq. (1).

Eq. (2), together with the assumed monotonic behavior of p, are fully justifiable for the pumping pulses being much longer that the relaxation times of the nonlinear media; pumping frequencies being far from any resonances; and the generated field being so weak that the induced polarization is largely determined by the pumping field. In real experiments, however, one can hardly expect all these conditions to hold. Deviation from them would be reflected in p being a complex function with an intensity- (and, therefore, coordinate-) dependent phase Ψ. Nevertheless, if Ψ is a *slow–varying* function of the intensity, one may neglect it. Then, if also $|p|$ decreases with $|E_1|$ rapidly and monotonically, Eq. (1) can still hold. For want of calculations of the single-atom response to strong biharmonic pumping, we have tried to verify these assumptions regarding Ψ and $|p|$ only for HHG. In all the three models of HHG [3] that have been most successfully compared with experimental data, Ψ *is* a relatively slow-varying function of intensity. At the same time, for most values of the pumping intensity, and on average, the amplitude of the dipole moment *is* a monotonic and rapidly increasing function of the pumping intensity. Moreover, in all the experiments on HHG and HDM, the pumping field is in a form of a pulse, so that its intensity at any given point changes in time. Obviously, for most of the pulse duration, we should expect p to be smaller at the points in space where the pumping intensity is lower. If our general assumptions hold also for HDM, the optimal phase-matching conditions, Eq. (1), would present a quite accurate approximation for nonperturbative HDM as well.

In conclusion, we have demonstrated that optimal phase-matching conditions for high-order harmonic generation and high-order difference-frequency mixing of weak fields remain largely intact beyond the perturbation limit under quite broad assumptions regarding nonlinear polarization. It allows one to use safely perturbative phase-matching conditions, Eq. (1), for both HHG and HDM in strong fields.

References

1 P. L. Shkolnikov, A. E. Kaplan, and A. Lago, Opt. Lett. 18, 1700 (1993).

2 A. L'Huillier, L. A. Lompre, G. Mainfray, and C. Manus, in *Atoms in Intense Laser Field*, Ed. M. Gavrila (Acad. Press, Inc., Boston, 1992), p.139.

3 W. Becker, S. Long, and J. K. McIver, Phys. Rev. A 41, 4112 (1990); A. L'Huillier, Ph. Balcou, S. Candel, K. J. Schafer, and K. C. Kulander, Phys. Rev. A 46, 2778 (1992); A. E. Kaplan and P. L. Shkolnikov, Phys. Rev. A 49, 1275 (1994).

A 3-D Numerical Code for the Propagation of Optical Pulses in Nonlinear Media

M.A. Franco, B.S. Prade, E.T.J. Nibbering, J.-Y. Vinet, and A. Mysyrowicz

Laboratoire d'Optique Appliquée, Centre National de la Recherche Scientifique URA 1406, ENSTA, Ecole Polytechnique, F-91120 Palaiseau, France

Abstract. A three-dimensional code for high-intensity nonlinear propagation through transparent media is presented in conjunction with experimental results. Methods for pulse phase retrieval and material n_2-measurements are presented as applications.

The understanding of the effects occuring during propagation of high-intensity ultrashort laser pulses through nonlinear transparent media is of utmost importance for careful transport of the laser beam to the targets and application of these pulses in experiments such as X-ray generation in solids [1] and high-harmonic generation in gases [2]. In the weak perturbation regime the nonlinear refractive index may cause self-focusing and self-phase modulation of the laser pulses. At higher intensities the onset of filamentation of a laser beam through nonlinear media may be induced, thereby facilitating channeling through laser-induced plasmas [3].

Only very simple situations of propagation of optical pulses through nonlinear media can be studied analytically. Simulation codes are thus indispensable to extend the field of applications. We have developed a numerical code that models the propagation of high-intensity femtosecond pulses through nonlinear media such as glass or air. This code is a fully three-dimensional adaptation of the well-known split-step algorithm used in the analysis of propagation through optical fibers [4]. In a first step linear dispersion and diffraction effects are calculated in the frequency domain over an arbitrarily small propagation distance. The second step consists of reconstructing the temporal pulse shape and calculating the nonlinear effects in the time domain over the same distance. Propagation through various successive media can easily be implemented and the method can encompass the group velocity dispersion to all orders in a natural way. The code has been developed for the axial-symmetric case (in which the linear dispersion and diffraction are calculated using a Hankel transform) as well as for the fully three-dimensional case. It uses the pulse spatial and temporal profile parameters as input. Possible ultrafast phenomena that can be simulated with the code include the description of dispersive effects and several kinds of nonlinearities such as gain, absorption and saturation. Various initial conditions can be introduced such as chirped pulses or sequences of pulses which are coupled inside the medium by nonlinear effects. The effect of different transverse beam shapes can be modeled enabling the study of self-focusing and self-guiding.

We have performed experiments to determine how high-intensity femtosecond laser pulses are influenced by propagation through transparent materials in the weakly nonlinear regime. The pulses were taken from a 100 mJ Ti:Sapphire-system using the chirped-pulse amplification technique in a multipass configuration [5]. Using a parallel beam with large transverse dimensions spectra were recorded after transmission through the nonlinear media. Since distortions of the beam phase plane are expected to occur more pronounced in the temporal domain than in the transverse spatial dimensions (leading to e.g. deterioration of the focus of the beam), we have adopted spectral analysis as the sensitive probe of nonlinear propagation effects [6]. In addition to spectral broadening due to self-phase modulation the appearance of significant fringe-like structures was observed (see Fig. 1). The broadening as well as the oscillatory structure are influenced in an asymmetric fashion with respect to the sign of the chirp.

Nonlinear propagation can be used to probe either the pulse or the material characteristics. With Phase Retrieval by High-Intensity Propagation (PREHIP) the known nonlinearity of a sample makes it possible to determine the phase parameters of the laser pulses. For example the results shown in Fig. 1 allow us to extract the large values for the cubic and quartic phase dispersion terms contained in the pulses delivered by our oscillator-amplifier system. These phase distortion terms are orders of magnitude larger than calculated for the laser set-up using a classical linear ray-tracing program. By changing the compressor alignment from its optimum

Editors: P. F. Barbara · W. H. Knox · G. A. Mourou · A. H. Zewail

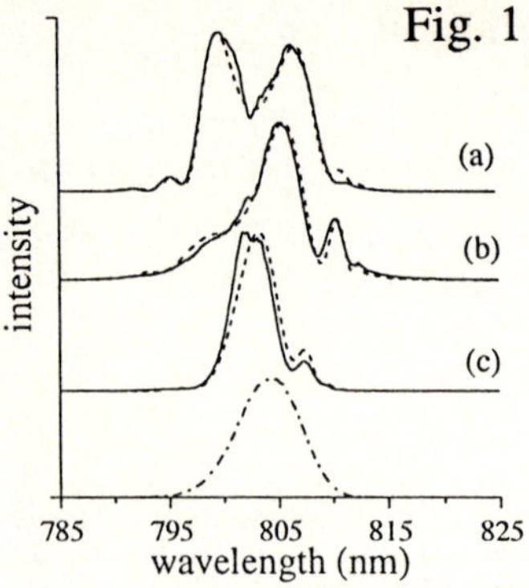

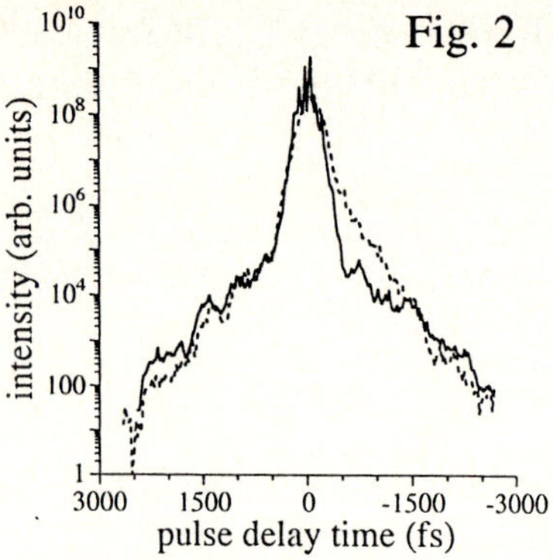

Fig. 1: (Left) Experimental spectra after propagation of 40 mJ 150 fs pulses through 1 cm BK7 (solid lines) are shown together with the dotted fits. The compressor settings correspond to positive (a), approximately zero (b) and negative (c) quadratic phase shift in the pulses, while the large cubic and quartic phase values remain identical. The reference spectrum is depicted as the dot-dashed trace.

Fig. 2: (Right) High-dynamic third-order pulse autocorrelations for the initial (dashed) and final (solid) compressor configurations corresponding to the shortest pulse.

in the absence of gain, it is possible to diminish the third-order phase dispersion term. This results in improved more symmetric pulse temporal profiles as reflected by the high-dynamic third- order autocorrelations (see Fig. 2) as well as the spectra after nonlinear propagation.

With High-Intensity Propagation with High-Order Phase (HIPHOP) on the other hand one imposes large high-order phase terms in the pulse, leading to strong asymmetric features in the nonlinear spectra. One can then easily determine the magnitude and sign of the nonlinear refractive index of a variety of different materials such as gases, liquids and solids by analysis of the propagated spectra. With HIPHOP a direct measure of the nonlinear effect is given by the relative magnitude of the secondary maximum compared to the main peak in the propagated spectrum. In addition the size of the samples can extend to values for which the beam profile does not change significantly, making HIPHOP suitable for measurements on materials with low n_2-values such as gases.

References.

[1]: M.M. Murnane, H.C. Kapteyn, M.D. Rosen and R.W. Falcone, Science **251**, 531 (1991); J.D. Kmetec, C.L. Gordon, III, J.J. Macklin, B.E. Lemoff, G.S. Brown and S.E. Harris, Phys. Rev. Lett. **68**, 1527 (1992).

[2]: A. L'Huillier, Ph. Balcou and L.A. Lompré, Phys. Rev. Lett. **68**, 166 (1992); A. L'Huillier and Ph. Balcou, Phys. Rev. Lett. **70**, 774 (1993); J.J. Macklin, J.D. Kmetec and C.L. Gordon III, Phys. Rev. Lett. **70**, 766 (1993).

[3]: A.B. Borisov, A.V. Borovskiy, V.V. Korobkin, A.M. Prokhorov, O.B. Shiryaev, X.M. Shi, T.S. Luk, A. McPherson, J.C. Solem, K. Boyer and C.K. Rhodes, Phys. Rev. Lett. **68**, 2309 (1992); C.G. Durfee and H.M. Milchberg, Phys. Rev. Lett. **71**, 2409 (1993); A. Sullivan, S. Gordon, H. Hamster, H. Nathel and R.W. Falcone, QELS'93 Technical Digest, 298 (1993).

[4]: R.A. Fisher and W.K. Bischel, J. Appl. Phys. **46**, 4921 (1975); E.T.J. Nibbering, M.A. Franco, B.S. Prade, J.-Y. Vinet, J.-P. Chambaret and A. Mysyrowicz, in preparation.

[5]: C. Le Blanc, G. Grillon, J.P. Chambaret, A. Migus and A. Antonetti, Opt. Lett. **18**, 140 (1993).

[6]: G.P. Agrawal, *Nonlinear Fiber Optics*, AT&T / Academic, New York (1989).

Time-Resolved Thermionic Electron Emission and Non-Thermal Ion Emission from Au Excited by Intense fs Pulses

D.M. Riffe, X.Y. Wang**, Y.-S. Lee, and M.C. Downer*

Physics Department, The University of Texas, Austin, TX 78712, USA
*Present address: Physics Department, Utah State University, Logan, UT 84322-4415, USA
**Present address: Columbia Radiation Laboratory, Columbia University, New York, NY 10027, USA

Abstract. Two-pulse-correlation space-charge-limited thermionic emission is used to monitor the cooling of electrons heated between 3,500 and 12,000 K in polycrystalline Au. The data show that thermal excitation of the Au *5d* electrons significantly alters the electron-phonon-coupling constant *G*, electronic specific heat C_e, and thermal conductivity κ from their RT parameterizations. Above the damage threshold superthermal, electrostatically accelerated H^+ ions are observed.

Electron and ion emission have long been used as probes of the interaction of visible laser pulses with solid-state materials [1]. For excitation by femtosecond pulses most of the focus has been on either the low excitation regime where multiphoton electron emission dominates and surface states play an important role or in the very intense regime where significant x-ray generation occurs. Notably absent are emission studies in the transition region near the damage threshold where peak electron temperatures T_p are ~1 eV, a temperature region where the relaxation dynamics of electrons are important in nanoscale electronic devices, ultrafast desorption of adsorbates, and fs-laser plasma physics [2].

We have recently shown, for heating timescales ≤10 ps, that space-charge-limited thermionic emission can be quantitatively characterized; and further, for T_p above ~0.3 eV, the total electron yield varies with T_p in a linear manner [3]. Here we extend those earlier one-laser-pulse measurements to time-resolved two-pulse measurements in order to directly access the cooling dynamics of electrons heated between 0.4 and 1.0 eV. Data obtained on polycrystalline Au with two (nearly equal intensity) 620 nm, 120 fs pulses are shown in Fig. 1. The measured electron cooling rates are much faster than that predicted with low-temperature characterizations of *G*, C_e, and κ. E.g., at an excitation level of 1.5×10^{12} W/cm^2, corresponding to the top curve in Fig. 1, RT characterizations of these parameters leads to T_p=1.4 eV (vs 0.95 eV measured) and cooling in 4 ps (vs 1.5 ps measured). However, at these high excitation levels the Au *5d* electrons become thermally excited, producing a strong temperature dependence to these parameters which causes the cooling to occur much faster than naively expected. The calculations of the temperature dependencies of *G*, C_e, and κ are described elsewhere [4]. By including this thermal excitation of the *5d* electrons we have successfully modeled the measured cooling rates and peak temperatures as shown by the dashed curves in Fig. 1.

Springer Series in Chemical Physics, Vol. 60 **Ultrafast Phenomena IX**
Editors: P. F. Barbara · W. H. Knox · G. A. Mourou · A. H. Zewail

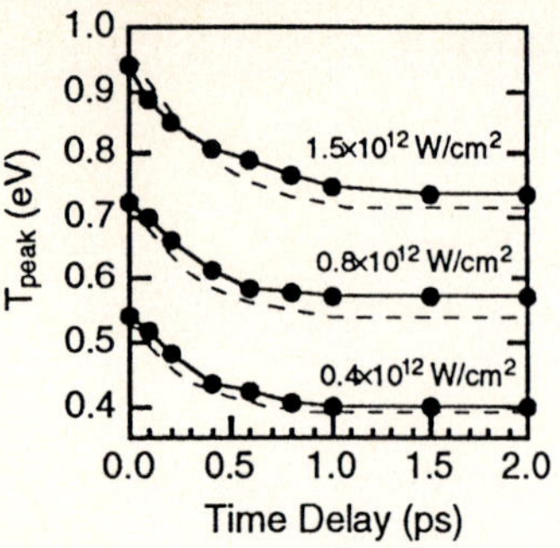

Fig. 1. T_p vs time delay between two (nearly) equal intensity laser pulses at indicated laser intensities. Solid circles—experiment, dashed lines—calculation.

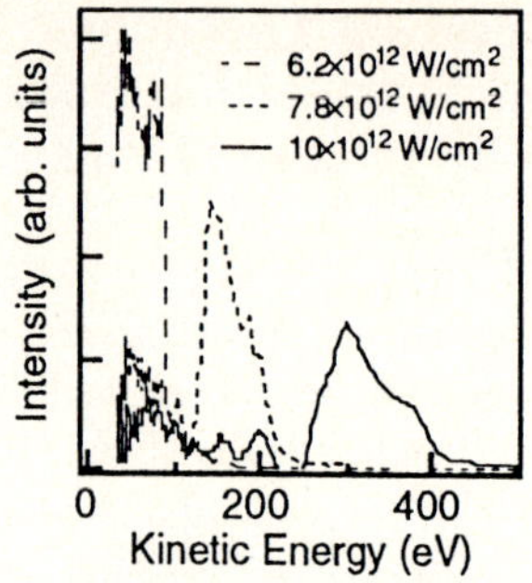

Fig. 2. Energy distributions of H^+ emitted from Au surface as a function of incident laser intensity.

Above the damage threshold we find that ion emission becomes increasingly important and that the total yield increases dramatically (similar to observations with ps pulses [5]). Figure 2 shows kinetic energy spectra of H^+ ions (from adsorbed H) obtained by time-of-flight spectrometry in this high-excitation region. The sharp distributions and large energies shown in Fig. 2 are surprising given the thermionic nature of the electron emission and a $T_p \approx 10$ eV at 10^{13} W/cm^2 laser intensity. However, we propose that it is just these thermionically emitted electrons which provide the force necessary to accelerate the ions to the observed kinetic energies. In the intensity regime studied here the electron density is so high that the Coulombic (space-charge) fields of the faster electrons (and their image charges) significantly suppress the slower electrons from escaping. Just below the damage threshold the suppression is so great that only one in $\sim 10^5$ electrons escape compared to the theoretical yield without space charge. Any surface-adsorbed H atom which is ionized near the surface experiences these same space-charge fields. However, due to its positive charge, it accelerates away from the surface. Preliminary calculations indicate that the force from the thermionic cloud of electrons is indeed sufficient to produce the measured ion energies. The fairly narrow distributions arise because all of the H^+ begin accelerating at nearly the same time from the surface. Application as a proton-injector source is intriguing.

This work supported by the AFOSR, NSF, and Robert A. Welch Foundation.

References

[1] See e.g., J.F. Ready, Phys. Rev. **137**, A620 (1965); E. Bernal, J.F. Ready, and L.P. Levine, IEEE J. Quantum Elect. **2**, 480 (1966).

[2] See, e.g., H. Shichijo and K. Hess, Phys. Rev. B **23**, 4197 (1981); M.M. Murnane, et al., Science **251**, 531 (1991); F. Budde, et al., Phys. Rev. Lett. **66**, 3024 (1991).

[3] D.M. Riffe, X.Y. Wang, M.C. Downer, D.L. Fisher, T. Tajima, and J.L. Erskine, J. Opt. Soc. Am. B **10**, 1424 (1993).

[4] X.Y. Wang, D. M. Riffe, Y.-S. Lee, and M.C. Downer (to be published).

[5] See, e.g., A.M. Malvezzi, H. Kurz, and N. Bloembergen, Appl. Phys. A **36**, 143 (1985).

Phase-Sensitive Characterization of Short-Scale-Length Plasmas

J.-P. Geindre[1], *P. Audebert*[1], *F. Falliès*[1], *A. Rousse*[1], *J.C. Gauthier*[1], *A. Antonetti*[2], *J.P. Chambaret*[2], *G. Grillon*[2], *and A. Mysyrowicz*[2]

[1]Laboratoire pour l'Utilisation des Lasers Intenses, Ecole Polytechnique, F-91128 Palaiseau, France

[2]Laboratoire d'Optique Appliquée, Batterie de l'Yvette, ENSTA, Ecole Polytechnique, F-91120 Palaiseau, France

Abstract: Using phase-controlled twin femtosecond optical pulses, we measure the dynamics of a short scale-length laser-produced plasma with high spatial and temporal resolution.

By irradiation of solid targets with intense femtosecond pulses it is possible to create microplasmas of very small gradient scale length $L/\lambda<0.1$. Useful information on the early evolution of these plasmas on a picosecond or even subpicosecond time scale can be obtained via optical detection. For instance, subpicosecond time-resolved Schlieren measurements can determine the location of the critical density layer of the plasma [1]. However diffraction effects limit the accuracy to a value of the order of the incident wavelength. A method to detect plasma velocities relies on the spectral analysis of the reflected probe beam at different delays [2]. Expansion velocities can be inferred from Doppler shifts. However, the large Fourier spectrum $\Delta\omega\approx1/\Delta t$ of short pulses makes it difficult to estimate frequency shifts much less than $\Delta\omega$. Furthermore, this method is sensitive to the detrimental shot-to-shot frequency and spatial fluctuations of the lasers.

We have implemented a method to determine the plasma front and expansion velocity with greatly improved sensitivity. In the experiments, the moving plasma target is probed with two successive identical femtosecond probe pulses separated in time by an external Michelson interferometer. It is well known [3] that the power spectrum of a double pulse sequence is modulated with a fringe period inversely proportional to the pulse separation. Any phase shift due the motion of the plasma critical layer occuring between the two pulses can be therefore detected directly in the reflected spectrum as a fringe displacement. The improvement of this method over single pulse spectrum analysis is manifold. First, the sharpness of fringes allows high precision measurements of small frequency shifts. Secondly, the inherent multiplexing of the information in all observed fringes for each laser shot further improves the signal to noise ratio. Finally, the method is largely insensitive to shot-to-shot frequency fluctuations. Indeed, besides phase shifts due to the signal, only phase fluctuations occuring inside the external interferometer during the time interval between both probe pulses affect the fringe position.

Two different regimes can be implemented. In the first regime, the plasma-inducing strong pump pulse lies between both probe pulses (figure 1a). This regime is well adapted to the absolute measurement of the location of the reflective plasma layer at a given delay. In the second regime (figure 1b) the pump pulse precedes both probe pulses. This regime yields with high accuracy the relative location of the critical layer at two different times and is therefore well suited for the precise measurement of the plasma axial expansion velocity.

We have used our method to study the expansion of a hot dense laser-induced plasma formed in aluminum.The pump pulse, $\lambda=620$ nm, duration 80 fs, had a peak intensity of 3×10^{15} W/cm^2 and was at normal incidence on the target. The two probe pulses of duration 50 fs and $\lambda=580$ nm were separated by 387 fs. The observed phase shift (left scale) or

Springer Series in Chemical Physics, Vol. 60 **Ultrafast Phenomena IX**
Editors: P. F. Barbara · W. H. Knox · G. A. Mourou · A. H. Zewail

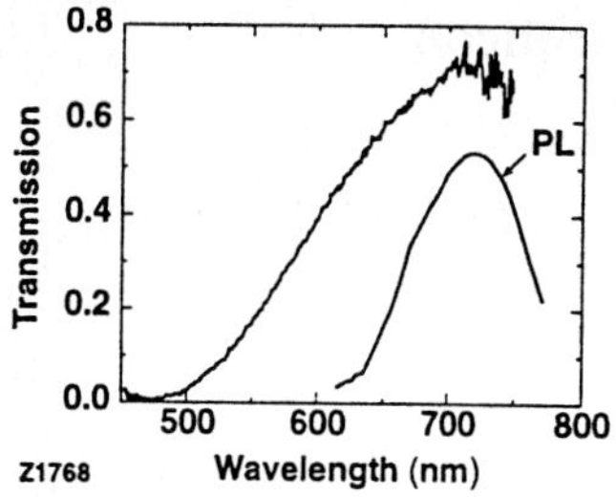

Figure 1
Transmission and photoluminescence spectra of a free standing oxidized mesoporous silicon film measured at room temperature.

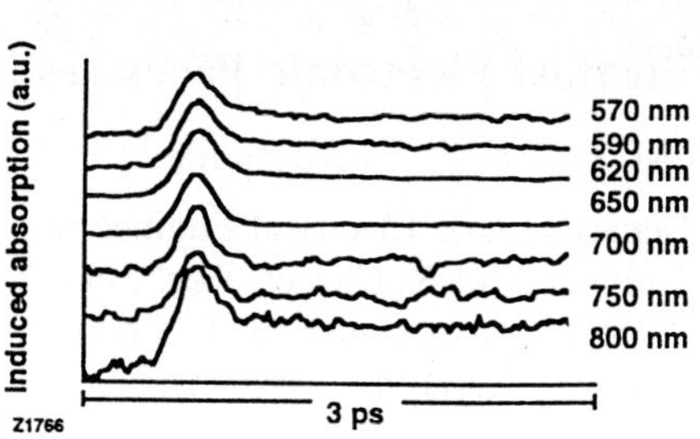

Figure 2
Induced absorption measured at different probe wavelengths after 2-eV excitation.

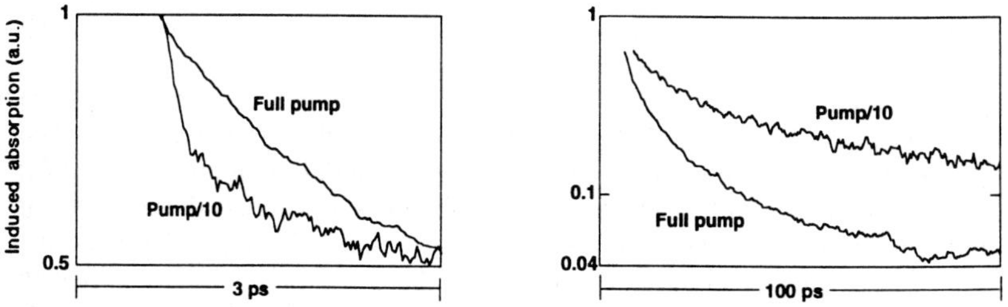

Figure 3:
Recovery of the induced absorption at intermediate pump intensity and high pump intensity. The initial recovery is faster at lower pump intensity while on a longer time scale, the recovery is faster at higher pump intensity. Note that the data are plotted on a semi-logarithmic scale.

within the distribution of surface states and recombination, either with photon emission (luminescence) or without photon emission (Auger processes). We believe that Auger recombination is the most likely candidate, since the slow recovery speeds up at higher pumping level and theoretical estimates of the Auger recombination rate in PSi are in qualitative agreement with our results.

This work was supported in part by grants from Rochester Gas & Electric, the New York State Energy Research and Development Authority, and Xerox.

References

1. L.T. Canham, Appl. Phys. Lett. **57**, 1046 (1990)
2. P.M. Fauchet et al, to appear in *Advanced Photonics Materials for Information Technology*, S. Etemad editor [SPIE Proc. vol 2144, (1994)]
3. P.M. Fauchet et al Ed., Mat. Res. Soc. Symp. Proc. **283** (Pittsburgh, 1993)
4. P.D.J. Calcott et al, J. Luminesc. **57**, 257 (1993)
5. F. Koch et al, in Ref. 3, p 197; J. Luminesc. **57**, 271 (1993)
6. M. Lannoo, private communication.

Terahertz Spectroscopy of Silicon-Micromachined Three-Dimensional Photonic Band-Gap Crystal

E. Özbay[1], *E. Michel*[1], *G. Tuttle*[1], *R. Biswas*[1], *K.M. Ho*[1], *J. Bostak*[2], and *D.M. Bloom*[2]

[1]Microelectronics Research Center and Ames Laboratory, Iowa State University, Ames, IA 50011, USA

[2]Edward L. Ginzton Laboratory, Stanford University, Stanford, CA 94305, USA

Abstract: Using micromachined silicon wafers, we have fabricated three-dimensional photonic bandgap crystals. Terahertz spectroscopy techniques have been used to measure bandgap frequencies over 500 GHz.

In analogy to electrons in a crystal, the propagation of electromagnetic (EM) waves in a three-dimensional dielectric structure can be forbidden for a certain range of frequencies. These three-dimensional structures that are called photonic band gap crystals, have recently received both theoretical and experimental attention[1-2]. The early research in the field has concentrated on possible optical frequency range applications, such as thresholdless semiconductor lasers and single-mode light-emitting diodes. The proposed applications were later extended to the millimeter and sub-millimeter wave regime, like efficient antennas, sources, waveguides and other components[3]. But, the difficulties associated with the fabrication of smaller scale structures along with the unfeasible three-dimensional structure designs, have restricted the experimental demonstration of the basic photonic band gap crystals to microwave frequencies (12-15 GHz).

We have recently designed a new three-dimensional structure that may alleviate some of the problems associated with earlier photonic band gap designs[4]. We have shown the feasibility of structures with 100 GHz photonic band gap frequencies by using semiconductor micromachining techniques[5]. In this paper, we demonstrate the scalability of the new structure to a band gap frequency of 500 GHz, which was measured by terahertz spectroscopy techniques. Such a performance offers a readily available structure for the demonstration of the proposed millimeter and sub-millimeter wave applications.

The new structure, shown in Fig. 1, is constructed of layers of dielectric rods. The stacking sequence repeats every four layers, corresponding to a single unit cell in the stacking direction. In order to build the 500 GHz crystal, we have used fabrication techniques similar to the one used to build 100 GHz crystals[6]. Fabrication consisted of defining stripes that were parallel to (111) planes and

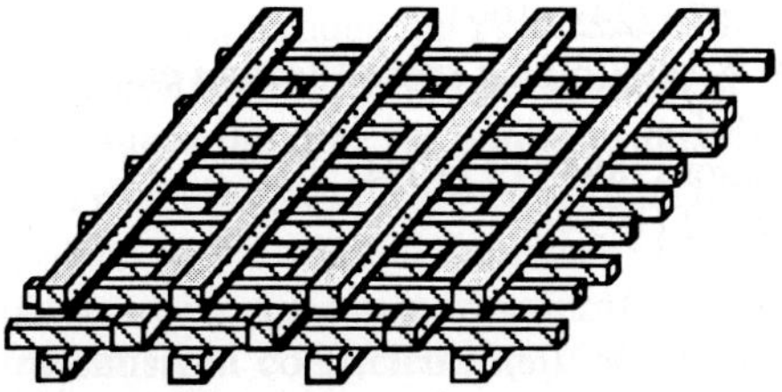

Figure 1. This figure schematically illustrates the design of the three-dimensional photonic bandgap crystal. The structure is built by an orderly stacking of dielectric rods and repeats every four layers in the stacking direction.

Springer Series in Chemical Physics, Vol. 60 **Ultrafast Phenomena IX**
Editors: P. F. Barbara · W. H. Knox · G. A. Mourou · A. H. Zewail

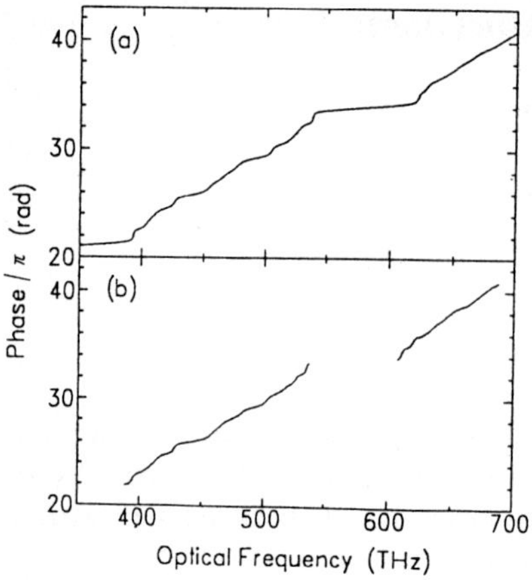

Fig. 1. The phase change of the light transmitted through the Fibonacci lattice obtained by (a) theoretical calculation, and (b) the Fourier-transform interferometry.

$\mathscr{F}\{C_C(t)\}/\mathscr{F}\{C_A(t)\} = |t(\omega)| \exp[i\phi(\omega)]$. Here, $C_A(\tau)$ and $C_C(\tau)$ are the autocorrelation and the cross-correlation, with τ being the delay time, and $\mathscr{F}\{\ \}$ denotes the Fourier transform of the function in the bracket. The transmissivity of light through the sample is the squared magnitude, $|t(\omega)|^2$, and the spectrum of the phase of the light is given by $\phi(\omega)$.

In Fig. 1, the spectrum of the phase of the transmitted light obtained by the Fourier transformation is shown, along with the theoretical curve. These spectrum can be regarded as the dispersion relation of photons in the finite-length Fibonacci lattice, since the phase, ϕ, is related to the wave vector, k, and the sample thickness, L, by $\phi = kL$. The obtained phase spectrum shows very good agreement with the calculated one, and clearly shows the self-similar structure of the band structure typical of Fibonacci lattices. Besides the positions of the two large gaps, smaller gaps are also clearly observed at the right positions, and the overall shape of the spectrum agrees very well with the calculated phase spectrum. Thus we could actually observe the photonic band structure in a one-dimensional quasiperiodic lattice experimentally.

The Fourier-transform interferometric technique presented here can be applicable to various time- and frequency-domain studies of coherent propagation of light in dispersive media [3]. The technique can also be utilized to the simultaneous measurement of the real and the imaginary parts of the refractive index of optical materials.

[1] T. Hattori, N. Tsurumachi, S. Kawato, and H. Nakatsuka, to be published.
[2] M. Kohmoto, B. Sutherland, and K. Iguchi, Phys.Rev.Lett. **58**, 2436 (1987).
[3] N. Tsurumachi, T. Fuji, S. Kawato, T. Hattori, and H. Nakatsuka, to be published.

Microscopic Study of Ferroelectrics with Ultrashort Phonon Polaritons

H.J. Bakker, S. Hunsche, and H. Kurz

Institut für Halbleitertechnik II, Sommerfeldstrasse 24,
D-52074 Aachen, Germany

Abstract : Impulsively generated phonon polaritons are used to study the THz dielectric response of $LiTaO_3$ and $LiNbO_3$. The extreme anharmonicity of the lattice vibrations results in low-frequency resonances that lead to the observation of phonon-polariton beats.

The low-frequency dielectric response of the ferroelectrics $LiTaO_3$ and $LiNbO_3$ is dominated by the lowest-energy lattice vibration of A_1 symmetry. This so-called ferroelectric mode leads to a strong absorption peak at 6 THz (200 cm^{-1}) in $LiTaO_3$ and at 7.5 THz (250 cm^{-1}) in $LiNbO_3$. We investigate the dielectric response of this mode via the excitation of coherent phonon polaritons with ultrashort laser pulses.

The phonon-polaritons are excited and probed in a three-beam pump-probe setup [1] using femtosecond laser pulses (τ_p = 60 fs, λ = 625 nm, E = 5 μJ). Two of the three pulses serve as excitation pulses and are focussed into the sample under a chosen angle to a common focus with a diameter of 200 μm. This angle defines the wavevector of the polariton. The polariton frequency that is phase matched for this wavevector is obtained out of the broad bandwidth of the laser pulses. The polariton dispersion is obtained by varying the angle between the pump beams. Due to the electro-optic effect, the polaritons are associated with a refractive-index grating from which the time-delayed third pulse (probe) is diffracted. Interference of the diffracted light with background scattered light leads to a signal that oscillates with the frequency of the polaritons. For both $LiTaO_3$ and $LiNbO_3$ we find that at most wave vectors the time-resolved signal of the phonon polaritons is formed by a regular oscillation. However, for $LiNbO_3$ we observe strong beating phenomena (polariton beats) for wave vectors near 1500 cm^{-1}, 3000 cm^{-1} and 4000 cm^{-1} (Fig. 1). These polariton beats indicate that in these wavevector regimens two or even three different polariton frequencies are coherently excited. This observation results from the presence of low-frequency resonances that lead to avoided crossings in the polariton dispersion. Near these avoided crossings, polaritons of the lower and upper branch of the dispersion can be excited with comparable efficiency, so that beating is observed. We find four avoided crossings in the polariton dispersion of $LiNbO_3$ (Fig. 2) due to resonances at 1.3 THz, 2.4 THz, 3.4, and 4.1 THz. For $LiTaO_3$, we find also observe polariton beats due to a weak resonance at 1 THz. None of these resonances have been found previously. The observed resonances can not result from transitions between anharmonically-coupled normal lattice vibrations, because the observed frequencies do not fit the energy differences of the optical phonons [2]. Hence, we conclude that the resonances must result from the strong anharmonicity of the ferroelectric mode *itself*.

The measured polariton dispersion can be used for an accurate fit of the shape of the microscopic potential of the ferroelectric mode. We find that this potential is extremely anharmonic and consists of three wells. The fundamental phonon frequency corresponds to transitions between vibrational states that are localized in the lowest well of the potential. The observed low-frequency resonances can all be assigned to transitions between excited vibrational states that are delocalized over the lowest and the middle well. The re-

Springer Series in Chemical Physics, Vol. 60 **Ultrafast Phenomena IX**
Editors: P. F. Barbara · W. H. Knox · G. A. Mourou · A. H. Zewail

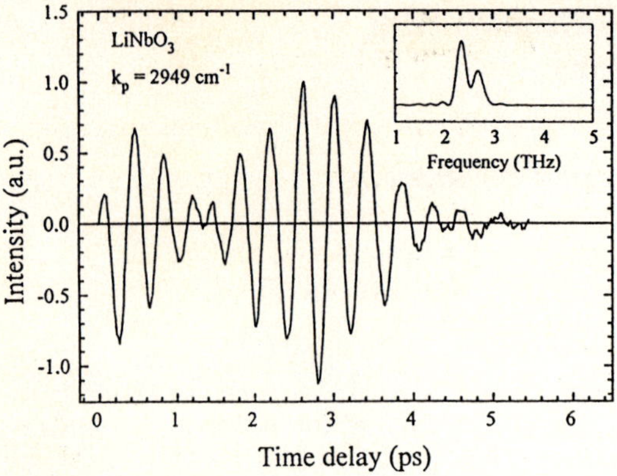

Figure 1: Time-resolved measurement of phonon-polariton beats in $LiNbO_3$. In the inset the Fourier transform of the signal is shown.

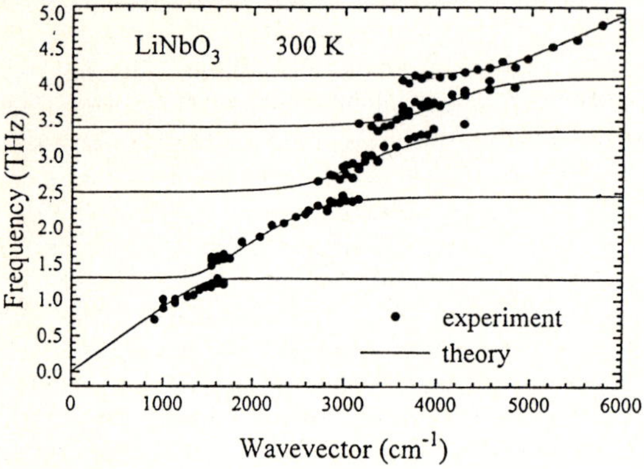

Figure 2: Phonon-polariton dispersion of $LiNbO_3$ at 300 K.

sonances at 1.3 and 4.1 THz in $LiNbO_3$ and at 1 THz in $LiTaO_3$ are tunneling resonances and correspond to the classically forbidden oscillation of the Li^+ ion through the barrier that separates the lowest and the middle well of the potential. We find that the triple-well potential also provides an accurate description of the ferroelectric phase transition and the temperature dependence of the dielectric response.

[1] T. P. Dougherty, G. P. Wiederrecht, and K. A. Nelson, J. Opt. Soc. Am. B **9**, 2179 (1992); H. J. Bakker, S. Hunsche, and H. Kurz, Phys. Rev. B **48**, 13524 (1993)
[2] A. S. Barker, Jr. and R. Loudon, Phys. Rev. **158**, 433 (1967)

Adsorbate Infrared Spectral Response Following Femtosecond Metal-Substrate Heating

A.G. Yodh[1], *J.P. Culver*[1], *M. Li*[2], *L.G. Jahn*[1,2], *and R.M. Hochstrasser*[2]

[1]Department of Physics and Laboratory for Research on the Structure of Matter, University of Pennsylvania, Philadelphia, PA 19104, USA
[2]Department of Chemistry and Laboratory for Research on the Structure of Matter, University of Pennsylvania, Philadelphia, PA 19104, USA

Ultrafast laser heating pulses create unique environments for surface chemistry on metal surfaces. Recent femtosecond experiments have shown, for example, that nonequilibrium substrate conditions can play a central role in laser induced desorption[1]. In these experiments extremely high electronic temperatures (4000 °K) are produced which modify adsorbates in a manner which is impossible to achieve via conventional thermal processes. Here we report the experiments probing the internal vibrations of CO adsorbate molecules as a function of time following the impulsive excitation of electrons in the underlying Cu(111) substrate by 300 fs visible light pulses[2]. We find that a low frequency CO mode couples to substrate electrons and phonons, and that the representative coupling rates can be separately determined[2,3]. Importantly, with increased excitation of the substrate, our measurements provide indication of stronger, temperature dependent couplings between the adsorbate vibrations and the substrate reservoirs[4].

Our experiment uses gated upconversion to follow the time evolution of the CO internal stretch vibration ($\omega_o = 2073.5\ cm^{-1}$, $\Gamma_o = 4.5\ cm^{-1}$) after femtosecond pulsed heating of the substrate. The metal substrate excitation is described by electron and lattice temperatures, T_e and T_l, having distinctly different temporal profiles (Fig. 1). Although the stretch mode is nominally populated through interaction with these substrate reservoirs, the dominant spectral change is a result of a temperature dependent shift of the oscillator complex frequency. The stretch mode is anharmonically coupled to the frustrated translation mode of the CO complex. As the frustrated translation mode is heated by the excited substrate, the stretch frequency absorption, broadens, and shifts to lower frequency. By measuring the reflectivity on the high frequency side of the CO stretch absorption at 2075 cm^{-1}, the time evolution of the low frequency mode population can be discerned (Fig. 2).

The change in temperature of the low frequency mode can be predicted using a rate equation with independent couplings to the electron (γ_e) and to the phonon (γ_l) reservoirs.

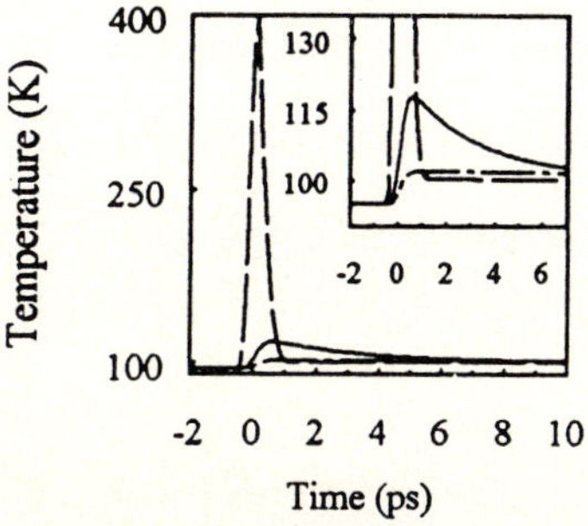

Fig. 1: Calculations of electron and phonon reservoir temperature as a function of time: electronic (dashed line), phonon (dotted dashed line): Adsorbate temperature (solid line) calculated using the best fit values of $\gamma_e = 167(24)\ GHz$ and $\gamma_l = 145(56)\ GHz$. Inset: same with expanded scale.

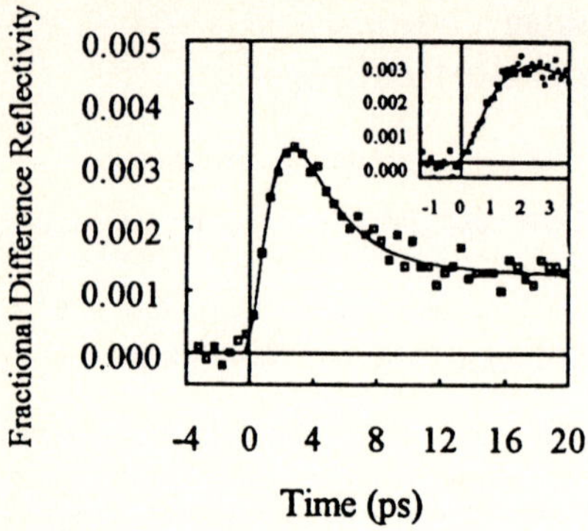

Fig. 2: $F_{abs} = 0.09\ mJ/cm^2$: Data (open squares), a best fit (solid line) using the model and values described in the text. Inset: rise of the signal.

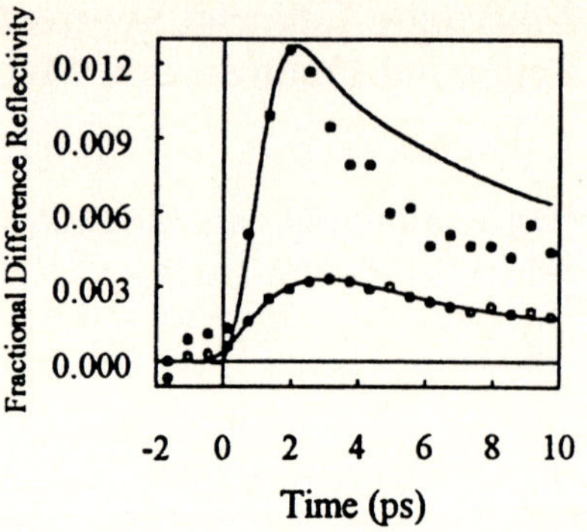

Fig. 3: $F_{abs} = 0.77\ mJ/cm^2$ (solid circles): $F_{abs} = 0.09\ mJ/cm^2$ (open circles). The lines represent fits using $\gamma_e = 167$ GHz and $\gamma_l = 145$ GHz.

The time evolution of the spectra was calculated using temperature dependent FTIR spectra to predict the corresponding changes in the stretch mode complex frequency. A best fit gives coupling rates $\gamma_e = 167(24)\ GHz$ and $\gamma_e = 145(56)\ GHz$. Couplings to both reservoirs are revealed to be necessary and as a result of relatively large differences in their respective temporal profiles, their respective contributions can be distinguished .

Experiments with eight times greater excitation of the substrate (Fig. 3) exhibit deviations from this model. As the molecule approaches excitation levels near the desorption threshold there are several mechanisms by which energy may be more quickly exchanged with the substrate. These new mechanisms, which include other anharmonic couplings between the adsorbate-like modes, temperature dependencies of γ_e and γ_l, and combined electron-lattice effects, have been discussed in recent molecular dynamics calculations for CO on Cu(100)[4]. Preliminary analysis of the data in Fig. 3 suggests that the deviations maybe indicative of such temperature dependent adsorbate-substrate coupling rates.

References

[1] R. R. Cavanaugh, D. S. King, J. C. Stephenson and T. F. Heinz, J. Phys. Chem. **97**, 786 (1993) and references therein.

[2] J. P. Culver, M. Li, L. G. Jahn, R.M. Hochstrasser, and A.G. Yodh, Chem. Phys. Lett. **214**, 431 (1993).

[3] T. A. Germer, J. C. Stephenson, E. J. Heilweil and R. R. Cavanagh, Phys. Rev. Lett. **71**, 3327 (1993).

[4] J. C. Tully, M. Gomez and M. Head-Gordon, J. Vac. Sci. Technol. **A11**, 1914 (1993).

Nonlinear Vibrational Spectroscopy of Liquids and Liquid/Air Interfaces with Ultrashort Laser Pulses

K. Wolfrum and A. Lauberau

Physik-Department der Technischen Universität München,
James-Franck-Strasse, D-85748 Garching, Germany

The availability of suitable fs-pulses has stimulated the study of molecular dynamics in recent years. An important prerequisite of our experimental investigations is parametric pulse generation. We use a pulsed operation mode of the optical parametric oscillator that demonstrated its superior properties for diffraction- and Fourier transform-limited pulses with wide tuning ranges.[1] Compared to cw OPO's the synchronous pumping by microsecond pulse trains allows to work with large output coupling of typically 50 % at a higher peak power level, notably insensitive to the frequency dependence of the parametric gain. Starting with 500 fs-pump pulses of a flash-lamp-pumped Nd:glass laser, bandwidth-limited pulses of 120 fs and 20 nJ were generated in the tuning range 900 - 1,500 nm with a single set of cavity mirrors.[2]

In addition, the simultaneous operation of a second fs-OPO was demonstrated, yielding two independently tunable, synchronized IR pulses. Cross correlation measurements indicate a jitter of the two OPO pulses of 50 fs and a time resolution of 75 fs for pump-probe experiments.[2]

Femtosecond 3-Colour CARS with Magic Polarization

Although time-resolved CARS and spontaneous Raman spectroscopy are interrelated by a one-to-one correspondence, different information is supplied in actual measurements. The superiority of the time domain technique could be demonstrated recently because of: (i) a remarkable improvement of experimental accuracy [3], and (ii) special polarization geometries providing novel information on vibrational and rotational dynamics.[4]

Different contributions to the third order nonlinear susceptibility $\chi^{(3)}$ were recognized in previous work but not yet exploited for time domain spectroscopy. For liquids and far off electronic resonances three components of $\chi^{(3)}$ can be distinguished with different temporal and tensor character: a nonresonant part, and the isotropic and anisotropic contributions of vibrational resonances.[4] Our novel polarization geometries to suppress individual parts are depicted in the top of Fig. 1. Parallel polarization of the excitation fields E_L, E_S is used whith the probe field E_P rotated by 60° and an analysing polarizer in the scattered beam. By polarization interference now one of the three scattering contributions is alternatively eliminated for the analyser oriented at the magic angles θ_A = -60°, -30° and 49.1°, respectively.[4]

A spectroscopic application is presented in Fig. 1. The symmetric CH_3-stretching mode at 2913 cm^{-1} of neat dimethyl-

Editors: P. F. Barbara · W. H. Knox · G. A. Mourou · A. H. Zewail

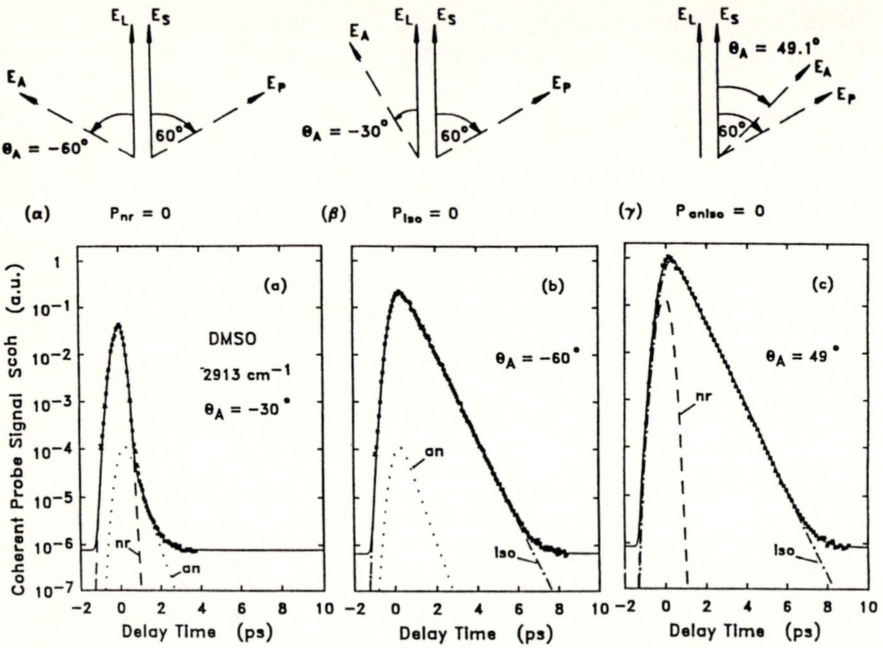

Fig. 1: Femtosecond 3-Colour CARS with magic polarization; top: polarization geometries suppressing the nonresonant (α), isotropic (β) and anisotropic (γ) scattering component, respectively; bottom: signal transients of the symmetric CH_3-mode of DMSO (295 K) for the magic polarization angles (a-c) yielding the three scattering components; experimental points, calculated curves. Time resolution 80 fs.

sulphoxide is investigated using the three special polarizations. The measured anti-Stokes scattering signal is plotted on a semi-log scale versus delay time. The three scattering contributions with different time evolutions can be uniquely determined in Figs. 1 a-c. The nonresonant signal transient is also available from other measurements. The measured time constants are the vibrational dephasing time T_2 = 982 ± 10 fs for the exponential decrease of the isotropic scattering and the decay time of the anisotropic component, τ_{an} = 654 ± 25 fs, the dynamics of the latter being accelerated by the reorientational motion.[5]

It is important to note that the magic angles allow to unravel the isotropic scattering part for small time delay. In this way the non-exponential relaxation dynamics at early times can be investigated. This behaviour is represented by the measured correlation time τ_c = 240 ± 50 fs, describing the delayed onset of the asymptotic dephasing rate $1/T_2$.[6]

Vibrational Sum-Frequency Spectroscopy at Interfaces

The method exploits single-frequency resonances of the second order nonlinear susceptibility $\chi^{(2)}$ using tunable infrared

(ω^{IR}) and visible pulses (ω^{vis}) for the generation of the radiated field $E(\omega^{SF} = \omega^{vis} + \omega^{IR})$.[7] Due to the symmetry breaking at the boundary of centrosymmetric media where the bulk-$\chi^{(2)}$ vanishes, only the molecules at the surface contribute (electric dipole approximation). The SF-signal observed with ultrashort pulses is the spatial and temporal integral over $|E(\omega^{SF})|^2$. The tensor elements of $\chi^{(2)}$ can be expressed as a superposition of a nonresonant part and the individual contributions of different vibrational modes ν with transition frequencies $\omega^{(\nu)}$ and amplitudes $A^{(\nu)}$.

The individual line amplitude depends on the electric transition dipole moment in the IR and the Raman polarizability, involving also an orientational average in the transformation from molecular coordinates to the laboratory frame. Analysing the experimental spectra taken for different polarization conditions for the coupled fields (parallel, p, or perpendicular, s, with respect to the plane of incidence) information on the molecular orientation can be obtained from the measured ratio of the line amplitudes [7,8].

Depending on the sign of the Raman polarizability and the orientational average, the $A^{(\nu)}$'s can be positive or negative [9]. As a result, destructive interference in the SF-signal may occur for opposite phases of the radiated field components of adjacent modes.

Destructive interference of adjacent lines. The long-chain molecule hexadecanol $CH_3(CH_2)_{15}OH$ is known for forming highly stable and ordered monolayers at the water:air interface. The measured SF-spectrum for ssp-polarization (i.e. s-SF, s-VIS, and p-IR) is depicted in Fig. 2.[9] Bandwidth-limited IR pulses of 3 ps and 5 cm^{-1} are used.[8]

The maximum at 2875 cm^{-1} is attributed to the symmetric CH_3-stretching vibration whereas the resonance at 2936 cm^{-1} is assigned as an overtone of the methyl bending vibration intensified by Fermi resonance with the methyl stretching mode. At 2958 cm^{-1}, where the degenerate CH_3-stretching vibration is expected, a distinct minimum is observed. A theoretical analysis shows that the amplitudes of the adjacent lines at 2936 cm^{-1} and 2958 cm^{-1} have opposite signs.

In Figs. 2 b, c the imaginary and real parts of the relevant tensor element $\chi^{(2)}_{yyz}$ for the resonances at 2875 cm^{-1}, 2936 cm^{-1}, and 2958 cm^{-1} are plotted (dashed, dash-dotted and dotted lines, respectively). It is readily seen that at 2958 cm^{-1} destructive interference between the symmetric overtone and the degenerate stretching vibration occurs, leading to a corresponding minimum in the SF-signal at this frequency position. The calculated spectrum for destructive interference is plotted in Fig. 2 a as a solid line. The dashed line also shown for comparison refers to the assumption of all $A^{(\nu)}$'s being positive. It is obvious that only the negative amplitude of the vibration at 2958 cm^{-1} accounts for the experimental data.

From the measured ratio of the amplitudes for ssp- and sps-polarization the orientational ordering in the monolayer is determined. The tilt angle of the molecular chain with respect to the surface normal is estimated to be 8.2 ± 1.8^{o} [9].

Chemical equilibrium at the interface. For the properties of surface molecules with ionizable head groups, the pH-value of the aqueous subphase plays an important role [10]. We have studied the SF-spectra of an octadecylamin monolayer (ODA,

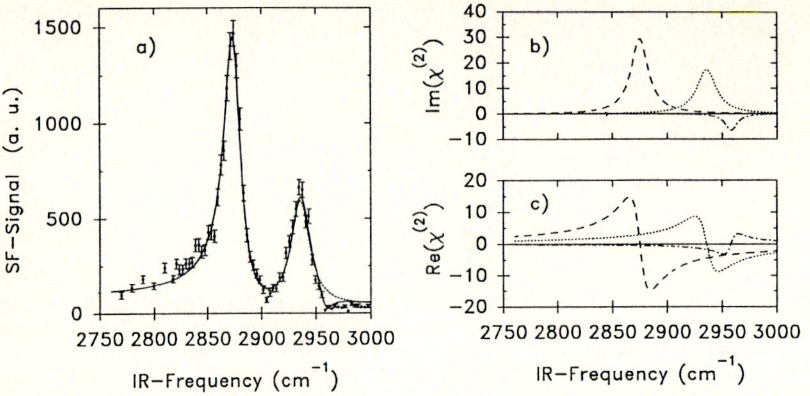

Fig. 2: a) SF-spectrum of a monolayer of hexadecanol on water (295 K) for ssp-polarization; experimental points and calculated signal curves $\propto |\chi^{(2)}_{yyz}|^2$; b), c) imaginary and real parts, respectively, of the individual lines contributing to the nonlinear susceptibility $\chi^{(2)}_{yyz}$: symmetric stretch (dashed line), symmetric overtone (dotted), and degenerate stretch with opposite phase (dash-dotted).

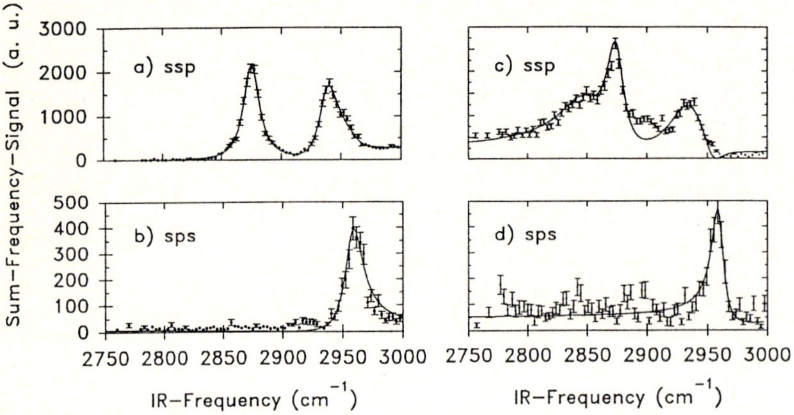

Fig. 3: SF-spectra of a monolayer of ODA on water (295 K) for bulk- pH = 13 (a, b; 0.1 m NaOH) and bulk- pH = 1 (c, d; 0.1 m HCl); experimental points, calculated solid curves $\propto |\chi^{(2)}_{yyz}|^2$ (a, c; ssp), and $\propto |\chi^{(2)}_{yzy}|^2$ (b, d; sps).

$CH_3(CH_2)_{17}NH_2$) in the bulk-pH range of pH = 1 to pH = 13.[11] In Figs. 3 c, d the SF-spectra for pH = 13 are shown for ssp- and sps-polarization. The major difference to the spectrum of hexadecanol (Fig. 2) is a weak, broad band around 3000 cm^{-1} assigned to NH_2 vibrations .

The spectra for pH_1 = 1 are depicted in Figs. 3 c and d. A shoulder at 2850 cm^{-1} appears, attributed to the symmetric CH_2-stretching vibration, which shows up when the chain symmetry is broken by trans-gauche transformations. Depending on the acid-base equilibrium the NH_2-group is neutral at high pH values and becomes ionized at low pH. On an acid subphase, therefore, electrostatic repulsion of the ionized head groups leads to a less closely packed film with structural changes as compared to an alkaline subphase with neutral headgroups and higher degree of ordering.

The normalized line amplitude of the symmetric CH_3-overtone may be used to monitor the concentration of neutral head groups. For bulk-pH > 11 the spectra are within experimental accuracy identical to Figs. 3 a and b indicating neutral head groups. The line amplitude notably decreases for bulk-pH < 10 providing evidence for ionized amino groups. From these observations the surface-pK value (chemical equilibrium constant) of ODA may be estimated [11].

References

1. R. Laenen, H. Graener and A. Laubereau, Optics Lett. 15 971 (1990); J. Opt. Soc.Am. B 8, 1085 (1991); K. Wolfrum, R. Laenen and A. Laubereau, Opt. Commun. 97 41 (1993).
2. R. Laenen and A. Laubereau, to be published.
3. M. Fickenscher, H.-G. Purucker and A. Laubereau, Chem. Phys. Lett. 191, 182 (1992).
4. W. Li, H.-G. Purucker and A. Laubereau, Opt. Commun. 94, 300 (1992).
5. H.-G. Purucker, V. Tunkin and A. Laubereau, J. Raman Spectrosc. 24, 453 (1993).
6. F. Lindenberger, C. Rauscher and A. Laubereau, to be published.
7. X. D. Zhu, H. Suhr and Y. R. Shen, Phys. Rev. B35, 3047 (1987); P. Guyot-Sionnest, J. H. Hunt and Y. R. Shen, Phys. Rev. Lett. 59, 1597 (1987).
8. K. Wolfrum, H. Graener, and A. Laubereau, Chem. Phys. Lett. 213, 41 (1993).
9. K. Wolfrum and A. Laubereau, Chem. Phys. Lett. in press.
10. G. L. Gaines jr., "Insoluble Monolayers at Liquid-Gas Interfaces" (John Wiley & Sons, New York 1966).
11. K. Wolfrum, J. Löbau and A. Laubereau, to be published.

Ultrafast Time-Resolved Dynamics of Hot-Electron Distributions in Silicon

*J.A. Prybyla and J.R. Goldman**

AT&T Bell Laboratories, 600 Mountain Avenue, Murray Hill, NJ 07974, USA
*Present address: Division of Applied Physics, Harvard University, Cambridge, MA 02138, USA

Ultrafast optical probing techniques uniquely provide a direct means for performing state-resolved studies of hot electron dynamics in materials. While ultrafast optical methods have been applied extensively to the study of carrier dynamics in GaAs, there have been very few such studies of Si because conventional optical pump-probe schemes are difficult to interpret for indirect gap materials. This is unfortunate since understanding hot electron behavior in Si is key to understanding and predicting the properties of next-generation Si devices. Here we report the first direct subpicosecond (150 fsec) time-resolved measurement of the energy relaxation of optically excited electrons in Si [1]. This is accomplished in a laser pump-probe format by analyzing the *photoelectron* spectrum generated by an ultrashort uv probe pulse. By this method, we obtain *directly* the time–evolution of the photo-excited electron distribution function (bulk and interface) in Si with 150 fsec resolution. These measurements would not be possible by either reflectivity/transmission or luminescence schemes.

We also report a new effect in two-photon photoemission unique to ultrashort laser pulses resulting in a sensitive new monitor of electron dynamics. We show that at high enough probe intensities excited conduction band (cb) electrons from the pump pulse can 'block' the dominant two-photon photoemission from the valence band, leading to an energy-dependent depletion, D(E), in the probe-induced photoemission. A simple model shows this depletion to be approximately proportional to the *fractional* occupation of intermediate (cb or surface) states.

The experiments were carried out using a 2 eV photon energy pump pulse followed by a 4 eV probe on a UHV-clean Si(100) sample. The cross-correlation of the 2 eV pump with the 4 eV probe was 150 fsec. The ultrafast time–resolved photoemission results are shown in Figures 1 and 2. Spectral assignment is identical to that of reference 2 for longer laser pulses. In Figure 1, we show the photoemission spectra N(E) (electron energy distribution) for delay times of 120 and 180 fsec. Conduction band electrons fit a *thermal* distribution by 120 fsec; this gives an upper bound on the electron-electron scattering time of <120 fsec under the conditions of our experiment ($\geq 10^{18}$carrier$/cm^3$). At this delay, a best-fit Boltzmann distribution gives a temperature of T$\sim$1600K. By 180 fsec, the distribution has cooled to $\sim$800K. This cooling rate is extremely fast ($\sim$10x faster than gold) but is broadly consistent with recently calculated electron-phonon scattering rates using a pseudopotentials approach [3]. We are presently in the process of quantitatively extracting the *energy* dependence of the scattering rates from our experiment through use of Monte Carlo modeling.

After 200 fsec, the cooling rate is much slower, as seen in the D(E) spectra of Figure 2. Again, D(E) gives the fractional occupation of the cb or surface states at E_{cb}, and is obtained by taking the difference between the probe and pump-probe

Springer Series in Chemical Physics, Vol. 60 **Ultrafast Phenomena IX**
Editors: P. F. Barbara · W. H. Knox · G. A. Mourou · A. H. Zewail

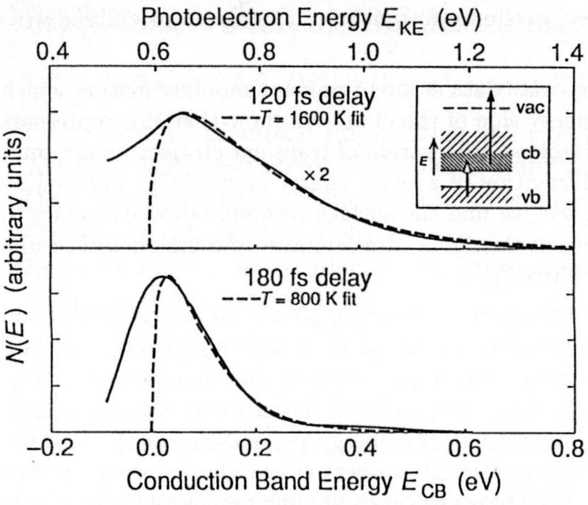

Figure 1. Time-resolved photoemission N(E) vs. photoelectron energy. The data are fit to thermal distributions (dashed lines).

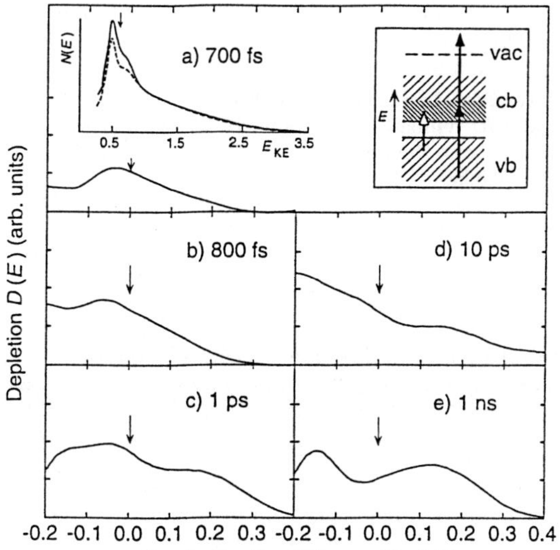

Figure 2. Time-resolved photoemission D(E) vs. photoelectron energy. The inset to (a) shows the probe-only (solid) and pump-probe (dashed) spectra at 700 fsec delay.

photoemission spectra normalized by the probe. This slower cooling rate at longer delay times is consistent with the scattering rate decreasing as electrons move toward lower energy. The D(E) spectra at 700 fsec delay shows a broad distribution of excited electrons extending from gap states to above the cb minimum (cbm). In going from 0.7 psec to 1 psec (Figs. 2(a)-(c)), there is an increase in electron occupation (D(E)) in the cbm region, E_{cb}=0. This indicates that the cb electrons had not yet equilibrated to the (bulk) lattice for these <1 psec delays. By 1 psec, the cb electrons have clearly reached equilibrium with the bulk lattice since beyond this delay relaxation *into* the

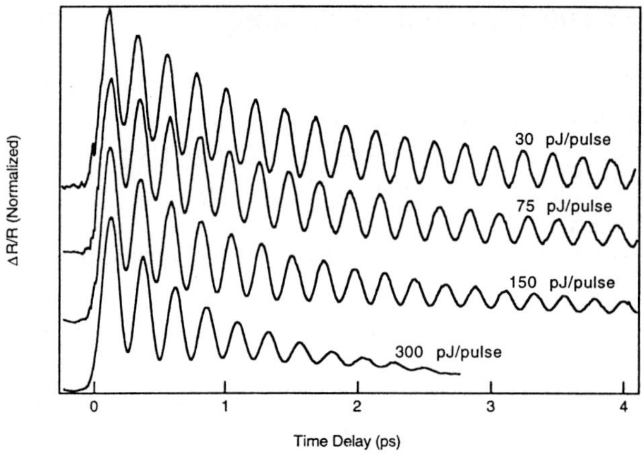

Figure 1: Pump intensity dependence of pump-probe data for single crystal antimony.

by a decay back to equilibrium. Superimposed on this relaxation are large oscillations which are the signatures of the coherent optical phonon. As the pump intensity increases, we observe that the peaks and valleys of the different traces do not coincide. In general, we observe a systematic increase in the phonon period as a function of increasing pump power. We can quantify this increase in oscillation period by piecewise frequency matching 0.5 ps blocks of the pump-probe data in Figure 1. In doing this, one observes that for increasing pump powers, the magnitude of the shift from the Raman Frequency (4.5 THz) increases near t=0 ($\Delta\nu$ = 0.4 THz or $\Delta\nu/\nu$ = 9 %, at 300 pJ/pulse). Furthermore, as the system relaxes back to the pre-pump state, the phonon frequency correspondingly relaxes to the Raman frequency on a picosecond timescale.

The transient shifts in the phonon frequency are directly related to the time-evolving curvatures of the ionic potential in the crystal. One possible mechanism for this evolving ionic potential is a plasma-induced effect caused by the pump-pulse absorption. The plasma screens the bare-ion potential of the solid yielding a "dressed" phonon with a reduced phonon frequency, analogous to the case of acoustic phonons in metals [10]. However, unlike metals, the plasma induced by the pump is highly non-equilibrium and will evolve as a function of time. By monitoring the phonon frequency, we directly measure the relaxation of the carrier-ion screening process.

In most femtosecond studies of carrier dynamics, the induced optical changes are the result of carrier distribution changes that evolve with energy and momentum loss to the lattice. In the present experiment, the induced optical changes are dominated by the lattice and we measure changes in the phonon frequency as affected by the ions' interaction with the excited carriers. In the most direct way possible, we observe electron-phonon interactions as experienced from the point of view of the lattice itself.

This research was sponsored in part by the Joint Services Electronics Program DAAL03-92-C-001 and the Air Force Office of Sponsored Research F49620-91-C-0901.

References

[1]T.K. Cheng, S.D. Brorson, A.S. Kazeroonian, J.S. Moodera, G. Dresselhaus, M.S. Dresselhaus, E.P. Ippen, Appl. Phys. Lett. **57**, 1004 (1990).
[2]T.K. Cheng, J. Vidal, H.J. Zeiger, E.P. Ippen, G. Dresselhaus and M. S. Dresselhaus, Appl. Phys. Lett. **59**, 1923 (1991).

[3]G. C. Cho, W. Kutt and H. Kurz, Phys. Rev. Lett. **65**, 764 (1990).
[4]W.A. Kutt, W. Albrecht and H. Kurz, J. Quant. Elect. **28**, 2434 (1992).
[5]J.M. Chwalek, C. Uher, J.F. Whitaker, G.A. Mourou and J.A. Agostinelli, Appl. Phys. Lett. **58**, 980 (1991).
[6]H.J. Zeiger, J. Vidal, T.K. Cheng, E.P. Ippen, G. Dresselhaus and M.S. Dresselhaus, Phys. Rev. **B45**, 768 (1992).
[7]T.K. Cheng, J. Vidal, H.J. Zeiger, E.P. Ippen, G. Dresselhaus and M.S. Dresselhaus, Appl. Phys. Lett. **62**, 1901 (1993).
[8]See, for example, E. P. Ippen and C. V. Shank, in *Ultrashort Light Pulses*, edited by S. L. Shapiro, (Springer-Verlag, Berlin, 1984), Chap. 3, p. 83.
[9]R. N. Zitter, in *The Physics of Semimetals and Narrow Gap Semiconductors* (D. L. Carter and R. T. Bates eds.), Pergamon Press, New York (1971) p. 285.
[10]For "dressed" acoustic phonons in metals, see D. Pines, *Elementary Excitations in Solids*, (Addison-Wesley, NY, 1963) p241.

Ultrafast Isomerization Dynamics at Interfaces by Time-Resolved Second-Harmonic Generation

E. Borguet, X. Shi, and K.B. Eisenthal

Department of Chemistry, Columbia University, New York, NY 10027, USA

Abstract. Second harmonic generation probes ultrafast dynamics of Malachite Green at air, alkane and silica/aqueous interfaces. Results suggest a new isomerization model, and a water structure near the solid/aqueous interface that results in an increase in friction there.

1. Introduction.

Ultrafast studies of molecular relaxation in the bulk have been of great value in revealing the dynamics of fundamental chemical and physical events. However, our knowledge of these processes at interfaces is still in its infancy. The interface is a unique environment, of interest in its own right, providing a further test of our understanding of elementary molecular processes. Time resolved second harmonic generation, by avoiding the complications of bulk contributions, enables surface species to be probed directly and on an ultrafast timescale [1,2]. Our results on the barrierless isomerization of an organic dye indicate novel and unanticipated behavior: a very large change in relaxation rates, differing by an order of magnitude from those in the bulk and surprisingly similar rates at very different interfaces, namely the air/liquid, liquid/liquid and liquid/solid interfaces.

As a prototype system we have explored the isomerization of Malachite Green at a number of aqueous interfaces; silica/aqueous, pentadecane/aqueous and air/aqueous. In the bulk the ultrafast relaxation dynamics of triphenylmethane dyes, such as Malachite Green, are sensitive to the nature of the molecular environment [3]. The fluorescence quantum yield and excited state relaxation rate show a strong dependence on solvent viscosity. The rate limiting step is believed to involve the rotation of the three aromatic groups about the bond between each one and the central carbon atom. The solvent drag forces opposing these rotations are manifested by the slowing down of the isomerization as the macroscopic bulk solvent viscosity increases.

2. Experimental.

A 10 Hz amplified CPM laser provides 130 fs pulses at 625 nm. The intense laser light is focused at the interface and the second harmonic radiation is separated from the reflected fundamental using filters and a monochromator. Polarizers enable distinct components of the surface nonlinear susceptibility to be probed. The results were independent of energy density in the range studied; 0.1mJ/cm^2 to 1mJ/cm^2.

Springer Series in Chemical Physics, Vol. 60 **Ultrafast Phenomena IX**
Editors: P. F. Barbara · W. H. Knox · G. A. Mourou · A. H. Zewail

3. Results and Discussion.

The dynamics of distinct nonlinear susceptibility elements were examined. Both χ_{zxx} and χ_{xzx}, which depend differently on molecular orientation, show similar dynamic behavior, indicating that molecular rotation is not significant in the timescale investigated. The dynamics of the SHG response were independent of the pump polarization (s or p). The polarization of the second harmonic fields radiated from the various interfaces differ only by about 10^{0}, suggesting similar molecular orientations. The relaxation dynamics are characterized by a single exponential decay (3.5±0.3 ps at the air/aqueous interface, 3.6±0.3 ps at the pentadecane/aqueous interface, and 6.2±0.5 ps at the silica/aqueous interface). These rates are significantly slower than the dominant decay component (0.7 ps) reported for Malachite Green in bulk water [4,5].

The generally accepted model for the photoisomerization of triphenylmethane dyes in bulk solution involves the intramolecular rotations (twisting) of the three aromatic groups about their respective bonds with the central carbon atom. This motion is initiated by the change in the charge distribution of the molecule following excitation to the first excited singlet state. The frictional resistance exerted by the solvent against the twisting of the aromatic moieties in the bulk isomerization dynamics is seen in the viscosity dependence of the kinetics. We expect that the isomerization at the interface will also be sensitive to the friction experienced along the twisting coordinates.

Let us now consider the molecule of interest, Malachite Green, at the air/aqueous interface. It is known from many studies that the hydrophobic part of a molecule, in this case the phenyl group, would be pointed away from the aqueous phase and the more hydrophilic part, in this case the partially charged dimethylaniline groups would be oriented into the aqueous phase [6]. At the air/aqueous and alkane/aqueous interfaces the phenyl group would project into the air and alkane phases, respectively. The twisting motions about the carbon-phenyl group bond are therefore expected to be very sensitive to the frictional properties of the phase into which the phenyl group is projecting. The air vs. the pentadecane phase should offer very different frictional resistances to the twisting of the phenyl group. Contrary to this expectation, we found that the isomerization dynamics were the same at these interfaces having the values of 3.5±0.3 ps and 3.6±0.3 ps at the air/aqueous and pentadecane/aqueous interfaces, respectively. We thus suggest that the photoisomerization dynamics in Malachite Green does not significantly depend on the

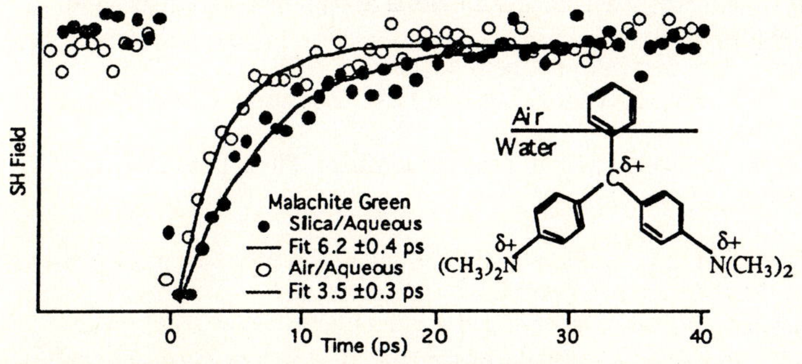

Figure 1. Interface Dynamics of Malachite Green.

rotation of the phenyl group. This is contrary to the generally accepted model which assumes that the twisting of all the aromatic moieties determines the rate of isomerization. Our results suggest that it is the nitrogen containing dimethylaniline moieties that determine the isomerization rate. Since these more hydrophilic moieties project into the aqueous phase, at approximately the same orientations, the twisting motions experience the same aqueous friction, and therefore yield the same isomerization dynamics at these various interfaces. It is of interest to note that at the silica/aqueous interface, where the phenyl group as well as the dimethylaniline groups are immersed in the water phase, the isomerization dynamics are slower than at the air/aqueous and alkane/aqueous interfaces. Although the phenyl group and the dimethylaniline groups are immersed in the aqueous phase, the isomerization dynamics are different (slower) than in the bulk aqueous phase. This result has a bearing on the considerable interest and some experimentally based inferences that the structure of water and its viscosity are different near solid/aqueous interfaces [7, 8]. The SH results reported here provide support for these ideas, which have important implications for aqueous interfaces in many physical, chemical and biological systems.

In summary, we have succeeded in measuring the dynamics of photoisomerization at various aqueous interfaces. The dynamics are not only slower than in the bulk solution but surprisingly are the same at the air/aqueous and alkane/aqueous interfaces, contrary to the current models of isomerization of triphenylmethane dyes. We suggest on the basis of the SH experiments that rotation about the central carbon-phenyl bond is not important in the structure change dynamics. The twisting motions of the dimethylaniline groups control the photoisomerization dynamics. The isomerization dynamics at the silica/aqueous interface are consistent with the idea that the friction and therefore the structure of water in the neighborhood of the solid (silica)/aqueous interface differs from that of bulk water.

References

1. E.V. Sitzmann and K.B. Eisenthal, J. Phys. Chem. **90**, 2831 (1989)
2. A. Castro, E.V. Sitzmann, D. Zhang and K.B. Eisenthal, J. Phys. Chem. **95**, 6752 (1991)
3. D. Ben-Amotz and C.B. Harris, J. Chem. Phys. **87**, 4856 (1987)
4. D.J. Erskine, A.J. Taylor and C.L. Tang , J. Chem. Phys. **80**, 5338 (1984)
5. A. Migus, A. Antonetti, J. Etchepare, D. Hulin and A. Orszag, J. Opt. Soc. Am. B **2**, 584 (1985)

6 K.B. Eisenthal, Ann. Rev. Phys. Chem **43**, 627 (1992)

7 T.S. Filligan, S.B. Zhu, S. Yao, J. Lee and G.W. Robinson, Chem. Phys. Lett. **161**, 444 (1989)

8 M. Yanagimachi, N. Tamai and H. Masuhara, Chem. Phys. Lett. **201**, 115 (1993)

Femtosecond Photoemission Studies of Electron Relaxation at Cu Surfaces

C.A. Schmuttenmaer[1], *R.J.D. Miller*[1], *M. Aeschlimann*[2], *J. Cao*[3], *Y. Gao*[3], *and D.A. Mantell*[4]

[1]Center for Photoinduced Charge Transfer, Department of Chemistry, University of Rochester, Rochester, NY 14627, USA
[2]Laboratory for Technical Chemistry, ETH Zürich, CH-8092 Zürich, Switzerland
[3]Department of Physics and Astronomy, University of Rochester, Rochester, NY 14627, USA
[4]Xerox Webster Research Center, Webster, NY 14850, USA

Abstract. Femtosecond photoemission has been used to fully characterize the non-radiative relaxation of hot electrons, which competes with induced desorption, at single crystal Cu surfaces.

A number of recent studies have illustrated the importance of photoexcited electrons in surface reactions at metal surfaces. The present work focuses on directly determining the relaxation dynamics for photoexcited electrons at Cu(100) surfaces. This particular surface is important as recent observations for photoinduced desorption of CO have found desorption yields as high as 0.1 ML, corresponding to a quantum efficiency of 10^{-3}/absorbed photon, for the hot electron reaction channel at high laser fluences.[1] In order to further understand surface photochemistry, it is essential to have a clear picture of the electron dynamics near metal surfaces. Equal pulse correlation studies employing 3.2 eV pulses (50-60 fs) with both parallel and crossed polarization and low energy (<100 nj/cm^2) have been used to characterize the intrinsic electron relaxation dynamics in the absence of excess lattice heating or transient collective effects. Dynamical information is obtained for the intermediate state, E_i, and the position of this state above the Fermi energy, E_F, is known through the relation: $E_i - E_F = E_{KE} + \Phi - h\nu$, where E_{KE} is the measured photoelectron kinetic energy, Φ is the work function [4.6 eV for Cu(100)], and $h\nu$ is the photon energy, 3.2 eV. These results are shown in Figure 1. The highest energy electrons give a pulse width limited response function which can be used to deduce the slower relaxation dynamics of the lower energy electrons. The slower relaxation dynamics are clearly resolved as a broadening in the FWHM of the two pulse correlation of the photoemitted electron at specific energies. The dyanmics at low energy were also confirmed by a red-blue 2 color experiment. The deconvolved relaxation times for the electron energy distribution are shown, and the solid line through the data is a fit to Fermi liquid theory for electron relaxation scaled by a factor of 2.45. The extent of this agreement for the electron relaxation dynamics at a single crystal surface is similar to the relaxation dynamics observed previously for polycrystalline gold surfaces at higher fluences.[2] Excited electronic states that are closer to the Fermi level are monitored by depositing a small amount of Cs (≈ 0.05

Springer Series in Chemical Physics, Vol. 60 **Ultrafast Phenomena IX**
Editors: P. F. Barbara · W. H. Knox · G. A. Mourou · A. H. Zewail

Real-Time Vibrational Dynamics of K_2

S. Rutz and E. Schreiber

Institut für Experimentalphysik, Freie Universität Berlin,
Arnimallee 14, D-14195 Berlin, Germany

Abstract. Using femtosecond pump&probe technique combined with mass-selective detection the dynamics of the excited potassium dimer's A-state $((1)^1\Sigma_u^+)$ could be investigated. The transient three photon ionization (3PI) spectrum was measured over $185ps$ period of time and exhibits a strong modulation superimposed by a pronounced beat structure. Fourier analysis of the spectrum yields a group of frequencies settled between $63cm^{-1}$ and $69cm^{-1}$ which can clearly be assigned to the vibrational motion of a wave packet in the bound A-state. The vibrational periods could be estimated to be $490fs \ldots 530fs$.

1. Introduction

In recent years diatomic molecules have been studied by multiphoton ionization employing a great variety of experimental and theoretical techniques [1]. The ionization process being in general well understood is predominantly due to resonance enhanced multiphoton ionization (REMPI). However, detailed time-resolved investigations on dynamical features as the laser induced excitation process as well as different energy transfer processes [2] are rarely done, although picosecond and femtosecond lasers have been around for some time. The ultrafast dynamics of molecules have been studied in the past few years through the advent of femtosecond pump&probe techniques pioneered by Zewail [3-5]. The group of Gerber studied the femtosecond time-resolved multiphoton ionization of the Na_2 molecule. Wavepacket motions in different molecular potentials showed that incoherent contributions from direct photoionzation of a singly excited state and from a doubly excited molecular state determine the observed transient ionzation signal [6]. The transient spectrum presents strong oscillations with superimposed modulations. The estimated periods of $T_A = 306fs$ and $T_\Pi = 363fs$ could be ascribed to the A- and Π-state, respectively. So the observed signal of Na_2 mirrors the coupled dynamics of the two involved molecular potentials.

Here, we report novel time-resolved studies of the related molecule K_2. The excitation and ionzation steps are quite similar, but as will be seen the involved Π-state is used as a Frank-Condon window, only. Over and above that the final ionization produces K_2^+ in the stable region of the ion's ground state. Therefore we observe the pure dynamics of the K_2 A-state.

Springer Series in Chemical Physics, Vol. 60 **Ultrafast Phenomena IX**
Editors: P. F. Barbara · W. H. Knox · G. A. Mourou · A. H. Zewail

2. Experimental

The temporal evolution of the dimer's ion signal was measured by use of a regeneratively modelocked titanium sapphire laser system pumped by an $10W$ Ar^+ laser. The repetition rate of the laser pulses amounted $82MHz$. Within the tuning range $(740 - 910nm)$ the laser pulses were nearly tranform-limited with a pulse width (FWHM) of $70fs$ and a bandwidth of $170cm^{-1}$. The pulse energies were about $20nJ$.

To measure the transient spectrum of K_2 pump&probe technique was applied. A first photon is used to create a wavepacket in the A-state [7]. To probe the dynamics in the A-state two further photons being optically delayed (delay time Δt) by an Michelson arrangement are involved. This two photon step ionizes K_2 using the $(2)^1\Pi_g^+$-state as a Frank-Condon window. The wavelength of the laser assures that the final $(1)^1\Sigma_g^+$-state of the K_2^+ is reached below its dissociation level [8].

Measurements were performed by setting the mass-spectrometer on the mass peak of $^{39}K_2$, while the ion signal is recorded as a function of delay time Δt. The K_2 molecules are produced by coexpansion of potassium vapor of about $100mbar$ partial pressure together with $5bar$ argon as an inert carrrier gas through a small cylindrical nozzle of about $70\mu m$ diameter.

3. Results and Discussion

We measured the transient 3PI spectrum of $^{39}K_2$ over a $185ps$ period of time. The first part $(60ps)$ of the observed spectrum with $\lambda_{pump} = \lambda_{probe} = 833.7nm$ is presented in Fig.1. The spectrum shows a distinct oscillation with period $T_A \approx 500fs$ being superimposed on a clear beat structure with period $T_S \approx 20ps$. The slight decay in Fig.1 is a consequence of a decreasing cluster beam

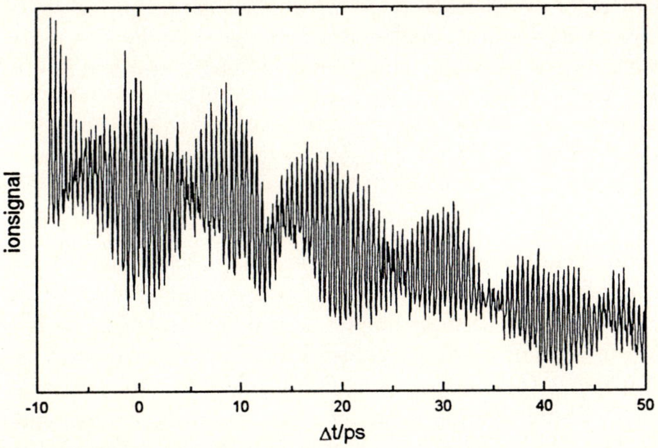

Fig.1 Time-resolved 3PI spectrum of K_2.

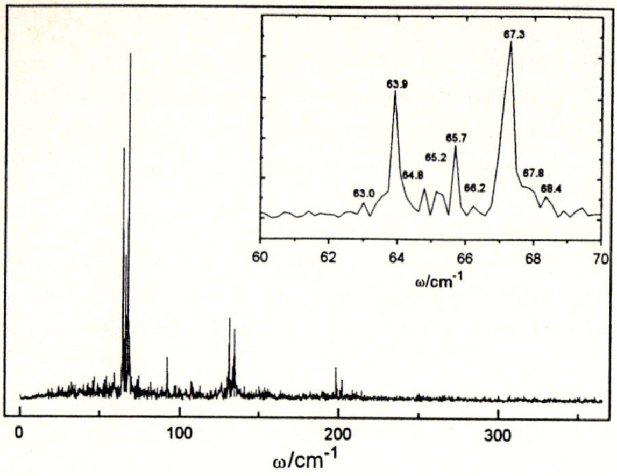

Fig.2 Fast Fourier Transform of experimental data.

intensity during the measurement. The fast oscillation (T_A) is shifted with respect to the zero-of-time with a 180° phase. This phase shift indicates that the ionization step takes place at the outer turning point of the wavepacket motion in the potential of the $^{39}K_2$ A-state.

More information concerning the involved frequencies in the transient 3PI signal is extracted by help of a Fourier analysis. The Fourier spectrum shown in Fig.2 is dominated by several frequencies centered around $65.5cm^{-1}$ which is depicted in more detail in the inset of this figure. The frequencies with higher amplitude at $63.9cm^{-1}$ and $67.3cm^{-1}$ (corresponding to $T_{A1} = 522fs$ and $T_{A2} = 496fs$, respectively) can be identified to be responsible for the beat oscillation with $T_S = 20ps$ in Fig.1.

It can be deduced from [8] that this frequency group is related to the molecular vibration of K_2 in its A-state ($(1)\ ^1\Sigma_u^+$). Each frequency is correlated to one difference of two vibrational modes being excited by the pump pulse. These differences vary with the vibrational quantum number due to anharmonicities of the A-state's potential curve.

4. Conclusion

We were able to measure selectively the vibration of $^{39}K_2$ in the electronically excited A-state. The oscillation periods of the molecule amount around $500fs$. There appear slightly different frequencies due to anharmonicities of the potential of the potassium dimer. The observed phase shift of 180° proves that the ionization step takes place at the outer turning point of the A-state potential.

Future work should illuminate the behavior of K_2 in similar experiments at different wavelenghts. Furthermore the examination of the heteronuclear $^{41}K^{39}K$ is under work.

References

1. S.J. Smith, P.L. Knight (Eds.): *Multiphoton Processes* (Camebridge Univ. Press, Oxford 1988)
2. J. Gaus, K. Kobe, V. Bonačić-Koutecký, H. Kühling, J. Manz, S. Rutz, E. Schreiber, L. Wöste: J. Phys. Chem. **97**, 12509 (1993)
3. R. Khunakar, A.H. Zewail: Ann. Rev. Phys. Chem. **41**, 15 (1990)
4. M. Dantus, M.H. Janssen, A.H. Zewail: Chem. Phys. Lett. **181**, 281 (1991)
5. M. Dantus, A.H. Zewail: J. Chem. Phys **89**, 6113 (1988)
6. T. Baumert, M. Grosser, R. Thalweiser, G. Gerber: Phys. Rev. Lett. **67**, 3753 (1991)
7. W.O. Crane, A. Christy: Phys. Rev. **36**, 421 (1930)
8. L.H. Jeung, A.J. Ross: J.Phys. B **21**, 1473 (1988)

Femtosecond Studies of Photoinduced Electron Dynamics at the Liquid–Solid Interface of Semiconductor Nano-Particles in Water

J.Z. Zhang, R.H. O'Neil, T.W. Roberti, J.L. McGowen, and J.E. Evans

Department of Chemistry and Biochemistry, University of California, Santa Cruz, CA 95064, USA

Abstract. We report direct measurements of the dynamics of photoinduced electrons at the liquid-solid interface of aqueous CdS colloids on the femtosecond time scale. Surface electron trapping was found to occur in <100 fs, followed by rapid decay of the trapped electrons primarily through electron-hole recombination.

Colloidal semiconductors with diameters of 10-100 Å provide a unique system for investigating interfacial electron dynamics because of their small size (thus transparent to light) and special optical as well as electronic properties due to quantum confinement [1-2]. CdS colloids present an excellent prototype for such investigations [3-5]. In this paper, we report a study of the interfacial electron dynamics in aqueous CdS colloids using femtosecond laser spectroscopy. We show new evidence of trapped electrons at the liquid-solid interface based on the sensitive dependence of the decay dynamics on the excitation intensity and the solvent.

The experiments were performed on aqueous CdS colloids using a pump-probe scheme with a regeneratively amplified, mode-locked femtosecond Ti-Sapphire laser. The CdS colloids were prepared following the procedure given in Ref. 2c. The details of the experiment and the laser system have been reported elsewhere [5].

The time evolution of the photoinduced electrons probed at 780 nm is shown in Fig. 1 The transient absorption has been assigned to electrons trapped at the liquid-solid interface [5]. The data fit well to a 100 fs exponential rise and a fast 2.5 ps exponential decay followed by a slower 50 ps exponential decay, plus an offset. The initial rise indicates that the trapped electrons are formed with a time constant of <100 fs. This observation is consistent with previous studies [3,4].

Immediately following the initial fast rise, the population of the trapped electrons begins to decay (Fig. 1). The double exponential decay components (2.5 ps and 50 ps) suggests that the trapped electrons disappear through a non-first-order decay process. The amplitudes for both decays are found to be sensitive to the pump power. The pump power used in recording the spectra in Fig. 1 is 3 μJ/pulse. When the power is varied from 3 μJ/pulse to 0.3 μJ/pulse (Fig. 2), the amplitude of the 2-3 ps decay component decreases almost linearly with the pump power and becomes negligible at 0.3 μJ/pulse, while the 50 ps decay component also decreases but is still weakly visible at 0.3 μJ/pulse. We have attributed both the fast and slow decay components to trapped electron-hole recombination [5]. At low pump power (Fig. 2a and 2b) the decay is dominated by a single exponential with a time constant of 50 ps. The amplitude, but not the rate, increases with the pump power. We assign this slow decay to geminate electron-hole recombination. At high pump

Springer Series in Chemical Physics, Vol. 60 **Ultrafast Phenomena IX**
Editors: P. F. Barbara · W. H. Knox · G. A. Mourou · A. H. Zewail

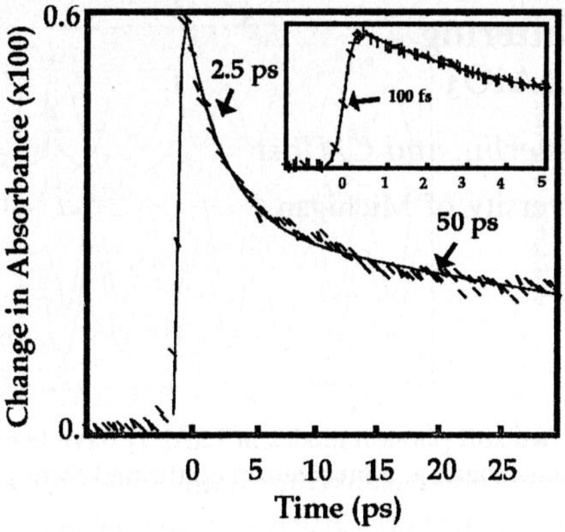

Fig. 1. The time evolution of the photoinduced electrons at the CdS-water interface probed at 780 nm following excitation at 390 nm. The pump fluence is 1.18 photons/Å^2.

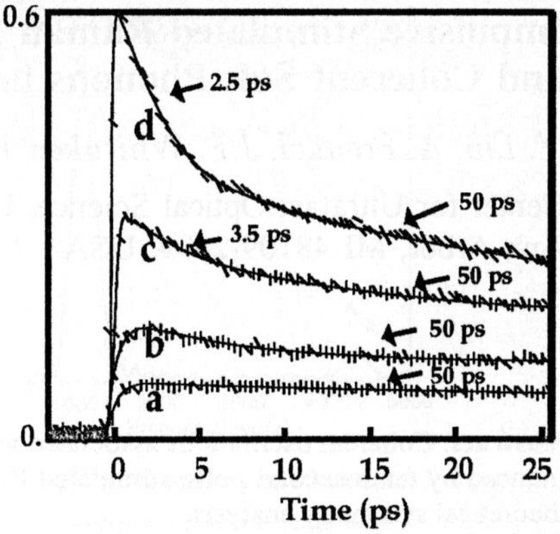

Fig. 2. Excitation intensity dependence of the dynamics probed at 780 nm. The photon fluence is as follows (photon/Å^2): a. 0.12; b. 0.20; c. 0.59; d. 1.18.

power (Fig. 2c and 2d) we observed enhancements both in the amplitude and in the rate of the fast decay component. We attribute this fast decay to non-geminate electron-hole recombination. We have also found that both the fast and slow decays are sensitive to species in the liquid that are likely to be adsorbed onto the particle surface, e.g. I^- and OH^-. The fast decay is also slower when CdS is coupled to TiO_2, indicating electron transfer from CdS to TiO_2 in <2 ps and thus slower electron-hole recombination.

References

1. (a) P. V. Kamat, *Chem. Rev.* **1993**, *93*, 267; (b) A. Henglein, *Topics in Curr. Chem.* **1988**, *143*, 113.
2. (a) N. Chestnoy, T. D. Harris, R. Hull, L. E. Brus, *J. Phys. Chem.* **1986**, *90*, 3393; (b) L. E. Brus, *J. Chem. Phys.* **1983**, *79*, 5566; (c) V. L. Colvin, A. N. Goldstein, A. P. Alivisatos, *J. Am. Chem. Soc.* **1992**, *114*, 5221.
3. (a) P. V. Kamat, T. W. Ebbesen, N. M. Dimitrijevic, A. J. Nozik, *Chem. Phys. Lett.* **1989**, *157*, 384; (b) R. Rossetti, L. E. Brus, *J. Phys. Chem.* **1986**, *90*, 558.
4. M. Kaschke, N. P. Ernsting, U. Muller, H. Weller, *Chem. Phys. Lett.* **1990**, *168*, 543.
5. J. Z. Zhang, R. H. O'Neil, T. W. Roberti, *J. Phys. Chem.* **1994**, *98, 3859.*

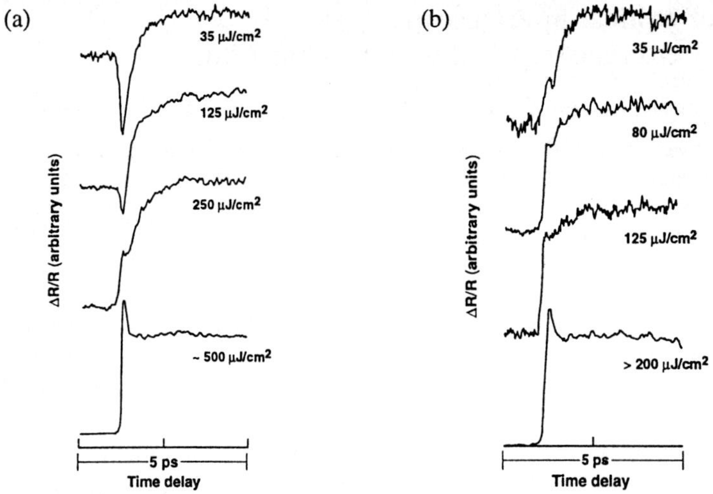

Fig. 2 Intensity dependence of the transient reflectivity traces measured on a 60-K YBCO film at two different regions (a) and (b).

We have measured the intensity dependence of the optical response by 2-eV pump/2-eV probe at two different regions of the same 60-K sample. Figures 2(a) and 2(b) display the measured ΔR/R traces at various pumping intensities. In region (a), a gradual change of sign is observed at t ~ 0^+. This clearly suggests that E_F, initially below 2 eV, is up-shifted with increasing the pump intensity. On the other hand, if we start with E_F above the energy of the probe beam [Fig. 2(b)], the signal is always positive. Thus, an increase of the pump intensity also leads to the up-shift of E_F. It is important to point out that the intensity dependence is fully reversible, i.e., reducing the pump intensity restores the initial trace. Hence, the possibility of any stoichiometric changes in the film, can be excluded.

The amplitudes of ΔR/R at t ~ 0^+ ps, corresponding to regions (a) and (b), are further evaluated by calculating the change of the state occupancies with the initial Fermi level as an adjustible parameter [3]. The good fit confirms that the thermomodulation model is indeed valid for partially oxygen-depleted YBCO. It also shows that the intensity dependence of ΔR/R can be explained by the up-shifting of the Fermi level with respect to the 2-eV-probe photon energy.

In conclusion, we have examined the validity of the thermomodulation model in 55-K and 60-K YBCO films. In addition, we have demonstrated that the change in reflectivity depends on the intensity of the pump laser near the Fermi energy.

References

1. S. L. Cooper *et al.*, Phys. Rev. B **47**, 8233 (1993).
2. S. D. Brorson *et al.*, Solid State Commun. **74**, 1305 (1990).
3. L. Shi *et al.*, unpublished.

Terahertz Emission by Coherent Lattice Vibrations in Tellurium

T. Dekorsy, H. Auer, C. Waschke, H.J. Bakker, H.G. Roskos, and H. Kurz

Institut für Halbleitertechnik, RWTH Aachen, D-52056 Aachen, Germany

Abstract. Coherently excited infrared-active phonons in Te are investigated by pump-probe and THz-emission experiments. We report on the first observation of the THz-wave emission from coherent phonons in a semiconductor.

THz-emission spectroscopy has been shown to be a powerful tool for the investigation of ultrafast polarization changes in different organic and anorganic materials. Up to now, oscillations observed in transient THz signals from semiconductors are of electronic origin, *e.g.* the coherent motion of electronic wavepackets at low temperatures [1]. Here, we report on room-temperature THz emission from coherently excited lattice vibrations from single crystal Tellurium. Additional to the THz emission experiments, we perform pump-probe experiments for a detailed time-resolved investigation of coherent infrared (IR) active phonons.

The observation of coherent phonons in transient reflectivity measurements employing femtosecond pump-probe techniques has been reported for several solid state materials [2,3]. To investigate the FIR emission of coherent phonons, we choose a "soft" material with phonon frequencies within the detection limit of a conventional dipole antenna used for time-resolved detection of THz emission. The lattice vibrations of Te consist of a fully symmetrical, only Raman-active A_1-mode (3.6 THz), two degenerate Raman- and IR-active E-modes ($E'_{TO/LO}$=2.76/3.09 THz, $E''_{TO/LO}$=4.22/4.26 THz) and one only IR-active A_2-mode ($A_{2,TO/LO}$=2.6/2.82 THz)[4].

In pump-probe experiments with 50 fs pulses from a CPM dye laser, coherent phonons are excited. The anisotropic contribution of the reflected probe beam is analyzed. In recent experiments, only the dominant A_1 mode could be observed in the isotropic reflectivity [3]. Fig. 1 shows the oscillatory contribution of the reflectivity changes $(R_x - R_y)/R_0$ of a Te crystal with the c-axis perpendicular to the surface. An instantaneous part has been omitted. The Fourier transform of the data reveals sharp peaks at the E_{TO} frequencies, which are associated with polarization along the x-y-plane (parallel to the surface). The modulation of the anisotropic part of the reflectivity by the E-modes can be derived from the Raman tensor. The A_2-mode with polarization along the c-axis is not observable in the optical experiment, since it is only IR-active, and the fully symmetric A_1-mode is obviously suppressed.

For the observation of the THz emission, the same crystal is excited with pulses derived from a Ti:Sapphire laser. The THz radiation emitted under 45° is focused on a gated dipole antenna. This technique allows the determination of amplitude and phase of the emitted electro-magnetic wave.

The THz emission from Te is composed of an instantaneous part and a persistent high-frequency part, that is clearly detectable over several ps (Fig. 2). The Fourier transform peaks at 2.85 THz, which is close to the $A_{2,LO}$ frequency. The A_2-mode has a polarization perpendicular to the excited surface, so that the polarization associated with the coherent LO-phonon radiates into the detected direction. When the surface parallel to the c-axis

Springer Series in Chemical Physics, Vol. 60 **Ultrafast Phenomena IX**
Editors: P. F. Barbara · W. H. Knox · G. A. Mourou · A. H. Zewail

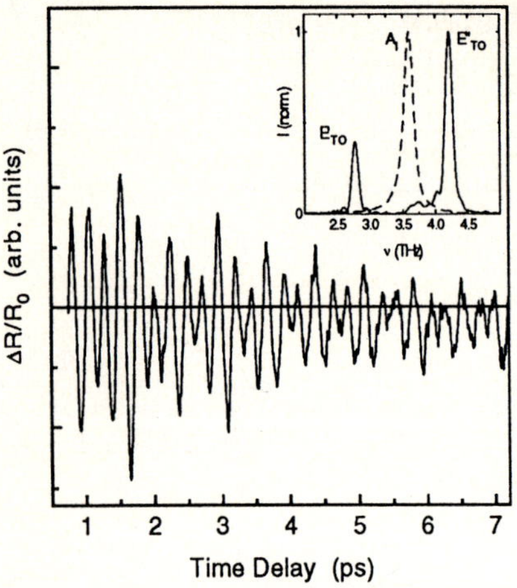

Fig. 1: Time-resolved anisotropic reflectivity changes $\Delta R = R_x - R_y$ from Te. The inset shows the Fourier transform of ΔR (E'_{TO} and E''_{TO}-mode, solid line) and of the isotropic reflectivity (A_1-mode, dashed line, not shown in time domain).

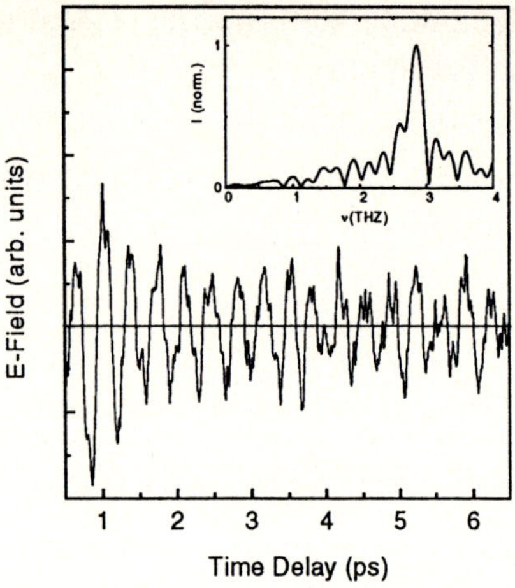

Fig. 2: THz emission from coherent phonons in Te. The inset shows the Fourier transform. Time delay zero is arbitrary.

is excited, the high frequency emission is observed at 3.2 THz, which is close to the E'_{LO}-phonon with polarization perpendicular to the c-axis. This mode is also observed in anisotropic reflectivity measurements at this surface. The emission intensity at the phonon frequency is weak due to the frequency cut-off of the antenna at 3.5 THz. The analysis of the Fourier transform of the whole THz signal reveals that the oscillations are driven by the exciting laser pulse and not by absorption of the instantaneous polarization. We therefore conclude that the oscillations observed are the emission of FIR radiation from coherent phonons.

We thank G.C. Cho, A.V. Kuznetsov and V. Wagner for useful discussions and P. Grosse for supplying high quality Te samples.

[1] H.G. Roskos M.C. Nuss, J. Shah, K. Leo, D.A.B. Miller, A.M. Fox, S. Schmitt-Rink, K. Köhler, Phys. Rev. Lett. **68**, 2216 (1992).
[2] W.A. Kütt, W. Albrecht, and H. Kurz, IEEE J. Quant. Electr. **QE-28**, 2434 (1992).
[3] T.K. Cheng, J. Vidal, H.J. Zeiger, G. Dresselhaus, M.S. Dresselhaus, and E.P. Ippen, Appl. Phys. Lett. **59**, 1923 (1991).
[4] P. Grosse, Springer Tracts in Modern Physics **48**, Springer (Berlin 1969).

Femtosecond Measurements of Ballistic and Interactive Transport of Non-Equilibrium Electrons in Thin Gold Films

T. Juhasz[1], *C. Suárez*[1], *W.E. Bron*[1], *and H.E. Elsayed-Ali*[2]

[1]Department of Physics, University of California, Irvine, CA 92717, USA
[2]Department of Electrical and Computer Engineering, Old Dominion University, Norfolk, VA 23529, USA

Abstract. The effect of the crystal structure on the transport of laser excited nonequilibrium electrons in gold films have been investigated.

Nonequilibrium electrons in metals can be excited by laser pulses with durations less than, or comparable to the excited-electron energy-loss lifetime (τ_e). Loss of electron energy in the excited volume is attributed to energy transfer to the lattice through electron-phonon interactions plus electron transport out of the excited region. It has been established on experimental and theoretical grounds that electron-grain boundary interactions impede electron transport in polycrystalline films [1]. In the present experiment we investigate the transport of femtosecond laser generated nonequilibrium electrons in poly- and single-crystalline gold films, through measurements of the time-of-flight across the film thickness.

A synchronously pumped femtosecond dye laser with output pulses of ~150 fs duration at 2 eV photon energy and 76 MHz repetition rate is used to excite the electrons with ~100 mJ/cm^2 fluence per pulse on the front surface of the films. The skin depth is $D \approx 15$ nm. Transport of the nonequilibrium electrons out of the excited volume occurs simultaneously with electron energy relaxation. Excited electrons which reach the back surface of the film introduce a change in the optical reflectivity which is probed with an appropriately delayed laser pulse.

The normalized transient thermoreflectivities, ($\Delta R/R(t)$), as measured on the back surface of the single-crystalline gold film samples with thicknesses of 25 nm, 100 nm, 200 nm, 300 nm, and 400 nm are displayed in Fig.1. Similar results are obtained in poly-crystalline films. As the thickness of the films is increased the traversal time of the electrons to cross the film, and therefore the delay in the transient $\Delta R/R(t)$, is increased. In order to measure the time required by the first wave of electrons to traverse the films, we measured the time to reach the signal onset of $\Delta R/R(t)$, relative to the excitation at t=0, for different sample thicknesses. For simplicity we assign the onset as 15% of the full signal level. The open symbols in Fig.2 display the results. The circles and triangles represent the poly- and single-crystalline samples respectively. For each film thickness, the delay of the peak of $\Delta R/R(t)$ corresponds to the time required for the bulk of the excited electrons to reach the surface. Measurements determining this delay (relative to t=0) in both poly- and single-crystalline gold films are displayed by the filled symbols in Fig. 2.

We interpret these results as follows. The transport of the excited electrons across the films with thicknesses less than, or equal to 100 nm is mostly ballistic in

Editors: P. F. Barbara · W. H. Knox · G. A. Mourou · A. H. Zewail

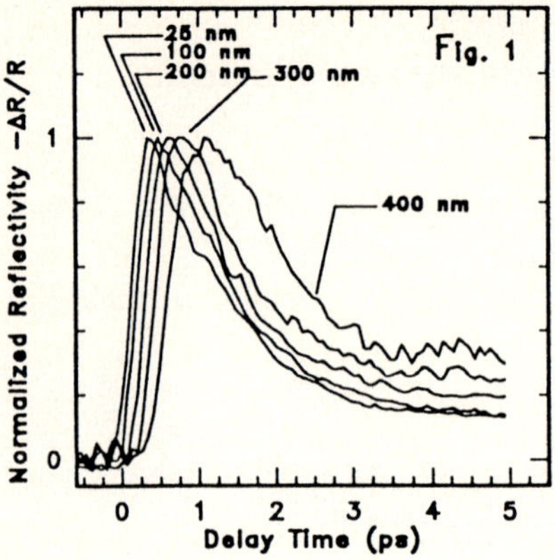

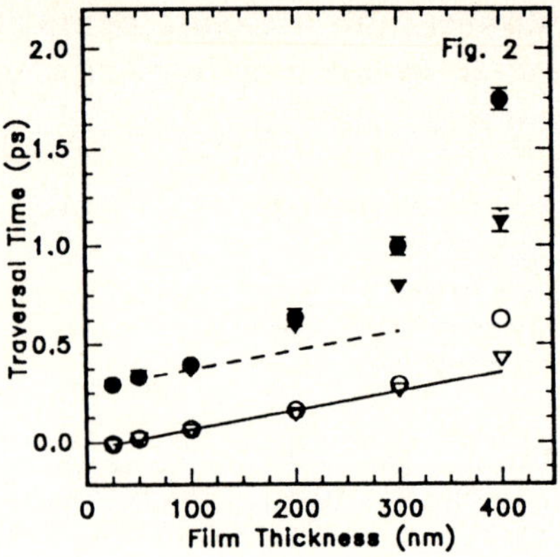

both single- and polycrystalline materials with a velocity of 10^6 m/s. As the thickness of the film increases the fraction of the ballistic component decreases, and the excess energy is carried by electrons interacting with other electrons or/and with the lattice. These electrons undergo one or more scattering events, and we refer to them as the interactive component of the transport. Interactive transport also includes the possibility of pure diffusion. Since the interactive electrons arrive at the back surface of the sample later than the ballistic electrons, the appearance of the peak of the transient $\Delta R/R(t)$ signal is somewhat delayed. The effect is observed to be stronger in the polycrystalline samples, which indicates that the grain boundaries cause a slowing of the transport of the excited electrons. In samples with 400 nm thicknesses the ballistic component is diminished in both poly- and single-crystalline samples. The absence of the ballistic component could imply the presence of a thermal equilibrium among the electrons which reach the back surface, although no direct evidence of this is available from the present experiment. Fitting the "two temperature model" [2] to the data obtained in the 400 nm films provides diffusion coefficient κ_s=310 W/mK and κ_p=180 W/mK in the case of single-crystalline and polycrystalline films respectively. The reflection coefficient of excited electrons at the grain boundaries is extracted from the experiment by analyzing the results in the terms of the Mayadas-Shatzkes model [1], and is estimated to be $r \cong 0.12$.

REFERENCES:

1. A. F. Mayadas and M. Shatzkes, Phys. Rev. B 1, 1382 (1970)
2. S. I. Anisimov, et al., Zh. Eksp. Teor. Fiz. 66, 776 (1974).

Observation of Coherent Surface-Optical Phonons by Time-Resolved Second-Harmonic Generation

L. Xu, Y.-M. Chang, and H.W.K. Tom

Department of Physics, University of California, Riverside, CA 92521, USA

We report the observation of coherent surface phonons at optical phonon frequencies. We excite the surface phonon by impulsive stimulation and detect the surface phonon by time-resolving the second-harmonic generation after the pump pulse. Several groups [1,2,3] have reported the observation of coherent bulk phonons in various media using short optical pulses to impulsively excite the vibrational mode. In contrast to linear reflectivity, SHG in reflection geometry is known to have decided surface sensitivity. SHG from surfaces is dipole-allowed but dipole-forbidden from the bulk of centrosymmetric media. Using the right excitation geometry, one can also measure SHG under conditions where the tensor elements are dipole-forbidden in the bulk of non-centrosymmetric media[4]. It should therefore be possible to observe coherent surface phonons using time-resolved surface SHG from almost any media. Indeed, we observe an oscillation in time-resolved SHG at 8.36 THz from a GaAs (100) excited with 30 fs, 850 nm pump pulses. The excitation conditions are similar to those under which bulk LO and TO coherent phonons are generated (as reported by Kurz and coworkers [2]). The bulk LO and TO phonon frequencies are 8.75 and 8.02 THz. Because SHG has a significant surface contribution and our observed oscillation at 8.36 THz is in good agreement with the oscillation of the Fuchs-Kliewer surface phonon of GaAs[5], we assign the 8.36 THz oscillation to the surface phonon.

Laser pump-probe experiments were performed on the surface of a GaAs (100) wafer which was n-doped with 3×10^{18} Si/cm^3. The sample was held in air and had a native oxide. The depletion field is estimated to be about 400 kV/cm. In general, the SH comes from the surface and the bulk. Even though the bulk nominally has $\bar{4}3m$ symmetry, both the surface and the near surface bulk have mm2 symmetry because the surface and the depletion field break symmetry normal to the surface.

In Fig. 1, we show the time-resolved SHG from the probe pulse after the pump pulse injects 5×10^{18} carriers/cm^3. The 856 nm s-polarized pump and the p-polarized probe were incident at 45° from the surface normal. During the pump pulse, the probe SHG increases rapidly as injected carriers screen the depletion field and returns the bulk symmetry to $\bar{4}3m$. There is also a change due to increasing the number of carriers. After the depletion field is screened, there is a slow decrease and significant oscillations caused by the generation of coherent phonons. These oscillations decay with ~ 1 ps time constant. The Fourier transform of the time-resolved SHG (see inset) shows that the oscillation has three frequencies: 7.65 THz, 8.36 THz, and 8.76 THz. The time-resolved experiment was also performed on n$^-$

Springer Series in Chemical Physics, Vol. 60 **Ultrafast Phenomena IX**
Editors: P. F. Barbara · W. H. Knox · G. A. Mourou · A. H. Zewail

Time-Resolved Studies of Bloch Oscillations in Semiconductor Superlattices

P. Leisching[1], *T. Dekorsy*[1], *C. Waschke*[1], *H.G. Roskos*[1], *K. Leo*[2], *H. Kurz*[1], *and H. Köhler*[3]

[1]Institute of Semiconductor Electronics, RWTH Aachen, Sommerfeldstrasse 24, D-52056 Aachen, Germany
[2]Institute of Applied Photophysics, Technische Universität Dresden, Mommsenstrasse 13, D-01062 Dresden, Germany
[3]Fraunhofer-Institute of Applied Solid State Physics, Tullastrasse 72, D-79108 Freiburg, Germany

Abstract: Bloch oscillations in GaAs/AlGaAs superlattices are investigated in a comparative study by time-resolved four-wave mixing, transmittive electro-optic sampling, as well as by THz-emission spectroscopy.

1. Introduction

The atomic-layer precision achieved in the MBE growth of GaAs/AlGaAs semiconductor superlattices (SL) has allowed to observe a variety of interesting novel coherent phenomena on a pico- and subpicosecond time scale upon resonant excitation with femtosecond laser pulses.

Since the discovery and exploration of coherent motion of wave packets in quantum-well structures, the evidence for the existence of Bloch oscillations (BO) moved into the center of interest. BO have been predicted in 1970 by Esaki and Tsu for SL [1] as the periodic motion of Bloch electrons in a homogeneous electric field F. This has been shown to be equivalent to the beating of coherently excited Wannier-Stark states of the SL [2]. The first observation and further detailed investigations of BO's succeeded by femtosecond degenerate four-wave mixing (DFWM) experiments [3,5]. More recently, further evidence was obtained with THz-emission experiments [4].

Here, we report on a comparative study of BO's [7] in a series of $GaAs/Al_{0.3}Ga_{0.7}As$ superlattices [5] for temperatures up to 200 K by time-resolved optical and THz experiments performed with 130 fs laser pulses emitted by a Kerr-lens modelocked Ti:sapphire laser. In the first part, the results of time-integrated and time-resolved DFWM experiments sensitive to *interband* polarizations are presented. The second part introduces the transmittive electro-optic sampling (TEOS) measurement technique and discusses the results of TEOS and THz-emission experiments sensitive to *intraband* polarizations. The paper is concluded with a brief summary.

Springer Series in Chemical Physics, Vol. 60 **Ultrafast Phenomena IX**
Editors: P. F. Barbara · W. H. Knox · G. A. Mourou · A. H. Zewail

2. Four-wave mixing experiments

In the time-integrated DFWM signals several field-dependent oscillations are visible if the excitation energy is peaked slightly below the spatially direct heavy hole (hh_0) excitonic Wannier-Stark transition. The experimentally observed field dependence of the oscillation frequency for samples with electron miniband widths from 12 meV to 46 meV corresponds excellently to the theoretical prediction for BO [5]. The oscillation frequency can be tuned by an external electric bias up to the frequency resolution limit of the experimental setup at 5 THz given by the 20 meV spectrum of the optical pulse. In the sample with 46 meV miniband width, BO's are observed up to 200 K [5].

With time-resolved DFWM experiments, a much deeper insight into the physics of the coherent motion of the *excitonic* electron-hole wavepackets can be obtained. In Fig. 1, the results of a time-resolved DFWM experiment with a 18 meV miniband width sample are summarized. At the applied bias of 7 kV/cm we excite only the hh_0 and the spatial indirect hh_{-1} Wannier-Stark transitions. The temporal evolution of the DFWM signal set up by two co-polarized input pulses #1 and #2 is mapped out by a third pulse #3 gating the emitted DFWM signal via upconversion. The data are plotted for different time delays τ between the two input pulses #1 and #2 and the time delay t of pulse #2 to pulse #3.

Due to the excitonic nature of the excited Wannier-Stark states, the diffracted signal is originates from the evolution of a polarization wave of excitonic dipoles. It is therefore retarded in time delay t by approximately the phase relaxation time T_2 as observed in earlier experiments on single quantum wells [6]. The delayed signal

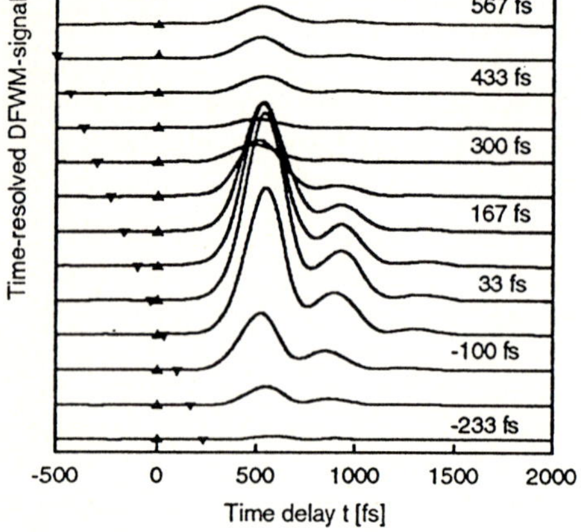

Figure 1: Time-resolved DFWM signal for different delays τ at an excitation density of $3*10^9$ cm^{-2} carriers per period and a lattice temperature of 10 K. The downward and upward triangles symbolize the arrival of pulse #1 and #2 respectivly.

exhibits BO's with a period of 400 fs in both time delays t and τ separately, proving that the beating is due to quantum beat interference. There is no evidence of a photon echo behavior, i.e. an inhomogeneous broadening of the transitions, that would produce a shift of the emitted signal in time delay t for positive time delays τ. This indicates a purely homogeneous broadening of the transitions and thus demonstrates the very high quality of our samples. At an excitation density close to 10^{10} cm^{-2} (not shown), a new "prompt" feature appears with a rise time in t and τ comparable to the duration of the exciting pulses. This can be attributed to a noninteraction induced polarization phenomena. The BO-induced DFWM polarization wave signal appears with about the same time delay t as at the lower density.

The final proof whether the excitonic wavepacket performing the BO's is associated to an oscillating dipole moment, as predicted by the BO theory, is accomplished by time-resolved TEOS and THz-emission experiments.

3. THz-emission and TEOS experiments

The amplitude and phase of the THz pulse emitted from the oscillating Bloch wavepacket is temporally resolved by an optically gated subpicosecond photoconductive dipole antenna [4,7]. In Fig. 2, the emitted field transients are plotted at different electrical bias levels. From flatband (+0.4 V) to a reverse bias of -1.6 V, the external electric field is completely screened. This is due to the accumulation of optically excited charge carriers at the ends of the superlattice structure [5]. At a reverse bias of -1.6 V, a strong BO related signal emerges, whose period decreases linearly with increasing applied voltage. Up to seven BO periods are visible at 10 K. At high oscillation frequencies, the detected signal decreases due to the limited bandwidth of the photoconductive dipole antenna. The initial part of the transients at high bias voltage is attributed to instantaneous polarization associated with the generation process of spatially separated electron-hole pairs (10^9 cm^{-2} carriers per period).

Complementary results are obtained by an alternative optical time-resolved technique: the transmittive-electro-optic sampling (TEOS). This method measures directly the internal electric field generated by the oscillating dipole moments [7]. TEOS is basically a pump/probe technique where a cirularly polarized probe beam is analysed polarization sensitive. The observed polarization changes of the transmitted probe beam result on birefringence induced by the macroscopic polarization associated with the intraband dipoles of the electron-hole separation performing BO's. The experimental measurement of the amplitude and phase of the excitonic wave packet oscillations launched are of central importance for the fundamental understanding of BO's.

Conventional DFWM experiments are only sensitive to the diffracted field intensity of the interband polarization. Important complementary information on the inter- and intraband polarizations are obtained by simultaneously performing DFWM and TEOS measurements under identical experimental conditions.

As an example for TEOS measurements, Fig. 3 depicts the periodic modulation of the polarization of the probe pulse transmitted through a SLS biased at different

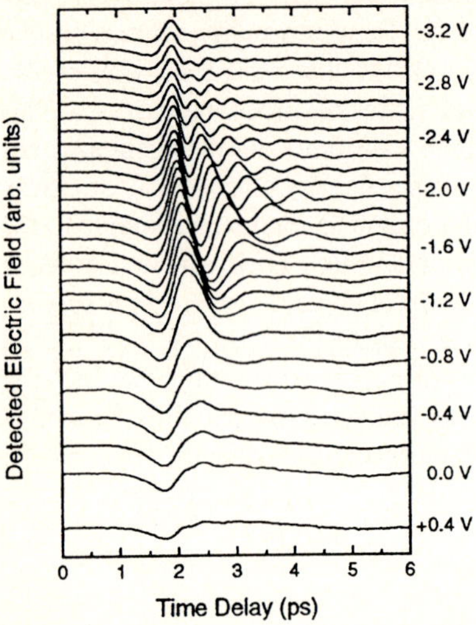

Figure 2: Measured coherent electromagnetic transients at a lattice temperature of 10 K for the 18 meV miniband width sample.

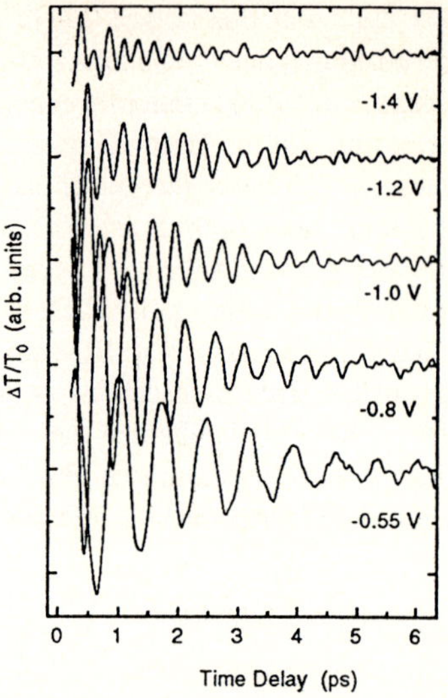

Figure 3: Oscillatory traces of TEOS signals for different voltages applied. An instantaneous part at zero time delay is omitted.

levels. The spectral position of the exciting laser pulse is identical to the DFWM experiments, at a reverse bias of -1.0 V we excite only the hh_0 and hh_{-1} transitions. Up to 10 periods of the BO are clearly detected over a time interval much longer than in the case of the DFWM experiments. For an external voltage of -0.55 V, the *interband* T_2 of 1.4 ps measured by DFWM (data not shown) is increased by a factor of about 2 to the *intraband* T_2 of 2.8 ps measured by TEOS. The frequency shift with bias voltage is clearly visible. The voltage dependence in the TEOS experiments in comparison to the THz experiments is shifted to lower reverse bias. This is due to the lower excitation density used in this experiment, wich corresponds to a reduced quasistatic screening offset [5]. The decrease of the amplitude at high reverse bias results from the increasing localization of the Wannier-Stark states leading to a reduction of the intraband dipole moment associated with BO.

4. Summary

In summary, these comparative studies provide a large body of essential information on Bloch oscillations of wave packets photo-injected into GaAs/AlGaAs superlattices. Time-resolved DFWM measures the excitonic interband polarization in contrast to TEOS and THz-emission that are sensitive to the intraband polarization. The interband and intraband T_2 measured simultaneously by DFWM and TEOS differ by

a factor of 2. DFWM is an powerful tool for the investigation of *excitonic interaction effects*. Among many other results, evidence for the density dependence of the excitonic polarization interactions is found. DFWM is not sensitive to the macroscopic electric fields induced by the oscillation of Bloch dipoles. In THz-emission and TEOS the instantaneous polarization, associated with the generation of electron-hole pairs is detected, and the dependence of the dipole moment of the Bloch oscillations on the bias field has been measured. The time-resolution of TEOS is limited by the optical pulse width, like in the case of DFWM, in contrast to the limitation to 4 THz in THz experiments to due to the antenna response.
TEOS and THz-emission should allow to answer the question, weather the phase of the dipole moment of the coherent wave packet can be reversed by tuning the wavelength of the excitation. Such a phase reversal is expected for selective excitation of the upper and lower branches of the Wannier-Stark ladder, respectively.

This work has been supported by the Deutsche Forschungsgemeinschaft, the Volkswagen foundation and the Alfried von Krupp foundation.

References

[1] L. Esaki, and R. Tsu, IBM J. Res. Dev. **61**, 61 (1970).

[2] G. Bastard and R. Ferreira, in *Spectroscopy of Semiconductor Microstructures*, NATO ASI Series (Plenum, New York, 1989), p. 333.

[3] J. Feldmann, K. Leo, J. Shah, D. A. B. Miller, J. E. Cunningham, S. Schmitt-Rink, T. Meier, G. von Plessen, A. Schulze, and P. Thomas, Phys. Rev. B **46**, 7252 (1992); K. Leo, P. Haring Bolivar, F. Brüggemann, R. Schwedler, and K. Köhler, Solid State Commun. **84**, 943 (1992).

[4] C. Waschke, H. G. Roskos, R. Schwedler, K. Leo, H. Kurz, and K. Köhler, Phys. Rev. Lett. **70**, 3319 (1993).

[5] P. Leisching, P. Haring Bolivar, W. Beck, Y. Dhaibi, K. Leo, H. Kurz, and K. Köhler, submitted to Phys. Rev. B

[6] D.S. Kim, J. Shah, T. C. Damen, W. Schäfer, F. Jahnke, S. Schmitt-Rink, and K. Köhler, Phys. Rev. Lett **69**, 2685 (1992).

[7] C. Waschke, P. Leisching, P. Haring Bolivar, R. Schwedler, F. Brüggemann, H. G. Roskos, K. Leo, H. Kurz, and K. Köhler, Solid-State Electron. **37**, 1321 (1994).

[8] T. Dekorsy, P. Leisching, K. Leo, H. Kurz, and K. Köhler, submitted to Phys. Rev. B.

Dynamics of Coherently Excited Semiconductor Microcavities

J.-K. Rhee[1], *T.B. Norris*[1], *Y. Arakawa*[2], *M. Nishioka*[2], *and C. Weisbuch*[3]

[1]Center for Ultrafast Optical Science, University of Michigan, 2200 Bonisteel Blvd., Ann Arbor, MI 48109, USA

[2]Institute of Industrial Science, University of Tokyo, 7-22-1 Roppongi, Minato-ku, Tokyo 153, Japan

[3]Thomson-CSF, Laboratoire Central de Recherches, Domaine de Corbeville, F-91404 Orsay Cedex, France

Abstract We have investigated ultrafast vacuum Rabi oscillations in semiconductor microcavities. Using an interferometric pump-probe technique, we have probed the coherence of the polarization of the exciton-cavity system.

In a semiconductor microcavity, when the cavity lifetime and transition dipole polarization decay rate are slow enough compared to the vacuum Rabi frequency, we can expect normal mode splitting of strongly coupled exciton-cavity system, similar to atom-cavity system [1–3]. Recently, Weisbuch *et al.* have reported the first observation of vacuum Rabi splitting (VRS) in 2-D-exciton/planar-DBR microcavities [4]. The observed splittings were as large as 6 meV in reflection and transmission spectra, implying 690-fs vacuum Rabi oscillations (VRO) of impulsively excited microcavities.

In this paper, we have performed an investigation of time-domain vacuum Rabi oscillations in a semiconductor microcavity using ultrafast techniques, in order to investigate VRO, including 2-D exciton-polariton behavior, phase, and coherence of the coupled-mode oscillation. AlGaAs-based microcavity sample used in the present work was the same as that used by Weisbuch *et al.* [4].

The microcavity sample was impulsively excited with 80-fs laser pulses from a Ti:sapphire oscillator tuned near exciton and cavity resonance with 3° incidence angle. The sample temperature was kept at 12 K so that the exciton dephasing time was about 1 ps [5]. The optical fluence of the excitation pulse is kept small, so as not to bleach the exciton absorption. The coupled exciton-photon mode (with the initial excitation corresponding to a photon mode) evolves in time with VRO, showing oscillations in the emitted radiation. In Fig. 1(a), the reflection spectra at three positions on the sample are shown. Fig. 1(b) presents the corresponding time-resolved emission signals from the sample using a standard upconversion technique. When the exciton and cavity modes are resonantly tuned (case 1), we can observe 600-fs VRO with 260-fs decay time. Note that the decay time is about twice the cavity lifetime (140 fs) due to linewidth averaging [2,3]. As further evidence of VRO, we observed the same oscillation period at lower optical fluences. When the cavity is slightly detuned (case 2), the VRO period is reduced to 580 fs due to larger normal mode splitting and the first dip of the emission signal is shifted by 60 fs. The VRO phase shifts with different detuning between laser center frequency and each normal mode. When the cavity and the exciton modes are out of resonance (case 3), VRO is weakened but the period is reduced to 370-fs.

In order to probe the coherence of the the polarization of the coupled exciton-cavity modes, we have developed an interferometric pump-probe technique of

Springer Series in Chemical Physics, Vol. 60 **Ultrafast Phenomena IX**
Editors: P. F. Barbara · W. H. Knox · G. A. Mourou · A. H. Zewail

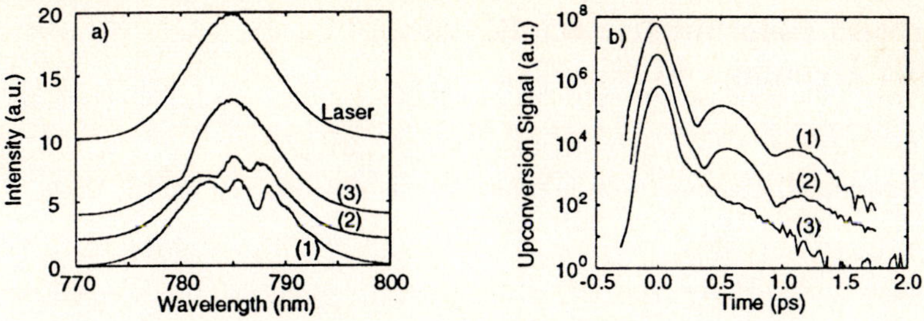

FIG. 1. Reflection spectra of pump lasers (a) and the upconversion experiment signal of the corresponding reflection beams (b) at various positions on the sample: (1) for exciton-cavity resonance case, (2) for 1.2-nm off-resonance case, and (3) for 5-nm off-resonance case. The thick curve in figure (a) shows the laser spectrum.

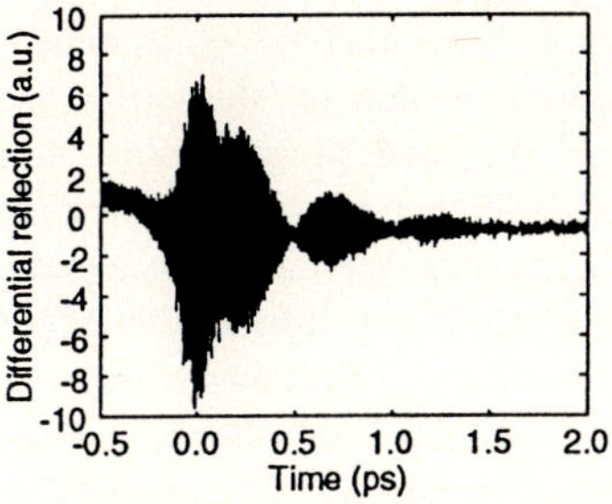

FIG. 2. Interferometric pump-probe experiment data for case 2.

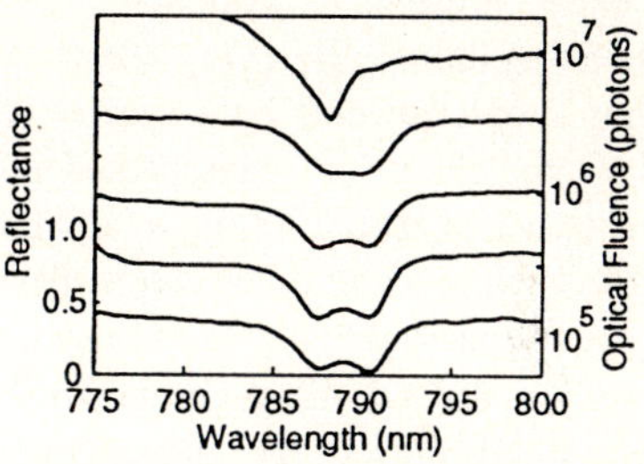

FIG. 3. Normalized reflectance spectra at high optical fluences. Focus spot diameter was $160\mu m$.

which signal has contributions from differential reflection due to pump-induced nonlinearity and interference between reflected probe and elastically scattered pump or coherent radiation from sample [6]. Fig. 2 shows the result of the interferometric pump-probe experiment with the same experimental condition for case 2. The slow envelope of the signal represents the VRO. The fine fringes after timing zero are the result of the coherent radiation from the sample.

In Fig. 3, the time-integrated nonlinear response of the microcavity sample is presented. With higher optical fluence, the reduction of excitonic transition oscillator strength is manifested as reduced VRS.

The work at the University of Michigan was supported by the NSF through the Center for Ultrafast Optical Science under STCPHY 8920108.

References

[1] G. S. Agarwal, Phys. Rev. Lett. **53**, 1732 (1984).

[2] M. G. Raizen, R. J. Thompson, R J. Brecha, H. J. Kinble, and H. J. Carmichael, Phys. Rev. Lett. **63**, 240 (1989).

[3] Y. Zhu, D. J. Gauthier, S. E. Morin, Q. Wu, H. J. Carmichael, and T. W. Mossberg, Phys. Rev. Lett. **64**, 2499 (1990).

[4] C. Weisbuch, M. Nishioka, A. Ishikawa, and Y. Arakawa, Phys. Rev. Lett. **69**, 3314 (1992).

[5] L. Schultheiss, J. Kuhl, A. Honold, and C. Tu, Phys. Rev. Lett. **57**, 1635 (1986).

[6] J.-K. Rhee, T. B. Norris, Y. Arakawa, M. Nishioka, and C. Weisbuch, in *Ultrafast Phenomena*, Vol. 7, 1994 OSA Technical Digest Series (Opt. Soc. Am., Washington, D.C., 1994), pp. 293–295.

Femtosecond Ellipsometric Study of Carrier Dynamics in $Si_{1-x}Ge_x$

X.F. Hu and M.C. Downer

Physics Department, University of Texas at Austin, Austin, TX 78712, USA

Abstract: Carrier dynamics of novel $Si_{1-x}Ge_x$ materials have been studied in the femtosecond time scale.

This paper presents a systematic study of the fs-time-resolved ellipsometric response of epitaxial $Si_{1-x}Ge_x$ ($0 \leq x \leq 1$) alloys as a first step in measuring and modelling subpicosecond bulk carrier dynamics. Pump and probe beams were derived from either an unamplified CPM laser (620 nm wavelength) or an unamplified KLM laser (800nm + 400nm second harmonic). We have developed a femtosecond ellipsometric technique, described elsewhere[1], to measure both $\Delta\varepsilon_1(t)$ and $\Delta\varepsilon_2(t)$ of epitaxial $Si_{1-x}Ge_x$ films on opaque substrates in reflectance mode.

Data on ten distinct alloy compositions spanning the range ($0 \leq x \leq 1$) have been acquired. Fig.1 shows the measured $\Delta\varepsilon_1(t)$ and $\Delta\varepsilon_2(t)$ data for a representative selection of alloy compositions and pump-probe wavelengths: 620 nm pump and probe, approximate pump fluence 10^{-4} J/cm^2 (Fig.1(a)-1(d)); 400 nm pump, 800 nm probe, several time smaller pump fluence (Fig.1(e)-1(h)). In all cases, the real part $\Delta\varepsilon_1(t)$ behaves according to the conventional Drude response of a diffusing electron-hole plasma. The imaginary part $\Delta\varepsilon_2(t)$, on the other hand, shows a more interesting and complicated dependence on alloy composition and pump-probe wavelength configuration. For all Si-like alloys (e.g. Bottom 2 rows of Fig.1), $\Delta\varepsilon_2(t)$ is positive, signifying induced absorption, and shows a remarkable similarity in the shape of its temporal development despite very different magnitudes, which approximately track α_{pump} (a similarly shaped response also occurs for pure Ge with 620 nm pump-probe (Fig.1(a)). In these cases the photon energy $h\nu_{probe}$ is on an indirect transition, and therefore insensitive to transient interband absorption saturation effects. For Ge-like alloys (including pure Ge), on the other hand, $\Delta\varepsilon_2(t)$ is sometimes negative and shows complicated temporal behavior which depends strongly on alloy composition and pump-probe configuration. In all these cases, $h\nu_{probe}$ is on a direct transition, and therefore sensitive to transient interband absorption saturation effects.

Modelling results for Ge at 620 nm pump-probe, derived from a kinetic model of carrier generation and transport developed by van Driel et al.[2], are presented in Ref.1. From the calculated time dependence of $N_{eh}(t)$, $T_e(t)$, $T_h(t)$, and $T_L(t)$ at the Ge surface, the calculated Drude contribution is shown by the lower dashed curve for $\Delta\varepsilon_1(t)$ in Fig.1 (a) and approximate nothing for $\Delta\varepsilon_2(t)$. The observed $\Delta\varepsilon_1(t)$ and $\Delta\varepsilon_2(t)$ are well explained by introducing band gap renormalization (BGR) by hot carriers. The dielectric functions $\Delta\varepsilon_{1,2}(t)$ are directly related to the exchange-

Springer Series in Chemical Physics, Vol. 60 **Ultrafast Phenomena IX**
Editors: P. F. Barbara · W. H. Knox · G. A. Mourou · A. H. Zewail

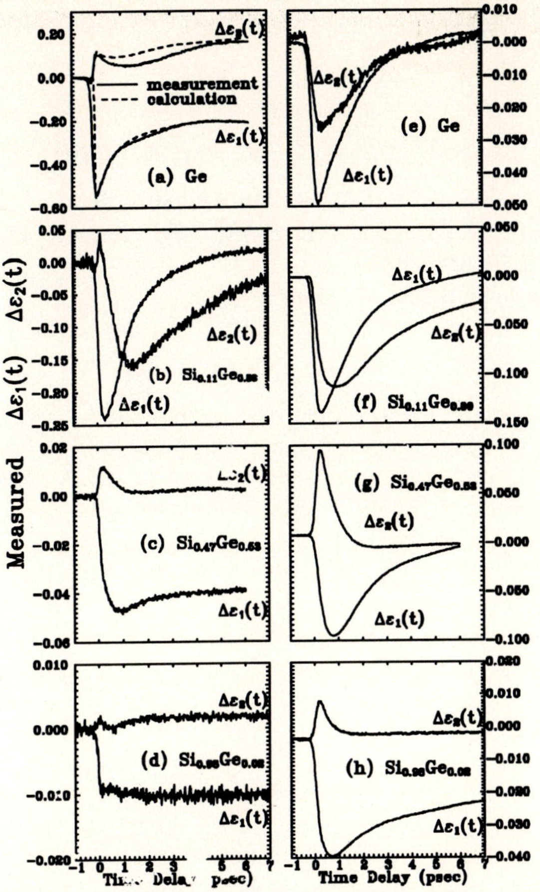

Figure 1 Femtosecond time-resolved changes in the real and imaginary parts of the dielectric function.

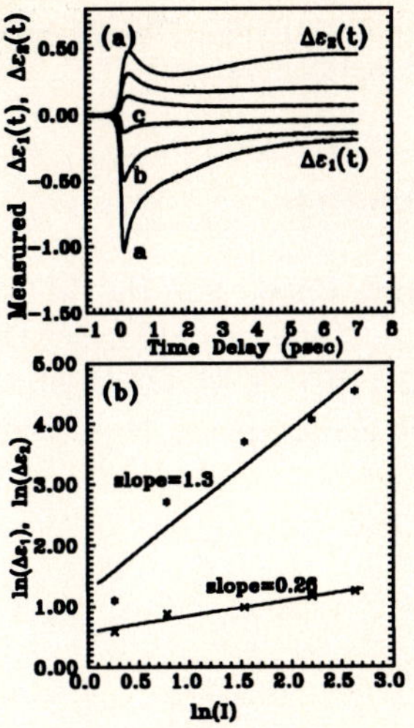

Figure 2: Dependence of dielectric response on pump intensity. (a)$\Delta\varepsilon_{1,2}$ for three pump intensities at 620nm with $I_a > I_b > I_c$. (b) log relation between peak $\Delta\varepsilon_{1,2}$ and pump intensity.

correlation energy ΔE_g, which equals the downward shift of the band gap when free carriers of density $N_{eh}(t)$ and temperature $T_{e\text{-}h}(t)$ are generated. Zimmermann[3] has derived a universal analytic formula for ΔE_g, which shows that ΔE_g increases approximately as $N^{1/4}$, but decreases with increasing carrier temperature. The revised $\Delta\varepsilon_2(t)$ from BGR is shown by the upper dashed curve in Fig.1(a). This model clearly captures the basic observed dynamics of $\Delta\varepsilon_2(t)$. Preliminary calculations show that a similar model explains the dielectric responses of all of the Si-like materials.

To further test this Zimmermann BGR model, we investigated the dependence of dielectric response on pump intensity (see Fig.2). The slope for $\Delta\varepsilon_2^{max}$ relative to N_{eh} is ~ 1/4, which agrees very well with Zimmermann BGR theory. The slope for $\Delta\varepsilon_1^{max}$ relative to N_{eh} is about 1.3, which agrees well with Drude model assuming linear absorption. The $Si_{1-x}Ge_x$ samples were provided by T.E. Haynes (Oak Ridge National Laboratory).

References

[1] H.R. Choo, X.F. Hu, M.C. Downer, and V.P. Kesan, APL **63**, 1507(1993).

[2] A. Otho os, H.M.van Driel, J.F.Young, and P.J.Kelly, Phys. Rev. B **43**, 6682(1991); H. M van Driel, *ibid* **35**, 8166(1986).

[3] R. Zimmermann, Phys. Status Solidi B **146**, 371 (1988).

Advances in Ultrafast Scanning-Probe Microscopy

S. Weiss, D. Botkin, D.F. Ogletree, M. Salmeron, and D.S. Chemla

Materials Sciences Division, Lawrence Berkeley Laboratory, Berkeley, CA and Department of Physics, University of California at Berkeley, Berkeley, CA 94720, USA

Abstract. We review the development of the ultrafast scanning tunneling microscope and present results on the ultrafast excitation of the tunneling gap.

Recently, we proposed a general technique to wed ultrafast laser spectroscopy and Scanning Probe Microscopies (SPM) to obtain simultaneous picosecond time and atomic space resolution [1]. Demonstration of this concept immediately followed [2]. In the experiment, a short (100 fs) laser pulse activated a photoconductive switch on a SOS transmission line, in order to generate short (650 fs) voltage pulses. While the tip was in tunneling range from one of the conductors of the transmission line, a second laser pulse gated the Scanning Tunnelng Microscope (STM) tip assembly. Thus, we could excite the tunneling gap of the STM with a short voltage pulse, and measure its response. We also used this instrument to image a surface with 2 ps time resolution and better than 50 Å spatial resolution. The details of the experimental arrangement are described eleswhere [2].

One of the most exciting prospects for USPM is the potential to create movies of surface dynamics on atomic scale, with atomic (ultrafast) time resolution. This can be done by collecting a series of STM images for increasing values of time delay and animating them. We beleive it should be possible to spatially and temporaly resolve phenomena such as vibronic motion on the atomic scale, charge transfer in biological materials, semiconductors and semiconductor devices, photochemical reactions on surfaces and more. To this end, understanding the dynamic response of the quantum point contact formed between the STM tip and the sample surface is essential.

Repeated time resolved tunnel current measurements suggest that a large portion of the tunneling signal comes from a capacitive-like response. The lineshape of such measurements is derivative-like with respect to the lineshape of a measurement done with the tip in ohmic contact with the sample [2]. Moreover, while in tunneling, the height of the correlation peak in each trace, as function of the corresponding DC current, shows a linear relation (Fig. 1). The dashed line is a linear regression to the data. It is clear from the figure that the AC part of the tunneling current has the same gap impedance, and hence distance, dependence as that of the DC current. In addition, when the tip is withdrawn from the surface by 50 Å, not only does the DC tunnel current vanish (as expected), but the AC part nearly vanishes as well. These observations, which have been reconfirmed many times, suggest that "quantum mechanical capacitance", associated with the tunneling proccess itself, is responsible for the time resolved tunnel current data.

Springer Series in Chemical Physics, Vol. 60 **Ultrafast Phenomena IX**
Editors: P. F. Barbara · W. H. Knox · G. A. Mourou · A. H. Zewail

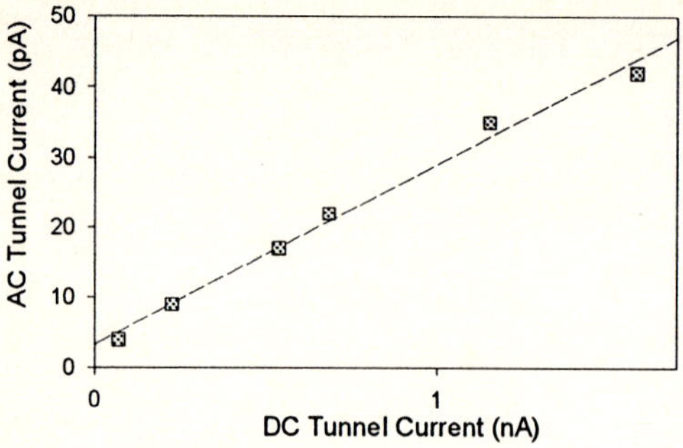

Figure 1. Amplitude of the time resolved tunnel current vs. DC tunnel current.

The tunneling gap height dependence of this response is very different from the one expected for the geometrical capacitance of the junction. In fact, the geometrical capacitance of the tip assembly contributes negligibly to the signal. The existance of an intrinsic capacitance associated with the tunneling process itself (quantum susceptance) has been discussed for instance in the context of the hotly disputed problem of "tunneling time" [3] and in the context of SIS tunnel junctions, where it is attributed to the noninstantaneous current voltage relations in these junctions which originates from "quantum sloshing" due to photon assitted tunneling [4]. In both cases, quantum susceptance is the result of time lag between excitation and the tunnel current itself.

A detailed study of the nature and the origin of the tunnel-gap capacitance is currently under way. Our current efforts are direced at comparing the predictions of the existing theories for quantum susceptance with our experimental results.

Simultanously, a second generation microscope, which is optimized for "movie making", has been completed. We will describe this microscope and the near future experiments which will take advantage of this new design.

Acknowledgements
This work was supported by a Laboratory Directed Research and Development grant from the Lawrence Berkeley Laboratory, under the US Department of Energy, contract DE-AC03-76SF00098 and by ONR/ARPA under contract #N000-14.93.105.36

References

[1] S. Weiss, D. Botkin and D.S. Chemla, OSA proceedings on *Ultrafast Electronics & Optoelectronics*, vol.14, San Francisco, January 25 1993, edited by J. Shah and U. Mishra, (OSA, Washington D.C. 1993), pp.162-165.

[2] S. Weiss, D.F. Ogletree, D. Botkin M. Salmeron and D.S. Chemla, Appl. Phys. Lett., **63**, 2567 (1993).

[3] For a review, see A.P. Jauho (edited by J. Shah): *Hot Carriers in Semiconductors Nanostructures*, (Academic Press, NY, 1992) p. 121.

[4] J.R. Tucker and M.J. Feldman, Rev. Mod. Phys., **57**, 1055 (1985).

Time-Resolved Nonlinear Near-Field Optical Microscopy of Semiconductor Microdisks

J.B. Stark, U. Mohideen, E. Betzig, and R.E. Slusher

AT&T Bell Laboratories, Murray Hill, NJ 07974, USA

Abstract. We use 40 femtosecond optical pulses and near-field scanning optical microscopy techniques to image the nonlinear optical response of GaAs quantum wells in the plane of a semiconductor microdisk.

We report the results of experiments that use nonlinear, time-resolved, near-field scanning optical microscopy techniques to study the local carrier dynamics and spatial distributions in optically thin GaAs microdisks that are a few wavelengths in diameter. In this microcavity limit, there is only one low-loss mode that interacts with the optically active material in the cavity: the whispering-gallery mode propagating within a few hundred nanometers of the disk edge [1]. In addition to very low threshold laser oscillation, there are many interesting phenomena associated with microcavities, including the control of the spontaneous emission rates and statistics of the emitted light, anomalous laser linewidths, and nonequilibrium effects in the active region. Time-resolved nonlinear spectroscopy, together with near-field microscopy, can be used to directly measure the nonequilibrium carrier dynamics of these microdisks, on a time scale of 40 fs, and length scale of 100 nm.

The near-field scanning optical microscope [2] consists of a tapered single-mode optical fiber that is coated with aluminum to leave a 100 nm aperture at the tip. Light injected into the fiber is coupled evanescently into the near field of this aperture, and propagates into the far field as a traveling wave. In the near field, the electric field is confined laterally by the aperture, so that the dielectric response of a near-field sample can be measured with sub-wavelength spatial resolution.

The microdisk consists of eight layers of 10 nm GaAs quantum wells and 5 nm barriers, with 15 nm $Al_{0.28}Ga_{0.72}As$ protective layers on the top and bottom of the disk. Microdisks, 150 nm thick and with diameters between 2 and 10 µm, are floated off of a GaAs substrate by dissolving an AlGaAs layer below the disks. They then settle onto a sapphire substrate.

The excitation and probe optical pulses are generated by a mode-locked Ti:Sapphire laser with output pulsewidths below 20 fs. The central wavelength of 800 nm overlaps the GaAs quantum well bandedge absorption at room temperature, while the 80 nm spectral bandwidth provides a complete measurement of the carrier dynamics near the bandedge. A grating precompensator disperses the pulses so that, after emerging from the near-field fiber, their pulsewidths are 40 fs. The relative time delay between the equal-amplitude pump and probe pulses is determined by a variable delay line. Crossed linearly polarized pump and

Springer Series in Chemical Physics, Vol. 60 **Ultrafast Phenomena IX**
Editors: P. F. Barbara · W. H. Knox · G. A. Mourou · A. H. Zewail

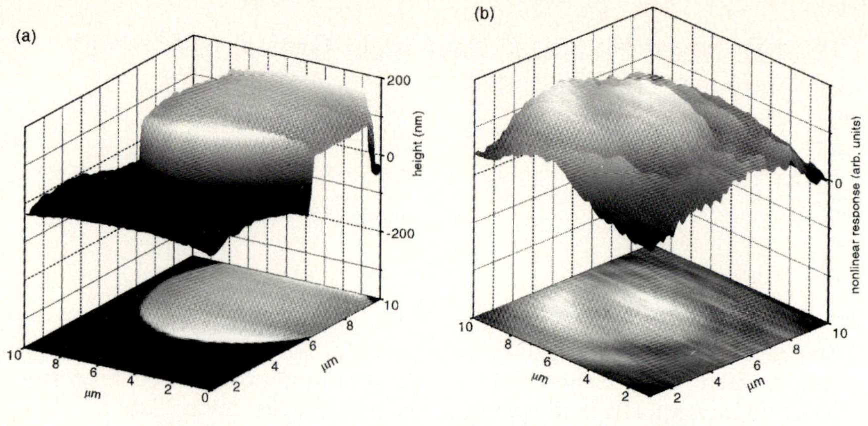

Figure 1. GaAs microdisk, shear-force image (a) and near-field nonlinear optical response (b) for pump-probe delay of 100 fs.

probe fields are used to minimize interference. The energy per pulse at the sample, with a 1 mW average input to the probe fiber and 50 dB insertion loss, is on the order of 1 $\mu J/cm^2$. This corresponds to 10^{11} electron-hole pairs/cm^2, near the saturation density of the GaAs quantum wells.

The fiber tip is used in shear-force feedback mode to image a 10 μm diameter microdisk, shown in Fig. 1(a). The gray-scale image below, and relief image above, depict the topography of the microdisk. A nonlinear transmission image is shown in Fig. 1(b), where the signal is the amplitude of the transmitted pump and probe pulses, measured at the sum of the frequencies used to chop the two beams. This nonlinear image represents the intensity of the probe, as the transmission of the sample is modulated by the pump. Structure in the nonlinear response, presumably due to the proximity of the surface, is visible at the edges of the microdisk, as well as in the center. Studies continue, to resolve the spatial, spectral and temporal details of this nonlinear response, and its relation to carrier dynamics.

Time-resolved nonlinear near-field optical microscopy provides a measure of the local nonlinear optical response of semiconductor microdisks. The field of high spatial- and temporal-resolution measurements, now in its infancy, is developing rapidly, and will soon find widespread application to the measurement of local dynamics in semiconductor, chemical and biological microstructures.

References

[1] U. Mohideen, W. S. Hobsen, S. J. Pearton, F. Ren and R. E. Slusher, Appl. Phys. Lett **64**, 1911 (1994).

[2] E. Betzig and J. K. Trautman, Science **257**, 189 (1992).

Ultrafast Dynamics in CdSe Nanocrystals

D.M. Mittleman[1], *S.J. Rosenthal*[1], *R.W. Schoenlein*[1], *A.T. Yeh*[1], *C.V. Shank*[1], *J.J. Shiang*[2], *V.L. Colvin*[2], *R.A. Grubbs*[2], *and A.P. Alivisatos*[2]

[1]Lawrence Berkeley Laboratory, 1 Cyclotron Road, MS 70-193A, Berkeley, CA 94720, USA

[2]Department of Chemistry, University of California at Berkeley, Berkeley, CA 94720, USA

The control and understanding of the dynamics of charge carrier motion within confined semiconductor systems is a subject of technical importance. In this paper we present experiments which probe and modify the confined wavefunction within chemically synthesized CdSe nanocrystals of 20-100Å in diameter. The shape of this wavefunction is controlled not only by the nature of the crystalline interior, but by the chemical nature of the surface as well. By performing time resolved femtosecond pump-probe and three pulse photon echo (TPPE) experiments, in combination with steady state spectroscopies such as Resonance Raman(RR), one can determine the size and time evolution of the optically excited wavefunction. In this paper, we discuss two complimentary experiments, Resonance Raman, and TPPE.

Both the apparatus and the synthesis of the nanocrystallites have been previously described.[1] The blue-green laser system[2] produces 20fs pulses which are tunable between 480nm and 580nm. In both the RR and TPPE experiments, the laser wavelength was tuned to the peak of the lowest excited state. The CdSe nanocrystals were prepared using the decomposition methods developed by Murray et al. The resulting nanocrystals are monodisperse to within 5% of the radius, and have highly crystalline interiors, with less than one defect per crystallite. The surfaces of the nanocrystal are chemically passivated with tri-octyl-phosphine-oxide (TOPO).

The RR and TPPE measurements probe similar features of the excited state surface, and our previous studies showed that both experiments may be analyzed within the framework of an offset harmonic oscillator model which has two parameters.[3] One is the electronic dephasing time,T_2 ,measured by the decay of the TPPE signal. The other is electron-optical phonon coupling strength, Δ, which appears both in the TPPE signal and as a change in the ratio of the RR fundamental to overtone intensity. A summary of the experimental results obtained at 15K for different size nanocrystals is given in Figure 1.

The results show two trends with size. The first is that the total rate of dephasing decreases as the size is increased. The rate of population decay from the initial state (T_1) is given by the pump-probe signal. Our results are consistent with the theoretical results reported by Nomura et al. which show a decrease in acoustic mode coupling as the size is increased [4]. The second result is that Δ ,the coupling to the optical modes , increases as the size is increased. In solids,

Springer Series in Chemical Physics, Vol. 60 **Ultrafast Phenomena IX**
Editors: P. F. Barbara · W. H. Knox · G. A. Mourou · A. H. Zewail

A Four-Wave-Mixing Study of Electron Scattering in Modulation-Doped GaAS Quantum Wells in High Magnetic Field

S. Bar-Ad and I. Bar-Joseph*

Department of Physics, The Weizmann Institute of Science,
Rehovot 76100, Israel
*Present address: Materials Sciences Division, Lawrence Berkely Laboratory,
1 Cyclotron Road, Berkeley, CA 94720, USA

Abstract. A decrease of the dephasing rate is observed as a high magnetic field is applied to modulation doped GaAs quantum wells. We attribute an enhancement of the polarization at the lowest Landau level to the Fermi edge singularity.

It has recently been shown by Kim et al. that the dephasing of the induced polarization in the vicinity of the absorption edge in modulation doped (MD) quantum wells (QW) is due to electron-electron (e-e) scattering.[1] Our goal in this work was to measure the dephasing rate at strong magnetic field and to determine the effect of such field on the e-e scattering process. To achieve this goal we have conducted time-resolved four wave mixing (FWM) measurements in two GaAs MD samples ($n= 1.8x10^{11}$ and $1.0x10^{11}$ cm^{-2}) in strong magnetic fields (B<8T) at low temperatures (T=4K).

The FWM signals from both samples show the same qualitative behavior. Figure 1a compares the FWM signals which are measured in the high density sample close to the Fermi energy, at B=0 and at B=6.5 T, when the Fermi level is in the lowest Landau level (LL). Two main findings are clearly seen: (i) the high field signal is substantially stronger than that at B=0; (ii) the high field signal rises and decays more slowly than the B=0 signal (it decays exponentially with a time constant of 1.5 ps.) As we tune our laser throughout the spectral width of the lowest LL, the same temporal behavior is measured.

Our measurements imply that the e-e scattering rate in high magnetic fields is lower, and is not as sensitive to excess energy as in the zero field case.[1] We can only set an upper limit for this scattering rate. It is very likely that impurity scattering is the limiting process which determines the dephasing time at high magnetic field in our samples. In fact, the 1.5 ps dephasing time at 6.5 T is roughly the Drude scattering time we determine from mobility measurements. The measured increase of the dephasing time for B > 0 could be understood if we consider the large inhomogeneous broadening. While at zero magnetic field a smooth random potential only slightly changes the density of states and does not affect the scattering rate significantly, in strong B severe limitations are imposed on the phase space for e-e scattering.[2]

We now turn to the spectral behavior of the FWM signal. Figure 1b illustrates the evolution of the FWM spectrum as the magnetic field is increased, on a logarithmic scale, for the low density sample. It can be seen that the FWM spectra

Springer Series in Chemical Physics, Vol. 60 **Ultrafast Phenomena IX**
Editors: P. F. Barbara · W. H. Knox · G. A. Mourou · A. H. Zewail

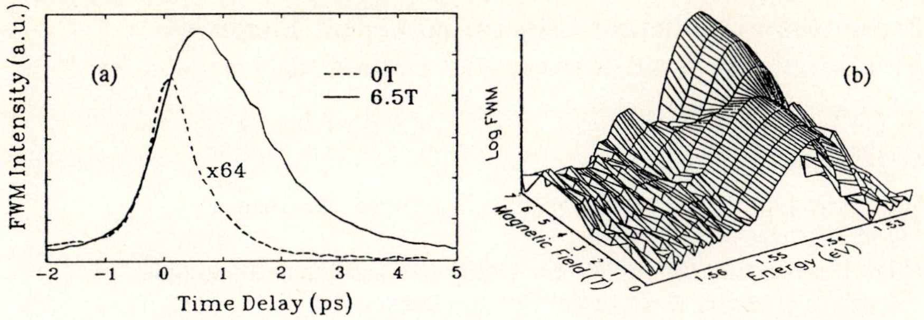

Figure 1: (a) Decrease of the dephasing rate close to the Fermi level at high B. (b) FWM spectra in magnetic field on a logarithmic scale.

are dominated by a single peak, which shifts as the Fermi level jumps from the second LL to the first LL. The signal from the second LL *decreases* with increasing magnetic field, opposite to what one finds in intrinsic samples. In high magnetic fields it is weaker by more than two orders of magnitude than the signal from the first LL. Also, it is surprising to note the apparent lack of any contribution to the FWM signal from light hole related transitions. The signal which is measured at the second LL shows two temporal components, one which follows the pulse and a slower component, which rises and decays on a time scale comparable to that of the dephasing in the first LL. Experiments are in progress to investigate this anomalous response.

It is evident from our results that the proximity of the Fermi level to the excitation energy strongly affects the observed FWM signal in the MD samples. Unlike the measurements in zero magnetic field,[1] where a strong dependence of the FWM intensity on the excess energy corresponded to a decrease of the dephasing time τ, our high magnetic field measurements do not exhibit a strong dependence of τ on excess energy. We attribute the observed spectral dependence of the signal to the Fermi edge singularity.[3] Indeed, the FWM peak intensity decreases exponentialy with temperature, with a decay constant of 17K (~1.4 meV), while no significant changes of τ are observed up to 30K. These temperature and energy scales are characteristic of the Fermi edge singularity.[4] We emphasize that the Fermi edge singularity is a manifestation of the scattering of conduction electrons by the photo-excited holes, and that this scattering mechanism should be considered particularly as e-e scattering diminishes. In conclusion, a better theory is needed, which would take into account e-e scattering, scattering by holes, and the effect of disorder.

References

[1] D. S. Kim et al., Phys. Rev. Lett. **68**, 2838 (1992).
[2] S. Bar-Ad et al., Phys. Rev. Lett. **72**, 776 (1994).
[3] G. D. Mahan, Phys. Rev. **153**, 882 (1967).
[4] G. Livescu et al., IEEE J. Quant. Electron. **QE 24**, 1677 (1988).

Femtosecond Coherent Generation of Hot Electrons Monitored Via Band-to-Acceptor Luminescence in GaAs

A. Leitenstorfer[1], *T. Elsaesser*[1,2], *A. Lohner*[1], *T. Kuhn*[3], *S. Haas*[4], and *F. Rossi*[4]

[1]Physik Department E 11, Technische Universität München, D-85748 Garching, Germany
[2]Max-Born-Institut für Nichtlinear Optik und Kurzzeitspektroskopie, Rudower Chaussee 6, D-12489 Berlin, Germany
[3]Institut für Theoretische Physik, Universität Stuttgart, D-70550 Stuttgart, Germany
[4]Fachbereich Physik, Universität Marburg, D-35032 Marburg, Germany

Abstract. Luminescence spectra observed after femtosecond excitation demonstrate initial electron distributions that are strongly influenced by the coherent coupling of the incident light field and the semiconductor.

Optical interband excitation of semiconductors results in an ultrafast electron-hole dynamics which is influenced by both coherent and incoherent phenomena. Until now, the influence of the coherent coupling between the femtosecond optical pulse and the semiconductor on the initial carrier distribution has not been clarified although the shape of this distribution is essential for the subsequent thermalization of the carriers. In this contribution, we provide experimental evidence that nonequilibrium electron distributions created by femtosecond excitation are strongly affected by the coherent light-matter interaction. Non-equilibrium electron distributions in p-type GaAs which are created by excitation with 150 fs pulses centered at 1.73 eV are monitored via time-integrated band-to-acceptor luminescence [1]. In Fig. 1 (a-c), we present hot luminescence spectra for excitation densities between 8×10^{13} and 2×10^{16} cm^{-3}. The first peak around

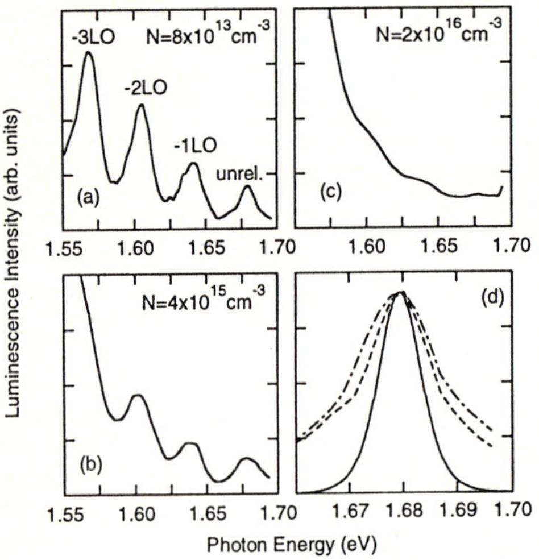

Figure 1 (a-c) Hot electron luminescence spectra of GaAs measured for different excitation densities N (lattice temperature 10 K). The spectra show a first peak (unrel.) that originates from conduction band states optically coupled by the excitation pulse and a series of phonon replicas. The lines exhibit an identical spectral width and broaden with increasing carrier concentration. (d) Spectral profile of the first luminescence peak for carrier densities of 8×10^{13} (dashed line) and 5×10^{14} cm^{-3} (dash-dotted line). The laser spectrum is given for comparison (solid line).

Springer Series in Chemical Physics, Vol. 60 **Ultrafast Phenomena IX**
Editors: P. F. Barbara · W. H. Knox · G. A. Mourou · A. H. Zewail

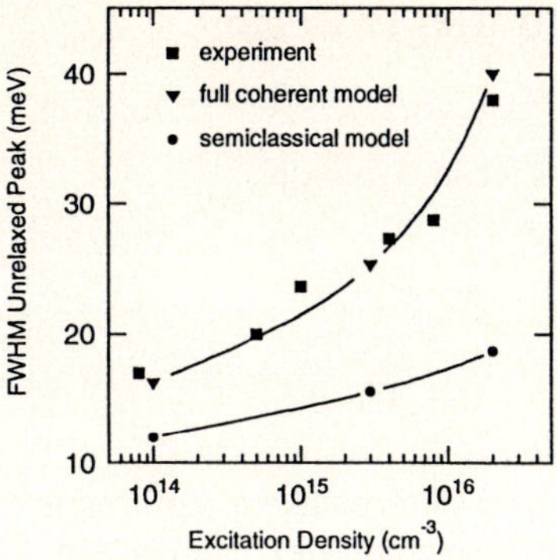

Figure 2 Spectral full width at half maximum (FWHM) of the first luminescence peak as a function of excitation density. The experimental results (squares) are compared to theoretical values calculated from a simulation including the coherent coupling of laser field and polarization in the sample (triangles) and from a semiclassical model of the incoherent carrier dynamics (circles).

1.68 eV originates from the recombination of electrons in conduction band states directly populated by laser excitation.

In Fig. 1 (d), this part of the spectrum is plotted on an extended scale. Even at low carrier concentration, the spectral width is substantially larger than the bandwidth of the laser pulses. At lower photon energies, a series of phonon replicas is observed which are due to electrons which have emitted one, two or three LO phonons and reflect the carrier dynamics at later times. With increasing carrier density, the luminescence maxima broaden due to the increasing strength of carrier-carrier (cc) scattering. At each density, the first ('unrelaxed') peak exhibits already the same spectral width as the phonon replicas occurring at later times. In Fig. 2, the spectral width of the unrelaxed peak is plotted as a function of excitation density (squares).

Our experimental results reveal the coherent nature of the excitation process. Phase breaking scattering events during excitation lead to a significant broadening of carrier generation in **k**-space, resulting in an energetic width larger than the laser bandwidth. The measured constant width of the different luminescence peaks is a clear consequence of the fact that the broadening is mainly introduced during carrier generation while the subsequent incoherent scattering processes are of minor importance.

This interpretation is supported by simulations of the coherent dynamics and of the carrier redistribution processes based on a Monte-Carlo solution of the semiconductor Bloch equations [2]. The linewidths calculated from this model are plotted in Fig. 2 (triangles) and reproduce the data very well whereas the behavior derived from a semiclassical model neglecting the coherent coupling are not in agreement with the data (circles). The full coherent calculation gives an identical width of the different luminescence peaks at a fixed density. These findings are related to the details of the cc-scattering processes : The efficiency of carrier-carrier interaction for dephasing is higher than that for electron redistribution. The decay of the interband polarization responsible for the linewidth is strongly enhanced by scattering among the holes created in the excitation process which is much faster than electron-electron scattering.

References

[1] J.A. Kash, Phys. Rev. **B 40**, 3455 (1989)
[2] F. Rossi, S. Haas, and T. Kuhn, Phys. Rev. Lett. **72**, 152 (1994)

Terahertz Study of Field Screening: 10–100 fs Regime

B.B. Hu, E.A. De Souza, W.H. Knox, M.C. Nuss, and J.E. Cunningham

AT & T Bell Laboratories, 101 Crawfords Corner Road, Holmdel,
NJ 07733-3030, USA

One of the unsolved problems in THz emission from semiconductor surfaces is to distinguish between the contributions caused by acceleration of carriers and by instantaneous polarization of electron and hole pairs under an electric field. Such a study requires a time resolution in the range of 10 fs, for the THz emission due to the ballistic acceleration of photocarriers quickly increase in the early time stage. Here, we present the first THz study on such a time scale. Our experimental results clearly separate the contributions of both effects in the different times.

The experimental setup is shown in Fig. 1. To achieve a time resolution of the optical pulse duration, we use correlation techniques. In this case, the time resolution is limited only by the optical pulse duration, not the detector response. We measure the total THz pulse energy created by optical pulse 1 under the influence of the prepulse (pulse 2) at different relative time delays. The change in THz energy then reflects the screening of the field by the pre-pulse.

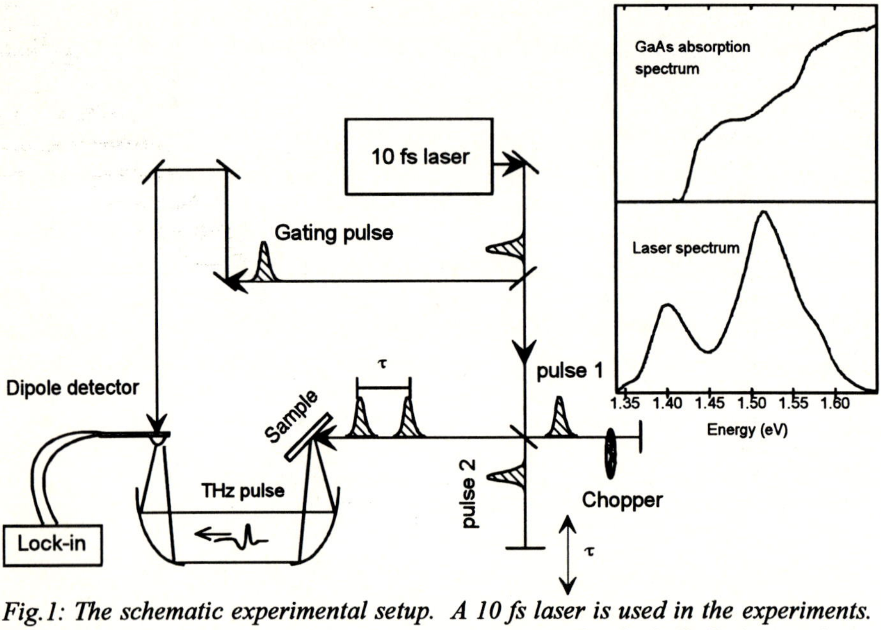

Fig. 1: The schematic experimental setup. A 10 fs laser is used in the experiments. Inset: The laser spectrum and the absorption spectrum for the bulk GaAs sample.

Springer Series in Chemical Physics, Vol. 60 **Ultrafast Phenomena IX**
Editors: P. F. Barbara · W. H. Knox · G. A. Mourou · A. H. Zewail

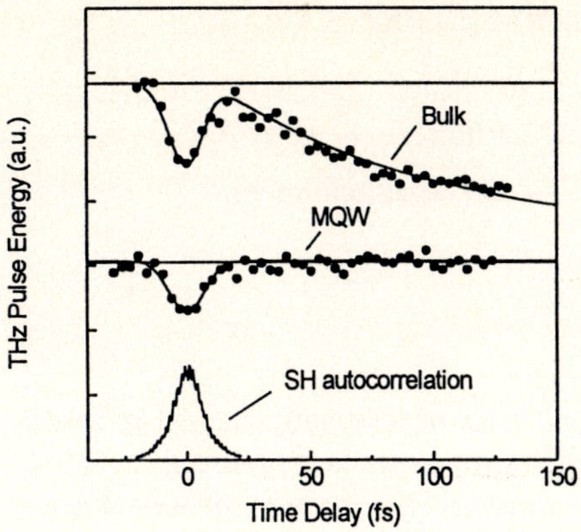

Fig.2 : The THz pulse energy as a function of the time delay relative to the prepulse. Bottom: Intensity autocorrelation of the 10fs optical pulse.

Both a bulk sample and a MQW sample are used in the experiments. The bulk sample has a 10 μm GaAs layer grown by MBE on a n^+ GaAs substrate. The MQW sample consists of 100 periods of GaAs/$Al_{0.3}Ga_{0.7}As$ (100Å/150Å). The built-in field in both samples is estimated to be 20 kV/cm. The experimental results are plotted in Fig. 2. All measurements have been repeated at different crystal azimuthal angles. No crystal orientation dependence is observed, indicating that the bulk $\chi^{(2)}$ effect in our measurements is not important.

The experimental results clearly demonstrate both the instantaneous polarization effect and the transport effect at different times. In a MQW structure, since the carrier transport motion is prohibited by the quantum confinement, the instantaneous polarization is the only source for THz emission [1]. We observe a sharp dip with a width comparable to the optical pulse duration. In the bulk sample however, both the instantaneous polarization and the carrier acceleration contribute to the THz emission. As shown in Fig.2, in addition to the sharp dip as in the MQW case, we also observe a relatively slow decay due to the field screening caused by the carrier transport motion. This decay has a time constant of 100 fs. In our measurement the photocarrier density is around $4 \times 10^{15} cm^{-3}$. In the initial 100 fs, the carrier transport motion is ballistic. We calculate from our experiment condition the decay time to be 120 fs which is close to the experimental value.

References

[1] P.C.M. Planken, M. C. Nuss, W. H. Knox, K. W. Goossen, D. A. B. Miller, and J. E. Cunningham, Appl. Phys. Lett. vol., **61**, 2009, 1992.

Extremely Fast Biexcitonic Radiative Decay in GaAs Quantum Wells

H. Wang, J. Shah, T.C. Damen, and L.N. Pfeiffer

AT&T Bell Laboratories, 101 Crawfords Corner Road, Holmdel, NJ 07733-3030, USA

In this paper we present the first femtosecond study of resonantly-excited excitonic as well as biexcitonic luminescence in GaAs quantum wells. By directly creating biexcitons through two-photon absorption, we are able to measure the spontaneous emission rate of biexcitons in GaAs quantum wells. The improved time resolution also demonstrates the crucial role of momentum relaxation in initial spontaneous emission of excitons in quantum wells, reflecting the fundamental difference between the exciton-photon and the atom-photon interactions.

The femtosecond time resolution for resonantly-excited luminescence is achieved by upconverting the luminescence in a nonlinear crystal with output from a synchronously pumped optical parametric oscillator. Excitons are resonantly excited with 120 fs pulses centered at 6 meV below the heavy hole resonance. The samples were kept at 10 K and the exciton densities used are estimated to be between 10^9 to $10^{10}/cm^2$.

Although decay of the excitonic luminescence has been studied in detail in GaAs quantum wells [1,2], no information is available on decay of the biexcitonic luminescence. We investigate the biexcitonic spontaneous emission by using linearly polarized excitation pulse and detecting the decay of the cross-polarized

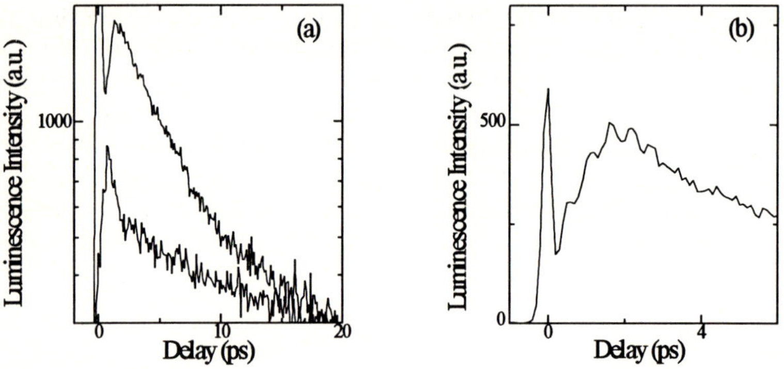

Fig. 1 Temporal evolution of resonantly-excited exciton and biexciton luminescence. The spike at t=0 is due to scattering of the laser pulse from the sample surface. (a) Excited with a linearly polarized field. Top curve: co-polarized luminescence. Bottom curve: cross-polarized luminescence. (b) Excited with a circularly polarized field.

Springer Series in Chemical Physics, Vol. 60 **Ultrafast Phenomena IX**
Editors: P. F. Barbara · W. H. Knox · G. A. Mourou · A. H. Zewail

luminescence. Because of the unique optical selection rule for the biexciton, in the absence of spin relaxation, the cross-polarized luminescence results from decay of the biexciton into a photon and an exciton. The biexcitons in this case are created resonantly through the biexcitonic two-photon absorption process. Figure 1a shows decay of both co-polarized and cross-polarized luminescence from a 135 Å quantum well. The ratio of the peak intensities of the cross-polarized over the co-polarized luminescence increases linearly with the input photon flux at low excitation levels, confirming the biexcitonic origin of the cross-polarized luminescence. The initial decay of the cross-polarized luminescence is extremely fast (3 ps), in sharp contrast to the exciton luminescence. The decay rate is much faster than both the exciton energy relaxation rate and the exciton recombination rate [1,2]. Below 20 K and an exciton density of $10^{10}/cm^2$, the initial decay is nearly independent of the excitation level and temperature. The extremely fast decay rate, however, is consistent with the giant oscillator strength for the decay of a biexciton into a photon and an exciton [3].

The reverse process for the biexcitonic decay discussed above is the conversion of an exciton into a biexciton by absorbing an additional photon. This process can be demonstrated by pump-probe measurements with circularly polarized fields, and should be greatly enhanced by the giant biexciton oscillator strength. As expected, a pronounced biexciton-induced resonance is observed in the transmission spectrum of the probe pulse when a cross-polarized pump is used.

The temporal evolution of the photoluminescence shown in Fig. 1b is obtained with a circularly polarized excitation field. The periods of the oscillation correspond to the heavy and light hole splitting of the sample. The observed quantum beats are the result of the excitation of a coherent superposition state of the heavy and light hole excitons. What is striking in Fig. 1b is that, unlike similar measurements in atomic vapors, the luminescence signal does not peak right after the excitation pulse but at 2 ps later. Conceptually, excitons created by the excitation pulse have a momentum $k_{||}$, the photon momentum along the plane of the quantum well. Because of momentum conservation., the exciton can not emit a photon in directions other than the transmitted or reflected direction of the excitation pulse before the exciton momentum is randomized through momentum scattering. Hence, the observed rise time of the exciton luminescence is determined by momentum relaxation. We also note that the rise time of the biexcitonic luminescence shown in Fig. 1a still remains to be explained.

We wish to thank Spectra-Physics Lasers, Inc. for the loan of an optical parametric oscillator.

References:

1) B. Deveaud *et al.*, Phys. Rev. Lett. **67**, 2355 (1991).
2) A. Vinattieri *et al.*, Solid State Comm. **88**, 189 (1993).
3) A.A. Golovin and E.I. Rashba, JETP Lett. **17**, 478 (1973).

Bound and Unbound Two-Exciton Contributions to the Nonlinear Optical Response of GaAs Quantum Wells

J. Kuhl[1], *E.J. Mayer*[1], *G.O. Smith*[1], *D. Bennhardt*[2], *T. Meier*[2], *A. Schulze*[2], *S.W. Koch*[2], *P. Thomas*[2], *R. Hey*[3], *and K. Ploog*[3]

[1]Max-Planck-Institut für Festkörperforschung, D-70569 Stuttgart, Germany
[2]Department of Physics, Philipps University, D-35032 Marburg, Germany
[3]Paul-Drude-Institut für Festkörperelektronik, D-10177 Berlin, Germany

The influence of exciton/exciton Coulomb interaction on the nonlinear optical response of the 2D exciton in GaAs quantum wells (QW's) has been demonstrated in several recent publications [1-3]. Generally this interaction involves a renormalization of the transition matrix elements, dephasing rates and frequencies of excitonic states and the formation of biexcitons. Although evidence for the existence of biexcitons in GaAs QW's has been derived from high excitation photoluminescence (PL), pump-probe studies and degenerate-four-wave-mixing (DFWM), their contribution to the 3rd order nonlinearity is still an open question.
We present a 3-pulse DFWM configuration which can differentiate between pure local field effects and biexcitonic contributions to the time-integrated (TI) signal. Experiments were performed on an almost homogeneously broadened 20-nm-GaAs/$Al_{0.3}Ga_{0.7}As$ single QW (PL linewidth 0.3 meV, homogenous linewidth 0.15 meV) in backward reflection at 10 K. The sample was excited by a sequence of three pulses (duration 1.1 ps) with equal intensity, wave vectors $\mathbf{k}_1$, $\mathbf{k}_2$, $\mathbf{k}_3$ and delays τ_{12}=1.3 ps and τ_{13} (variable). The signal was monitored in the direction $\mathbf{k}_s=\mathbf{k}_1+\mathbf{k}_2-\mathbf{k}_3$. Analysis of the signal's dependence on the polarization of the exciting pulses is the crucial point of our investigations.
Figs. 1a and b depict the TI signal for 7 combinations of circularly and linearly polarized pulses versus $\tau_{13} = \tau$. The signal for the $\sigma^+\sigma^+\sigma^-$-geometry is 100 times smaller and is probably due to a 5th order process since 3rd order contributions are zero according to the theoretical model. The data show remarkable differences of the peak amplitude for the different polarization geometries. Several curves exhibit a pronounced modulation of the signal for $\tau > 0$ with a frequency corresponding to the biexciton binding energy Δ=1.2 meV, which has been independently determined from density dependent measurements of the PL line profile.
The DFWM signal is calculated by solving the optical Bloch equations for a model system which represents single-exciton and two-exciton contributions to the optical response by discrete transitions with adjustable oscillator strengths, dephasing rates and frequencies. The different mechanisms of exciton/exciton interaction are included phenomenologically, preserving the structure of the full semiconductor Bloch equations. According to this theoretical model, variations of the signal peak amplitude can originate from either excitation-induced dephasing (EID) or biexciton formation, whereas pure local field effects (LFE) have no influence on the amplitude but slow down the decay rate of the signal for τ>0 to twice the dephasing

Springer Series in Chemical Physics, Vol. 60 **Ultrafast Phenomena IX**
Editors: P. F. Barbara · W. H. Knox · G. A. Mourou · A. H. Zewail

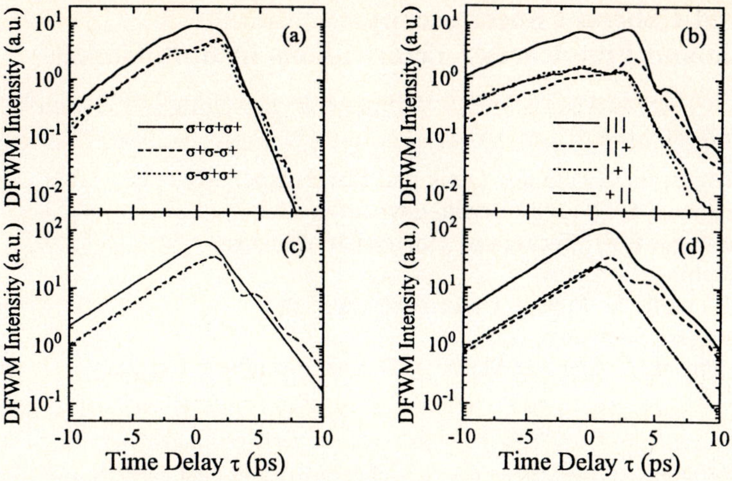

Fig. 1 The DFWM intensity versus delay τ for various polarization combinations. Exp.: (a) circular (b) linear. Theoretical: (c) circular and (d) linear polarizations (EID=0, LFE=1.2 meV, BI Δ=1.2meV, same oscillator strengths for the transitions from the single-exciton states to the two-exciton states and biexciton.).

rate of the exciton [4]. The unique feature for biexcitons (BI) is the appearance of quantum beats (QB) between BI and two-exciton scattering states for $\tau > 0$.
Comparison of the theoretical curves in Figs.1c and d with the experimental data proves that biexcitons provide a significant contribution to the 3rd order optical nonlinearity. Their presence is further confirmed by time-resolved DFWM [5]. At present, a quantitative determination of the individual contributions is impossible since signals of almost the same relative magnitude can se simulated with slightly different parameter sets. Finally, it should be mentioned that biexcitonic effects are more pronounced (strong QB appear for both positive and negative delays) at higher excitation densities ($>6\times10^8$ cm^{-2}) where 5th order nonlinear processes contribute to the diffracted signal.

[1] K. Bott et al., Phys. Rev. B **48**, 17418 (1993).
[2] H. Wang et al., Phys. Rev. Lett. **71**, 1261(1993).
[3] T. Saiki et al., Phys, Rev. B **49**, 7817 (1994).
[4] M. Wegener et al., Phys. Rev. A **42**, 5675 (1990).
[5] E.J. Mayer et al., see this book.

Femtosecond Coherent Spectroscopy of Excitonic and Free-Carrier Polarizations in Bulk GaAs

T. Elsaesser[1,2], *A. Leitenstorfer*[2], *A. Lohner*[2], *P. Leisching*[2], *T. Kuhn*[3], *F. Rossi*[4], *and W. Stolz*[4]

[1]Max-Born-Institut für Nichtlinear Optik und Kurzzeitspektroskopie, Rudower Chaussee 6, D-12489 Berlin, Germany

[2]Physik-Department E 11, Technische Universität München, D-85748 Garching, Germany

[3]Institut für Theoretische Physik, Universität Stuttgart, D-70550 Stuttgart, Germany

[4]Fachbereich Physik, Universität Marburg, D-35032 Marburg, Germany

Abstract. Third-order polarizations due to simultaneously excited excitons and free electron-hole pairs in GaAs are studied by femtosecond four-wave-mixing.

Femtosecond spectroscopy of coherent optical polarizations gives insight into the dynamics and the coupling of elementary excitations in semiconductors. In this paper, we present spectrally and temporally resolved four-wave-mixing (FWM) studies of undoped and p-doped bulk GaAs, providing new information on the excitonic and free-carrier polarization close to the bandgap. The experimental results are analyzed by theoretical calculations based on the optical Bloch equations for the FWM geometry including Monte Carlo simulations of the carrier dynamics.

In our experiments, 0.5 μm thick layers of undoped and p-doped GaAs (doping concentration 10^{18} cm^{-3}) clad by AlGaAs barriers are studied in a degenerate FWM geometry using 100 fs pulses from a mode-locked Ti:sapphire laser. The spectrum and the temporal characteristics of the FWM signals diffracted from the transient grating are measured with the help of a frequency-selective detection system and parametric up-conversion in a nonlinear crystal, respectively.

In undoped GaAs, excitonic and free-carrier components both occurring within the bandwidth of the laser pulses are distinguished for the first time [1]. In Figs. 1 (a,b), the spectrally resolved FWM signals are plotted for different excitation densities and time delays T_{12} between the two pulses generating the transient grating. The excitonic part dominates at carrier densities below 10^{16} cm^{-3} and follows a picosecond kinetics governed by many-body effects. The excitonic contribution persists up to an excitation density of 6×10^{16} cm^{-3}. At carrier densities above 10^{16} cm^{-3}, non-thermalized free carriers give rise to an additional polarization with a spectrum resonant to the laser pulse. With increasing T_{12}, the spectrum of this contribution shifts towards smaller photon energies while the excitonic enhancement disappears (Fig. 1 b). The free-carrier component shows a femtosecond photon-echo like time behavior [1].

The nonlinear response at the bandgap of the p-type material is governed by excitons which are bound to impurities and - consequently - show a slower dephasing kinetics than free excitons. In the time resolved FWM signal plotted in Fig. 2, a prompt component due to phase space filling and a second delayed contribution related to exciton-exciton interaction [2,3] are clearly distinguished.

The two components of nonlinear polarization and their change with carrier density are quantitatively reproduced by the theoretical calculations. The MC simulations give insight into the momentary carrier distributions and show that the photoexcited carriers

Springer Series in Chemical Physics, Vol. 60 **Ultrafast Phenomena IX**
Editors: P. F. Barbara · W. H. Knox · G. A. Mourou · A. H. Zewail

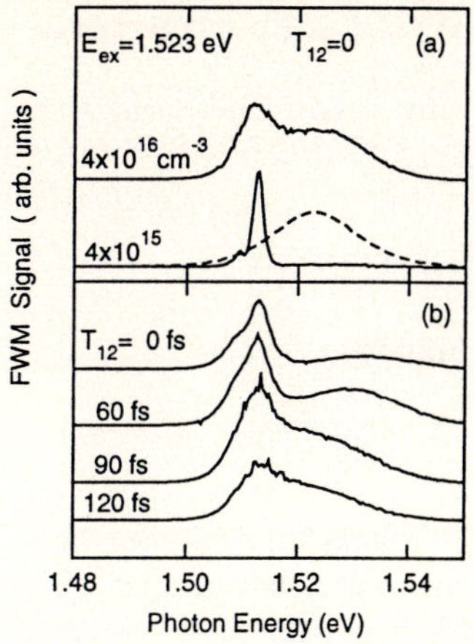

Figure 1 Spectrally dispersed FWM signals from undoped GaAs. (a) Spectral profiles found for zero delay (T_{12}=0) between the two pulses generating the transient grating in the sample (dashed line : spectrum of the 100 fs pulses centered at E_{ex} = 1.523 eV). (b) Spectrally resolved FWM signals for different delay times T_{12} (E_{ex}=1.531 eV, excitation density 1.5×10^{16} cm^{-3}).

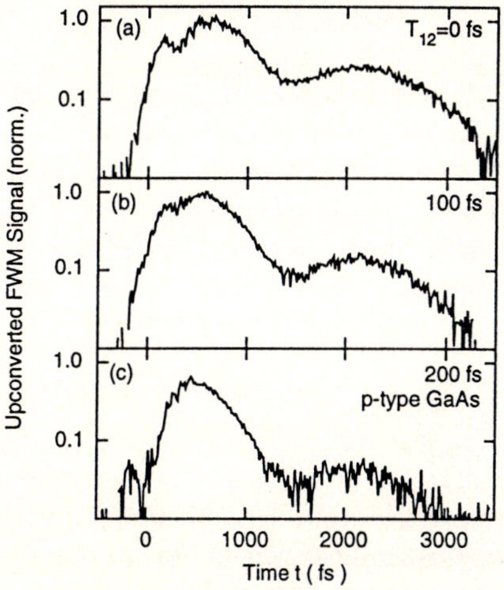

Figure 2 Temporal profiles of the FWM signal from p-type GaAs. The diffracted intensity is plotted as a function of the delay time t between the second pulse forming the grating and the (third) up-conversion pulse. Data are shown for different delays T_{12} (excitation density 6×10^{15} cm^{-3}).

thermalize on a time scale of 200 fs to Fermi distributions with different electron and hole temperature. The broad carrier distributions result in a reduced screening of the attractive electron-hole interaction and in an elevated Mott density of excitons. For increasing T_{12}, the diffracted intensity due to many-body effects in the carrier system is reduced, leading to a decrease of the excitonic component of the FWM signal.

References

[1] A. Lohner, K. Rick, P. Leisching, A. Leitenstorfer, T. Elsaesser, T. Kuhn, F. Rossi, and W. Stolz, Phys. Rev. Lett. **71**, 77 (1993); Phys. Rev. **B 49**, in press (1994)

[2] S. Weiss, M.-A. Mycek, J.-Y. Bigot, S. Schmitt-Rink, and D.S. Chemla, Phys. Rev. Lett. **69**, 2685 (1992)

[3] D.S. Kim, J. Shah, T.C. Damen, W. Schäfer, F. Jahnke, S. Schmitt-Rink, and K. Köhler, Phys. Rev. Lett. **69**, 2725 (1992)

Carrier Transport in 1.5 μm Quantum-Well Laser Structures

S. Marcinkevičius[1], *U. Olin*[2], *K. Fröjdh*[1], *J. Wallin*[3], *and G. Landgren*[3]

[1]Department of Physics II, Royal Institute of Technology, S-100 44 Stockholm, Sweden
[2]Institute of Optical Research, Royal Institute of Technology, S-100 44 Stockholm, Sweden
[3]Semiconductor Laboratory, Royal Institute of Technology, S-164 40 Kista, Sweden

Abstract. Carrier transport in step-graded InGaAsP/InP quantum well laser structures is studied by time-resolved photoluminescence using upconversion. In all investigated structures ambipolar carrier motion proceeds with a constant velocity of $(1.5\pm0.2)\times10^6$ cm/s. The experimental results are discussed in terms of step-driven and ordinary diffusion.

Carrier transport time is one of the possible factors limiting the high-speed performance of quantum well (QW) lasers. Therefore, investigations of transport mechanisms are essential for the improvement of laser characteristics. In this paper we examine the carrier transport in the 1.5 μm InGaAs(P)/InP graded-index separate-confinement heterostructure (GRINSCH) laser structures by means of time-resolved photoluminescence (PL). Such measurements produce a direct picture of the carrier motion and allow to draw conclusions about the transport mechanisms.

The MOVPE grown laser structures consisted of a *n*-InP buffer layer, a quaternary GRINSCH, a *p*-InP spacer, a 40 nm grating layer (E_g = 0.95 eV) and a 10 nm *p*-InP cap layer. The thickness of the InP spacer was different for various structures: 250 nm for structure A, 170 nm for structures B and C, and 25 nm for structure D. The GRINSCH of the structures is schematically shown in Fig. 1. The PL was measured at room temperature with the upconversion technique using a Ti:sapphire laser for excitation (pulse duration 100 fs, excitation wavelength 770 nm). The photoexcited carrier densities were of the order of 10^{18} cm^{-3} .

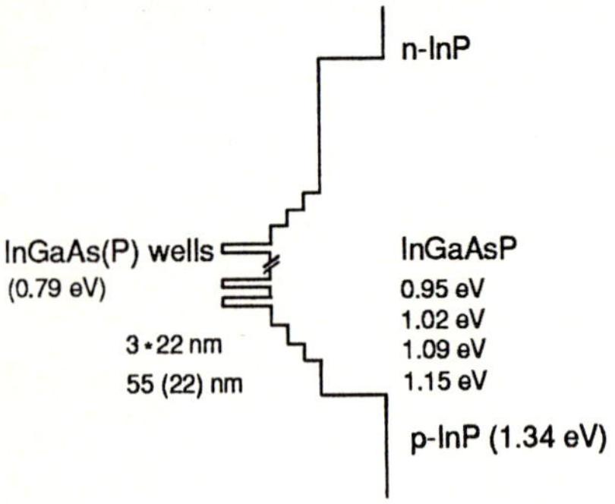

Fig. 1. Schematic diagram of the laser diode structures in the GRINSCH region. Given energy values denote room temperature band gaps for various layers.

Springer Series in Chemical Physics, Vol. 60 **Ultrafast Phenomena IX**
Editors: P. F. Barbara · W. H. Knox · G. A. Mourou · A. H. Zewail

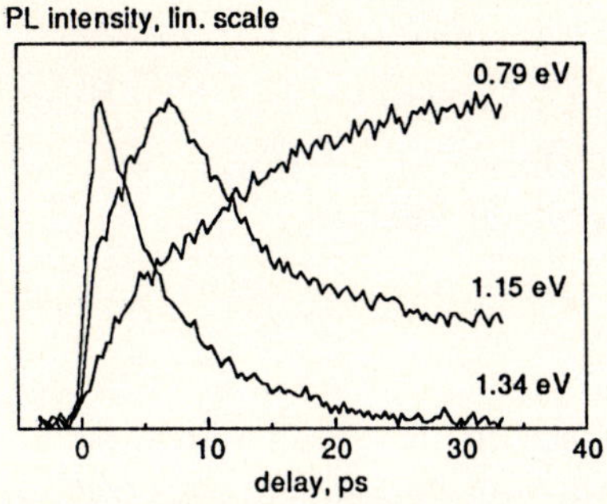

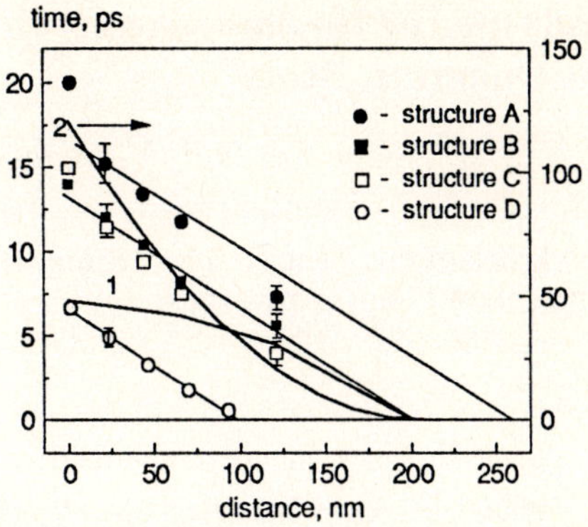

Fig. 2. (left) Temporal transients of the PL for the structure A measured at the band gap energies of certain layers in the GRINSCH.

Fig. 3. PL rise times as a function of a layer distance from the first QW on the *p* side of the structure. Calculated curves correspond to the step-driven (1) and ordinary (2) diffusion models for the structures B and C.

The PL spectra of the laser structures (not shown) exhibit a shift towards lower energies with increasing time after the excitation pulse, indicating carrier transfer from the InP spacer to the bottom of the GRIN structure and the QW. In more detail, the carrier transfer across the GRINSCH is traced measuring temporal transients of the PL at energies, corresponding to the band gaps of the GRINSCH layers. Several dependencies of this type are shown in Fig. 2. The decay time of the InP PL corresponds to the rise time of the 1.15 eV energy gap step. The same correlation holds for several subsequent steps. This allows us to associate the PL rise time with the carrier transfer time to a given GRINSCH layer.

Fig. 3 sums up the carrier transfer times for all the structures as a function of the carrier travel distance. For better comparison between various structures, the origin of the distance axis is chosen at the position of the first QW on the *p* side of the structure. For all the structures, the increase of the transfer time with distance is linear and has the same slope (straight lines in Fig. 3). The corresponding velocity of the carrier packet is $(1.5\pm0.2)\times10^6$ cm/s.

For large photoexcited carrier densities, ambipolar diffusion is the most viable transport mechanism. However, ordinary diffusion is too slow to describe the carrier transport in our structures (Fig. 3, right time axis). Instead, the model of step-driven diffusion [1] which takes into account carrier sink at the boundaries of structure layers with different band gaps, fits fairy well our experimental data, especially in the region of larger band gap layers (Fig 3). A slow-down of the carrier motion, as compared to this model, for the smaller band gap steps may be due to the partial filling of the bottom of the GRINSCH, and smaller potential steps in the bands. Then, the lower band gap steps do not act as effective carrier sinks, and in this region both, the step-driven and ordinary diffusion have to be considered.

[1] B. Deveaud, J. Shah, T. C. Damen, B. Lambert, and A. Regreny, Phys. Rev. Lett. **58**, 2582 (1987).

Influence of Inhomogeneous Broadening on Quantum Beats

J. Erland[1], K.-H. Pantke[1], V. Mizeikis[1], V.G. Lyssenko[2], and J.M. Hvam[2]

[1]Fysisk Institut, Odense Universitet, DK-5230 Odense M, Denmark
[2]Mikroeletronik Centret, The Technical University of Denmark, DK-2800 Lyngby, Denmark

Quantum beats in an inhomogeneously broadened three-level system are studied by spectrally resolved, transient four-wave mixing. The important parameters are the ratio of inhomogeneous to homogeneous broadening and the correlation between the two transitions.

Current issues in the exploration of light-matter interactions are coherence and quantum interference. With ultrafast nonlinear spectroscopy, such as transient four-wave mixing (TFWM), the coherence of the optical excitations in e.g. GaAs multiple quantum wells has been studied. We study the influence of inhomogeneous broadening on quantum beats observed by spectrally resolved TFWM. In the experiments, the light-hole exciton (lhx) and the heavy-hole exciton (hhx) in a 100 Å GaAs multiple quantum well are excited simultaneously with 150 fs pulses from a self mode-locked Ti:sapphire laser. The laser profile is shown in the inset of Fig. 1, where our experimental results are summarized.

The TFWM signal is modulated with T_{beat} = 460 fs corresponding to a hhx-lhx splitting of 8.9 meV. The phase of the beats is constant through the resonances showing the quantum beat nature of the modulation[1]. In both resonances a clear dependence on detuning δ is observed, as the signals decay faster for increasing detuning. This dependence is stronger for the hhx than for the lhx-resonance.

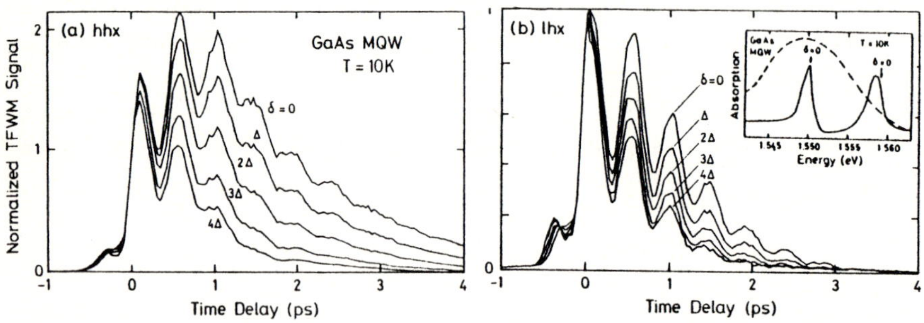

Fig. 1. Experiment with spectrally resolved TFWM at hhx (a) and at lhx (b) in a GaAs multiple quantum well. The curves are for different detunings δ, in steps of $\Delta = 0.21$ meV, and all curves are normalized at zero delay, $\tau = 0$. The inset shows the absorption spectrum together with the laser profile used in the experiments. In the inset, the arrows indicate zero detuning from the nonlinear resonances.

Springer Series in Chemical Physics, Vol. 60 **Ultrafast Phenomena IX**
Editors: P. F. Barbara · W. H. Knox · G. A. Mourou · A. H. Zewail

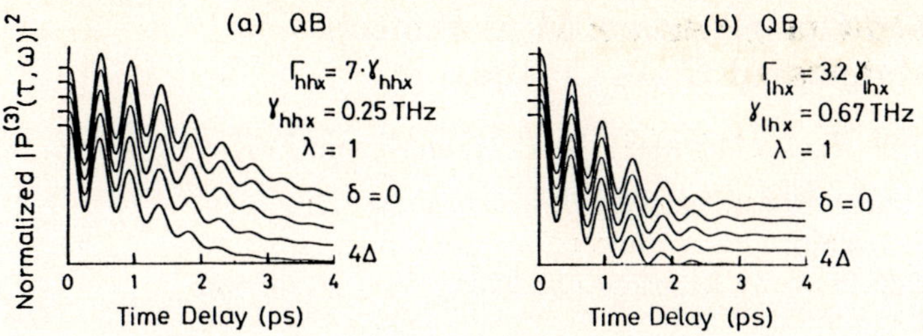

Fig. 2. Calculated TFWM-signal as function of delay τ and detuning δ, in steps of $\Delta = 0.21$ meV, for the hhx (a) and the lhx (b). The curves have been normalized and are shown offset by 0.1. For this calculation we chose full correlation, $\lambda=1$.

The third-order polarization $P^{(3)}_{hom,QB}(t,\tau)$ for a homogeneously broadened (γ) three-level system has been calculated from the Optical Bloch equations[2]. We introduce a correlated (λ) distribution, $g_N(\omega_{10},\omega_{20},\lambda)$, of the two transition frequencies to describe the inhomogeneous broadening (Γ) in a three-level system[3]. We calculate the third-order polarization from an inhomogeneously broadened three-level system

$$P^{(3)}_{inh,QB}(t,\tau) \propto \iint g_N(\omega_{10},\omega_{20},\lambda) P^{(3)}_{hom,QB}(t,\tau,\omega_{10},\omega_{20}) d\omega_{10} d\omega_{20} \tag{1}$$

We apply spectral resolution of the nonlinear signal, which corresponds to the absolute square of the Fourier Transform (FT) of Eq. (1). This can be done analytically[4]. Fig. 2 shows examples of such a calculation

From the reasonable agreement between the data presented in Fig. 1 and the calculation in Fig. 2, we estimate $\beta_{hhx} \approx 7$ and $\beta_{lhx} \approx 3.2$. The corresponding intensity HWHM-linewidths are: γ(hhx) = 0.16 meV, γ(lhx) = 0.44 meV, Γ(hhx) = 0.41 meV and Γ(lhx) = 0.49 meV. The full correlation ($\lambda \approx 1$) gives maximum number of beats in the signal. For small correlation ($\lambda \approx 0$) the modulation of the signal only persists on a delay-time scale given by the inverse inhomogeneous broadening, $\tau < 1/\Gamma$. We can therefore conclude, that the inhomogeneous broadenings of the hhx and the lhx are highly correlated.

References

1. V.G. Lyssenko et al., *Phys. Rev. B* **46**, 5720 (1993).

2. J. Erland and I. Balslev, *Phys Rev. A* **48**, 1765 (1993).

3. Private communication with S.T. Cundiff.

4. J. Erland, K.-H. Pantke, V. Mizeikis, V.G. Lyssenko, and J.M. Hvam, submitted.

Time-Resolved Optoelectronic Measurements of Nitrogen-Implanted GaAs Crystals

H. Shen[1], *Y. Jin*[1], *G.A. Wagoner*[1], *X.-C. Zhang*[1], *and L. Kingsley*[2]

[1]Physics Department, Rensselaer Polytechnic Institute, Troy, NY 12180-3590, USA

[2]Pulse Power Center, Army Research Laboratory, Fort Monmouth, NJ 07703, USA

We compare the measurement of optically induced THz radiation and time-resolved photoreflectance (ΔR/R) from nitrogen-ion implanted GaAs samples, where surface implantation densities ranged from $5 \cdot 10^{11}$ cm^{-2} to $5 \cdot 10^{15}$ cm^{-2}.

Understanding the effects of electrochemical passivation on the electronic properties of semiconductor surface states is crucial to semiconductor device fabrication. Previous results indicated that the static electric field across the surface of a semiconductor is substantially modified by hydrogen plasma passivation. As an example, after 30 minutes of passivation, the surface field from n(p)-type GaAs samples has been shown to increase by up to a factor of 2(10).

Herein, we report the results of our recent measurements of the transient optoelectronic properties of nitrogen-ion implanted GaAs, where the nitrogen ions served as isoelectronic centers and surface implantation densities ranged from $5 \cdot 10^{11}$ cm^{-2} to $5 \cdot 10^{15}$ cm^{-2}. Optically induced THz electromagnetic radiation and time-resolved photoreflectance (ΔR/R) measurements were made and the results of these two different optoelectronic techniques was compared. Both measurements show a strong dependence on implantation density and either technique is a good diagnostic of surface implantation density.

GaAs substrates were cleaved from a single semi-insulating GaAs wafer. The measured resistivity and steady state mobility were over $1 \cdot 10^7$ Ωcm and $4 \cdot 10^3$ cm$^2 \cdot$V$^{-1} \cdot$s^{-1}, respectively. The substrates were then implanted with nitrogen ions. One GaAs substrate remained unimplanted serving as a reference. The implantation was performed on a Varian 350R Medium-Energy & Medium-Current ion implanter. The beam current ranged from 0.18 μA to 100 μA, increasing proportionally with dose level. Typical sample temperature was kept below 80 °C and no annealing was performed after ion-implantation.

The energy of the incident nitrogen ions was set at 180 keV. Nitrogen ion surface densities ranged from $5 \cdot 10^{11}$ cm^{-2} to $5 \cdot 10^{15}$ cm^{-2}, with incremental steps alternating between factors of 2 and 5 successively. The mean value of the damage depth in the samples was approximately 0.6 μm which is comparable to the optical absorption length (α^{-1} = 0.7 μm at 810 nm).

The optical source was a mode-locked Ti:sapphire laser, operating at a central wavelength of 810 nm. During the measurement of optically-induced THz radiation emitted from the wafers, the waist of the normally-incident, unfocused optical beam was approximately 4 mm. Further, the wafers were subjected to an external magnetic field of roughly 0.3 T, perpendicular to the direction of optical propagation and applied via pair of permanent magnets. The forward THz radiation was

Springer Series in Chemical Physics, Vol. 60 **Ultrafast Phenomena IX**
Editors: P. F. Barbara · W. H. Knox · G. A. Mourou · A. H. Zewail

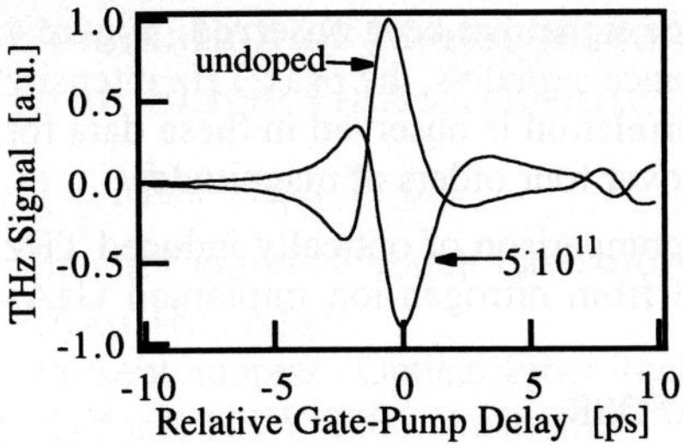

Figure 1. THz emission from undoped and implanted samples. Implanted samples exhibit a polarity flip associated with a reversal in direction of the surface field.

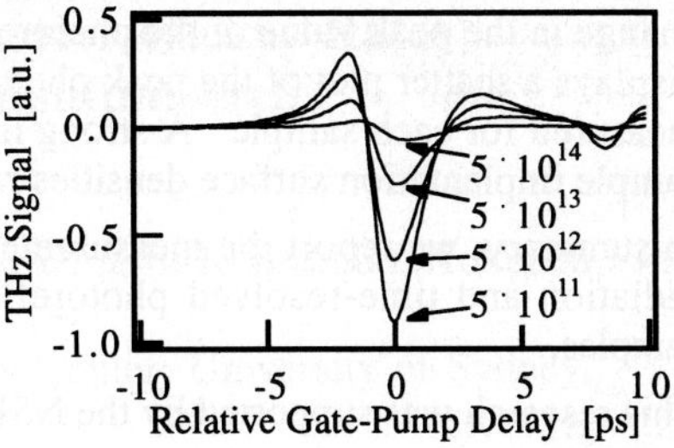

Figure 2. THz emission from samples with various implantation surface densities. Suppressed emission with increasing implantation associated with a reduction in the surface field and/or carrier mobility.

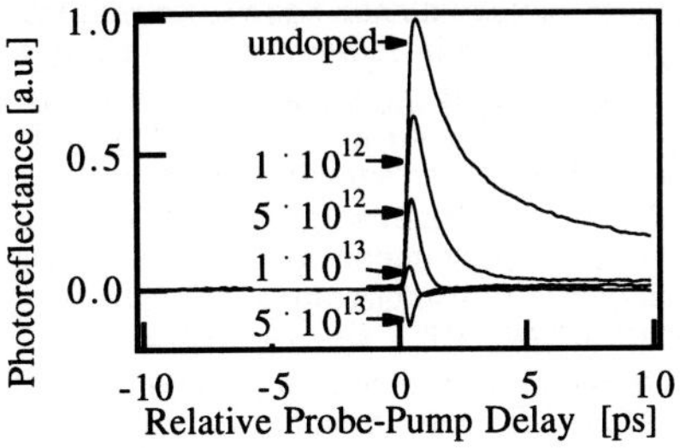

Figure 3. Time resolved photoreflectance signal at various implantation densities.

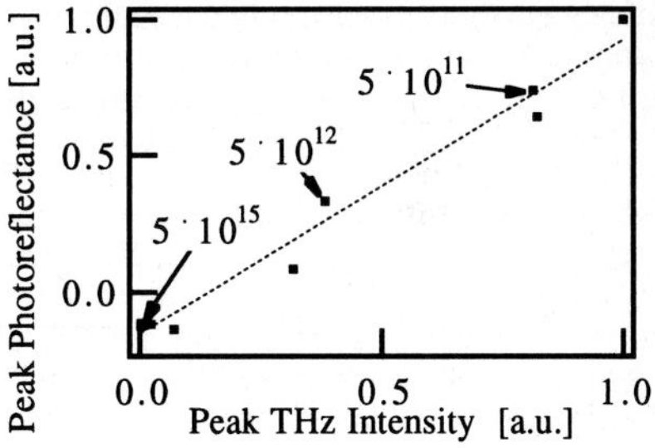

Figure 4. Peak photoreflectance vs. peak THz intensity.

detected by an ultrafast photoconducting antenna. In the time-resolved photoreflectance experiment (pump/probe), the laser beam was focused to spot size of approximately 100 μm. Both experiments were performed at room temperature.

The unimplanted GaAs wafer was slightly n-type and the direction of its surface depletion field was outward. However, after implantation, the field direction reversed. The THz radiation waveforms emitted from a GaAs wafer prior to, and after implantation at $5 \cdot 10^{11}$ cm^{-2} are shown in Figure 1. The observed polarity flip is associated with a reversal in the direction of the surface depletion field. Figure 2 displays THz radiation waveforms from several ion-implanted samples with implantation surface densities ranging from $5 \cdot 10^{11}$ cm^{-2} to $5 \cdot 10^{14}$ cm^{-2}. A decrease in emitted THz intensity with increasing surface implantation density is observed. This may be a result of a reduction in the surface depletion field strength and/or photocarrier mobility due to the implanted nitrogen ions.

Figure 3 shows the temporal waveforms of the photoreflectance signals, ($\Delta R/R$) from both implanted and unimplanted samples. Corresponding with increased implantation density, there is a significant decrease in both the amplitude and decay time of the photoreflectance waveforms. For samples with ion dose levels greater than $1 \cdot 10^{13}$ cm^{-2}, decay times as short as 0.35 ps have been measured and a sign

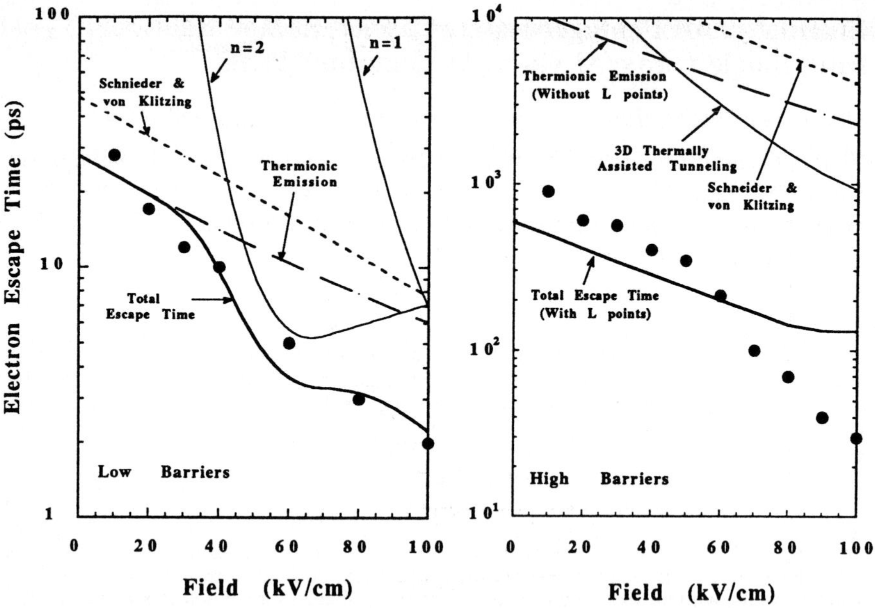

Figure 1 Electron escape times as a function of applied field from wells with a) low (x=0.2) barriers (left) and b) high (x=0.4) barriers (right). The solid circles are experimental data from Ref[1]. The light solid lines in a) are thermally assisted tunneling times from the n=1 and n=2 levels and the results labelled Schneider and von Klitzing are from Ref[6].

In conclusion, we have carried out theoretical calculations of photogenerated carrier escape times for single GaAs/AlGaAs quantum wells and compared the redults directly with recent experiments. Agreement for electrons is extremely good, when thermionic emission, thermally assisted tunneling, the contribution of the L point conduction band minima, and the 2D density of states are all included.

1. J.A.Cavailles, D.A.B.Miller, J.E.Cunningham, P.Li Kam Wa, and A.Miller, IEEE Journal of Quantum Electronics (JQE), 28 2486 (1992).
2. T.H.Wood, J.Z.Pastalan, C.A.Burrus Jr., B.C.Johnson, J.L.de Miguel, U.Koren, and M.G.Young, Applied Physics Letters (APL) 57 1081 (1990).
3. D.Moss, D.Halliday, S.Charbonneau, G.Aers, D.Landheer, F.Chatenoud, and D.Conn, SPIE Quantum Well and Superlattice Physics IV, Somerset NJ, **1675** 418 (1992).
4. B.W.Takasaki, J.S.Preston, J.D.Evans, J.G.Simmons, S.Charbonneau, and D.J.Moss, Applied Physics Letters 62 2525 (1993).
5. D.A.B.Miller,Opt. and Quantum Electronics 22 S61-S98 (1990).
6. H.Schneider and K. von Klitzing, Physical Review B 38, 6160 (1988).
7. P.J.Stevens, M.Whitehead, G.Parry, and K.Woodbridge, IEEE JQE, 24 2007 (1988).
8. D.J.Moss, T.Ido, and H.Sano, IEEE Journal of Quantum Electronics, April (1994).

Picosecond Optical Time-Domain Reflectometer for Optoelectronic IC Device Characterization

Y. Beaulieu[1], *R. Normandin*[2], *P. van der Meer*[2], *F. Chatenoud*[2], *A. Delage*[2], *and B.K. Garside*[1]

[1]Opto-Electronic Inc., Unit 9, 2538 Speers Road, Oakville, Ontario, Canada
[2]Optoelectronic Devices, Institute for Microstructural Sciences, National Research Council, Ottawa, Ontario, Canada

We demonstrate a real-time nonlinear guided wave optical time domain reflectometer for discrete and integrated optoelectronic device high resolution spatial probing.

1. Introduction

Several years ago Stegeman and Normandin predicted, and subsequently demonstrated [1], the nonlinear mixing of oppositely propagating guided waves. The resultant second harmonic field was coupled to radiation modes and propagated in a direction perpendicular to the waveguide surface, in the case of equal fundamental frequencies. By using the $Ga_{1-x}Al_xAs$ alloy, $\lambda/2$ layers of high conversion efficiency (low Al content) can alternate with transparent $\lambda/2$ layers of low conversion efficiency (high Al content). The in-phase and out-of-phase contributions have different intensities and the second harmonic generation (SHG) efficiency in the waveguide can be enhanced by many orders of magnitude [2].

When two optical pulses interact in a nonlinear waveguide, the second harmonic polarisation source field will reflect their convolution. If the signal from each element of the waveguide surface is integrated over the signal overlap time, the output corresponds to the square of the convolution of the two waveforms with a time compression of 2. By using a detector array that integrates the total light output at each position of the waveguide, it is therefore possible to obtain directly the temporal convolution of the two input pulses [3]. This detection scheme can then be used for ultrahigh resolution optical time domain reflectometry applications.

2. Experimental Aspects

The instrumental set-up is shown schematically in figure 1. Pulses are generated by a Nd-YAG laser (100 ps at 1.06 μm) and compressed to 2.5 ps FWHM with a standard fiber and grating pair set-up. The light beam is first split into two beams propagating along different optical paths: a reference path, consisting of a variable optical delay, and a signal path, which is used to probe a discrete or integrated optoelectronic device in transmission (T) or reflection (R) mode. Light from the two paths is then fed into opposite faces of a SHG waveguide. By properly adjusting the optical delay in the reference path, pulses from both paths can be made to meet in the SHG waveguide, thereby producing spots of light that can be directly observed above the nonlinear waveguide. This SHG waveguide consisted of 6 repetitions of $\lambda/2$ layers of $Ga_{1-x}Al_xAs$, where the Al content alternates between 70 and 90%, grown on a buffer layer deposited on a GaAs substrate.

Springer Series in Chemical Physics, Vol. 60 **Ultrafast Phenomena IX**
Editors: P. F. Barbara · W. H. Knox · G. A. Mourou · A. H. Zewail

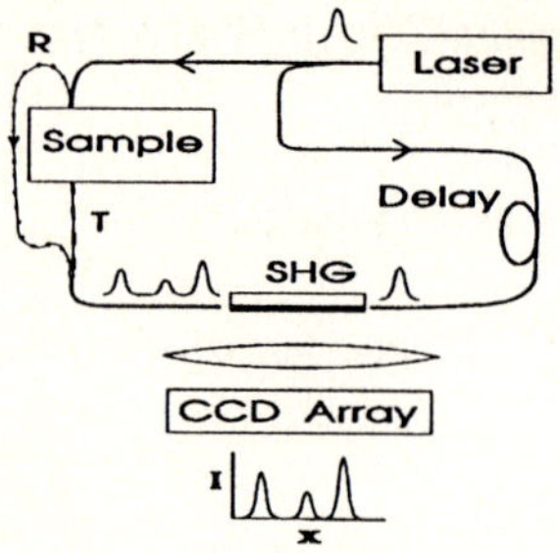

Figure 1. Instrumental set-up used for optical time domain reflectometry.

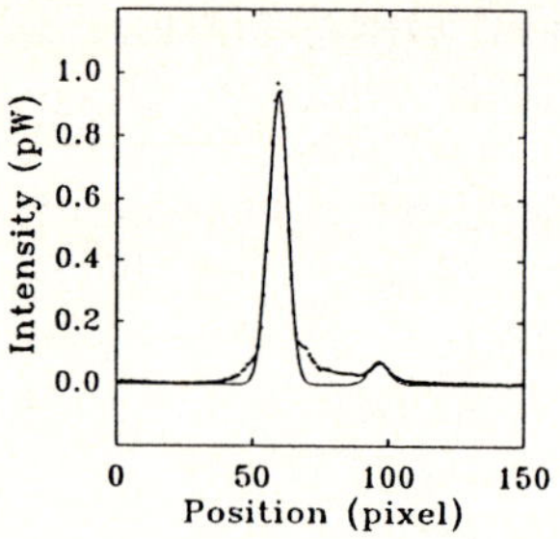

Figure 2. SHG intensity across the waveguide surface when probing a 0.65 mm thick GaAs substrate.

3. Results

In our proof-of-concept experiment, we used a piece of bulk GaAs (about 0.65 mm thick) as a sample, observed in transmission mode. The distribution of the light output intensity (dots) across the SHG waveguide surface (3.02 mm long) and the least-square fit (line) to a Gaussian pulse shape are shown in figure 2. An intense spot corresponding to the pulse transmitted through the sample and a second spot corresponding to the pulse reflected once inside the sample can be identified. Knowing the index of refraction of the SHG waveguide and of GaAs, this separation (37 pixels) is calculated to be 0.65 mm, in good agreement with the measured thickness. The two spots are very easily resolved in this case, since the resolution is about 0.15 mm in the sample with our 2.5 ps pulses.

4. Conclusion

We have demonstrated a stable OTDR for optical ranging without the use of interferometric techniques or the need of Fourier transforms. The set-up has no moving parts and is fiber compatible. Efficiencies are sufficient for real time CCD monitoring. The 0.15 mm interfeature resolution and the 50 μm position uncertainty are presently limited by the laser pulse FWHM of about 2.5 ps.

References

1. R. Normandin and G.I. Stegeman, Optics Lett. **4**, 58 (1979).
2. R. Normandin, S. Létourneau, F. Chatenoud and R. L. Williams, IEEE Journal of Quantum Electronics **27**, 1520 (1991).
3. D. Vakhshoori and S. Wang, Appl. Phys. Lett. **53**, 347 (1988).

Upper Satellite Conduction-Band Population Dynamics in GaAs

M.A. Cavicchia and R.R. Alfano

Institute for Ultrafast Spectroscopy and Lasers, Center for Advanced Technology for Photonic Materials and Applications, Physics Department, The City College and the Graduate School of the City University of New York, Convent Avenue, 138th Street, New York, NY 10031, USA

Abstract: Population dynamics of the *upper* satellite, X_7, conduction band valley via intervalley scattering with the *lower*, X_6 and central, Γ_6, valleys, was **directly** measured in GaAs.

Summary

A Monte Carlo calculation has shown that the *upper* satellite, X_7, conduction band valley plays an important role in the high-field transport properties of GaAs.[1] We present for the first time an experimental study of the intervalley relaxation dynamics of a hot electron population in the *upper* satellite, X_7, conduction band in GaAs using time-resolved UV pump - IR probe absorption spectroscopy.

Fig. 1 shows the transient change in IR absorption as a function of pump- probe delay time for λ_{pump} = 293-nm and λ_{probe} = 4.7-μm. The temporal behavior is characterized by a rapid rise in absorption, followed by a fast decay component (~ 4 ps), and finally a slow decay. For λ_{probe} = 4.7-μm only free carrier absorption contributes to the induced absorption. The peak observed in Fig. 1 at early delay times is not seen when λ_{pump} = 585-nm and λ_{probe} = 4.7-μm. This result, shown in Fig. 2, is characterized by a rapid rise in absorption and only a slow decay component.

For λ_{pump} = 293-nm, there are electron populations in both the X_6 and X_7 valleys (see inset Fig. 1). On the other hand, for λ_{pump} = 585-nm, the photoexcited electrons do not obtain sufficient kinetic energy to undergo intervalley scattering to the X_7 valley (see inset Fig. 2). Therefore, the observed 4-ps decay of the peak in Fig. 1 is attributed to depopulation of the upper satellite, X_7, valley.

The temporal evolution of the X_7 valley electron population (Fig. 3) is obtained from the excess absorption at early delay times. The X_7 valley electron population peaks at 1.7-ps and has a decay time of 3.4-ps. The buildup and decay times of the nonequilibrium electron population in the X_7 valley in GaAs have been resolved for the first time. For λ_{pump} = 293-nm, the electrons are excited to an excess energy of 1.58-eV and it is estimated an electron will undergo intervalley scattering back and forth between the X_7 valley and the X_6 and Γ valleys 19 times before its energy falls below the X_7 valley minimum (see inset of Fig. 3).[2] Consequently, an upper limit for the intervalley scattering time for an electron in the X_7 valley to either the X_6 or the Γ valley, $t(X_7 \rightarrow X_6,\Gamma)$, is determined to be 180-fs in GaAs.

Springer Series in Chemical Physics, Vol. 60 **Ultrafast Phenomena IX**
Editors: P. F. Barbara · W. H. Knox · G. A. Mourou · A. H. Zewail

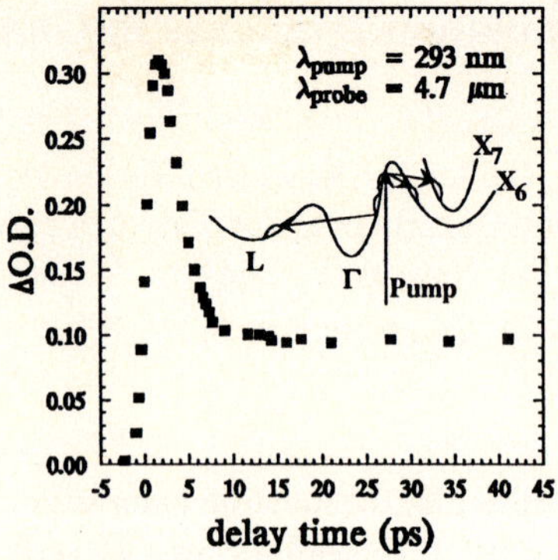

Fig. 1 Transient IR absorption where λ_{pump} = 293-nm and λ_{probe} = 4.7-µm. Inset shows the initial relaxation channels for UV photoexcitation.

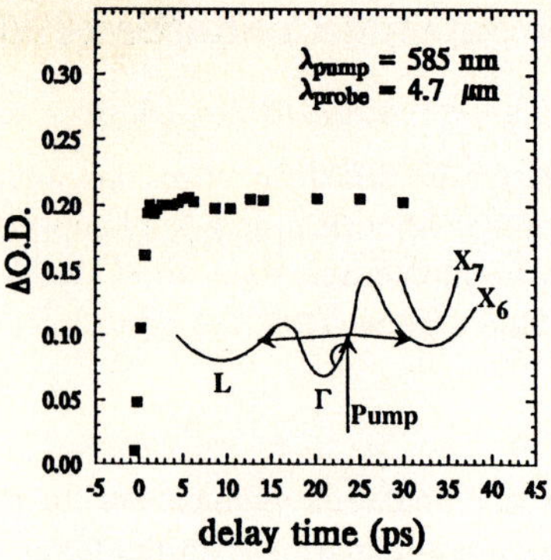

Fig. 2 Transient IR absorption where λ_{pump} = 293-nm and λ_{probe} = 4.7-µm. Inset shows the initial relaxation for visible photoexcitation.

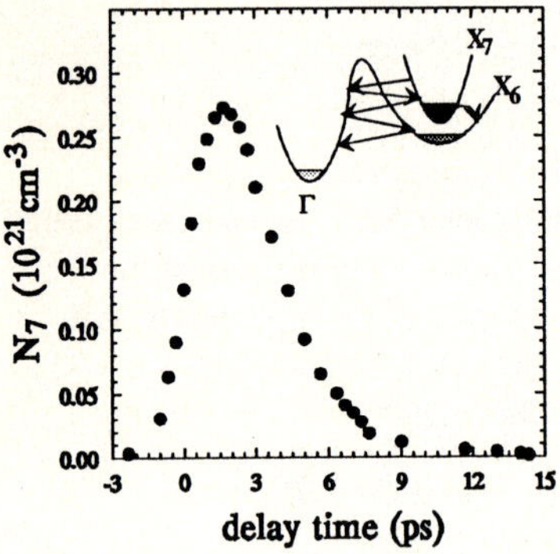

Fig. 3 The temporal evolution of the X_7 valley electron population. Inset shows the intervalley relaxation which depopulates the X_7 valley.

This work is supported by NASA.

References

[1] T. González Sánchez, J. E. Velázquez Pérez, P. M. Gutiérrez Conde, and D. Pardo Collantes, Semicond. Sci. Technol. **6**, 862 (1991).

[2] M. A. Cavicchia and R. R. Alfano, Phys. Rev. B **48**, 5696 (1993).

Nonlinear Optical Studies of Microstructured GaAs/AlAs

S.T. Cundiff[1], *E.O. Böbel*[1], *V.P. Kochereshko*[2], *P.S. Ko'pev*[2], *and N.N. Ledentsov*[2]

[1]Fachbereich Physik und Wissenschaftliches Zentrum für Materialwissenschaften, Philipps-Universität, Renthof 5, D-35032 Marburg, Germany
[2]A.F. Ioffe Physico-Technical Institute, 194021 St. Petersburg, Russia

Abstract. The exciton dynamics in a microstructured GaAs/AlAs sample are studied using transient four-wave-mixing. The excitons in this sample are confined to micro-crystallites that form during growth. Contributions to the signal due to acceptor-bound and free excitons are distinguished and the excitons are shown to have dynamics that are attributed to the size of the crystallites.

Structures exhibiting high degrees of quantum confinement, i.e., quantum wires and dots, have attracted considerable attention, both as a natural extension of investigations of quantum wells and because of theoretical predictions of new physics [1]. Despite difficulties with fabricating sufficiently small structures through direct methods (e.g, etching), there has been some success in observing new phenomena [2]. Nevertheless the difficulties have spurred efforts to develop techniques that result in the formation of low dimensional structures during the growth. We present here transient-four-wave-mixing (TFWM) experiments on such a sample in which GaAs micro-crystallites naturally form during the growth process.

The sample is a GaAs/AlAs superlattice grown on a non-(100)-oriented faceted surface (311A), with the GaAs micro-crystallites introduced in the central AlAs layer [3]. The micro-crystallites are approximately 20 Å by 35 Å in cross-section and 400 Å long, with significant length fluctuation. The exciton wave function will extend into the neighboring GaAs regions, and hence is expected to be ~100 Å across.

We perform 2-pulse TFWM in the reflected geometry using ~150 fs pulses. The sample containing micro-crystallites displays a strong, polarized TFWM signal that is spectrally coincident with the PL attributed to the micro-crystallites, and no PL or TFWM from the superlattice, whereas a sample without micro-crystallites displays a weaker, unpolarized TFWM from superlattice (Fig. 1(a)). Note that the PL from both samples is polarized due to the faceting. Both samples show similar dephasing times (~10 ps) and quantum beats. By examining the spectral dependence for long delays (Fig 1(b)), and high incident intensity we assign the quantum beating to the presence of acceptor-bound (BE), which dominate at long delays, and free excitons (FE), which dominate at high intensity [4].

The dephasing rate of the BE is linear with temperature at low temperature (Fig. 2(a)). This is a surprising result for bound excitons, which usually display a nonlinear temperature dependence, and we attribute it to the small size of the micro-crystallites that suppresses hopping. Consequently other processes must dominate the dephasing.

The density dependence of the BE dephasing is examined by exciting the sample with a variable intensity prepulse (Fig. 2(b)), which arrives several dephasing times early. The dephasing time initially increases with increasing prepulse intensity, and then saturates at an incident intensity of ~10 mW. The FE component shows a

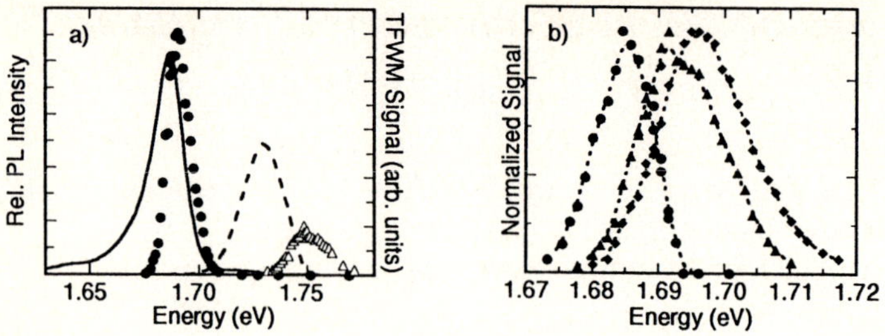

Fig. 1. (a) PL (solid line) and TFWM (•) for sample with crystallites and without (dashed,△). (b) TFWM signal for short delay, low intensity (▲), long delay, low intensity (●) and short delay, high intensity (♦), dashed lines are guide to the eye.

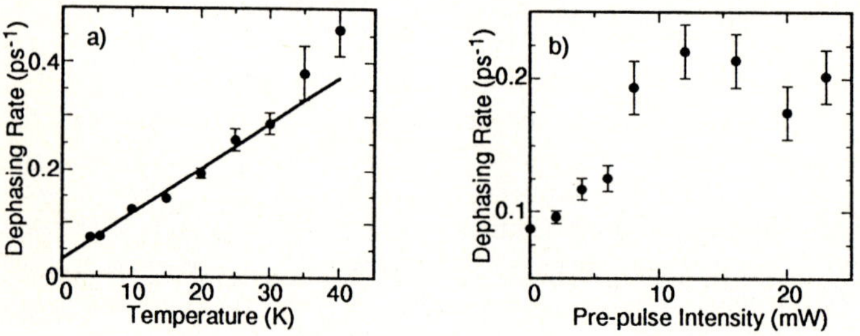

Fig. 2. (a) Dephasing rate as a function of temperature. (b) Dephasing rate as a function of prepulse density. Both (a) & (b) are for micro-crystallite excitons.

similar behavior (not shown). The orgin of this behavior is currently unclear, interaction between excitons on separate micro-crystallites, or a scattering cross-section enhancement due to biexciton formation are both possible mechanisms.

References

[1] see, e.g., K. Kash, J. Lumin. **46**, 69 (1990); H. Sakaki, Jap. J. App. Phys. **19**, L735 (1980).

[2] see, e.g., M. Oestereich, *et al.*, Phys. Rev. Lett. **70**, 1683 (1993); E.J. Mayer, *et al.*, Phys. Rev. B **49**, 2993 (1993).

[3] R. Nötzel, *et al.*, Phys. Rev. B. **45**, 3507 (1991); Z.I. Alferov, *et al.*, Sov. Phys. Semi. **26**, 959 (1992).

[4] M. Koch, *et al.*, Phys. Rev. B **47**, 1532 (1993).

Non-Equilibrium Dynamics of Holes in p-Type Germanium Studied by Femtosecond Infrared Spectroscopy

M. Woerner[1,2], *T. Elsaesser*[1,2], *C. Ludwig*[1], *M.T. Portella*[1], *W. Frey*[1], and *W. Kaiser*[1]

[1]Physik Department E 11, Technische Universität München, James-Franck-Strasse, D-85748 Garching, Germany
[2]Max-Born-Institut für Nichtlineare Optik und Kurzzeitspektroskopie, Rudower Chaussee 6, D-12489 Berlin, Germany

Abstract. Thermalization processes in a pure hole plasma are directly monitored by femtosecond pump-probe spectroscopy in the mid-infrared. We observe hole thermalization by inelastic carrier-carrier scattering on a time scale of 700 fs.

The ultrafast thermalization of nonequilibrium carriers in semiconductors, i.e. the relaxation to a Fermi-like distribution, has mainly been investigated by femtosecond absorption and luminescence studies of photoexcited electron-hole plasma where the optical response of the material is mostly governed by the electron dynamics. In this paper, we present the first time resolved observation of thermalization in a (single component) hole plasma, providing specific insight into the subpicosecond relaxation of holes in p-type germanium.

In our experiments, holes are promoted to the split-off band by excitation with femtosecond pulses around 3 μm (E_{ex}=430 meV). The subsequent redistribution dynamics is monitored in temporally and spectrally resolved measurements of inter-valence band absorption from the heavy hole (HH) to the split-off (SO) band. We investigate a Ga-doped germanium crystal with a total hole concentration of 3×10^{17} cm^{-3}, the excitation density has a value of about 10^{16} cm^{-3}. Femtosecond pump and probe pulses of 250 fs duration that are independently tunable in the wavelength range from 2.5 to 5 μm, are generated by parametric difference frequency mixing [1].

In Fig. 1 (a), the stationary absorption spectrum due to transitions between the HH and the SO band is plotted versus photon energy for lattice temperatures of T_L = 10K and 80K. In Figs. 1 (b-d), the transient increase of inter-valence band absorption is presented for three photon energies E_{pr} of the probe pulses. At the different E_{pr}, the enhanced absorption rises with a common delay time of approximately 700 fs (solid lines: calculated monoexponential response, dashed line in Fig. b: time integrated cross-correlation function of pump and probe) and decays subsequently on a slower picosecond time scale by carrier cooling [2]. The present data and additional measurements at other spectral positions [3] exhibit no spectral hole burning, even at delay zero and for identical pump and probe wavelengths.

The absence of a bleaching signal around delay zero shows that within our time resolution of 100 fs (i) holes excited to the SO band are scattered out of the optically coupled states and (ii) HH states depleted by the pump pulse are essentially repopulated by equilibration among unexcited holes. The holes in the SO band scatter back to the HH band mainly by emission of optical phonons via the deformation potential within the first 100 fs. These nonequilibrium carriers populating high-lying states loose energy by emission of optical phonons and - more important - by heating the cold plasma of unexcited holes via inelastic carrier-carrier collisions, resulting in an equilibration of the hole distribution. The concomitant rise of occupation probability of HH states leads to

Editors: P. F. Barbara · W. H. Knox · G. A. Mourou · A. H. Zewail

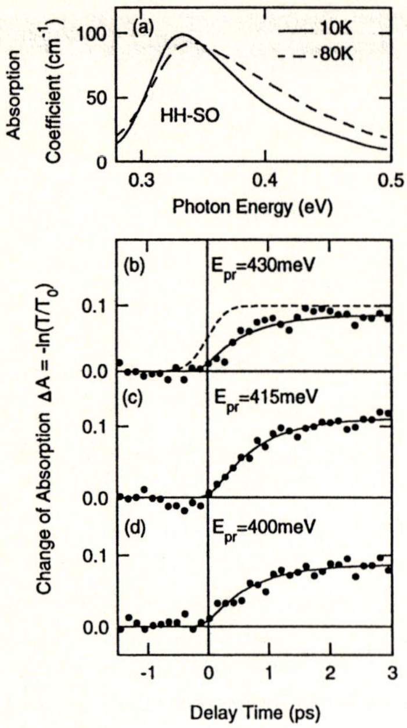

Figure 1 (a) Intervalence band absorption of p-type germanium at lattice temperatures of 10 and 80 K. (b-d) Transient increase of intervalence band absorption of p-type Ge after femtosecond excitation at E_{ex} = 430 meV ($T_L = 10$ K). The absorption change $\Delta A = -\ln(T/T_0)$ is plotted versus the delay time between pump and probe pulses for three different probe frequencies E_{pr} (T_0,T : transmission before and after excitation). The dashed line in (b) gives the time-integrated cross-correlation of pump and probe.

the observed increase of absorption with rise times around 700 fs, i.e. the thermalization of photoexcited holes is directly monitored in our experiments.

The rates of energy loss by inelastic carrier-carrier scattering and by emission of optical phonons are derived from a theoretical simulation including a dynamically screened Coulomb interaction [3,4]. For the excess energies of holes excited to the SO band, inelastic scattering of hot carriers with the cold hole distribution represents the main relaxation mechanism. Due to the weak dynamical screening in the carrier system, we derive thermalization times of 0.5 ps that are in good agreement with the experimental results.

References

[1] C. Ludwig, W. Frey, M. Woerner, and T. Elsaesser, Opt. Commun. **102**, 447 (1993)

[2] M. Woerner, T. Elsaesser, and W. Kaiser, Phys. Rev **B 45**, 8378 (1992)

[3] M. Woerner, W. Frey, M.T. Portella, C. Ludwig, T. Elsaesser, and W. Kaiser, Phys. Rev. B, in press (1994)

[4] J.F. Young, N.L. Henry, and P.J. Kelly, Solid State Electron. **32**, 1567 (1989)

Excitation-Induced Phase Shifts of Heavy- and Light-Hole Quantum Beats in Multiple Quantum Wells

W. Sha, A.E. Paul, S. Patkar, and A.L. Smirl

Center for Laser Science and Engineering, University of Iowa, 144 Iowa Advanced Technology Laboratories, Ioawa City, IA 52242, USA

Abstract. The phase of quantum beats from GaAs/AlGaAs multiple quantum wells is shown to be a strong function both of the fluence and of the wavelength of the exciting radiation. The fluence and wavelength dependence can be explained only if many body effects, such as local field corrections and excitation-induced dephasing, are included.

In recent years, transient four-wave mixing (FWM) techniques have been used to investigate the dephasing times of non-equilibrium carriers in semiconductor heterostructures by monitoring the decay of the macroscopic polarization. When excitons with slightly different energies are excited, oscillations (or beats) have been observed, for example, between light- and heavy-hole excitons (quantum beats, or QBs) and between excitons in wells of different widths (polarization interference). Both the spectral behavior [1] and the temporal behavior [2] of the FWM signal have been shown to allow the distinction between polarization interference and QBs. In addition, it has been demonstrated that the QBs produced by incident pulses with parallel polarizations are exactly out of phase with the beats produced by orthogonally polarized pulses. In each case [1-3], the period and phase of the beats and the polarization were explained by using a six-level model, which excluded many-body effects, to describe the J=1/2 conduction and J=3/2 valence states. Very recently, however, it has been shown that a complete description of the polarization selection rules for the FWM signal requires excitation-induced dephasing [4] or disorder-induced coupling of σ_+ and σ_- excitonic transitions [5].

Here, we report an excitation-dependent shift in the phase of the QBs from GaAs/AlGaAs multiple quantum wells (MQWs). The phase is observed to shift dramatically as the incident fluence is varied or as the central frequency of the incident pulses is tuned relative to the excitonic line centers. We demonstrate that the phases of the beats are determined by many-body effects, such as local field corrections and excitation-induced dephasing. While the role of many body effects in describing the formation and decay of the macroscopic polarization and in determining the polarization of the FWM signal is now established, to our knowledge, this is the first demonstration that many body effects are necessary for an accurate description of the QBs.

In our experiments, we generate the coherent polarization by using a conventional two-beam mixing geometry and monitor its decay by time resolving the FWM signal. We use ~100 fs pulses produced by a mode-locked Ti:Sapphire laser. The FWM signal traveling in the direction $2\mathbf{k}_2$-$\mathbf{k}_1$ generated by the two temporally overlapped pump pulses ($\mathbf{k}_1$ and $\mathbf{k}_2$) with parallel polarizations is cross-correlated with a third pulse. Here, we focus on the results from a MQW sample composed of 50 periods of 100Å-GaAs/75Å-$Al_{0.3}Ga_{0.7}As$. The sample is cooled to ~80 K.

The decay times and the phases of the QBs as a function of incident fluences extracted from the time-resolved signals are summarized in Fig. 1. The central frequency of the incident pulses is tuned at the HH excitonic line. Notice that the decay time initially decreases with increasing excitonic density then saturates at higher densities, behavior that

Springer Series in Chemical Physics, Vol. 60 **Ultrafast Phenomena IX**
Editors: P. F. Barbara · W. H. Knox · G. A. Mourou · A. H. Zewail

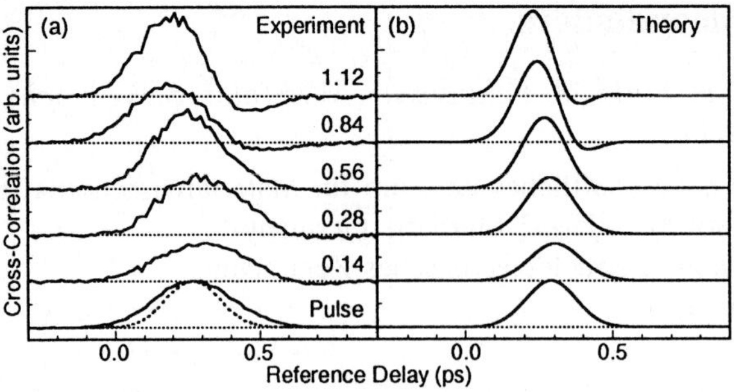

Fig. 1. Experimental (a) and theoretical (b) cross-correlation signal as a function of reference pulse delay and pulse intensity (traces offset vertically for clarity, indicated intensities in GW/cm^2 apply to experiment and theory). The bottom trace is the pulse cross correlation (dotted line in (a) shows pulse without propagation through sample).

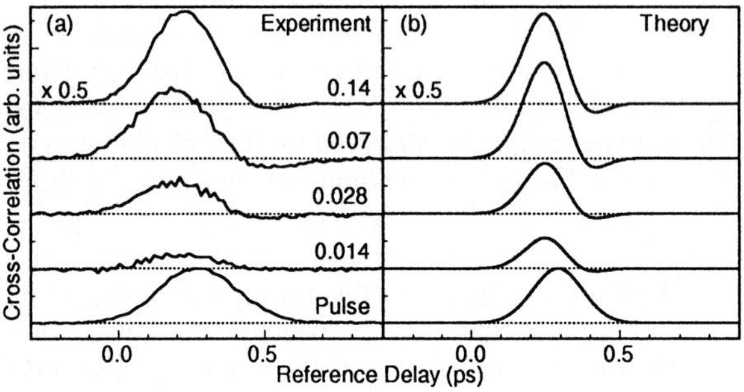

Fig. 2. As Fig. 1, except as a function of prepulse intensity.

References

[1] N. Peyghambarian, S.W. Koch and A. Mysyrowicz, *Introduction to Semiconductor Optics* (Prentice Hall, Englewood Cliffs, 1993).
[2] R. Binder, *et al.*, Phys. Rev. Lett. **65**, 899 (1990).
[3] H. Haug and S.W. Koch, *Quantum Theory of the Optical and Electronic Properties of Semiconductors* (World Scientific, Singapore, 1993) 2nd ed.

Femtosecond Blue-Continuum Generation and Its Application to the Study of ZnS:Mn Nanocrystals

T. Sosnowski[1], *P.B. Klein*[1,2], *T.B. Norris*[1], *R.N. Bhargava*[3], *and D. Gallagher*[4]

[1]Center for Ultrafast Optical Science, University of Michigan, 2200 Bonisteel Blvd., Ann Arbor, MI 48109-2099, USA
[2]Naval Research Laboratory, Code 6874, Washington, DC 20375-5347, USA
[3]Nanocrystals Technology, P.O. Box 820, Briarcliff Manor, NY 10510, USA
[4]Philips Laboratories, 345 Scarborough Road, Briarcliff Manor, NY 10510, USA

Abstract. This paper describes the generation of a continuum created with the frequency doubled output of a regeneratively amplified Ti:sapphire laser and the application of this source to the time resolved study of Mn doped ZnS nanocrystals.

Recently, the achievement of high quantum efficiency light emission from semiconductor nanocrystals using transition metal doping was reported [1] in the ZnS:Mn system. While this system is of considerable interest both scientifically and technologically, the excitation dynamics in doped nanocrystals are still not well understood. In order to probe such a system on the picosecond or sub-picosecond time scale, a tunable ultrafast UV source is required. We have developed such a source by using the frequency-doubled output of a 250kHz Ti:sapphire regenerative amplifier[2] to produce the blue continuum.

The pump pulses were created by chirped-pulse amplification in a cw-argon-pumped Ti:sapphire regenerative amplifier. The system produced 2-μJ, 85-fsec pulses with a repetition rate of ≥ 250 kHz. The 800nm pump pulses were frequency-doubled by weakly focusing into a 0.5 mm BBO crystal. Approximately 20% conversion efficiency is achieved, resulting in $\approx$0.4μJ, 400 nm pulses. The 400 nm pulses were focused at f/10 into a 3 mm sapphire crystal to create a continuum. Sapphire was chosen in part because its UV transmittance extends down to 155 nm.

Fig. 1a shows the continuum spectrum generated with the 400nm, 0.4μJ pulses. The spectrum is centered around the 400 nm pump with a range of $\leq$250 nm to $\geq$680 nm. The spectrum was measured with a multialkali photocathode photomultiplier tube and a grating spectrometer. The spectrum measurement is not corrected for the system response, and is limited on the short wavelength side by scattered light in the monochromator. The threshold pulse energy for blue continuum generation in this configuration was as low as 0.08μJ, which is 4-5 times lower than the threshold for 800 nm continuum generation in the identical configuration.

The crystal field split $^4T_1$ to $^6A_1$ transition of Mn^{2+} is responsible for the long-lived (1.8 msec) 584 nm emission that is characteristic of the bulk phosphor. Recently, Bhargava et al.[1] have shown that ZnS nanocrystals as small as 30Å can be doped with Mn, and exhibit high radiative efficiencies ($\geq$18%). This was a dramatic observation, as semiconductor nanocrystals are notoriously inefficient because of enhanced surface recombination[3]. In contrast to the slow decay of the bulk ZnS: Mn^{2+} photoluminescence (PL), Bhargava et al. observed a fast (3.7 nsec) decay in the nanocrystals. The much faster decay and enhanced radiative efficiency in the nanocrystals was interpreted as resulting from carrier localization at the Mn center. Attempts by Bhargava et al. to observe the energy transfer process by monitoring the rise of the luminescence signal was limited by the 0.5 nsec response of their system.

Springer Series in Chemical Physics, Vol. 60 **Ultrafast Phenomena IX**
Editors: P. F. Barbara · W. H. Knox · G. A. Mourou · A. H. Zewail

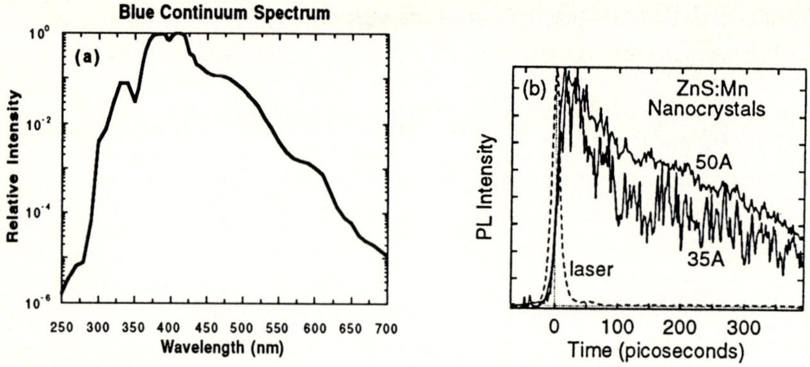

Figure 1 (a) Spectrum of the blue continuum produced from doubled Ti:Sapphire, (b) PL response of 35 Å and 55 Å ZnS:Mn nanocrystals

In the present work we have extended those measurements to the picosecond regime by exciting the nanocrystal above the ZnS band edge with the filtered blue continuum source described above. The blue continuum was passed through a Schott UG-11 filter, or a solar blind filter, resulting in a broad band excitation centered at 370 nm or 330 nm respectively. The PL emission was imaged onto a streak camera and detected by a CCD array. The PL temporal response in the nsec range for 30Å crystals was observed to exhibit a bi-exponential dependence, with decay components of 3.3 and 0.4 nsec. The slower decay corresponds to that observed by Bhargava et al., while they were unable to resolve the faster component. The rise of the PL was system limited. Fig. 1b shows similar data on a shorter time scale for 30Å and 55Å nanocrystals. Again, the PL rise is limited by the system response of ≈17 psec, while the decay is in the 300 psec range and appears to have more than one component in the 55Å nanocrystals. This may be of some significance, as different crystal sizes are expected to exhibit different decay times.

Spectral studies reveal several Mn^{2+} emission bands in the 400 - 650 nm region, which suggests that the energy transfer process leaves the Mn in a higher lying excited state. Nonradiative relaxation down the ladder of d-states will lead to population of the lowest $^4T_1$ excited state. Thus the PL risetime will not reflect the energy transfer process but rather the nonradiative relaxation, which has been observed to be several picoseconds in ZnS:Ni [4]. We are presently attempting to measure the energy transfer itself by a pump-probe experiment in which a probe pulse monitors the internal $^6A_1$-$^4T_1$ absorption while a UV pump pulse excites the ZnS nanocrystal above the band gap.

References

1. R.N. Bhargava, D. Gallagher, X. Hong and A. Nurmikko, Phys. Rev. Lett. **72**, 416 (1994).
2. T.B. Norris, Opt. Lett. **17**, 1009 (1992).
3. L.E. Brus, IEEE J. Quantum Electron. QE-**22**, 1909 (1986).
4. R. Heitz, A. Hoffman and I. Broser, J. Lum. **53**, 359 (1992).

Terathertz Radiation from Delta-Doped GaAs

D. Birkedal[1], *J.M. Hvam*[1], *and S.R. Keiding*[2]

[1]Mikroeletronik Centret, The Technical University of Denmark, DK-2800 Lyngby, Denmark
[2]Kemisk Institut, Aarhus Universitet, DK-8000 Aarhus C, Denmark

Abstract. In this work we study the THz-emission from four delta-doped samples having different spacings between the doping layer and the surface. The polarity of the emitted THz pulses is found to be opposite that of pulses emitted from samples having bulk *n*-type doping. For increasing depth of the doped-layer, we observe a decrease in the THz pulse amplitude.

The presence of a delta-doped layer in the vicinity of the surface creates a very large field at the surface. Electric fields up to 100 kV/cm have recently been inferred from the observation of Franz-Keldysh oscillations in the photoreflectance spectra of delta-doped GaAs [1]. These characteristics of the delta-doped structures make them very interesting as sources of THz-radiation, where both the high field and the high mobility should make the emitted THz amplitude greater than that emitted from bulk GaAs surfaces.

The delta-doped GaAs samples used in this work were grown by molecular beam epitaxy on Cr-doped semi-insulating GaAs substrates. The 4 delta-doped samples have the doped layer positioned at a distance, *d*, of 25 nm, 50 nm, 100 nm, and 200 nm from the surface. The delta-doping was of *n*-type with Si as dopant. The same Si dose was used for each sample. A 1 μm buffer layer of undoped GaAs with a background *p*-type concentration of approximately 2×10^{14} cm^{-3}, was grown between the substrate and the delta-doped layer. An identical layer of GaAs was used to cap the sample. As reference a sample for the THz experiments, we used a semi-insulating liquid-encapsulated Czochralski (LEC) GaAs sample.

The experimental setup used in the THz experiments is described in Ref. [2]. The laser source was a mode-locked titanium sapphire laser producing ≈ 200 fs pulses at 840 nm. The beam was gently focused to a diameter of approximately 1 mm at the sample surface. The pump power was 100 mW, corresponding to a sheet carrier density of 1×10^{12} cm^{-2}. The THz-signal generated from the surface propagates to a receiving dipole antenna defined on ion implanted silicon on sapphire. The receiver has a bandwidth extending to 2 THz, mainly limited by the dielectric loss in the sapphire substrate.

THz signals radiated from the delta-doped samples upon laser pumping at 840 nm are shown in Fig. 1. The curves are labeled to indicate the type of sample. Also shown are pulses radiated from the LEC GaAs reference sample. It is observed that the THz-pulses from the delta-doped samples show the opposite polarity of the THz-pulse emitted by the LEC GaAs sample.

Springer Series in Chemical Physics, Vol. 60 **Ultrafast Phenomena IX**
Editors: P. F. Barbara · W. H. Knox · G. A. Mourou · A. H. Zewail

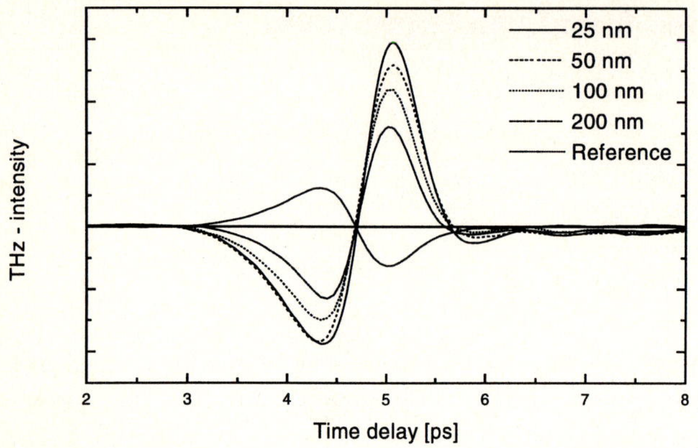

Figure 1: THz-pulse radiated from the 4 delta-doped samples and the GaAs reference sample with the Ti:sapphire laser operating at 840 nm.

A decrease of the THz amplitude as a function of delta-layer depth is also observed, and is explained in the following way: The source current responsible for the radiated THz-pulse originates from both sides of the delta-doped layer. Photoreflectance measurements showed that the field on the surface side of the delta-doped layer is homogeneous and well approximated with $F = -V_s/d$, where V_s is the surface potential equal to 0.6 V. The field on the substrate side of the delta-doped layer is almost homogeneous and equal to 9.1 kV/cm within the absorption length. Hence, the surface and substrate side contributions have opposite sign; the total THz signal is the sum of these contributions.

The sign of the electric field radiated from the surface, and correspondingly the direction of the photo-current can be determined from the measured pulses by calibrating the response of the detection system with a biased THz-antenna, where the direction of the current is controlled by the polarity of the applied bias. We find that the observed polarity of the pulses radiated from the delta-doped samples corresponds to a net current directed toward the the substrate. This is consistent with the majority of the THz signal arising from the substrate side of the delta-doped layer. For increasing depth of the delta-doped layer an increasing contribution to the THz-radiation of opposite polarity is generated, thus, reducing the total detected signal.

We would like to acknowledge O. Hansen, C. B. Sørensen, and S. D. Brorson for valuable contributions.

References

[1] D. G. Liu *et al.*, J. Appl. Phys. **72**, 1468 (1992). T. M. Hsu *et al.*, J. Appl. Phys. **72**, 1065 (1992).

[2] J. E. Pedersen, I. Balslev, J. M. Hvam, and S. R. Keiding, Appl. Phys. Lett. **62**, 1392 (1992).

Transient Reflective-Polarization Spectroscopy of Hidden Anisotropy in Cubic Crystals

N.I. Zheludev[1], *A.R. Bungay*[1], *S.V. Popov*[1], *and Yu.P. Svirko*[2]

[1]Department of Physics, University of Southampton, Highfield, Southampton SO17 1BJ, UK

[2]General Physics Institute, 38 Vavilov Street, 117333 Moscow, Russia

Abstract. Time resolved reflective polarisation spectroscopy allows for the first time observation of the anisotropy and non-locality terms in the nonlinear optical response of GaAs.

We report the first observation of a new fine, but quite pronounced pump-probe polarization sensitive transient effect in reflected light which results directly from the anisotropy of the zinc-blende structure. If a (001) surface of a zinc-blende crystal is stimulated with linearly polarized light propagating along the [001] direction, this affects the polarization state of the probe wave reflected from the surface (see fig.1). If the pump and the probe are initially identically linearly polarized, only hidden crystal anisotropy could lead to the alteration of the reflected probe polarization. With the following material equation $D_i = \varepsilon_{ij}E_j + \gamma_{ijk}\nabla_k E_j + \chi_{ijkl}E_j E_k E_l + \Gamma_{ijklm}E_j E_k \nabla_m E_l$ the alteration of the polarization azimuth, $\delta\alpha_r$, of the probe, induced by the linearly polarized incoherent pump is given by

$$\delta\alpha_r = \frac{|E_{pump}|^2\ [Re(\chi_{xxxx} - 2\chi_{xxyy} - \chi_{xyyx})\sin 4\Phi - 4kn^2\ Im(\Gamma_{xyyyz})\cos 2\Phi]}{2n(1-n^2)(1+n^2)}$$

indicating the involvement of the "cubic anisotropy" term ($\Delta\chi = \chi_{xxxx} - 2\chi_{xxyy} - \chi_{xyyx}$, this combination is equal to zero in any isotropic media) and the "cubic nonlocality" terms (Γ_{xyyyz}). Here Φ is the angle that the initial polarization direction of the probe makes with the [100] axis of the crystal.

In the configuration shown in fig.1, with a 2ps Ti:Sapphire laser working at the wavelength of 750nm we have clearly observed both the "cubic anisotropy" and

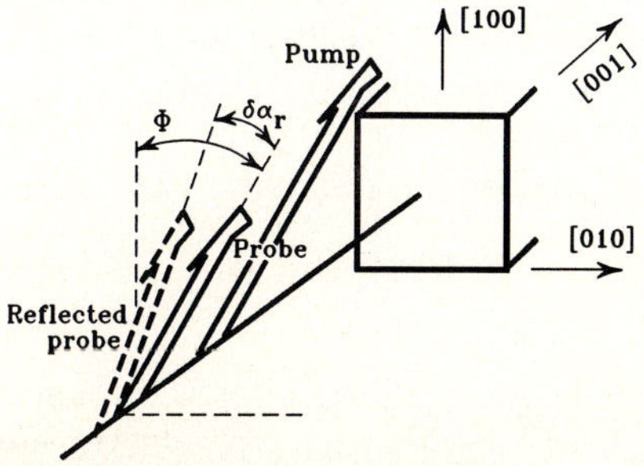

Figurc 1. Obscrvation of specular induced anisotropy in a zinc-blende crystal.

Editors: P. F. Barbara · W. H. Knox · G. A. Mourou · A. H. Zewail

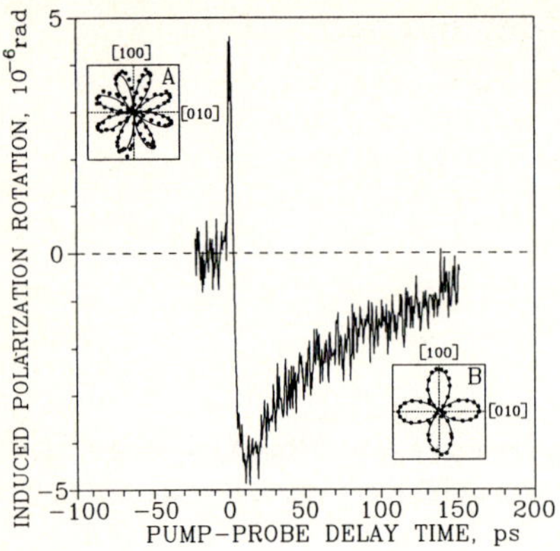

Figure 2. Induced reflected probe polarization rotation as a function of pump-probe delay, intensity $=100 MW/cm^2$.

"cubic nonlocality" terms of a GaAs crystal optical response. This experiment was performed with a 1µrad-sensitive polarimeter. At this wavelength GaAs is completely opaque with a penetration depth of about $\ell = 1\mu m$ ($\hbar\omega \approx 1.65 eV$, $E_g \approx 1.43 eV$). We have studied the reflected probe light polarization rotation as a function of the relative pump-probe delay, pump intensity and crystal orientation. It was found that the time-domain dependence of the rotation has two components with different orientational dependencies (see fig.2). The observed rotation is approximately 20 times weaker than in the optical Faraday effect [1].The first component gives rise to the initial sharp peak on fig.2. When the crystal is rotated around the [001] direction this fast component oscillates as $\sin 4\Phi$. This anisotropic nonlinearity is due [2] predominantly to valence-conductivity band transitions involving the heavily-warped heavy-hole sub-band, i.e the difference in the heavy-hole effective mass for momentum in the [111] and [100] directions. The second component decays within 50ps. It follows a $\cos 2\Phi$ orientational dependence and may be attributed to the second term in the above formula. This effect is due to the absence of an inversion centre in zinc-blende crystals and broken Kramer's degeneracy, associated with appearance of **q**-linear terms in the band structure.

References

[1] A.R.Bungay, A.Malinowski, S.V.Popov, Yu.P.Svirko, R.E.Worsley, N.I.Zheludev. Transient Coherent and Incoherent Specular Inverse Faraday Effect: Measurements of Surface Cubic Optical Nonlinearities. IQEC, Anaheim, May 8-13, 1994.

[2] M.G.Dubenskaya, T.M.Ilyinova, A.V.Trukhov, A.A.Fortigin. Bull. Russ. Ac. Sci., Phys. Sci., **57,** 1192 (1992).

Dominant Biexcitonic Nonlinear Response in GaAs Quantum Wells

H. Wang, J. Shah, T.C. Damen, and L.N. Pfeiffer

AT&T Bell Laboratories, 101 Crawfords Corner Road, Holmdel, NJ 07733, USA

Abstract. Polarization-dependent four-wave-mixing measurements show that biexcitonic effects are one of the leading contributions to the nonlinear optical response in GaAs quantum wells.

We present in this paper polarization dependent four-wave-mixing (FWM) measurements in GaAs quantum wells (QW). We show for the first time the dominant effects of the biexciton-induced two-photon coherence (TPC) on polarization dependent FWM. The biexcitonic effect also explains the anomalous polarization dependence in FWM [1]. These results indicate that the current theory based on the semiconductor Bloch equation is inadequate. Although biexcitons have attracted considerable efforts recently [2-3], the importance of the biexcitonic contribution to the nonlinear response has not been appreciated.

Our measurements were carried out at 10 K with pulses of 400 fs duration and 4.5 meV spectral-width. The samples contain 15 periods of 175 Å GaAs well and 170 Å $Al_{0.3}Ga_{0.7}As$ barrier. Figure 1a shows self-diffracted FWM responses. It is evident that both $I_{\|}$ (obtained with $\mathbf{E}_1 \| \mathbf{E}_2$) and $I_{\perp}$ (obtained with $\mathbf{E}_1 \perp \mathbf{E}_2$) feature significant contributions at negative delay ($t_2<t_1$). In GaAs quantum wells, both biexciton-induced two-photon coherence (TPC) and polarization scattering lead to signals at negative delay [2].

To understand the origin of the negative-delay signal, we show in Fig. 1b the FWM response obtained with $\mathbf{E}_1$ and $\mathbf{E}_2$ having the same circular polarization. The TPC does not contribute to the response in Fig. 1b because the creation of the TPC requires both σ+ and σ- polarized fields. Since the negative-delay signal in Fig. 1b is much smaller than that in Fig. 1a, we conclude that the negative-delay signal in both $I_{\|}$ and $I_{\perp}$ results primarily from the TPC. This signal remains nearly independent of the polarization change since creation of the TPC does not depend on the polarization direction as long as the applied field is linearly polarized.

We further confirm the biexcitonic origin of the TPC using circularly-polarized three-pulse FWM. The solid line in Fig. 2a shows the FWM signal obtained with all three beams σ+ polarized. The dashed line is obtained at the same condition except that $\mathbf{E}_2$ is σ- polarized. The signal at $t_3>t_1$ is due to the interaction of $\mathbf{E}_3^*$ with the TPC created by $\mathbf{E}_1 \cdot \mathbf{E}_2$ with $\mathbf{E}_1$ and $\mathbf{E}_2$ having opposite circular polarization. This signal vanishes when all three beams are co-circularly polarized as shown in Fig. 2a. Spectrally resolving the FWM emission gives a biexciton binding energy of 1.2 meV.

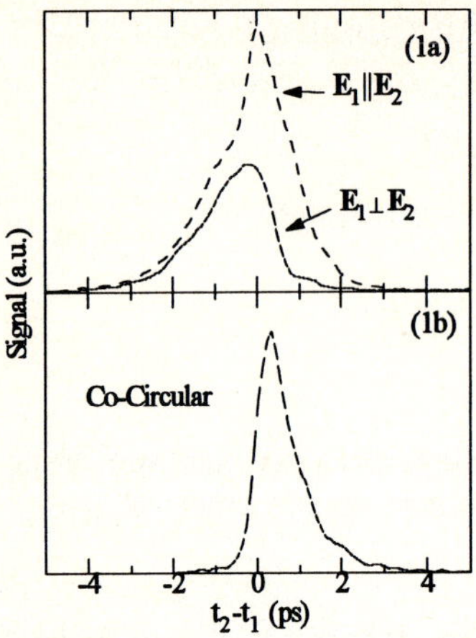

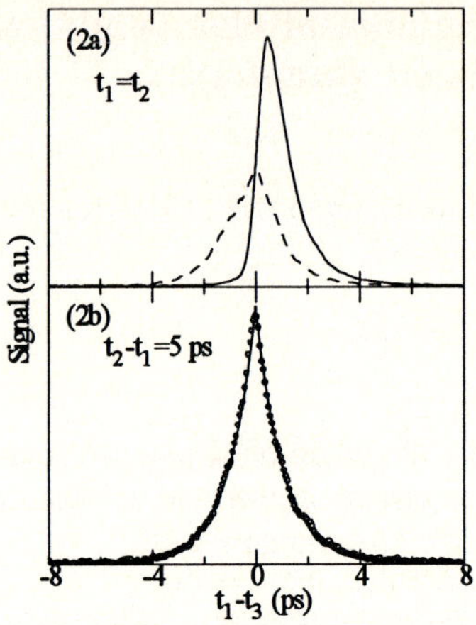

Fig. 1 Self-diffracted FWM responses in the direction of $2\mathbf{k}_2$-$\mathbf{k}_1$. (a) $\mathbf{E}_1$and $\mathbf{E}_2$ are linearly polarized. (b) $\mathbf{E}_1$and $\mathbf{E}_2$ are co-circularly polarized.

Fig. 2 Three pulse FWM responses in the direction of $\mathbf{k}_1$+$\mathbf{k}_2$-$\mathbf{k}_3$. (a) For the solid curve all three fields are co-circularly polarized. For the dashed curve $\mathbf{E}_2$ has the circular polarization opposite to $\mathbf{E}_1$ and $\mathbf{E}_3$. (b) The same as the dashed curve in (a) except that t_2 is delayed by 5 ps with respect to t_1.

Figure 2b shows the FWM response when we delay t_2 by 5 ps with respect to $t_1$to eliminate the contribution from the TPC. The response is symmetric, showing that the sample is homogeneously broadened. The solid line is a fit with a single decay rate for both the positive and negative delay.

We have extended the above result to inhomogeneously broadened systems by using the optical Bloch equations that include a biexciton resonance. Our calculation shows that the TPC-induced FWM signal does not rephase and behaves like a free polarization decay instead of an echo in an inhomogeneously broadened system. Decay of this signal can be much faster than the intrinsic biexciton dephasing. These results also account for the surprising polarization dependence in FWM of inhomogeneously broadened systems [1].

References:

1) D. Bennhardt et al. Phys. Rev. B47, 13485 (1993) and references cited there.
2) B.F. Feuerbacher, J. Kuhl, and K. Ploog, Phys. Rev. B**43**, 2439 (1991).
3) S. Bar-Ad and I. Bar-Joseph, Phys. Rev. Lett. **68**, 349 (1992); D.J. Lovering *et al.*, Phys. Rev. Lett. **68**, 1880 (1992); K.H. Pantke *et al.*, Phys. Rev. B**47**, 2413 (1993).

Enhancement of Coherent Charge Oscillations in Coupled Quantum Wells by Femtosecond Pulse Shaping

A.M. Weiner

Purdue University, School of Electrical Engineering, West Lafayette, IN 47907-1285, USA

Abstract. We analyze the use of femtosecond pulse shaping to enhance quantum mechanical charge oscillations and terahertz emission in coupled quantum wells.

Considerable interest has arisen in coherent control of quantum mechanical motions in atomic and molecular systems. Recently coherent charge oscillations have also been observed in quantum wells [1,2] and superlattices [3]. Here we analyze the use of femtosecond pulse shaping to enhance coherent quantum mechanical charge oscillations in GaAs/GaAlAs coupled quantum wells.

We consider coupled quantum wells consisting of a wide and a narrow well separated by a thin barrier. An electric field is applied in order to align the lowest electron energy levels, leading to an energy level splitting and a delocalization across the two wells. Short pulse excitation from the valence band in one well into a superposition state in the conduction band results in an oscillation of the electron between the two wells, which has been observed, e.g., by terahertz (THz) emission.

Simple experiments using pulse pairs [4] and sequences [5] to control these oscillations have been reported. These experiments did demonstrate control over the oscillation dynamics but did not increase the peak oscillation amplitude over that observed with the original (single) pulse. This raises the following question: can laser pulse shapes be designed that would enhance the charge oscillation amplitude?

Let us first explain why these experiments produce roughly the same peak THz emission as single pulse experiments. With dephasing ignored, the second order polarization responsible for THz emission from a coupled quantum well is given by

$$P^{(2)}(t) \sim \mathrm{Re}\left\{e^{i\omega_{21}t}\int^{t} dt'\, e_L(t')\exp(i\Delta_{13}t')\int^{t} dt'\, e_L^{*}(t')\exp(-i\Delta_{23}t')\right\} \qquad (1)$$

Here the Δs are the detunings of the laser from the interband transitions, and ω_{21} is the charge oscillation frequency. The shaped pulse $e_L(t)$ is given by the Fourier transform of the input spectrum multiplied by a pulse shaping filter $H(\omega)$ [6]. *After the end of the shaped laser pulse*, the coherent polarization is proportional to the product of the shaped laser spectral amplitudes at the two interband transitions. For passive pulse shapers, the spectral amplitudes are bounded by those of the original input pulse. Thus, for a fixed laser pulse energy, pulse shaping cannot increase the coherent polarization (or THz emission) left behind after the laser pulse has finished!

There is, however, the possibility of enhancing the oscillation amplitude *while the laser waveform is still on.* In this scenario the oscillation initially grows, peaks significantly above the level corresponding to the original single laser pulse, and then decreases during the final portion of the pulse to reach a final value identical to the single pulse case. By rewriting eq. (1) using an operator formalism, we have identified pulse shaping filters $H(\omega)$ that can optimize the peak coherent polarization.

Editors: P. F. Barbara · W. H. Knox · G. A. Mourou · A. H. Zewail

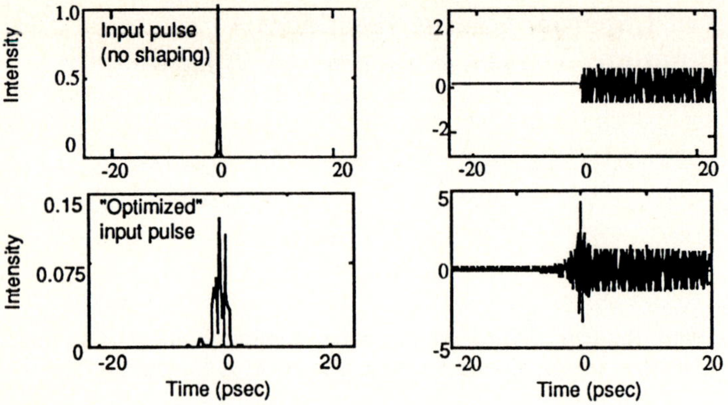

Fig. 1 Computed results for the coherent polarization. **(Top)** An input unshaped laser pulse (left) and the resultant induced polarization (right). **(Bottom)** Shaped input pulse (left) and resultant induced polarization (right) showing a strong enhancement in the peak polarization. Note the different vertical scales.

One example of our computed results is shown in Fig. 1. The top traces show the input (unshaped) laser pulse and the resultant induced polarization. As expected, the polarization turns on at t=0 and then oscillates with a constant amplitude at frequency ω_{21} (2 THz in these simulations). The lower trace shows one of our derived waveforms which causes the polarization to build up to a strong peak at t = 0, with the peak amplitude enhanced by 4.2× compared to no pulse shaping. The waveform shown, although complicated, can be created readily using a previously demonstrated 128-element phase modulator within a pulse shaping apparatus [6].

These initial theoretical results indicate the exciting possibility of strongly influencing charge motions in semiconductor coupled quantum wells by using ultrafast coherent control ! In the future we wish to include dephasing effects and investigate control of wave packets in three-well systems and superlattices.

References

1. K. Leo, J. Shah, E.O. Gobel, T. Damen, S. Schmitt-Rink, W. Schafer, and K. Kohler, *Phys. Rev. Lett.* 66, 201 (1991).
2. H.G. Roskos, M.C. Nuss, J. Shah, K. Leo, D.A.B. Miller, A.M. Fox, S. Schmitt-Rink, and K. Kohler, *Phys. Rev. Lett.* 68, 2216 (1992).
3. C. Waschke, H. G. Roskos, R. Schwedler, K. Leo, H. Kurz, and K. Köhler, Phy. Rev. Lett. 70, 3319 (1993).
4. P.C.M. Planken, I. Brener, M.C. Nuss, M.S.C. Luo, and S.L. Chuang, *Phys. Rev.* B48, 4903 (1993).
5. I. Brener, P.C.M. Planken, M.C. Nuss, L. Pfeiffer, D.E. Leaird, and A.M. Weiner, *Appl. Phys. Lett.* 63, 2213 (1993).
6. A. M. Weiner, D. E. Leaird, J. S. Patel, and J. R. Wullert, *IEEE J. Quantum Electron.* 28, 908 (1992).

Femtosecond Spectral Hole Burning in InGaAs/AlGaAs Strained-Layer Single-Quantum-Well Diode Lasers

C.-K. Sun[1,2], *B. Golubovic*[2], *H.K. Choi*[3], *C.A. Wang*[3], *G.D. Sanders*[4], *C.J. Stanton*[4], *and J.G. Fujimoto*[2]

[1]Division of Applied Science, Harvard University, Cambridge, MA 02138, USA
[2]Department of EECS, Research Laboratory of Electronics, Massachusetts Institute of Technology, Cambridge, MA 02139, USA
[3]Lincoln Laboratory, Massachusetts Institute of Technology, Lexington, MA 02173, USA
[4]Department of Physics, University of Florida, Gainesville, FL 32611, USA

Studies of the high speed modulation of semiconductor lasers, show strong relaxation oscillation damping and suggest the presence of an ultrafast nonlinear gain process[1]. Femtosecond pump-probe investigations performed using identical wavelengths[2,3] reveal characteristic time constants of 1.0 ps and 200 fs. The longer time constant was attributed to carrier heating effects, and the shorter time constant was suggested to arise from delayed carrier heating in the gain region and spectral hole burning in the loss region[3]. Further experiments are needed to better understand the role of spectral hole burning in the gain region.

In order to distinguish spectral hole burning in the gain region, we used a novel heterodyne multiple wavelength pump-probe technique to investigate gain dynamics in strained-layer InGaAs SQW diode lasers. The strain induced heavy and light hole band splitting allows experimental conditions to be chosen where only a single-transition contributes. This avoids the ambiguity of multiple spectral holes and simplifies the interpretation of the experimental data. In the gain region, spectral hole burning was directly observed for the first time.

Studies were performed using a modelocked Ti:Al_2O_3 laser. The pulse spectrum was first broadened in an optical fiber and spectrally windowed to produce pump and probe pulses[4]. A novel heterodyne detection technique[5] was used to separate the probe from the pump. The bandgap of the laser diode was near 960 nm and 910 nm for heavy and light hole transitions, respectively. The pump was chosen to be at 935 nm and the probe was varied over a range of ± 25 nm from the pump without allowing the light hole band transitions. The pulse duration was 200 fs with an 8 nm bandwidth. The diode was operated at high bias, but under the lasing threshold (18 mA, corresponding to $\sim 5 \times 10^{12} cm^{-2}$), so that all pump-probe wavelengths were far away from the transparency point in the gain region. The temperature and carrier population transient effects were minimized, and the gain at different probe wavelengths were roughly the same. The study was performed in the perturbative limit, so that differential transmission changes were directly proportional to the changes in the gain spectrum.

Figure 1 shows differential transmission changes measured at different probe frequencies as a function of time delay. Around zero delay, a sharp transmission decrease was observed with time resolution limited recovery.

Editors: P. F. Barbara · W. H. Knox · G. A. Mourou · A. H. Zewail

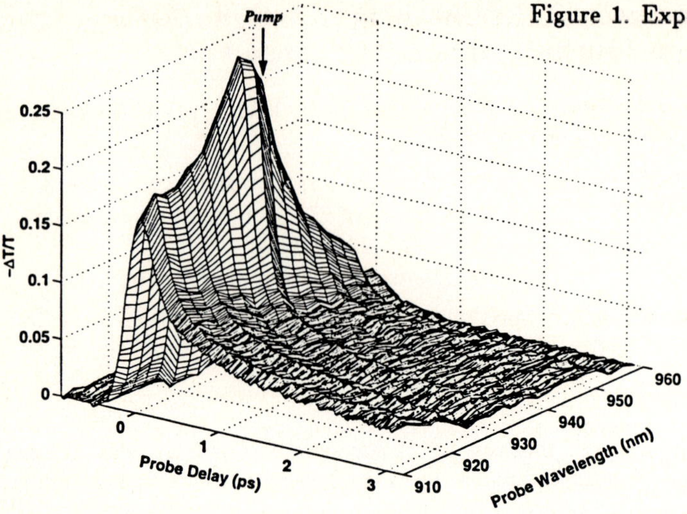

Figure 1. Experimental data.

This rapid transient was attributed mainly to spectral hole burning and two photon absorption. As the probe frequency is tuned close to the pump, this transmission transient increases. This wavelength dependent characterization provides direct evidence of spectral hole burning effects. The peak of the transmission decrease is blue-shifted 5 nm with respect to the pump center wavelength. This is probably due to the relaxation of the spectral hole toward higher energies. Compared to the work by Hall *et al.*[3], the strong spectral hole burning effect in our data might be a result of the operating regime. The device was operated at the bottom of the gain region where the occupation number is close to 1 and the temperature effects are less pronounced. The previous study was done close to the transparency point where the occupation number is close to 1/2 and the temperature effects are also much stronger.

We would like to acknowledge the scientific collaboration of Prof. E. P. Ippen. This research is supported in part by the ONR FEL program N00014-91-J-1956, JSEP contract DAAL03-92-C-0001, AFOSR contract F49620-91-C-0091, and the Dept. of the Air Force.

References

[1] R. S. Tucker, J. Lightwave Technol. **3**, 1180 (1985).

[2] M. P. Kesler and E. P. Ippen, Appl. Phys. Lett. **51**, 1765 (1987).

[3] K. L. Hall, G. Lenz, E. P. Ippen, U. Koren, and G. Raybon, Appl. Phys. Lett. **61**, 2512 (1992).

[4] C.-K. Sun, H. K. Choi, C. A. Wang, and J. G. Fujimoto, Appl. Phys. Lett. **62**, 747 (1993).

[5] K. L. Hall, G. Lenz, E. P. Ippen, and G. Raybon, Opt. Lett. **17**, 876 (1992).

Optical Enhancement of Terahertz Emission from Bulk Semi-Insulating GaAs

B.B. Hu, P.C.M. Planken, M.C. Nuss, K.W. Goossen, J.E. Cunningham, W.Y. Jan, H.T. Chiu, and C. Mendonca

AT&T Bell Laboratories, 101 Crawfords Corner Road, Holmdel, NJ 07733-3030, USA

Abstract: We investigate the optical enhancement of THz emission from bulk semi-insulating GaAs wafers.

We present an observation that the THz signal induced from semi-insulating bulk GaAs wafers by a femtosecond optical pulse is *enhanced, and sometimes even reversed in polarity* by a prepulse. This is in contrast to the regular screening case in which the prepulse screens the field, causing a reduction of the signal from the second optical pulse. We attribute this unexpected behavior to the preferential capture of photocarriers by fast traps at the surface which can greatly change the surface charge density, and therefore the surface field.

In our experiments, two optical pulses of equal power are used to induce the THz signal. Figure 1 shows the THz wave forms emitted from a Liquid-Encapsulated-Czochralski (LEC) grown semi-insulating (SI) GaAs substrate for different time delays between the two pulses. The pulse which comes later in time is shown to produce *more* signal (the dashed curves) than the first pulse. The inset shows the amplitude of the THz signal due to each pulse at a relative time delay of 10 ps for different temporatures. The enhancement is thermally activated with a thermal activation energy of 10-15 meV. The enhancement effects are reduced after the sample is desorbed in MBE chamber, and completely disappear after we cap the sample with a 2000 Å $Al_{0.3}Ga_{0.7}As$ layer. Most of the SI (including LEC grown undoped and Cr doped) samples show the enhancement; while MBE samples only show screening behavior.

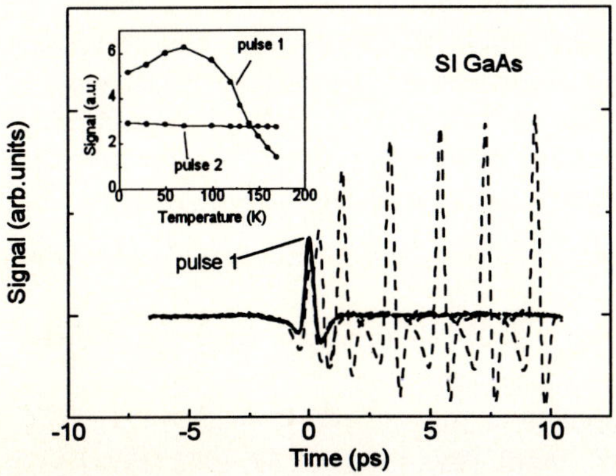

Fig. 1: THz wave forms from a LEC SI GaAs sample. The dashed curves are THz signals by the second pulse at different time delays relative to the first pulse. Inset: The amplitude of the THz signal from both pulses at different temperatures.

Springer Series in Chemical Physics, Vol. 60 **Ultrafast Phenomena IX**
Editors: P. F. Barbara · W. H. Knox · G. A. Mourou · A. H. Zewail

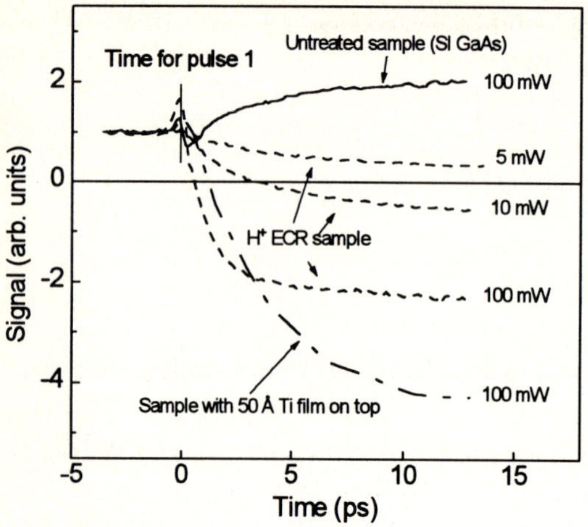

Fig.2 : The amplitude of the THz signal from the second pulse as a function of the relative time delay for three different samples at different powers of the pre-pulse.

The enhancement of the THz signal indicates the increase of the surface field which is determined by the surface charge density. For undoped SI GaAs, it is estimated that the Fermi-level is pinned at 0.82 V below the conduction band and the surface charge density is around 10^{12} cm^{-2}, resulting in a surface field on the order of 10^4 V/cm [1]. For high quality MBE samples, the surface charge density is not affected on a picosecond time scale. This is not true for commercial wafers because of the large number of shallow traps at the surface. The occupancy of these trap states can be greatly affected by the preferential capture of photocarriers, resulting in a big change in surface charge density on the same order as the surface trap density, even when the prepulse is weak [2,3]. This change increases with the photocarrier density, and saturates when the free carrier density approaches the total number of the surface traps. For SI commercial wafers, this effect causes more negative charges at the surface, therefore enhances the field as shown in our THz signal.

On the other hand, when more positive charges accumulate at the surface, the field will be reduced or even reversed in polarity. This is also observed in our measurements after we treat the sample using H^+ ECR (Electron Cyclotron Resonance), or evaporate a thin metal film on the surface. Fig. 2 shows the amplitude of the THz signal for different powers of the prepulse. When the power of the pre-pulse is small, the signal generated by the second optical pulse is reduced. When the power of the first pulse reaches about 8 mW, the prepulse flips the polarity of the THz signal generated by the second optical pulse. Further increase of the power enhances the amplitude of the signal.

References

[1] Robert B. Darling, Phys. Rev. B. vol. 43, 4071 (1991).

[2] J.-C. Manifacier and H. K. Henisch, Phys. Rev. B , vol. 17, 2648 (1978).

[3] S. E. Ralph, and D. Grischkowsky, Appl. Phys. Lett. vol. 59, 1972 (1991).

Pulse-Doped GaAs Nonlinear Transmission Lines for 630 fs, 1 V Transients

D. van der Weide[1] *and P. Brunemeier*[2]

[1]Max-Planck-Institut für Festkörperforschung, Heisenbergstrasse 1, D-70569 Stuttgart, Germany
[2]Power Spectra Inc., 919 Hermosa Ct., Sunnyvale, CA 94086, USA

Abstract We report nonlinear transmission lines (NLTLs) having 630 fs, 1 V edges (measured at T = 300 K), realized on GaAs with a pulse-doped profile for maximum capacitance-voltage nonlinearity, and measured with on-chip sampling circuits strobed by identical NLTLs. This epitaxial profile provides $f_c \approx 7$ THz abrupt-junction diodes by a simple etching process, enabling us to combine two essential requirements for an ultrafast electronic circuit onto one monolithic IC.

Recently, increasingly faster nonlinear transmission lines have been reported in the literature, with subpicosecond results now being observed for the first time from electronic circuits [1,2,3]. Electronic pulse generation has lagged behind electro-optic switching, although each method has been employed for producing ultrafast test signals used both for advanced instrumentation and for probing the physics of carrier transport within the devices themselves. The gap is now closing: since this report we have measured substantially faster edges with 3× larger amplitudes [4].

While the speed of an electro-optic switch is limited primarily by carrier lifetimes, the speed of a nonlinear transmission line is limited by the rate at which the depletion edge of a Schottky diode can move through the semiconductor. This edge speed involves both the electromagnetic environment of the diode and the carrier transport limitations of the semiconductor. We are investigating these limitations by designing and building nonlinear transmission lines on pulse-doped GaAs, combining the desirable properties of large *C(V)* nonlinearity with low doping for fast depletion and long mean-free paths through a simple etching step.

Nonlinear transmission lines are synthetic structures of series inductors (approximated by sections of high-impedance transmission line) with varactor diodes periodically placed as shunt elements. On this structure a voltage step develops from a sinusoidal input because the propagation velocity u is modulated by the diode capacitance, $\Delta u(V) = 1\big/\sqrt{LC(V)}$, where L is the inductance and $C(V)$ the sum of the diode and parasitic capacitance of the line, all per unit length. Limitations of the NLTL arise from its periodic cutoff frequency, $\omega_p = 2\big/\sqrt{LC(V)}$, waveguide dispersion, interconnect metallization losses, and diode resistive losses. Diode samplers can be integrated in the fabrication process with NLTLs to measure the output waveforms accurately.

As shown in Fig. 1, the pulse-doped profile combines two distinct diode types in one circuit through a simple etch step. Just below the Schottky contact lies a 500 Å layer of 5E16 cm^{-3} material, followed by an 80 Å pulse of carriers at 4E18 cm^{-3}. The 5E16 cm^{-3} N^- layer is 0.27 µm thick, followed by a 1 µm thick N^+ layer for a backside contact. Also shown is an abrupt-junction diode formed by etching

Editors: P. F. Barbara · W. H. Knox · G. A. Mourou · A. H. Zewail

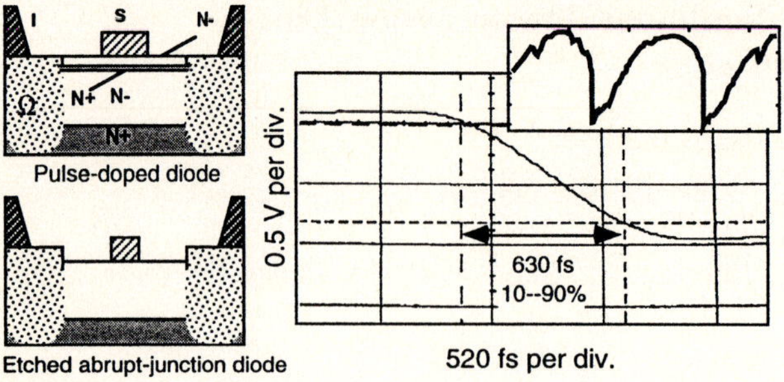

FIG. 1. Diode profiles (left) with semiconductor layer designations and (I)nterconnect, (S)chottky, and Ohmic (Ω) metal. NLTL output (right) measured by integral diode sampling bridge; inset shows two full cycles of the waveform.

through the pulse. The smallest diodes in the NLTL/sampler circuit have RC_{j0} cutoff frequencies ≈ 7 THz, as estimated from auxiliary measurements.

To make the measurements shown in Fig. 1, we packaged a monolithic NLTL/sampler chip with additional NLTLs and suitable microwave connectors. The NLTL test port and sampler strobe inputs were offset by 0.5 Hz by a rotating phase shifter to sample the NLTL output in equivalent time (with < 20 fs RMS jitter). A root-sum-squares deconvolution of the measured fall time with the sampler aperture time (assuming that the two times are equal) gives ~ 450 fs.

We thank D.M. Bloom (Stanford University) for fabrication support and several members of the technical staff at Watkins-Johnson Co. for useful discussions.

References

[1] D.W. van der Weide, J.S. Bostak, B.A. Auld, D.M. Bloom, Appl. Phys. Lett. **62**, 22-24 (1993).

[2] M.S. Shakouri, A. Black, B.A. Auld, D.M. Bloom, Electron. Lett. **29**, 557-558 (1993).

[3] S.T. Allen, U. Bhattacharya, M.J.W. Rodwell, Electron. Lett. **29**, 2227-8 (1993).

[4] D.W. van der Weide, submitted to Appl. Phys. Lett., Mar. 1994.

Generation of Tunable Narrow-Band Free-Space Terahertz Radiation

A.W. Weling, B.B. Hu, N.M. Froberg*, and D.H. Auston*

Microelectronics Sciences Laboratories, Room 1001 Schapiro CEPSR, 530 West 120th Street, Columbia University, New York, NY 10027, USA
*Present address: AT&T Bell Laboratories, 101 Crawfords Corner Road, Holmdel, NJ 07733, USA

We report a novel technique for generating narrowband far- infrared (FIR) radiation in free space continuously tunable from dc to > 1THz by mixing two linearly chirped optical pulses with a variable time delay on a THz emitter such as a photoconducting antenna.

Recently, a variety of techniques have been developed to generate ultrashort pulses of FIR radiation in free space using femtosecond optical excitation. The broadband nature of this THz radiation and the lack of frequency control limits the applicability of these devices to microwave communication systems and FIR spectroscopy. Previous efforts to optoelectronically generate narrowband microwave radiation with broad tunability include the use of cw laser heterodyne in photodetectors and HEMT's [1].

Our experimental scheme for producing quasi-sinusoidal optical modulation at THz frequencies involves a Michelson interferometer arrangement (fig. 1), the input to which is a linearly chirped optical pulse whose instantaneous carrier frequency is linearly swept in time at a rate μ. At any instant the frequency content of the two split-off pulses in the arms of the interferometer differs by a constant beat frequency f_o, which is directly proportional to the time delay τ between them. The cross term in the interferometer output $I_{total}(t, \tau) = |E(t)+E(t+\tau)|^2 = I_1(t) + I_2(t+\tau) + I_{cross}(t, \tau)$ contains the sinusoidal modulation at the beat frequency $f_o \sim \tau\mu/2\pi$. Photoconducting antennas or non-linear crystals can be used as optical mixers which convert this optical modulation into THz radiation with a center frequency f_o which can ideally be tuned out to the available optical bandwidth by varying the delay τ. For a finite emitter bandwidth Δf_{max}, the bandwidth of the THz radiation is given by $\Delta f \sim \Delta f_{max}\sigma/\sigma_n$ where σ is the transform limited pulsewidth for the given optical bandwidth, and σ_n the FWHM of the chirped pulse.

We used a regenerative amplifier based on chirped pulse amplification in Ti:sapphire [2] to produce linearly chirped pulses at 800 nm with an FWHM of 22.5 ps and 400 μJ pulse energy. A large aperture photoconducting antenna with 4 mm spaced Al electrodes on semi-insulating GaAs biased at 2.0 kV/cm was used as a THz emitter. The

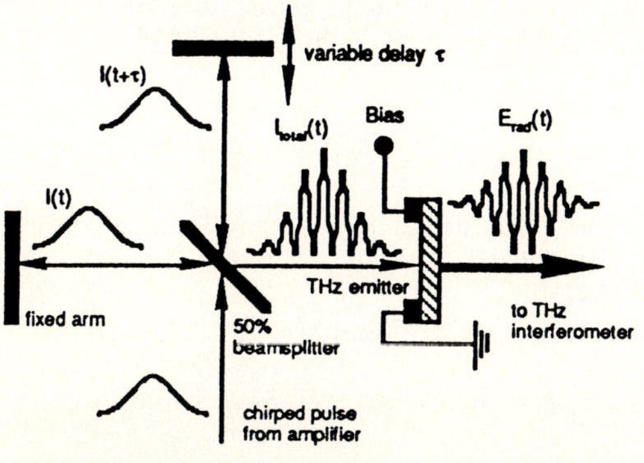

Fig.1: Schematic of experimental technique showing temporal overlap of the two linearly chirped pulses in the interferometer to produce quasi-sinusoidal modulation.

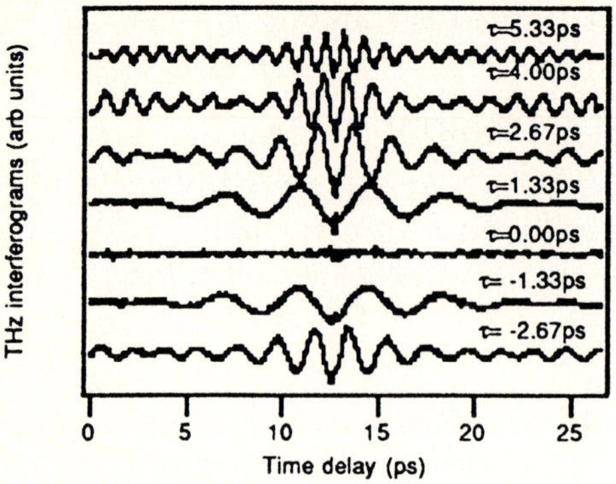

Fig.2: THz interferograms from beating of 22.5 ps chirped pulses at various delays τ; effects of satellite pulses at ± 12 ps due to multiple reflections in the emitter are visible.

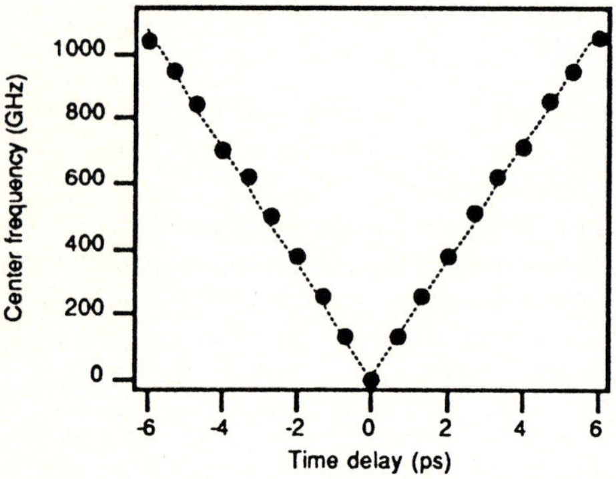

Fig.3: Tuning of THz frequency f_o with delay τ (dotted line shows linear fit using a Gaussian spectrum of 11.8 nm FWHM close to the measured value of 10.3 nm).

resultant free space THz radiation was characterized by a non-polarizing Michelson interferometer using a helium-cooled Si bolometer detector [3]. Fig. 2 shows the THz interferograms obtained for various time delays between the two chirped pulses. The results of our frequency tuning experiment along with a linear fit obtained by using a Gaussian σ of 80 fs are shown in fig. 3.

Optical difference frequency mixing in a non-linear crystal such as $LiTaO_3$ can be used to extend the tuning range to the optical bandwidth available (~ 4 THz). The above scheme of chirped pulse beating can be modified to spatially scan the narrowband THz radiation by having the two chirped pulses mix on the emitter surface in a non-collinear configuration and controlling the angle δθ between them. This research was supported by the Air Force Office of Scientific Research under grant no. F49620-92-J-0036.

References

1. G.J. Simmons and K.G. Purchase, IEEE Trans. Microwave Theory Tech., 38, 667 (1990); D.V. Plant, D.C. Scott, H.R. Fetterman, L.K. Shaw, W. Jones, and K.L. Tan, IEEE Photon. Technol. Lett., 4, 102 (1992).
2. P. Maine, D. Strickland, P. Bado, M. Pessot, and G. Mourou, IEEE J. Quantum Electron., 24, 398 (1988).
3. B.I. Greene, J.F. Federici, D.R. Dykaar, R.R. Jones, and P.H. Bucksbaum, Appl. Phys. Lett., 59, 893, (1991).

AC-Stark Effect of the Fermi-Edge Singularity: Observation of "Excitonic Polarons?"

E.I. Perakis[1,2], *I. Brener*[2], *W.H. Knox*[2], *and D.S. Chemla*[3]

[1]Department of Physics, Rutgers University, Piscataway, NJ 08855, USA
[2]AT&T Bell Laboratories, 101 Crawfords Corner Road, Holmdel, NJ 07733, USA
[3]Physics Department, Univerisity of California at Berkeley, Berkeley, CA 94720, USA

Intense below-bandgap pump optical pulses can dress semiconductors by inducing virtual charge fluctuations. As a result, the exciton absorption blue-shifts and decreases [1], analogous to the AC-Stark effect in atoms. However, excitonic resonances originate from Coulomb correlations and many-body effects must be addressed [2]. In modulation-doped quantum wells, the absorption edge resonance is called Fermi Edge Singularity (FES); its bleaching is much smaller than for undoped excitons (Fig. (1)). To clarify the role of the time-dependent many-body processes, we introduced a new theoretical approach [3]. The pump induces a fast virtual e-h polarization cloud, with adiabatically adjusting size. As in the polaron problem, the interacting electrons become heavier excitonic polarons, determined by the instantaneous pump amplitude and Coulomb correlations. The weak probe induces optical transitions in the dressed system, with reduced amplitudes and blue-shifted energies due to Pauli blocking among the cloud e-h pairs. The FES results from the e-h Coulomb scattering of large numbers of electrons across the Fermi surface [4]; sharp excitonic bound states do not exist. Such many-body processes in the dressed system involve time-dependent excitonic polarons. Unlike in atoms, excitonic resonances strongly depend on the masses. For long pulses, the polaronic effect increases the exponent of the FES or decreases the exciton Bohr radius; this overcompensates the Pauli decrease in the transition amplitude. For ultrashort pulses, the low-lying Fermi sea excitations are *slower* than the

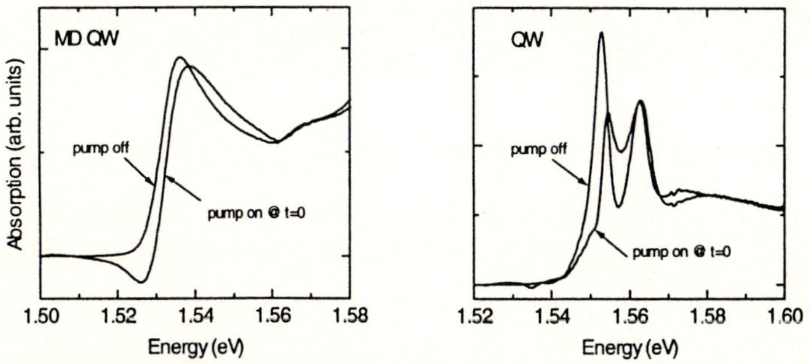

Figure 1: Nonlinear absorption spectra.

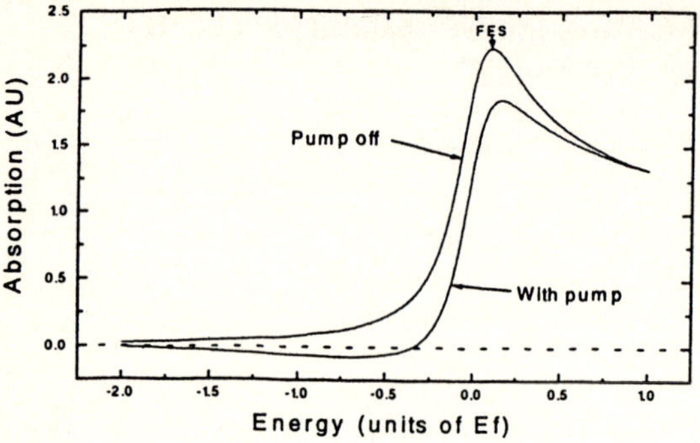

Figure 2: Calculated absorption spectrum of the FES.

time-dependent renormalizations, unlike interband e-h pairs; thus the Pauli reduction is only partially compensated. Fig. (2), obtained for sub-bandgap square pump and delta-function probe pulses in the limit of small scattering renormalizations, displays the qualitative features of the experiment. For large detunings, the dressed parameters and the resonance blue-shift are comparable in undoped and doped semiconductors. The bleaching is different for two reasons. The bound state semiclassical trajectory does not close in the dressed undoped semiconductor; the Bohr radius changes during a single Bohr period. Also, the pump induces indirect transitions [3], which broaden the FES.

D. S. C. is supported by the Office of Basic Energy Sciences- D. M. S. of U.S D.O.E., No. DE-ACO3-76SF00098.

References

[1] A. Mysyrowicz, D. Hulin, A. Antonetti, A. Migus, W. Masselink, and H. Morcoc, Phys. Rev. Lett. **56**, 2748 (1986); W. Knox, D. S. Chemla, D. A. B. Miller, J. Stark, and S. Schmitt-Rink, Phys. Rev. Lett. **62**, 1189 (1989).

[2] S. Schmitt-Rink and D. S. Chemla, Phys. Rev. Lett. **57**, 2752 (1986);

[3] I. E. Perakis and D. S. Chemla, to be published in Phys. Rev. Lett.

[4] K. Ohtaka and Y. Tanabe, Rev. Mod. Phys. **62**, 929 (1990); I. E. Perakis and Yia-Chung Chang, Phys. Rev. B **44**, 5877 (1991); ibid **47**, 6573 (1993).

Femtosecond Carrier Dynamics at InP/Liquid Interfaces in the Presence of Electric Fields

Y. Rosenwaks[1], *B.R. Thacker*[1], *A.J. Nozik*[1], *R.J. Ellingson*[2], *K.C. Burr*[2], *and C.L. Tang*[2]

[1]National Renewable Energy Laboratory, Golden, CO 80401, USA
[2]Materials Science Center, Cornell University, Ithaca, NY 14853, USA

Recently there has been considerable interest in studying photogenerated carrier dynamics in semiconductors under the influence of electric fields[1]. It is expected that these fields may have a crucial role in electron transfer (ET) processes at semiconductor electrodes. In this work we conduct fs time-resolved PL measurements in the presence of large electric fields controlled with an externally applied bias.

The photoelectrode is p-InP (Zn doped, 1.5×10^{17} cm^{-3}) single crystals with(100) surface orientation (from Nippon Mining). They were etched for 30 seconds in 2% Br_2/MeOH and then mounted in a sealed oxygen-free photoelectrochemical (PEC) cell. The electrolyte used was sodium sulfate and the redox couple was $Fe(CN)_6^{3-/4-}$; all the potentials were controlled with a potentiostat and were measured relative to a sodium saturated reference electrode (SSCE). Time-resolved PL (TRPL) spectra were obtained by the technique of sum frequency generation. Photoexcitation was with a Ti: sapphire laser that provided 80-fs pulses at a wavelength of 780 nm and 82 MHz repetition rate. The laser spot size was ~ 20 μ in diameter, and the temporal resolution of the system was 100 fs. The band-edge TRPL was measured at different potentials applied to the InP photoelectode.

Figure 1 shows the TRPL spectra of the InP electrode measured in air at two different laser intensities corresponding to excess carrier concentrations of $I_o=5\times10^{18}$ and 2×10^{17} cm^{-3}. The solid lines were calculated from our model based on a numerical solution of the drift-diffusion equations described previously[2].

The calculations of the theoretical PL decay curves (solid lines) in Fig. 1 were done using parameters based on a recent study of InP[3]. A $\delta(t)$ laser pulse was assumed, and V_B (the surface band bending) was 0.5 V which is typical for air/p-InP. The injection level, (defined here as the the initial (at t = 0) excess carrier concentration relative to

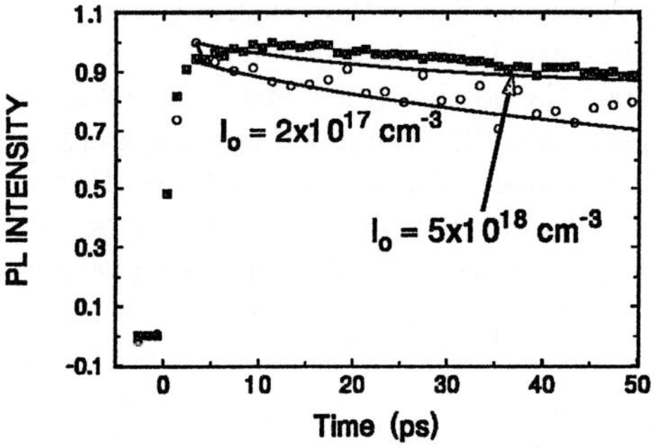

Figure 1. Experimental and calculated (solid lines) PLdecay curves for p-InP in air under high and moderate injection conditions.

Observation of Biexcitons in Time-Resolved Degenerate Four-Wave Mixing on GaAs Quantum Wells

E.J. Mayer[1], *G.O. Smith*[1], *J. Kuhl*[1], *D. Bennhardt*[2], *T. Meier*[2], *A. Schulze*[2], *P. Thomas*[2], *S.W. Koch*[2], *R. Hey*[3], *and K. Ploog*[3]

[1]Max-Planck-Institut für Festkörperforschung, Heisenbergstrasse 1, D-70569 Stuttgart, Germany
[2]Department of Physics, Philipps University, Renthof 5, D-35032 Marburg, Germany
[3]Paul-Drude-Institut für Festkörperelektronik, D-10177 Berlin, Germany

Although the importance of exciton/exciton interactions for the nonlinear optical response of GaAs quantum wells (QW) is well established by many experimental observations, the role of biexciton states in these processes is still an open question. Evidence for the formation of biexcitons has been independently derived from pump-probe studies [1], from the density dependence of the photoluminescence (PL) line shape [2] and from the shape and polarization dependence of the degenerate-four-wave-mixing (DFWM) signal [3,4,5]. Here, we report on time-resolved DFWM with parallel (PP) and cross-polarized (CP) pulses. Depending on the polarization geometry, the signals reveal strong quantum beats which correspond to the splitting between either the heavy hole (hh) and light hole (lh) exciton or the hh exciton and the biexciton.

In the experiments, a single 20nm-GaAs/$Al_{0.3}Ga_{0.7}As$ QW (photoluminescence linewidth 0.3 meV, homogeneous linewidth 0.15 meV) kept at 8 K is excited by two 120 fs pulses from a mode-locked Ti:Sapphire laser using the backward reflection geometry for the signal detection. Time-resolution of the signal diffracted in the direction $2\mathbf{k}_2$-$\mathbf{k}_1$ is accomplished via upconversion with a reference pulse in a 2 mm thick $LiIO_3$ crystal.

Figure 1 presents 3D plots of the time-resolved signal for parallel and perpendicular polarization of the exciting fields. Since the short laser pulse duration enables simultaneous excitation of the heavy and light hole exciton transition, the signal traces display a rich quantum beat structure. Careful analysis of the beating phenomenon leads to the peculiar result that the beat frequencies appearing for PP and CP orientation differ and correspond to energy splittings of 3.7 meV and 1.2 meV, respectively. Whereas the beat frequency in case of the PP configuration agrees fairly well with the hh-lh splitting measured by PL excitation spectroscopy, the beating is distinctly slower for the CP geometry and the corresponding splitting agrees with the biexciton binding energy of 1.2 meV derived from the appearance of a second luminescence peak on the low energy side of the hh exciton transition. At higher excitation densities, the modulation can be attributed to quantum beats between bound and unbound two exciton states in our sample. The dependence of the beat structure on the polarization of the exciting fields is explained in the frame of a nine-level scheme, consisting of the ground

Springer Series in Chemical Physics, Vol. 60 **Ultrafast Phenomena IX**
Editors: P. F. Barbara · W. H. Knox · G. A. Mourou · A. H. Zewail

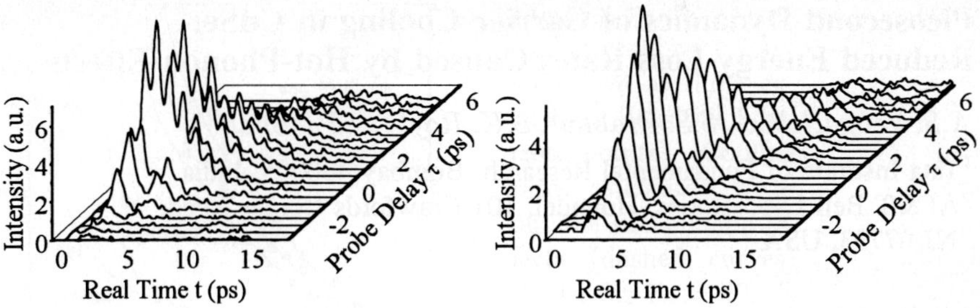

Fig. 1 The time-resolved DFWM intensities for parallel (left) and perpendicular (right) polarizations.

state, two hh and two lh single-exciton states representing excitons excited by x- and y- polarized light and five levels representing two-exciton states. The first two represent two-exciton scattering states consisting of either two hh or two lh excitons, the next two scattering states built up by one hh- and one lh- exciton. The final two-exciton state plays a specific role since it is due to a bound state between two hh-excitons with opposite spins, which is separated from the corresponding scattering state by the biexciton binding energy. For the PP geometry and probe delay $\tau > 0$, identically polarized hh and lh single-exciton states yield the dominant contribution to the signal. In this case, we observe strong hh-lh-exciton beating. For the CP geometry, generation of the DFWM signal necessarily involves the bound and unbound two-exciton levels, since this mechanism provides the only possible polarization path for this configuration. De-excitation of the two-exciton states then produces a beat frequency corresponding to the biexciton/unbound-two-exciton splitting. For the shorter dephasing time of the two-exciton states (T_2 = 3.6 ps) in comparison to that of the single-exciton state (T_2 = 5.3 ps), a smaller time-integrated DFWM signal is observed in the CP-configuration. Close inspection of the signal traces for PP and $\tau < 0$, shows the superposition of two beating phenomena. This observation is in accordance with the polarization of the 9-level scheme since, under this condition, both single- and two-exciton transitions contribute to the 3rd order polarization.

[1] S. Bar-Ad and I. Bar-Joseph, Phys. Rev. Lett. **68**, 349(1992).

[2] R.T. Phillips, D.J. Lovering, G.J. Denton, and G.W. Smith, Phys. Rev. B **45**, 4308(1992).

[3] K.-H. Pantke, D. Oberhauser, V.G. Lyssenko, and J.M. Hvam, Phys. Rev. B **47**, 2413(1993).

[4] K. Bott, O. Heller, D. Bennhardt, S.T. Cundiff, P. Thomas, E.J. Mayer, G.O. Smith, R. Eccleston, and J. Kuhl, Phys. Rev. B **48**, 17418 (1993).

[5] T. Saiki, M. Kuwata-Gonokami, T. Matsusue, and H. Sakaki, Phys. Rev. B **49**, 7817(1994)

acoustic phonon bottleneck [4], screening and phonon renormalization [5], etc. which are not considered here.

In summary, we find significantly reduced hot carrier cooling rates in CdSe using picosecond upconversion luminescence spectroscopy. This can be explained to be due to the hot phonon effects provided the optical phonon lifetime τ_{LO} is 6 to 9 ps, depending upon the excitation density. This may be compared with the corresponding value of 7 ps for GaAs obtained using Raman measurements [6]. It would be interesting to have a direct experimental determination of τ_{LO} in CdSe.

References

1. K. Kash, J. Shah, D. Block, A.C. Gossard and W. Wiegmann, Physica 134B, 189 (1985).
2. See for example, J. Shah and R.F. Leheny in 'Semiconductors Probed by Ultrafast Laser Spectroscopy', edited by R.R. Alfano (Academic Press, 1984) and references therein.
3. W. Potz and P. Kocevar, Phys. Rev. B28, 7040 (1983).
4. R. Baltramiejunas and A. Zukauskas, Phys. Stat. Solidi B149, 337 (1988).
5. S. DasSarma in 'Hot Carriers in Semiconductor Nanostructures' edited by J.Shah (Academic Press, 1992).
6. D. von der Linde, J. Kuhl and H. Klingenburg, Phys. Rev. Lett. 44, 1505 (1980).

Ultrafast Magnetoexciton Dynamics in GaAs

M.-A. Mycek, U. Siegner, and D.S. Chemla

Department of Physics, University of California at Berkeley and
Material Sciences Division, Lawrence Berkeley Laboratory, MS 2-300,
1 Cyclotron Road, Berkeley, CA 94720, USA

Abstract: Contrary to experiments in quantum wells, we observe evidence of Coulomb interactions between magnetoexcitons in GaAs at high fields.

The application of a magnetic field to a 3-D system quantizes the energy levels of states with momentum perpendicular to B and, in contradistinction to experiments performed on 2-D quantum well systems, leaves a 1-D continuum of states with momentum parallel to B. Time integrated (TI), time-resolved (TR) [1], and power spectrum measurements of four wave mixing (FWM) emission from GaAs were performed using co-circularly polarized 100fs pulses from a Ti:Sapphire laser tuned to excite the lowest order HH and LH magnetoexcitons. The TI-FWM signal is shown in Fig. 1. With field, in addition to an obvious nonperiodic beating modulation on the TI signal, there is a large signal at negative Δt which increases significantly with increasing field strength and disappears at higher excitation densities. These are all indications of the presence of Coulomb interactions between magnetoexcitons. TR-FWM measurements taken for negative Δt, where interactions dominate the signal, confirm this. In order to better understand the nonperiodic TI beat modulation, we measured the evolution in time of the FWM power spectrum (see Fig. 2).

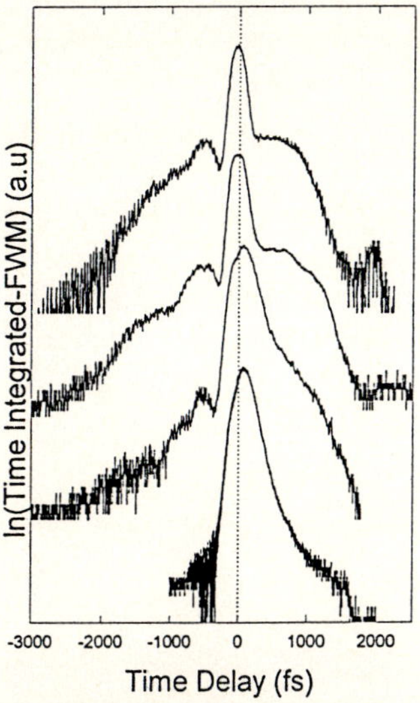

Figure 1: TI-FWM at T=1.6K and $N_x \approx 10^{16} cm^{-3}$ for B=10T (top), 6T, 2T, and 0T (bottom).

Springer Series in Chemical Physics, Vol. 60 **Ultrafast Phenomena IX**
Editors: P. F. Barbara · W. H. Knox · G. A. Mourou · A. H. Zewail

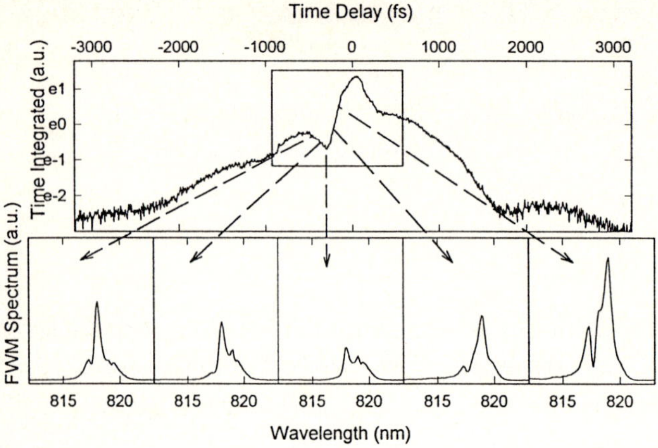

Figure 2: For B=6T, $N_X \approx 10^{16} cm^{-3}$, and T=1.6K: TI-FWM (top) and FWM Power Spectra (bottom) for Δt = -560fs, -400fs, -320fs, -240fs, and -160fs, left to right.

Two of the spectral components present in the FWM emission are from the LH and HH exciton, as expected, while the remaining components cannot be identified from lines in the linear absorption. At present, a two-photon process is a candidate to explain this feature. From these data, we measure the dynamics of each component individually and find that as they evolve in time, they decay and/or beat asynchronously with distinct dephasing times, thus giving rise to the nonperiodic beating observed temporally.

Thus, FWM emission measurements demonstrate that Coulomb interactions between magnetoexcitons persist at high fields in bulk GaAs. This is fundamentally different from 2-D quantum well systems, where the interactions become quenched at high fields. [2]

This work was supported by the US DOE under contract No. DE-AC03-76SF00098.

References: [1] M.-A. Mycek, S. Weiss, J.-Y. Bigot, S. Schmitt-Rink, D.S. Chemla, Appl. Phys. Lett. 60, 2666 (1992); [2] M. Hartmann, W. Schafer, S. Schmitt-Rink, Int. Meeting on Optics of Excitons in Confined Systems, Inst. Phys. Conf. Ser. No.123, p.265

Part VIII

Biology: Primary Dynamics, Electron and Energy Transfer

Direct Observation of Vibrational Dephasing in Functional Proteins

J.-L. Martin[1], M.H. Vos[1], J.-C. Lambry[1], J. Breton[2], M.R. Jones[3], and C.N. Hunter[3]

[1]Laboratoire d'Optique Appliquée, INSERM U275, CNRS URA 1406, Ecole Polytechnique, ENSTA, F-91120 Palaiseau, France

[2]SBE/DBCM, CEN de Sacly, F-91191 Gif-sur-Yvette Cedex, France

[3]Krebs Institute for Biomolecular Research and Robot Hill Institute for Photosynthesis, Department of Molecular Biology and Botechnology, University of Sheffield, Western Bank, Sheffield S10 2UH, UK

Abstract. The full spectral evolution of the stimulated emission band of functional membrane-bound reaction centers of the antenna-deficient *Rhodobacter sphaeroides* RCO1 mutant is reported at low temperature. The initial time-dependant Stokes shift of this band and its subsequent oscillatory behaviour are shown to be due to coherent vibrational motion on the excited state potential energy surface. It is concluded that the primary electron transfer takes place from a vibrationally unrelaxed excited state.

1. Introduction

In recent years ultrafast spectroscopy has been used to study coherent motions in condensed matter, including biological macromolecules. In such studies, a short pump pulse prepares a nuclear wavepacket on the potential energy surface of an electronic state and the time evolution of this wavepacket can be monitored subsequently with a probe pulse in resonance with an electronic transition from this state. In studies on bacterial reaction centers, we have demonstrated that vibrational motion in the excited state P^* can be observed in this way [1,2]. The role of the activated coherent nuclear motion in the ultrafast primary electron transfer step is still not understood. This paper deals with coherent excited state vibrational motions occurring during electron transfer. Wavepacket motion on the ground state potential energy surface can also be observed, via impulsive stimulated Raman excitation. Studies of this type have been performed on bacteriorhodopsin [3] and very recently on heme-proteins [4].

Nuclear dynamics play an essential role in electron transfer reactions as they allow the system to reach the nuclear configuration where the energy gap between the donor state and the charge separated state vanishes and where curve crossing of the donor and product state potential energy surfaces may take place. For vibrationally relaxed states the nuclear configuration space is thermally populated and electron transfer is brought about by stochastic dynamics and can be described phenomenologically with a simple rate constant. A different regime applies for ultrafast electron transfer reactions if the initial phase of vibrational motion is not lost on the timescale of electron transfer.

Springer Series in Chemical Physics, Vol. 60 **Ultrafast Phenomena IX**
Editors: P. F. Barbara · W. H. Knox · G. A. Mourou · A. H. Zewail

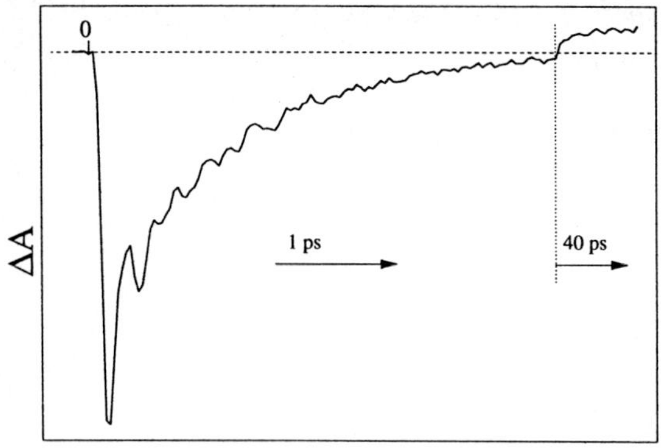

Figure 3. *Kinetics of stimulated emission at 938 nm. Note the different timescales.*

electron transfer reaction, a regime of vibrational coherence applies for the modes coupled to the (physiological) $P \rightarrow P^*$ optical transition. This result may be a clue to the extreme efficiency and speed of the reaction. It remains to be determined whether the reaction coordinates are along the concerted vibrational motions synchronized by the pump pulse.

A qualitatively similar spectral evolution and vibrational behaviour was observed at room temperature (not shown). In particular, the damping of the oscillations is not enhanced at room temperature. This indicates that the same regime of vibrational coherence applies under the normal functional conditions of the reaction center.

3. Acknowledgments

M.H.V. was supported by CEN, CNRS and Fondation pour la Recherche Médicale during the course of this work. M.R.J. and C.N.H. acknowledge support from the Wellcome Trust and the European Community. M.R.J. is an AFRC Senior Research Fellow.

4. References

1 Vos, M.H., Lambry, J.-C., Robles, S.J., Youvan, D.G., Breton, J. & Martin, J.-L. *Proc. Natl. Acad. Sci. USA* **88**, 8885-8889 (1991).
2 Vos, M.H., Rappaport, F., Lambry, J.-C., Breton, J. & Martin, J.-L. *Nature* **363**, 320-325 (1993).
3 Dexheimer, S.L., Wang, Q., Peteanu, L.A., Pollard, W.T., Mathies, R.A. & Shank, C.V. *Chem. Phys. Lett.* **188**, 61-66 (1992)
4 Zhu, L., Li, M., Huang, M., Sage, J.T. & Champion, P.M. *Phys. Rev. Lett.* **72**, 301-304 (1994).
5 Jones, M.R., Visschers, R.W., van Grondelle, R. & Hunter, C.N. *Biochemistry* **31**, 4458-4465 (1992).
6 Vos, M.H., Jones, M.R., McGlynn, P., Hunter, C.N., Breton, J. and Martin, J.-L. *Biochim. Biophys. Acta*, in press.
7 Vos, M.H., Jones, M.R., Hunter, C.N., Breton, J., Lambry, J.-C. and Martin, J.-L. *Biochemistry*, in press.

Femtosecond Vibrationally Coherent Photochemistry in the Primary Event in Vision

Q. Wang[1,2], *R.W. Schoenlein*[1], *L.A. Peteanu*[2], *S.J. Rosenthal*[1,2], *R.A. Mathies*[2], *and C.V. Shank*[1,2]

[1]Lawrence Berkeley Laboratory, MS 70-193A, Berkeley, CA 94720, USA
[2]Department of Chemistry, University of California, Berkeley, CA 94720, USA

ABSTRACT. We demonstrate that the first step in vision, the 200 femtosecond *cis-trans* photoisomerization of the rhodopsin chromophore, is a vibrationally coherent photochemical reaction.

One of the fastest chemical reactions is the first step in vision, the 11-*cis* -> all-*trans* isomerization of the retinal chromophore in rhodopsin (Fig. 1). We have demonstrated that this reaction is essentially complete in only 200 fs.[1,2] We report here the observation of vibrational coherence in the photoproduct of the isomerization reaction in rhodopsin. The presence of vibrational coherence in the photoproduct indicates that a nonstationary wavepacket is carried along a nearly barrierless nonadiabatic surface from the excited state of the reactant to the ground state of the photoproduct.

Rhodopsin is excited using 35 fs pump pulses at 500 nm (Fig. 1). Transient changes in absorption are measured from 450 to 640 nm using 10 fs blue probe pulses and 10 fs red probe pulses. Fig. 2 presents differential transmission measurements and Fourier transform power spectra at 500, 530, 550, 580 and 610 nm. Measurements near 500 nm ($\Delta T/T>0$) indicate the bleaching of the ground-state reactant while measurements near the photoproduct (bathorhodopsin) absorption maximum at 550 nm ($\Delta T/T<0$) indicate the formation of photoproduct within 200 fs.[1,2] Measurements at 500, 530, 580 and 610 nm show clear evidence of vibrational coherence with a period of ~550 fs (~60 cm^{-1}), which persists for several picoseconds. The phase analysis of the Fourier transforms shows a 180-degree phase change near 550 nm (Fig. 3), the photoproduct absorption maximum. In addition, the amplitude of the ~60 cm^{-1} oscillation (Fig. 2A) is small at the edges of the photoproduct absorption (500 and 610 nm), and the oscillations are largest when the probe

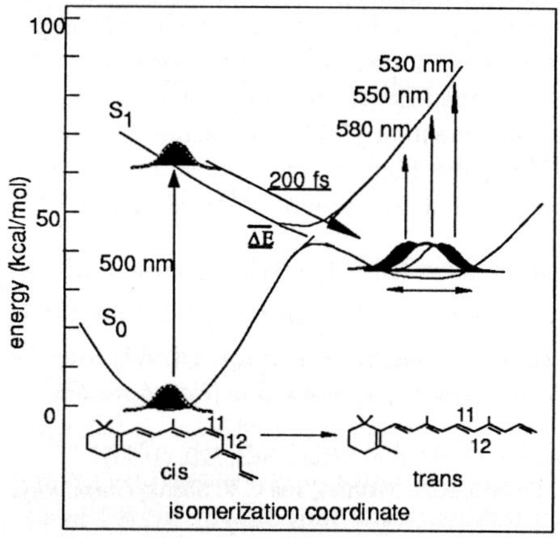

Figure 1. Schematic ground-state and excited-state potential energy surfaces for the 11-*cis* » all-*trans* isomerization in rhodopsin. The coherent vibrational motion of the initial ground-state photoproduct is indicated.

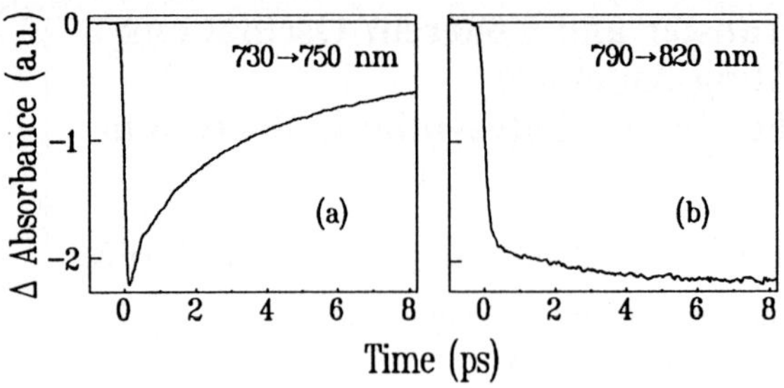

Figure 1. Two-color absorption difference profiles of chlorosomes for (a) 730→750 nm and (b) 790→820 nm pump-probe experiments.

BChl *c* absorption wavelengths 730 and 750 nm, respectively. Like all profiles obtained in this wavelength range, it exhibits *no* rise behavior; its decay components exhibit lifetimes of 101 fs, 1.33 ps, and 10.6 ps. However, rise behavior is obtained when the pump and probe wavelengths are < 800 nm and > 800 nm, respectively, as shown in Fig. 1(b). In this case, the 790→820 nm two-color profile exhibits PB/SE rise components of 103 fs, 2.08 ps, and 9.93 ps. These risetimes approximately mirror the decay components that are found at shorter pump/probe wavelengths, cf. Fig. 1(a). Broadband pump-probe experiments reveal that nearly the entire BChl *c* antenna Q_y band is uniformly bleached, and shows relatively little spectral evolution over the first several hundred fs. This suggests that the 740 nm BChl *c* spectrum arises from a sequence of exciton transitions, rather than from distinct BChl *c* spectral forms. In a dynamic scenario consistent with our results, much of the excitation deposited within the BChl *c* antenna equilibrates rapidly (within ~ 100 fs) among BChl pigments with peak absorption wavelengths ranging from ~730 to ~790 nm. Two kinetic scenarios can account for the fact that the long-wavelength (~810 nm) BChl *a* species apparently receives excitation with triexponential kinetics: a rod-limited model in which the slowest (~10 ps) components arise from energy transfer between BChl *c* rod elements, and a baseplate-limited model in which they arise from energy transfer between BChl pigments absorbing at ~790 and ~810 nm.

Many of the one- and two-color profiles for pump/probe wavelengths in the BChl *c* region exhibit coherent oscillations, which are damped within 2 ps (Fig. 2). These oscillations differ in several respects from the vibrational coherences observed in bacterial reaction center mutants [1]: the apparent phase of the intermediate-frequency oscillations varies with wavelength in *one-color* experiments, and their Fourier-transform spectra are more complex. These features may stem from the fact that ESA, PB, and SE all contribute to the pump-probe signal in the BChl *c* antenna, whereas SE strongly dominates the signal in the reaction centers.

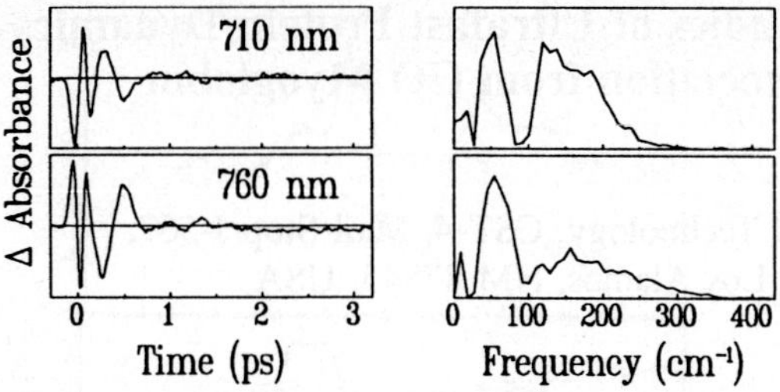

Figure 2. Oscillating parts of one-color BChl c antenna absorption difference signals at 710 and 760 nm (left) and their Fourier-transform spectra (right). The respective signals are dominated by ESA and PB/SE.

References

1. M. H. Vos, F. Rappaport, J.-C. Lambry, J. Breton, and J.-L. Martin, Nature **363**, 320 (1993).

other. The bleach at 1664 cm^{-1} in the difference spectrum is in the region assigned to turn structures; bands at 1676 cm^{-1} (a transient absorption in Fig. 1) have been assigned to either turn or extended chain conformations. The other main absorption at 1645 cm^{-1} is in the area assigned to unordered structure. Bands at lower frequency have been assigned to general β-type structures [6]. Specific structural assignments to the spectral features in Fig. 1 will require pH-dependent studies, selective isotopic labeling or site-directed mutagenesis. It is clear, however, that the major IR difference features are due to changes in the polypeptide backbone conformation. We conclude that these changes correspond to the global relaxation of the proximal helices of the protein. This relaxation is almost complete on the 50 ps timescale, with at least some motions having an 8 ps time constant. The changes in secondary conformation measurable in the amide I band are complete by 100 ns. In conclusion, we have demonstrated that TRIR in the amide I region gives structural information regarding protein conformational changes in real time. Assignment of many of the amide I peaks to specific amide or side chain structures will require much additional effort.

3. References.

1. Stoutland, P. O.; Dyer, R. B.; Woodruff, W. H. *Science* **1992**, *257*, 1913.
2. Anfinrud, P. A.; Han, C.; Hochstrasser, R. M. *Proc. Natl. Acad. Sci. USA* **1989**, *86*, 8387; Martin, J. L. *Chem. Phys.* **1989**, *131*, 31-47.
3. Findsen, E. W.; Friedman, J. M.; Ondrias, M. R.; Simon, S. R. *Science* **1985**, *229*, 661-665.
4. Richard, L.; Genberg, L.; Deak, J.; Chiu, H.-L.; Miller, R. J. D. *Biochemistry* **1992**, *31*, 10703-10715.
5. Xie, X.; Simon, J. D. *Biochemistry* **1991**, *30*, 3682-3692.
6. Surewicz, W. K.; Mantsch, H. H. *Biochim. Biophys. Acta* **1988**, *952*, 115-130.

Femtosecond Coherence Spectroscopy of Heme Proteins

L. Zhu, P. Li, J.T. Sage, and P.M. Champion

Department of Physics, Northeastern University, Boston, MA 02115, USA

1. Introduction

A great deal of information concerning heme proteins has been obtained using the technique of resonance Raman scattering. Here we explore the analogous technique of femtosecond transmission coherence spectroscopy, using pump and probe frequencies in resonance with the Soret band [1].

In previous work we have suggested that heme doming plays an important role in both the optical properties [2] and the ligand binding [3] of heme proteins. The present experiments reveal impulsively excited low frequency vibrational oscillations, subsequent to the photolysis of MbNO. The frequency and phasing of these oscillations suggests that heme doming as well as iron-histidine motion is coupled to the ligand photolysis event. The vibrational coherence of the product species appears to be excited by the electronic forces that develop during the bond breaking event. The ultrashort decay times that are usually associated with heme excited electronic states, along with the development of coherent motion in the product species, suggest that there is a relatively direct electronic route between the optically accessed MbNO excited state and the electronic ground state surface of Mb.

2. Results

Figure 1 demonstrates an experiment using 75fs pump and probe pulses centered near 435 nm with monochromator detection at 430 nm. In the lower panel, the equilibrium Mb sample shows the expected bleaching recovery signal for times less than 300fs, followed by a slower bleaching recovery arising from the red-shifted nature of the Mb* excited state absorption bands [4]. In contrast to Mb, the MbNO sample shows an anti-bleaching signal subsequent to the rapid 300fs recovery. This is expected if the signal probes MbNO photolysis, since the MbNO/Mb isosbestic point lies to the blue of the 430 nm detection wavelength, rather than to the red (at ~441 nm) as for Mb/Mb*. Moreover, the oscillations corresponding to the Fe-His vibration at 220 cm^{-1}, delineated with the dotted lines, appear out of phase when the two samples are compared. We suggest that this is due to the compressive nature of the forces that develop during photolysis as electrons are lost from the Fe-His antibonding d_{z^2} iron orbital. Taken together, these observations suggest that the

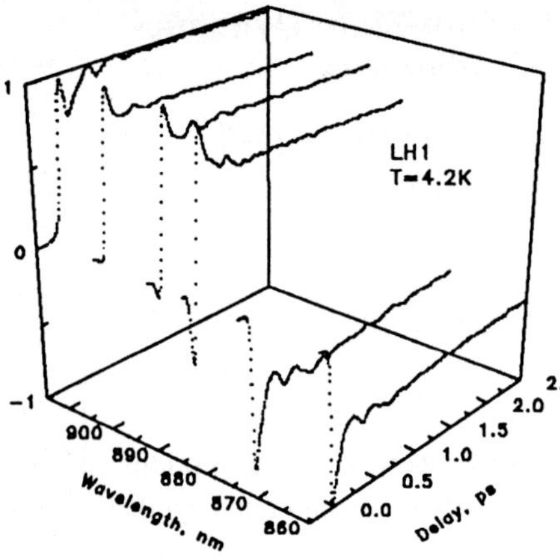

Fig. 1. Transient absorption kinetics of Rb. sphaeroides LH1 measured at several wavelengths at 4 K.

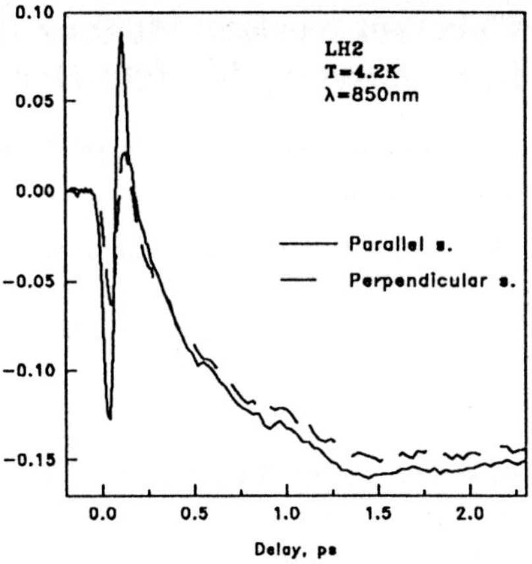

Fig.2. Transient absorption kinetics of Rb. sphaeroides LH2 measured at 850 nm with a 40 fs pulse.

were interpreted as inter exciton state relaxation [5]. Theoretical estimates have suggested very fast (10-100 fs) relaxation from upper to lower exciton states (-)→(+). Using ~40 fs transform-limited infrared pulses to excite and probe the transient absorption kinetics in the blue wing of the absorption spectrum of the 850 nm absorption band of LH2 we observe a very fast highly polarized relaxation at wavelengths which coincide with the expected upper exciton state of a B850 aggregate (835-845 nm, at 4 K). The kinetics displayed in Fig.2 exhibits an induced absorption which relaxes with a time constant of ≤20 fs to a bleaching signal which in turn relaxes with a 100-200 fs time constant. We interpret the absorption immediately following excitation by the 40 fs pulse as excited state absorption from the upper (-) exciton state which relaxes very rapidly (≤20 fs) to the lower (+) exciton state. The slower 100-200 fs relaxation time represents energy transfer to the (+)-state of other B850 units which absorb light at a slightly different energy due to different interactions with other pigment molecules and the environment (inhomogeneous spectral broadening of the B850 absorption band).

3. References

1. F.G. Zhang, R. van Grondelle and V. Sundström, Biophys. J. 61(1992)911.
2. S. Hess, F. Feldshtein, A. Babin, I. Nurgaleev, T. Pullerits, A. Sergeev and V. Sundström, Chem. Phys. Lett., 216(1993)243.
3. L. Zhu, P. Li, M..Huang, J.T. Sage and P. M. Champion, Phys. Rev. Lett. 72(1994)301.
4. M.H. Vos, F. Rappaport, J-C. Lambry, J. Breton and J-L. Martin, Nature 363(1993)320.
5. N.R.S. Reddy, R. Picorel and G.J. Small, J. Phys. Chem., 96(1992)6458.

Direct Observation of the Accessory Bacteriochlorophyll in the Primary Electron Transfer in Bacterial Reaction Centers

W. Zinth, S. Schmidt, T. Arlt, J. Wachtveitl, H. Huber, and T. Nägele

Institut für Medizinische Optik der Universität München, Barbarastrasse 16/IV, D-80797 München, Germany

Abstract. Subpicosecond experiments on native and modified, pheophytin containing reaction centers prove that the accessory bacteriochlorophyll is a real electron carrier in primary photosynthesis.

Early femtosecond experiments on bacterial reaction centers (RC) [1] performed nearly one decade ago have shown, that the primary electron donor (the special pair P) remains in its exited state for about 3 ps and that the electron arrives at the bacteriopheophytin (H) in practically the same time. These studies did not exhibit transient absorption features related to the reduction of the accessory bacteriochlorophyll (B) located between the special pair P and the bacteriopheophytin H. The absence of such a component led to the conclusion that B is not a real electron carrying intermediate. However in later experiments [2,3] we found an additional weak subpicosecond component. The spectral characteristics of this component and additional experiments on mutated and modified RC suggested that B is a real electron carrier [4]. This interpretation was not generally accepted in the literature. One main objection came from the fact that the subpicosecond kinetic component was only observed in "congested" spectral regions where other intermediates caused strong absorption changes and that no long lasting P^+B^- state was observed in mutated RC. It is the purpose of this contribution to present new experimental results proving unambiguously that the accessory bacteriochlorophyll is a real electron carrier.

The new aspect of the experiments is the use of modified RC containing pheophytin instead of bacteriopheophytin and the extension of the spectral range up to probing wavelengths of 1100 nm. In this range (around 1.000 nm) bacteriochlorophyll anions in vitro have a pronounced absorption band. In addition there is no absorption from ground state B, H as well as from H^- and special pair bands.

A first set of experiments treats unmodified RC from *Rb. sphaeroides R 26.1* in the spectral range of $\lambda > 920$ nm. At the shorter wavelengths, 920 - 980 nm, there is strong stimulated emission decaying with an average time constant of about 3 ps. When probing at the longer wavelengths the strong 3 ps component becomes weaker and practically vanishes at $\lambda > 1.020$ nm. From 980 nm on, one finds a pronounced contribution of a 0.9 ps component which is strongest around 1.020 nm (see Fig. 1, broken line). Analysing the data within the stepwise reaction model yields the expected spectrum of intermediate P^+B^- (with a peak at 1.020 nm) as well as the expected orientation of the transition dipole moment of B^-. In the frame of a superexchange model with direct electron transfer to H there would be no explanation for the strong 0.9 ps kinetic component observed in the 1.000 nm range.

A second set of experiments treated modified RC containing pheophytins instead of bacteriopheophytins. The replacement increases the energy of the radical pair state P^+H^- bringing it much closer to the state P*. Assuming a stepwise electron transfer via B, one would expect a long-lasting population of the radical pair state P^+B^- in

Editors: P. F. Barbara · W. H. Knox · G. A. Mourou · A. H. Zewail

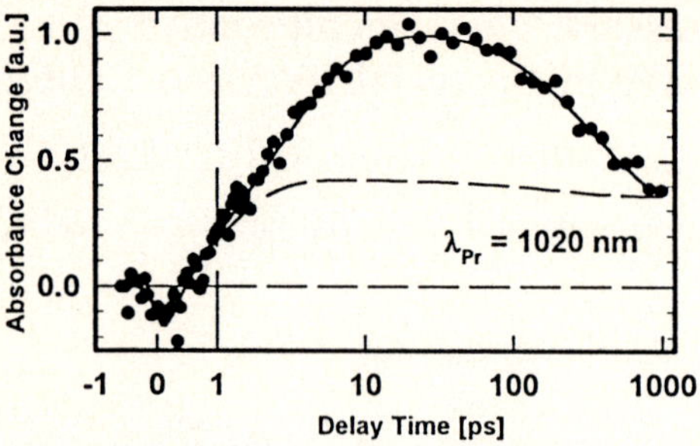

Fig. 1: Transient absorption changes for native (broken line) and modified (solid line) RC at 1.020 nm. The absorption increase with ~ 3 ps and a late decrease with a 370 ps time constant in the modified RC reflect the presence of a bacteriochlorophyll anion in equilibrium with the pheophytin anion.

thermal equilibrium with P^+H^-. Indeed the signal trace at 1.020 nm (see Fig. 1, solid line) shows a long-living increased absorption which can only be assigned to the bacteriochlorophyll anion B^-. Combining these results with data from transient emission studies on the same type of RC we can deduce that the intermediate P^+B^- is about 450 cm^{-1} below P*.

In conclusion: The new experimental data prove that the accessory bacteriochlorophyll B is a real electron carrier in the primary reaction of native bacterial reaction centers. The superexchange model does not apply to the primary reaction at room temperature.

References

1. J.-L. Martin, J. Breton, A.J. Hoff, A. Migus, A. Antonetti; Proc. Natl. Acad. Sci. USA 83 (1986) 957.

2. W. Holzapfel, U. Finkele, W. Kaiser, D. Oesterhelt, H. Scheer, H.U. Stilz, W. Zinth; Chem. Phys. Lett. 16O (1989)1.

3. W. Holzapfel, U. Finkele, W. Kaiser, D. Oesterhelt, H. Scheer, H.U. Stilz, W. Zinth; Proc. Natl. Acad. Sci. USA , 87 (1990) 5168.

4. W. Zinth, W. Kaiser; "The Photosynthetic Reaction Center", vol. II, eds. J. Deisenhofer and J.R. Norris (1993) Academic Press, p. 71

Femtosecond Infrared Spectroscopy of the Photosynthetic Reaction Center

G.C. Walker, S. Maiti, G.D. Reid, K. Wynne, C.C. Moser, R.S. Pippenger, B.R. Cowen, P.L. Dutton, and R.M. Hochstrasser*

Department of Chemistry and Johnson Research Foundation,
University of Pennsylvania, Philadelphia, PA 19104, USA
*Present address: Department of Chemistry, University of Pittsburgh, Pittsburgh, PA 15260, USA

The photosynthetic reaction center (RC) from the purple bacterium *Rb. sphaeroides* has been studied by ultrafast infrared spectroscopy. In the present experiment, spectral kinetics measured in the region between 1590-1760 cm^{-1} access cofactor and side chain carbonyls and backbone amide absorptions of the protein. Fig. 1(a) depicts the difference signal obtained from a representative vibrational band around 1702 cm^{-1}. The main contribution in this region comes from the 9-keto carbonyls of the special pair, P, which shift their resonance frequency from 1683 cm^{-1} to 1702 cm^{-1} when P becomes P^+. The evolution of the band at early times clearly manifests vibrational dephasing.[1,2] The three dimensional data therefore affords a measurement of the bandwidth of the metastable vibrational transitions in a protein.

The evolution of the bleach of the special pair ground state (9-keto, at 1682 cm^{-1}) occurs only on the time scale of electron transfer from P^* (See Fig. 1(b), dashed line is electronic absorption, dotted line is vibrational bleach). This proves that the vibrations of P and P^* have essentially the same frequencies and dephasing times in this region. A detailed analysis of the coherent vibrational spectra around 1665 cm^{-1} has revealed the presence of a transient $Bchl^-$ state.[1]

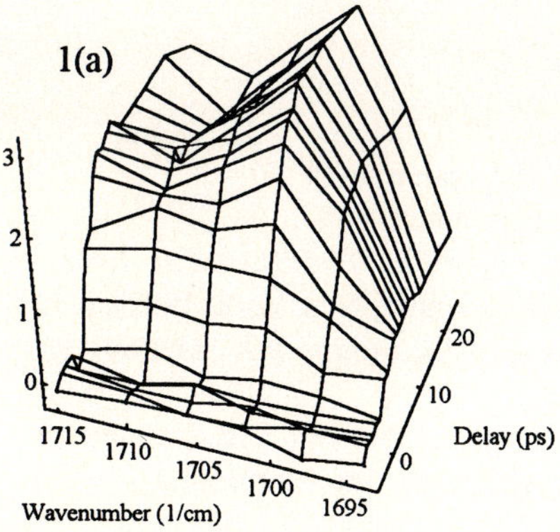

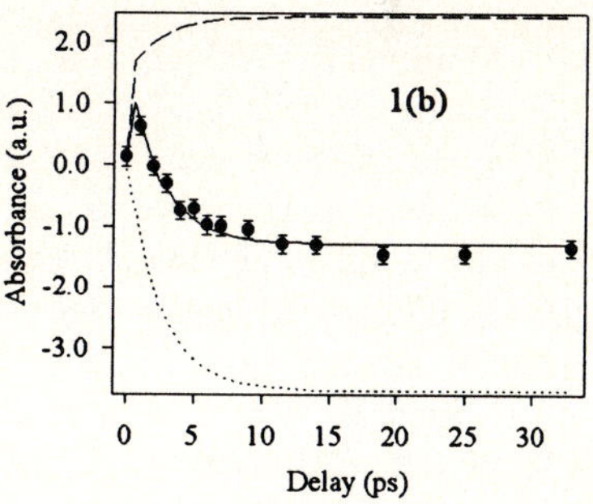

We have resolved broad bands (Fig. 2(a)) that we assign to low energy electronic transitions at ca. 6000 and 2000 cm^{-1} of the P^* and P^+ states. The lowest energy P^* electronic transition is identified with the interexcitonic transition of the special pair based on earlier predictions made from theoretical calculations. We suggest that the high energy transition locates the charge resonance state. The P^+

Editors: P. F. Barbara · W. H. Knox · G. A. Mourou · A. H. Zewail

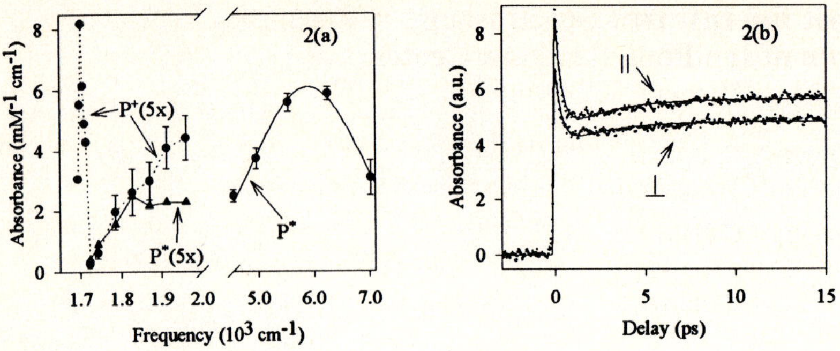

electronic band is assigned to a transition between hole transfer states of P, in accordance with steady state FTIR measurements at low temperatures.[3] The measurements of anisotropy around 1900 cm^{-1} yield identical values for all transitions.[4,5] This has been interpreted in terms of a simple model for the electronic nature of P^* involving excitonic and charge transfer states and provides a constraint for the theoretical calculation of the electronic wave functions for the ground and excited states of P, including the amount of charge shifted. Fig. 2(b) shows the transient absorption signal with excitation at 800 nm with 35 fs pulses[6] followed by a probe pulse at 1.215 μm with parallel and perpendicular polarization. Energy transfer from the initially excited B^* is seen as the fast decay with 201±4 fs in Fig. 2(b). The slow rise in the signal is assigned to P^+ absorption. This research was supported by grants from NSF and NIH.

References

1 S. Maiti, G.C. Walker, B.R. Cowen, C.C. Moser, P.L. Dutton, R.M. Hochstrasser, Proc. Natl. Acad. Sci. USA (submitted).

2 R.M. Hochstrasser, Acc. Chem. Res. (submitted).

3 W.W. Parson, L. Nabedryk, J. Breton, in *The Photosynthetic Reaction Center II*, Eds. J. Breton, A. Vermeglio (Plenum, New York, 1992), p.79.

4 G.C. Walker, S. Maiti, B.R. Cowen, C.C. Moser, P.L. Dutton, R.M. Hochstrasser, J. Phys. Chem. (in press).

5 G.D. Reid, K. Wynne, R.M. Hochstrasser, Chem. Phys. Lett. (submitted).

6 K. Wynne, G.D. Reid, R.M. Hochstrasser, Opt. Lett. (in press).

Ultrafast Energy Transfer in the Core Light-Harvesting Complex of the Photosynthetic Bacterium *Rhodobacter Sphaeroides* Observed by Fluorescence Upconversion

S. Bradforth[1], *R. Jimenez*[1], *V. Fidler*[1], *G. Fleming*[1], *S. Nagarajan*[1,2], *J. Norris*[1,2], *F. van Mourik*[3], *and R. van Grondelle*[3]

[1]Department of Chemistry, University of Chicago, Chicago, IL 60637, USA
[2]Chemistry Division, Argonne National Laboratory, Argonne, IL 60439, USA
[3]Department of Biophysics, Vrije Universiteit, Amsterdam, The Netherlands

Abstract: A study is presented of the room temperature sub-picosecond energy transfer of electronic excitation between bacteriochlorophyll pigments and between carotenoid and bacteriochlorophyll in the core light harvesting complex (LH I).

The LH I complex is thought to be constructed from aggregates of small building blocks containing transmembrane polypeptides and associated carotenoid and bacteriochlorophyll *a* (BChl*a*) pigment moieties[1]. There is evidence for a high degree of ordering of the pigments in the basic "minimal unit" although no crystal structure has been so far determined for this or any membrane bound bacterial light harvesting complex. The precise mechanism of energy transfer from carotenoid to BChl*a* and between BChl*a* pigments is not well understood. Both energy transfer processes are highly efficient, and are both expected to occur on sub-picosecond timescales. LH I makes an excellent system for study of ultrafast energy transfer in a biological system because of the existence of only a single BChl spectral pool and one type of carotenoid.

Two experimental arrangements were used in the time-resolved study of LH I. To monitor the fluorescence evolution of the Q_y BChl*a* band (B870), a single color experiment was performed using the 870 nm 76 MHz mode locked output of a Ti:S oscillator. The IR beam is split; part excites the B870 band and the remainder is used as a gate pulse to upconvert the emission from the sample in a $LiIO_3$ crystal[2]. For two-color experiments where the carotenoid band is excited, the 505 nm excitation pulse is selected from the white light continuum output of a 250 kHz regenerative amplifier; the gate pulse derives from the 820 nm amplifier fundamental. The excitation energy incident on the stirred sample is ~0.5 nJ in both experiments. The excitation / gate pulse cross-correlation is 190 fs FWHM for 870 nm / 870 nm and 250 fs for 505 nm / 820 nm. For the 870 nm experiments a chromatophore preparation of the *Rb. sphaeroides* M2192 mutant [3] was used; for the 505 nm experiment complexes isolated using the detergent BOG were studied. Neither preparation contains LH II or reaction centers.

Figure 1 shows the fluorescence observed when the BChl*a* band is excited at 870 nm with polarization parallel and perpendicular with respect to the polarization of the detected fluorescence. Simultaneous fitting of both curves recovers a 50% amplitude 1.3 ps isotropic decay component with the remaining 50% of the isotropic emission showing no decay over 10 ps. The fit also yields a 320 fs decay component for the fluorescence anisotropy from an initial value of 0.28 to 0.09. This final anisotropy is

Editors: P. F. Barbara · W. H. Knox · G. A. Mourou · A. H. Zewail

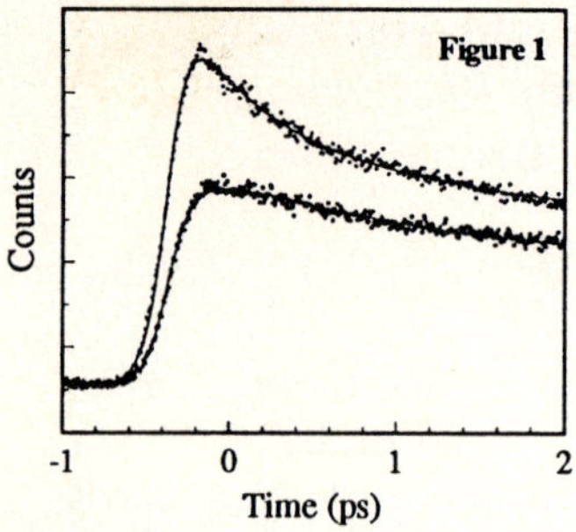

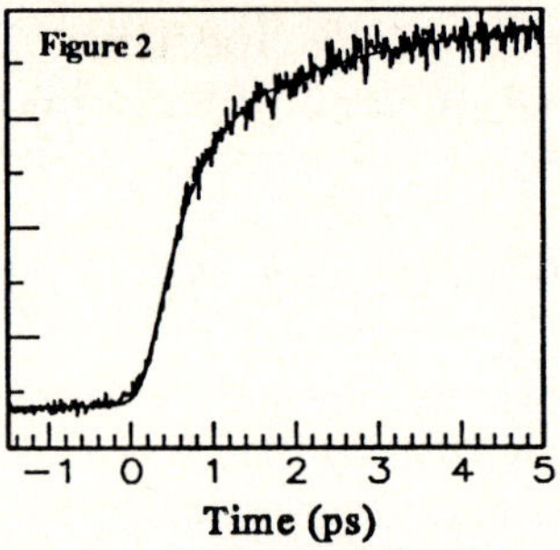

Figure 1: Fluorescence time evolution from B870 in LH I (excitation and detected emission wavelengths are 870 and 945 nm respectively). Upper curve (parallel) and lower curve (perpendicular).
Figure 2: Isotropic fluorescence (910 nm) from B870 in LH I; carotenoid band excited at 505nm.

close to the observed steady state value[4]. The short excited state lifetime component may result from singlet - triplet annihilation as a consequence of the high laser repetition rate establishing a significant triplet steady state population. The anisotropy decay, however, is due to energy transfer. It is consistent with a ~600 fs hopping rate assuming a model containing a basic unit of two BChl*a* dimers exchanging energy via an incoherent Forster-type mechanism between the lower exciton transition moments of each dimer at a separation of 20 Å.

Figure 2 shows the isotropic fluorescence at 910 nm when the carotenoid pigment in LH I is excited. The BChl*a* emission rises with two components (τ_1= 450 fs, 64% and τ_2 = 2.5 ps, 36%). In a study by Shreve et al. on the related antenna system LH II, a 200 fs component was assigned to energy transfer between carotenoid and B850[5]. These authors find evidence suggesting that carotenoid to BChl*a* energy transfer occurs from both carotenoid S_2 and S_1 states. Furthermore, the energy transfer from the carotenoids may occur via coupling with either BChl Q_x or Q_y transition dipoles. Both of these factors may be used to explain the observation of a multi-exponential rise in this dataset.

We thank Coherent for the loan of the regenerative amplifier used in this study.

[1] V. Sundstrom and R. van Grondelle in *Chlorophylls*, Hugo Scheer ed., p. 1097, CRC Press, Bacon Rouge, FL, 1991

[2] M. Du, S. Rosenthal, X. Xie, T. DiMagno, M. Schmidt, J. Norris and G. R. Fleming, *Proc. Natl. Acad. Sci. USA* **89**, 8517 (1992)

[3] C. N. Hunter, R. van Grondelle and R . J. van Dorssen, *Biochim. Biophys. Acta* **973**, 383 (1989)

[4] H. J. M. Kramer, J. D. Pennoyer, R. van Grondelle, W. H. J. Westerhuis, R. A. Niederman and J. Amesz, *Biochim. Biophys. Acta* **767**, 335 (1984)

[5] A. P. Shreve, J. K. Trautman, H. A. Frank, T. G. Owens and A. C. Albrecht, *Biochim. Biophys. Acta* **1058**, 280 (1991)

Ultrafast Mid-IR Spectroscopy of Carbonmonoxymyoglobin: The Dynamics of Ligand Motion

M. Lim, T.A. Jackson, and P.A. Anfinrud

Department of Chemistry, Harvard University, Cambridge, MA 02138, USA

Abstract. The rotational dynamics of CO in photolyzed MbCO have been investigated by time-resolved polarized mid-IR spectroscopy. The spectra reveal ultrafast spatial localization of the ligand in the heme pocket.

1. Introduction

Myoglobin (Mb) is a ligand-binding heme protein whose active binding site is located within the hydrophobic interior of the protein. It is well known that CO binds strongly to Mb. Yet, when MbCO is photolyzed at 300 K, less than 2% of the dissociated ligands rebind geminately. The protein evidently provides an efficient pathway for shuttling CO into the surrounding solvent. We have investigated the rotational dynamics of CO using fs mid-IR absorption spectroscopy and have found that the first step of CO expulsion involves localization in a position near the heme iron with the bond axis oriented approximately 62° from the heme normal.

2. Experimental Method

A time-resolved mid-IR absorption spectrometer has been developed which provides spectral coverage of ≈ 1800 - $2500\ cm^{-1}$, spectral resolution of $3\ cm^{-1}$, time resolution of < 200 fs, and a noise level of $< 5 \times 10^{-4}$ rms in absorbance units after 1 second of signal averaging. The fs mid-IR probe pulse is derived by difference-frequency mixing a fs optical pulse with an amplified and compressed continuum in a nonlinear $LiIO_3$ crystal. The mid-IR probe is split into two portions: one portion passes through the photolyzed volume of the sample and the other portion, a reference, bypasses the sample. The sample and reference pulses are routed through an IR monochromator and detected with amplified InSb photodetectors. The two signals are digitized and their ratio is computed on a shot-by-shot basis in order to recover the transmission of the sample. Signal averaging at the 1.5 kHz repetition rate of the laser contributes to the very high signal-to-noise ratio attained.

3. Dynamics of Ligand Motion in the Heme Pocket

Figure 1 shows time-resolved polarized mid-IR absorption spectra of horse MbCO photolyzed at 590 nm. The features near $2130\ cm^{-1}$ are assigned to CO confined within the heme pocket and are denoted B-states. The B-state spectra can be modeled with two Gaussian features with time-dependent amplitudes and frequencies but time-independent widths. A spectrum characteristic of longer times emerges by 320 fs and grows in amplitude with a ~1ps time constant. The center

Springer Series in Chemical Physics, Vol. 60 **Ultrafast Phenomena IX**
Editors: P. F. Barbara · W. H. Knox · G. A. Mourou · A. H. Zewail

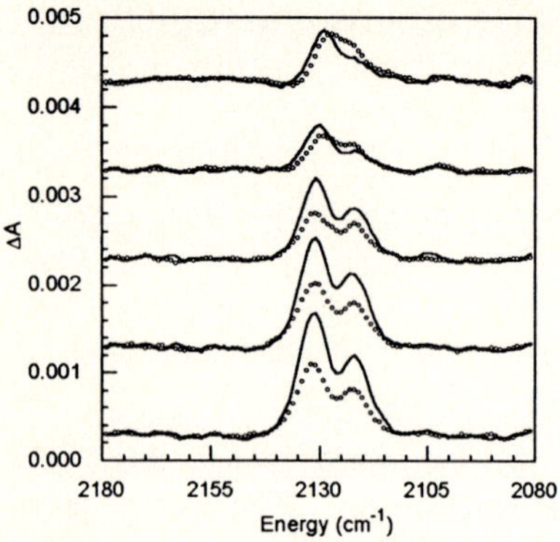

Figure 1. Parallel (solid line) and perpendicular (open circles) polarized mid-IR absorbance spectra of photolyzed horse MbCO at 200 fs, 320 fs, 1ps, 3.2 ps, and 10 ps (top to bottom).

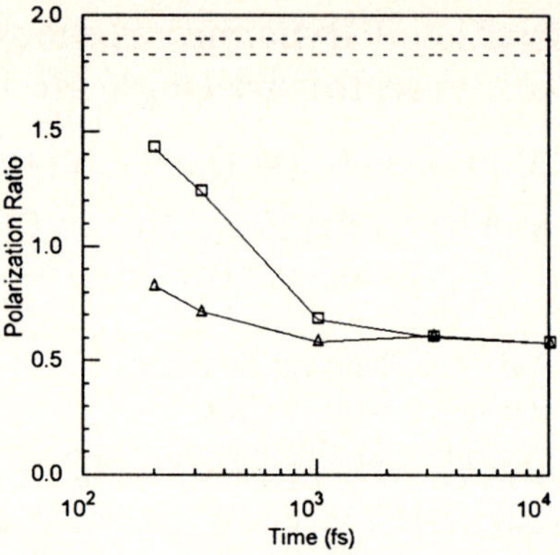

Figure 2. Polarization ratio of B-states. The open triangles (squares) correspond to the higher (lower) energy feature. The dashed lines indicate the polarization ratio of bound CO.

frequency of the larger feature blue-shifts nonexponentially with time while the center frequency of the smaller feature is essentially constant in time. The nonexponential shift evidently arises from conformational relaxation of the protein, a process which has been shown to evolve nonexponentially in time [1]. The ~1ps growth of the B-states is likely a consequence of heme cooling and the concomitant reduction in the librational amplitude of spatially localized CO. The ratio of the perpendicular and parallel polarized spectra ($\Delta A^{\perp}/\Delta A^{||}$) can be related to the orientation of the CO bond axis relative to the heme normal [2]. The polarization ratios obtained by modeling the spectra of Figure 1 are plotted in Figure 2. The change in the polarization ratio corresponds to rotation from ~20° to ~60° relative to the heme normal (angles determined by photolysis at 527 nm, where the heme is a *circular* absorber). The fact that the polarization ratio evolves to a static value by ~1 ps suggests that the ligand rotates less than one revolution before becoming trapped within the heme pocket.

4. Conclusions

These results show that the primary motion of CO following dissociation from Mb involves ultrafast rotation and spatial localization in the heme pocket. Even though CO is trapped near the iron, its bond axis is bent far from that required for binding, which may contribute to the low yield for CO geminate recombination.

References

1. M. Lim, T.A. Jackson, and P.A. Anfinrud, *Proc. Natl. Acad. Sci. U.S.A.* **90**, 5801 (1993); T.A. Jackson, M. Lim, and P.A. Anfinrud, *Chem. Phys*, **180**, 131 (1994).
2. J.N. Moore, P.A. Hansen, and R.M. Hochstrasser, *Proc. Natl. Acad. Sci. U.S.A.* **85**, 5062 (1988); P. Ormos, D. Braunstein, H. Frauenfelder, M.K. Hong, S.L. Lin, T.B. Sauke, and R.D. Young, *Proc. Natl. Acad. Sci. U.S.A.* **85**, 8492 (1988).

Vibrational Dynamics at the Active Site of Myoglobin: Picosecond Infrared Free-Electron-Laser Experiments

K.A. Peterson[1], *J.R. Hill*[2], *A. Tokmakoff*[3], *B. Sauter*[3], *D. Zimdars*[3], *D.D. Dlott*[2], *and M.D. Fayer*[3]

[1]Hansen Experimental Physics Laboratory, Stanford University, Stanford, CA 94305, USA
[2]School of Chemical Sciences, University of Illinois at Urbana Champaign, Urbana, IL 61801, USA
[3]Department of Chemistry, Stanford University, Stanford, CA 94305, USA

Summary: We have made direct measurements of vibrational energy transfer at the active site of a protein using intense tunable mid-IR pulses from a free-electron laser to monitor the loss of vibrational energy from CO bound to a water soluble heme and CO bound to the active sites of different conformational substates of myoglobin. Vibrational relaxation (VR) rates are found to depend on protein conformational substate and porphyrin substitution. Examination of data obtained on a variety of systems containing protoheme suggests the existence of a linear correlation between increasing CO stretching frequency and decreasing VR rate, which indicates that VR of excited CO becomes more efficient as external influences lower the CO stretching frequency. Significant contributions toward understanding protein-ligand interactions will come from additional studies of this type which investigate the effects of protein residue mutation, proximal ligand substitution and porphyrin structural changes.

Experimental: Glycerol:water (60:40) samples were ~30 mM horse heart Mb (Sigma) buffered to pH = 7.0 with sodium phosphate, held in a 0.1 mm path cell with 1 mm thick sapphire windows. Poly(vinyl alcohol) (PVA) samples were prepared by dissolving Mb-CO in a solution of 10% by weight PVA in 1 mM phosphate buffer at pH = 7.0. The PVA solution was spread on a sapphire window and allowed to dry under 1 atm of CO. The concentration is estimated at 30 mM. Iron tetraphenyl-porphyrinsulfate (Fe:TPPS) was synthesized from tetraphenyl-porphyrin by addition of H_2SO_4, followed by metallation. The Fe:TPPS-CO sample consisted of 30 mM Fe:TPPS in glycerol:water (75:25) adjusted to pH = 10.0 by addition of NaOH. Mid-IR spectra of Mb-CO in the region of the CO stretching frequency for the glycerol:water and PVA samples are shown in Fig. 1. Following Frauenfelder's nomenclature [1], we designate the bands as different A-states. The substantial background is caused by window reflection and solvent and protein absorption. The IR spectrum of Fe:TPPS was also obtained. Only one band in the CO stretch region was observed and can be accurately fit to a gaussian lineshape centered at 1963.7 cm^{-1} with a FWHM of 27 cm^{-1}. The IR pump-probe apparatus using the Stanford free-electron laser (FEL) has been described in detail previously [2]. The IR pulse autocorrelation was ca. 2 ps transform limited gaussian. FEL pulse spectra are also shown (lower curves) in Fig. 1. Experimental details will be given in a future publication [3].

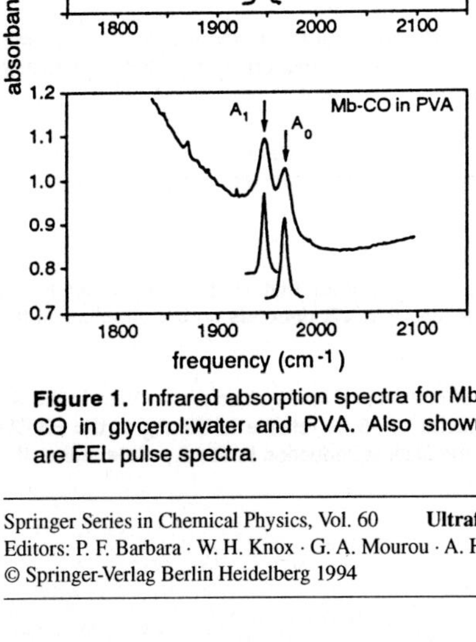

Figure 1. Infrared absorption spectra for Mb-CO in glycerol:water and PVA. Also shown are FEL pulse spectra.

Ultrafast Isomerization Dynamics of 11-*cis*-Locked 8-Membered Rhodopsin

H. Kandori[1,4], *H. Sasabe*[1], *K. Nakanishi*[2], *T. Yoshizawa*[3], *T. Mizukami*[4], and *Y. Shichida*[4]

[1]Frontier Research Program, The Institute of Physical and Chemical Research (RIKEN), Wako, Saitama 351-01, Japan

[2]Department of Chemistry, Columbia University, New York, NY 10027, USA

[3]Department of Information System Engineering, Faculty of Engineering, Osaka Sagyo University, Daito-shi, Osaka 574, Japan

[4]Department of Biophysics, Faculty of Science, Kyoto University, Kyoto 606-01, Japan

Abstract. The primary isomerization dynamics of a rhodopsin analog having flexible 8-membered ring retinal (Rh8) was studied by picosecond (ps) transient absorption and femtosecond (fs) fluorescence up-conversion spectroscopies. The former revealed product formation, suggesting that isomerization occurs in spite of fixation of the 11-ene. The latter determined the fluorescence lifetime to be 60 fs, which was probing wavelength independent. Such ultrafast processes ($\phi_f \sim 10^{-6}$) imply that cis-trans isomerization in vision is a coherent process along the barrierless potential surface of the excited state of rhodopsin.

Rhodopsin (Rh), a photoreceptor protein present in our eyes, is an excellent molecular switch to convert light signal to the electrical response of the photoreceptor cells. Elucidation of the primary photochemical event in Rh has been one of the central issues in vision as well as ultrafast spectroscopy [1]. We have studied the primary isomerization mechanism by applying ps spectroscopies to Rh analogs possessing 11-cis-locked chromophores. By observing a long-lived excited state (85 ps) and no product formation of a Rh analog having an 11-cis-locked 5-membered retinal (Rh5), we provided a direct evidence that cis-trans isomerization is a primary event in vision [2]. On the other hand, a ground-state product which resembles photorhodopsin (the first photointermediate of native Rh [3]) was observed on excitation of a Rh analog having an 11-cis-locked 7-membered retinal (Rh7) [2]. These results suggest that the flexibility along the 11-ene is closely related to the primary processes in Rh. Thus, we extended the study by measuring the primary processes of a Rh analog having an 11-cis-locked 8-membered retinal (Rh8), which has a more flexible ring structure including the 11-ene.

First, we applied ps transient absorption to Rh8. The Rh8 sample was prepared from 11-cis-locked 8-membered retinal (Ret8) and bovine opsin [4,5]. Upon excitation of Rh8 with a 532 nm, 21 ps pulse, we observed two products in ps regime, whose absorption spectra were similar to the primary intermediates of the native Rh (i.e. photorhodopsin and bathorhodopsin) [4]. Thus, an appearance of early intermediates of Rh is dependent on the flexibility of the 11-ene, and we eventually obtained the scheme shown in Fig. 1. One interesting aspect in the ps measurements was that the product formation yield of Rh8 was almost identical with that of the native Rh, while that of Rh7 was about 5-times lower. This result suggested that the excited-state dynamics of Rh8 is similar to that of the native Rh. Therefore, we next investigated the excited-state dynamics of Rh8 using a fs fluorescence up-conversion technique, by which we were able to detect an extraordinary weak fluorescence with fs resolution [6].

Typical fluorescence decay kinetics of Rh8 is shown at 578 nm in Fig. 2. The fluorescence signal decays very rapidly, which is comparable with the instrumental response (broken line in Fig. 2). Simultaneously, the small fluorescence signal (~10 % of the peak) is observed between 400~600 fs. The signal exists even at 20 ps, indicating its lifetime to be longer than 20 ps. In comparison to the transient absorption results [4], the slow component is not coupled to the processes to form the photorhodopsin-like product, but probably originates from any impurity. The lifetime of the fast component was determined to be 60 ± 10 fs by the deconvolution procedure (Fig. 2). The rapid decay process is due to the conformational change of the chromophore, probably cis-trans isomerization. From the obtained lifetime, we estimated the fluorescence quantum yield of Rh8 to be in the order of 10^{-6} ($\phi_f \sim 6 \times 10^{-6}$) by using the lifetime (60 fs) and the natural radiative lifetime of Rh8 (~10 ns). It should be noted

Springer Series in Chemical Physics, Vol. 60 **Ultrafast Phenomena IX**
Editors: P. F. Barbara · W. H. Knox · G. A. Mourou · A. H. Zewail

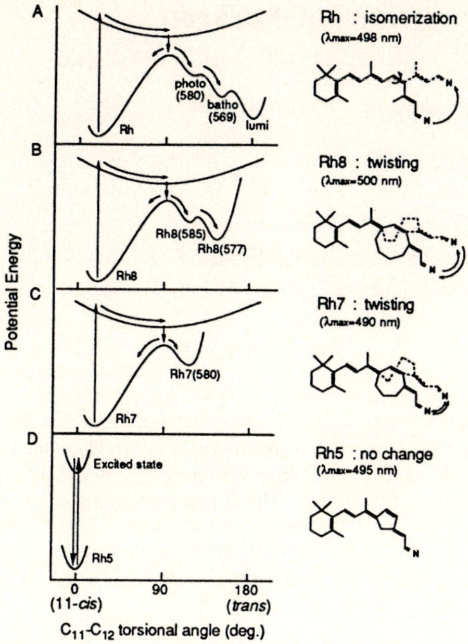

Fig.1

Schematic drawing of potential surface of native and 11-cis-locked rhodopsins.

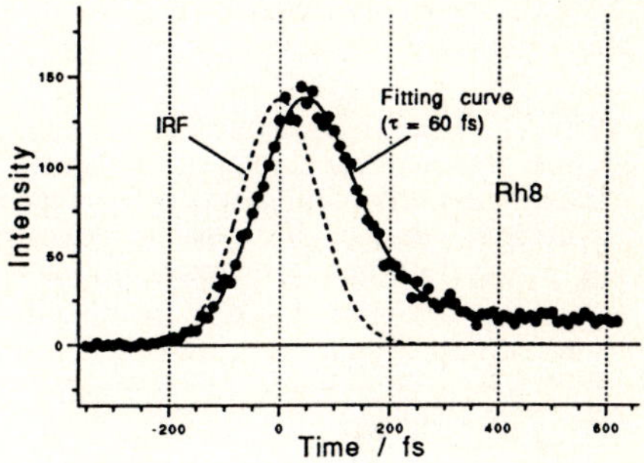

Fig.2

Typical fluorescence decay of Rh8 measured at 578 nm. IRF: instrumental response function.

that the kinetic profiles recorded at different wavelengths (551-664 nm) displayed similar features to that at 578 nm (Fig. 2), indicating that the process is wavelength-independent (no spectral evolution). Since the lifetime (60 fs) corresponds, for example, to a few period of C-C stretching vibration, the isomerization possibly occurs faster than the vibrational dephasing time, suggesting that the isomerization is a coherent process.

The recent fs transient absorption spectroscopy of the native Rh revealed that the appearance of photorhodopsin completes within 200 fs [7]. The present fs emission results on Rh8 are consistent with those by transient absorption on the native Rh. On the assumption that the excited-state dynamics of Rh8 is similar to that of the native Rh, we can speculate the primary reaction dynamics of Rh as follows:

"The excited wave-packet moves from fluorescent to non-fluorescent state in 60 fs along a barrierless potential surface of the C_{11}-C_{12} rotational coordinate. Visual cis-trans isomerization is a coherent process in its excited state and the isomerized photoproduct (photorhodopsin) is formed within 200 fs."

This work was supported in part by Grant-in-Aid for Scientific Research and Research from the Japanese Ministry of Education, Culture and Science, by a Human Frontier Science Program, and by NIH Grant GM 36564.

1. Y. Shichida, *Photochem. Photobiol.* **52**, 1179 (1990); T. Yoshizawa and H. Kandori, *Progress in Retinal Research* **11**, 33 (1992).
2. H. Kandori, S. Matuoka, Y. Shichida, T. Yoshizawa, M. Ito, K. Tsukida, V. Balogh-Nair and K. Nakanishi, *Biochemistry* **28**, 6460 (1989).
3. Y. Shichida, S. Matuoka and T. Yoshizawa, *Photobiochem. Photobiophys.* **7**, 221 (1984).
4. T. Mizukami, H. Kandori, Y. Shichida, A.-H. Chen, F. Derguini, C.G. Caldwell, C.F. Bigge, K. Nakanishi and T. Yoshizawa, *Proc. Natl. Acad. Sci. USA* **90**, 4072 (1993).
5. S. Hu, P.J. Franklin, J. Wang, B.E. Ruiz Silva, F. Derguini and K. Nakanishi, *Biochemistry* **33**, 408 (1994).
6. H. Kandori and H. Sasabe, *Chem. Phys. Lett.* **216**, 126 (1993).
7. R.W. Schoenlein, L.A. Peteanu, R.A. Mathies and C.V. Shank, *Science* **254**, 412 (1991); L.A. Peteanu, R.W. Schoenlein, Q. Wang, R.A. Mathies and C.V. Shank, *Proc. Natl. Acad. Sci. USA* **90**, 11762 (1993).

Direct Determination of the S_2 Lifetime of β-Carotene by Femtosecond Time-Resolved Fluorescence Spectroscopy

H. Kandori[1,*], *H. Sasabe*[1], *and M. Mimuro*[2]

[1]Frontier Research Program, The Institute of Physical and Chemical Resesarch (RIKEN), Wako, Saitama 351-01, Japan
[2]National Institute for Basic Biology, Okazaki 444, Japan
*Present address: Department of Biophysics, Faculty of Science, Kyoto University, Sakyo-ku, Kyoto 606-01, Japan

Abstract. Femtosecond fluorescence up-conversion was applied to directly determine the S_2 lifetime of β-carotene at room temperature. Upon excitation of β-carotene in *n*-hexane with a 425-nm pulse, the lifetime was determined to be 195 ± 10 fs throughout the whole emission wavelength. Absence of a dynamic Stokes shift suggests that intramolecular relaxations occur within 50 fs, followed by internal conversion from the S_2 to S_1 state in 195 fs.

Carotenoids have important physiological functions in photosynthesis. They work as light-harvesting pigments due to a large oscillator strength in the visible region and efficient singlet energy transfer to chlorophylls. Further they work as a quencher of chlorophyll triplet state and of singlet molecular oxygen due to their low-lying triplet state. Carotenoids are, in general, polyenes belonging to the C_{2h} point group [1]. Two energetically low-lying singlet states are expected; one is closely related to the 2^1Ag state which is dipole forbidden from the ground state by parity, and the other is related to the 1^1Bu state which is optically allowed. The Bu-based state is responsible for a strong visible absorption of carotenoids, while the Ag-based state is in a lower energy level (S_1-state) than the Bu-based state (S_2-state). According to the above assignment, we can expect the following relaxation processes of carotenoids; optical excitation to the S_2-state induces an internal conversion to the S_1-state, followed by relaxation to the ground state. Singlet energy transfer to chlorophyll is a competitive relaxation process from the both states, and hence the accurate estimation of their lifetime is very important. The S_1-lifetime was estimated by the ground state recovery, which was in a range of ~10 ps for β-carotene [2,3]. An estimation of the S_2 lifetime was carried out by the sub-ps transient absorption on β-carotene [3]. In their study, complicated kinetics including two relaxation processes were analyzed, which eventually provided the S_2 lifetime to be 250 or 200 fs in ethanol or CS_2, respectively, and the S_1 lifetime, 9.5 or 11 ps, respectively. However, a direct measurement of the S_2 lifetime is necessary to characterize the excited-state dynamics, and thus we report here the first accurate measurements of the S_2 lifetime.

Since a conventional time-correlated single photon counting apparatus or a streak camera system is not sufficient in the time-resolution, we adopted fluorescence up-conversion technique to determine directly the S_2 lifetime of β-carotene [4]. In our system, combination of the sum frequency generation and single-photon counting detection ensured a high sensitivity for the fluorescence in fs time regime. The SHG (425 nm) of the mode-locked Ti:Sapphire laser was used to excite a sample, and fluorescence and a fundamental gate pulse (850 nm) were focused onto a phase-matched thin BBO crystal to generate ultraviolet sum-frequency light. The polarization angle between the excitation and gate pulses was normally set to be 54.7°, while parallel and perpendicular polarizations were applied to obtain the time-resolved anisotropy change. The present instrumental response function, obtained by the cross-correlation measurement between the excitation and gate pulses, had a Gaussian shape of 200 fs (FWHM, Fig. 1, broken line). The steady-state fluorescence spectrum of β-carotene dissolved in *n*-hexane showed that the fluorescence predominantly came from the S_2 state [5]. Based on the spectrum, we chose 6 wavelengths covering its whole emission (488-624 nm) and applied the fs up-conversion. The excitation energy was 80 mW with a 0.1 mm diameter, thus about 0.6 % of the molecule were excited per pulse [5].

Upon excitation of β-carotene in *n*-hexane with a 425 nm pulse, the fluorescence was observed only in the fs regime at all probing wavelengths. Typical kinetics at 513 nm were

Springer Series in Chemical Physics, Vol. 60 **Ultrafast Phenomena IX**
Editors: P. F. Barbara · W. H. Knox · G. A. Mourou · A. H. Zewail

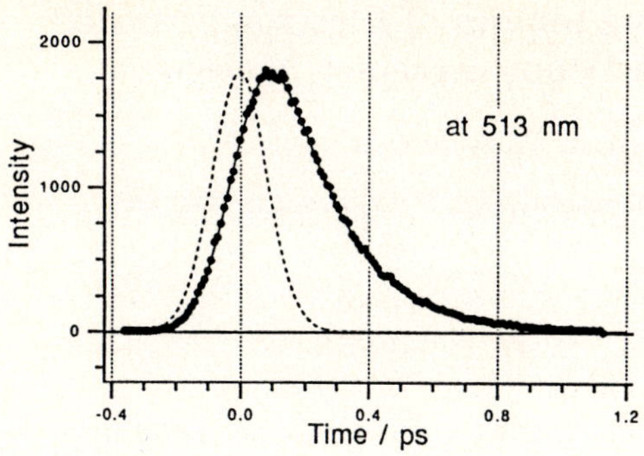

Fig. 1

Typical fluorescence decay kinetics of β-carotene in *n*-hexane.

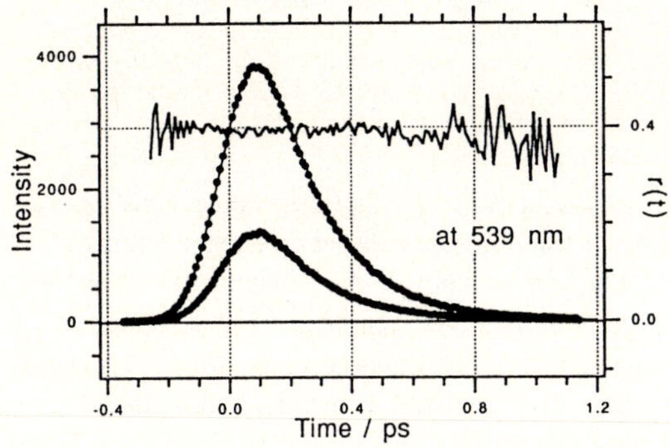

Fig. 2

Anisotropic fluorescence decays and anisotropy change of β-carotene in *n*-hexane.

shown in Fig. 1, where the deconvolution gave a lifetime of 195 ± 5 fs. It is noted that the fluorescence appeared without any delay and decayed to almost zero within 1 ps. The kinetic feature was essentially wavelength-independent; at 6 probing wavelengths, the S_2 lifetimes of β-carotene in *n*-hexane were in a range of 190-200 fs. Fluorescence anisotropy was measured at 539 nm (Fig. 2); it gave a high value, r = 0.39 ± 0.02, throughout the emission time, indicating that fluorescence directly originated from the S_2 state. The estimated fluorescence quantum yield was 1.7×10^{-4}. The 425 nm excitation yields the 0-2 vibrational transition of the S_2 state with ~2500 cm^{-1} excess energy. However the observed kinetics gave a unique lifetime for the whole emission wavelengths, indicating no spectral evolution. The absence of dynamic Stokes shift suggests that the intramolecular relaxation in the S_2 state occurs faster than the present time-resolution, and we estimated it shorter than 50 fs.

It is known that the relaxation processes from S_2 state follow the energy gap law of internal conversion [6]. Thus, a static energy gap between S_2 and S_1 states, which is determined by polarizability of solvents, and solvation process are very important. These will be analyzed through the up-conversion method, thus leading to general understanding of optical properties of polyenes.

1. B.S. Hudson, B. E. Kohler and K. Schulten, *Excited States* **6**, 1 (1982).
2. M.R. Wasielewski and L.D. Kispert, *Chem. Phys. Lett.* **128**, 238 (1986).
3. A.P. Shreve, J. K. Trautman, T. G. Owens and A. C. Albrecht, *Chem. Phys. Lett.* **178**, 89 (1991).
4. H. Kandori and H. Sasabe, *Chem. Phys. Lett.* **216**, 126 (1993).
5. H. Kandori, H. Sasabe and M. Mimuro, *J. Am. Chem. Soc.* **116**, 2671 (1994).
6. M. Mimuro, Y. Nishimura, S. Takaichi, Y. Yamano, M. Ito, S. Nagaoka, I. Yamazaki, T. Katoh and U. Nagashima, *Chem. Phys. Lett.* **213**, 576 (1993).

Femtosecond Transient-Grating Studies of Electron Transfer in Porphyrin and Chlorophyll Donor–Acceptor Molecules

G.P. Wiederrecht, W.A. Svec, and M.R. Wasielewski

Chemistry Division, Argonne National Laboratory, 9700 S. Cass Avenue, Argonne, IL 60439, USA

Abstract. Polarization sensitive transient grating experiments are performed on artificial photosynthetic systems. Separation of overlapping spectral features, as well as structural determinations and dynamic solvent interactions are observed.

1. Introduction

Transient absorption experiments on artificial photosynthetic systems have proven to be a valuable technique for the study of phenomena related to charge transfer[1,2]. However, even simplified model systems can be complicated because they possess moieties that have overlapping spectral features. We report on the application of polarization selective transient grating experiments to electron transfer. Transient grating experiments can be utilized to separate contributions from two chromophores if their transition dipoles are oriented in different directions. They are also utilized to probe processes associated with electron transfer, such as solvent reorientation, that may not result in a spectral change.

The transient grating technique has been described in detail previously[3]. Since values of $\chi^{(3)}$ are unequal for differing physical processes, the polarization selective nature of transient grating experiments can be utilized to determine the nature of a particular kinetic component of the material response to the light pulses[4,5]. This includes the ability to determine the anisotropy of the excited state grating, which can be utilized to provide valuable structural information about the relation between the excitation transition dipole and the probed transition dipole. The ability to separate absorption gratings with different anisotropies also enables this technique to be applied to charge transfer model systems with success in separating overlapping spectral components.

2. Results

The molecule utilized for the charge transfer studies was Zn 13^1 desoxomethylpyropheophorbide a - pyromellitimide (ZnDMPPH a-PI), illustrated in Fig. 1. The laser system is an amplified femtosecond dye laser that produced pulses of 100 fs duration at 645 nm with 20μJ of energy per pulse at 1 kHz. The transient grating data is shown in Fig. 2 with a probe wavelength of 730 nm. Through appropriate polarization techniques, the anisotropy of the absorption grating was

Springer Series in Chemical Physics, Vol. 60 **Ultrafast Phenomena IX**
Editors: P. F. Barbara · W. H. Knox · G. A. Mourou · A. H. Zewail

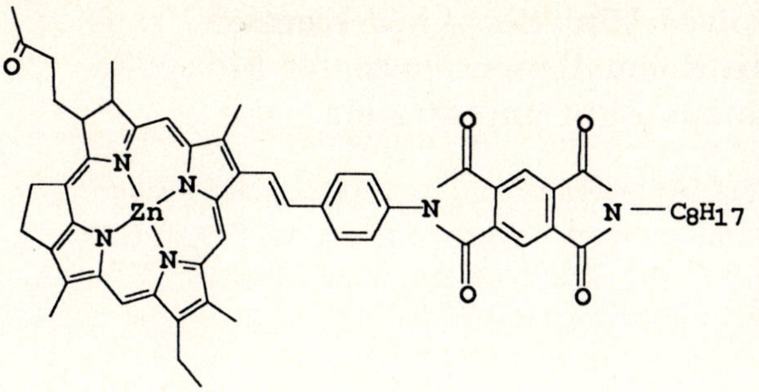

Figure 1. An illustration of ZnDMPPh a-PI. The PI anion radical has a large absorption at 730 nm that is well separated from the Zn DMPPh a bands. Optical excitation results in electron transfer to PI.

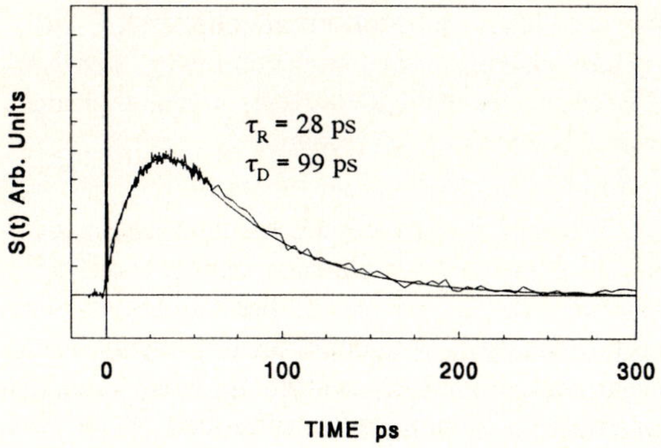

Figure 2. Optical excitation results in charge transfer in 28 ps and charge return in 99 ps. A polarization geometry that minimizes nonresonant nuclear contributions to the signal was utilized.

utilized to determine an angle between the donor and acceptor transition moments of 43°. This is in excellent agreement with theoretical results.

The authors wish to acknowledge the support of the Division of Chemical Sciences, Office of Basic Energy Sciences, US Department of Energy under contract W-31-109-Eng-38.

3. References

1. M. R. Wasielewski, M. P. Niemczyk, W. A. Svec and E. Bradley-Pewitt, J. Am. Chem. Soc. **107**, 5562 (1985).
2. G. P. Wiederrecht, S. Watanabe, and M. R. Wasielewski, Chem. Phys. **176**, 601 (1993).
3. Y. Yan and K. A. Nelson, J. Chem. Phys. **87**, 6240 (1987).
4. J. Etchepare, G. Grillon, J. P. Chambaret, G. Hamoniaux and A. Orszag, Opt. Commun. **63**, 329 (1987).
5. F. W. Deeg and M. D. Fayer, J. Chem. Phys. **91**, 2269 (1989).

Picosecond Time-Resolved Vibrational Spectrum of Batho-Rhodopsin at Room Temperature: Effect of Changing Rhodopsin Concentration

A. Popp, L. Ujj, and G.H. Atkinson

Department of Chemistry, University of Arizona, Tucson, AZ 85721, USA

Abstract. The first vibrational spectrum of batho-rhodopsin (batho) at room temperature, an intermediate formed during the initial 50 ps of the visual process of rhodopsin (Rh), has been measured by picosecond time-resolved coherent anti-Stokes scattering (PTR/CARS). The batho CARS-spectrum is obtained through a third-order susceptibility ($\chi^{(3)}$) analysis of the measured PTR/CARS data. The CARS spectrum of batho reveals structural changes in the retinal chromophore which have not previously been recognized.

1. Introduction and Experimental

The visual pigment Rh is responsible for the conversion of light into nerve impulses resulting in vision. The early steps following photon absorption by Rh are currently under discussion[1]. It is recognized, however, that only batho and Rh are present 200-ps after excitation. The vibrational degrees of freedom of Rh and batho are examined here by PTR/CARS. The instrumentation, experimental procedure, and $\chi^{(3)}$-analysis are described in detail elsewhere[2]. Laser wavelengths (λ_1= 568 nm, λ_s=623±5nm) are selected to resonantly enhance only the retinal chromophore within Rh and batho.

2. Results and Discussion

PR/CARS spectra in the C=C stretching region for different concentrations of Rh are presented in Fig. 1. All spectra are fitted with the same band parameters and only the concentration of Rh is changed. The PR/CARS data from Rh are in good agreement with RR data [3]. In Fig. 2, the PTR/CARS spectrum of a mixture of batho and Rh recorded 200 ps after excitation of Rh is presented. To accurately fit these PTR/CARS data, the relative contribution from Rh must be quantitatively known. Based on the Rh-concentration data (Fig. 1), three different Rh contributions to the PTR/CARS data can be considered (Fig. 2). A quantitative $\chi^{(3)}$ analysis of PTR/CARS data from the Rh/batho mixture shows that major changes occur in both near 1542 cm^{-1} and near 1651 cm^{-1}. It is evident (Figs. 1 & 2) that these spectral changes are not related only to Rh concentration differences. These are the first vibrational data assignable to room temperature batho and are significantly different than previously published RR-spectra of batho obtained at $\leq$120K[4]; *e.g.*, the C=C vibration of batho at room temperature is only 3 cm^{-1} lower than that of Rh instead of 12 cm^{-1} lower as observed at low temperature. These differences emphasize the importance of room temperature measurements of the vibrational degrees of freedom of the early Rh photo-intermediates to elucidate an accurate mechanism describing the molecular processes of vision.

Springer Series in Chemical Physics, Vol. 60 **Ultrafast Phenomena IX**
Editors: P. F. Barbara · W. H. Knox · G. A. Mourou · A. H. Zewail

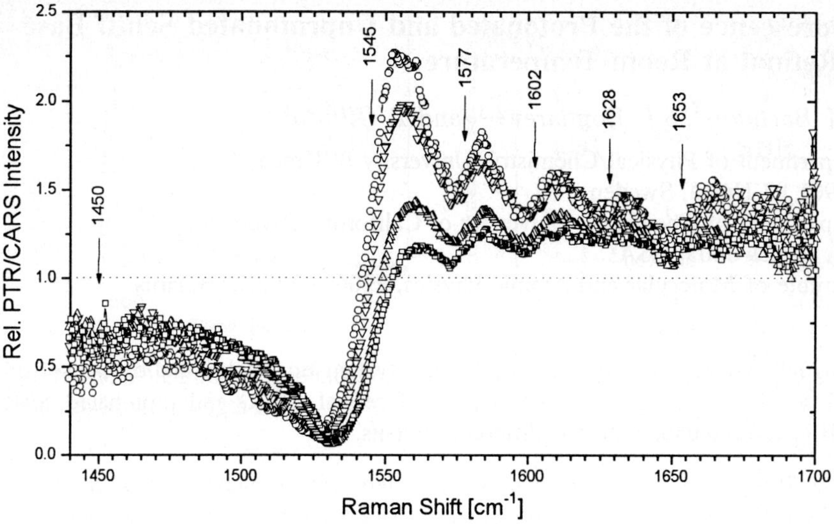

Fig. 1: PR/CARS data from four different concentrations of ground-state Rh. Wavenumber positions of band origins derived from $\chi^{(3)}$-analysis are shown as vertical arrows. The dependence of band shapes on the Rh concentration is derived from these data.

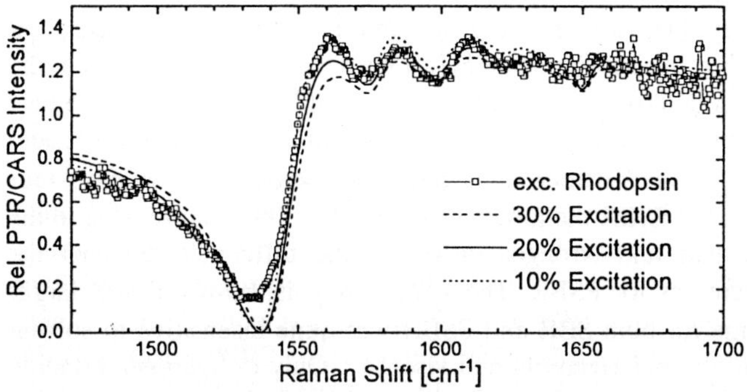

Fig. 2: PTR/CARS from the room temperature Rh photo-reaction at a 200-ps delay (□). Three different degrees of optical excitation of Rh are presented.

3. References

1. (a) Y. Shichida *Photochem. Photobiol.* (1990) **52** 1179-1185. (b) L.A. Peteanu, RW. Schoenlein, Q. Wang, R. Mathies, C. Shank *Proc. Natl. Acad. Sci.* (1993) **90** 11762-66.
2. (a) L. Ujj, B. Volodin, A. Popp, J. Delaney, G.H Atkinson *Chem. Phys.* (1994) **181** in press. (b) L. Ujj, G.H. Atkinson *Isr. J. Chem.* **33** 207-14. (c) G.H. Atkinson, A. Popp, L. Ujj in: *Time Res. Vibr. Spectr. VI* (eds.: A. Lau, *et al.*) p. 153 (1994) Springer, Berlin.
3. R. Mathies, A.R. Oseroff, L. Stryer *Proc. Natl. Acad. Sci. USA* (1976) **73** 1-5.
4. H. Deng, R.C. Callender *Biochemistry* (1987) **26** 7418-7426.

Picosecond Laser-Induced Electric-Field Modulation of Carotenoid Absorption Bands

D. Gosztola, H. Yamada, and M.R. Wasielewski

Chemistry Division, Argonne National Laboratory, Argonne, IL 60439, USA

Abstract. Photoinduced charge separation within an electron donor - acceptor pair is shown to induce large electrochromism in a nearby carotenoid molecule.

1. Introduction

Photoinduced electron transfer reactions are often used to generate long-lived ion pairs that mimic natural photosynthesis[1]. The dipolar ion pair products of these reactions generate large electric fields that can significantly perturb the electronic structures of adjacent molecules[2]. In molecule **1** we use the electric field generated by the ion pair product that results from oxidation of the lowest excited singlet state of a Zn porphyrin, ZP, covalently attached to a pyromellitimide acceptor, PI, to perturb the electronic properties of an adjacent carotenoid. The ZP^+ - PI^- ion pair possesses an 8.4 Å center-to-center distance that is restricted by the covalent attachment, thus generating a 40 D dipole moment. This electric field generating ion pair is, in turn, attached to a calix[4]arene spacer that exists in its cone conformation. Attachment of a carotenoid probe molecule to the calix[4]arene about 7 Å away from the field generator restricts the overall average orientation of the electric field generator relative to that of the probe molecule. Thus, the direction of the electric field relative to the optical transition moment direction of the probe molecule is maintained even though the entire supramolecular assembly rotates in liquid solution.

2. Results and Discussion

Upon photo-excitation of either **1** or reference molecule **2** lacking the carotenoid with 200 fs, 412 nm laser pulses[3], the ZP^+-PI^- ion pair is formed with a 64 ps time constant (97% quantum yield), and decays with a 3.7 ns time constant. Transient absorption spectra of these molecules at 150 ps following excitation show characteristic absorption bands of ZP^+ and PI^- at 660 nm and 720 nm, respectively. Changes in the ground state optical absorption spectrum of the carotenoid due to the presence of the electric field produced by ZP^+ - PI^- within **1**, were determined from the difference spectrum calculated by subtracting the transient absorbance spectrum of reference molecule **2** from that of **1** at 150 ps following excitation. Figure 1 shows the resulting difference spectra. We assign the derivative-shaped ΔA change observed for **1** in toluene (ε = 2.4) to a 15 nm electrochromic red shift of the carotenoid absorption band. This assignment is supported by the corresponding

Springer Series in Chemical Physics, Vol. 60 **Ultrafast Phenomena IX**
Editors: P. F. Barbara · W. H. Knox · G. A. Mourou · A. H. Zewail

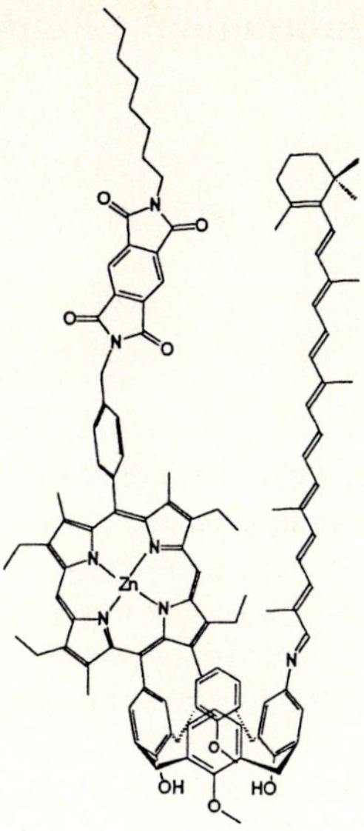

1

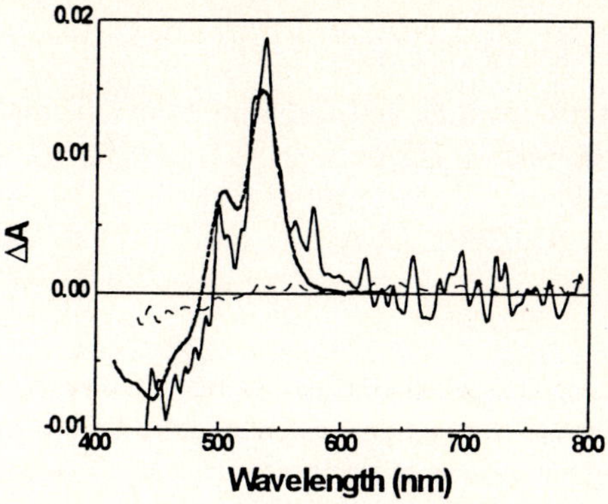

Figure 1. Electrochromically shifted ground state spectra of the carotenoid molecule within **1** at 150 ps following production of ZP^+ - PI^- in toluene (—) and benzonitrile(— —). The simulated 15 nm bandshift(----) of the carotenoid ground state spectrum is superimposed on the data.

spectrum in benzonitrile (ε=25). Figure 1 shows that no electrochromic shift is observed in benzonitrile. The electric field strength is inversely proportional to the static dielectric constant of the medium. The electrochromic red shift in **1** was simulated by recording the ground state spectrum of the carotenoid alone, calculating a spectrum shifted by 15 nm, and taking the difference between these two spectra, Figure 1. The electrochromic shift $\Delta\lambda$ for a band with an absorption maximum λ is given by[4]:

$$\Delta\lambda = \frac{\lambda^2}{hc}(\Delta\mu \cdot E \cdot \cos\theta + \frac{1}{2}\Delta\alpha \cdot E^2 \cdot \cos^2\phi) \qquad (1)$$

where $\Delta\mu$ is the change in the permanent dipole and $\Delta\alpha$ is the corresponding change in polarizability between the ground and excited state of the molecule. For $\Delta\mu$ = 2-4 X 10^{-29} Cm, $\Delta\alpha = 10^{-37}$ $CV^{-1}m^2$, $\Delta\lambda$ = 15 nm and λ= 480 nm, solving equation 1 gives E = 2-3 MV/cm at the carotenoid. (Supported by the Div. of Chem. Sci., BES of the US DOE under contract W-31-109-Eng-38.)

3. References

1 Wasielewski, M. R. *Chem. Rev.* **1992**, *92*, 435.
2 Boxer, S. G. in *The Photosynthetic Reaction Center*; Vol. 2, J. Deisenhofer, J. R. Norris, eds. Academic Press: San Diego, 1993; p 179.
3 Frank, H. A.; Farhoosh, R.; Gebhard, R.; Lugtenburg, J.; Gosztola, D.; Wasielewski, M. R. *Chem. Phys. Lett.* **1993**, *207*, 88.
4 Liptay, W. *Z Naturforsch. A* **1965**, *20*, 1441.

Time-Resolved Fluorescence Anisotropy in Chromophore Pairs

A. Matro and J.A. Cina*

James Franck Institute and Department of Chemistry, University of Chicago, Chicago, IL 60637, USA

*Present address: Department of Chemistry, University of Rhode Island, Kingston, RI 02881, USA

Abstract. Calculations of time-resolved fluorescence anisotropy from a pair of chromophores coupled by an excitation transfer interaction are presented for the purpose of studying the effects of nuclear motion on the energy transfer and anisotropy. An illustrative model is developed that provides each chromophore with a vibrational mode. Non-instantaneous excitation and time- and frequency-resolved detection are taken into account. Vibrational relaxation is included using the Redfield theory of relaxation.

1. Introduction

Electronic excitation transfer has been the subject of numerous experimental and theoretical studies. Recent experimental measurements of time-resolved fluorescence and pump-probe anisotropy have revealed sub-picosecond excitation transfer processes, seen as decays in the anisotropy. In recent fluorescence anisotropy measurements on Allophycocyanin (APC) and C-Phycocyanin (C-PC), Xie et al.[1] found that the ultrafast decay component in the anisotropy showed a strong dependence on excitation/detection wavelength. We have developed a quantum mechanical model for the chromophore pairs which accurately represents the ultrafast electronic excitation, excitation transfer, and detection. Each chromophore has two electronic states and a single vibrational mode. The singly excited states of the chromophore pair are coupled by an excitation transfer interaction. The detector consists of a collection of two-level systems with a Gaussian distribution of transition frequencies. The chromophore pair is first excited with a laser pulse and later returns to its ground state through an interaction with the detector. The emission rate is calculated from the change in excited state population of the detector over a time interval corresponding to the temporal resolution of the detection process. The dipole moment of the detector is set parallel or perpendicular to the polarization of the excitation pulse and the time-resolved fluorescence anisotropy $R(t) = \dfrac{S_{||}(t) - S_{\perp}(t)}{S_{||}(t) + 2S_{\perp}(t)}$, is calculated from $S_{||}(t)$ and $S_{\perp}(t)$, the parallel and perpendicular signals, respectively. This model allows us to include excitation and detection frequency dependence by choosing the center frequency of the excitation pulse and the center frequency (as well as bandwidth) of the detector.

2. Results

A test calculation (not shown) neglecting all but the lowest singly excited states and the lowest three vibrational levels of the ground state of the chromophore pair

Springer Series in Chemical Physics, Vol. 60 **Ultrafast Phenomena IX**
Editors: P. F. Barbara · W. H. Knox · G. A. Mourou · A. H. Zewail

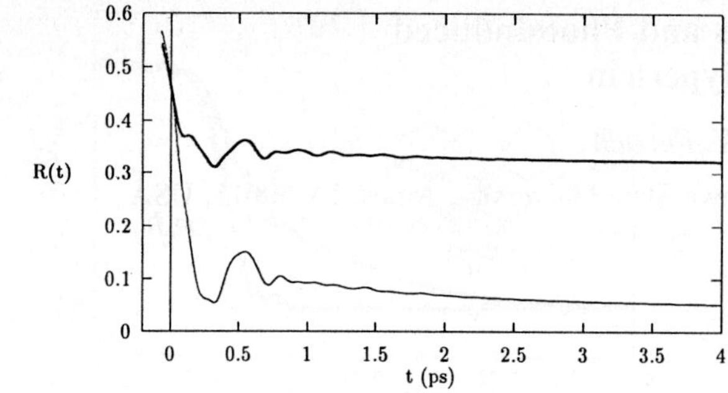

Figure 1. Time-resolved anisotropy calculated with 1.5 ps^{-1} vibrational relaxation rate. Upper (thin) curve corresponds to excitation at bare transition frequency of lower-energy chromophore, lower (thick) curve corresponds to excitation 100cm^{-1} above the bare transition frequency of higher-energy chromophore. Detection for both cases is 250cm^{-1} below the excitation frequency.

illustrates the fact that the time-resolved fluorescence anisotropy is strongly dependent on the nature of the initial and final states of the chromophore pair. Exciting from the lowest vibrational level of the electronic ground state, the anisotropy resulting from emission back to that state can take values greater than 0.4 if the transition moments of the individual monomers are not parallel, while the maximum anisotropy resulting from emission to the higher-lying levels is 0.4 and is independent of the angle between the two transition moments. This is in agreement with a theoretical study of pump/probe anisotropy by Wynne et al. [2] We have carried out numerical calculations of the time-resolved anisotropy including non-zero excitation pulse duration and non-zero time resolution of the detector. Vibrational (or optically active collective) motion is assumed harmonic with 100 cm^{-1} frequency, and the two chromophores differ by 150 cm^{-1} in their bare electronic transition frequency. The model parameters correspond roughly to the those of the strongly coupled chromophores in C-PC. [3] Figure 1 shows the calculated anisotropies for two excitation/detection conditions. The intra-chromophore vibrations are coupled to a thermal bath using the Redfield relaxation theory. [4] The anisotropies in Figure 1, calculated with a vibrational relaxation rate of 1.5 ps^{-1}, roughly twice the vibrational period, show the decay of the oscillations due to the coherences among the excitonic states. An optical dephasing rate need not be introduced, as our approach is to explicitly include any optically active collective modes in the intra-chromophore dynamics. [5] Vibrational relaxation is left out during the first 150 fs, while the excitation pulse and the detection window overlap. After the coherences have decayed, the anisotropy continues to decay due to the thermal equilibration of the populations in the excitonic states. Excitation wavelength dependence similar to that in experiments is observed. [1] This work was supported by the National Science Foundation and the Camille and Henry Dreyfus Teacher-Scholar Award Program (J. A. C.).

References

[1] X. Xie, M. Du, L. Mets and G.R. Fleming, SPIE Vol 1640, Time-Resolved Laser Spectroscopy in Biochemistry III (1992).
[2] K. Wynne and R.M. Hochstrasser, Chem. Phys. **171**, 179 (1993).
[3] K. Sauer and H. Sauer, Biochim. Biophys. Acta., **936**, 157 (1988).
[4] J.M. Jean, R.F. Friesner and G.R. Fleming, J. Chem. Phys. **96**, 5827 (1993).
[5] A. Matro and J. A. Cina, submitted.

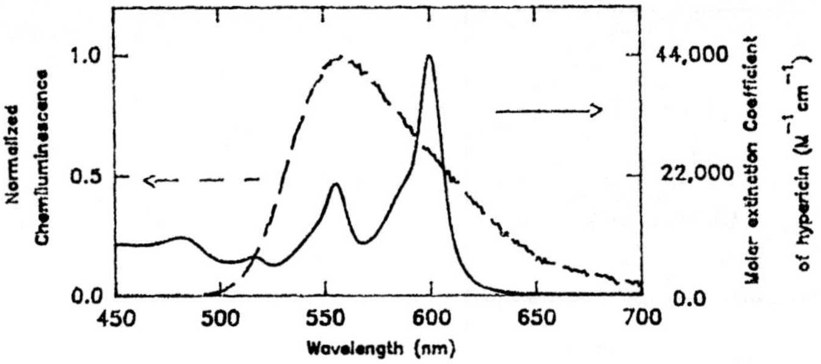

Figure 2. Spectral overlap between luciferin chemiluminescence and the visible absorption of hypericin.

In recent years, photodynamic therapy has emerged as a promising tool in both antiviral and cancer chemotherapy. Our approach is to combine a chemiluminescent light source with the antiviral agent, hypericin, so that photodynamic therapy can be extended to all regions of the body. We employ a chemical light source that has a broad emission spectrum in the region where the photoactive molecule absorbs. A natural light source is the luciferin molecule. We have demonstrated that hypericin reduces the infectivity of cell-free stocks of equine infectious anemia virus (EIAV), a lentivirus closely related to HIV, by 99.99%. The antiviral activity of hypericin is completely dependent on the presence of light [2]. This antiviral activity does not depend on the presence of oxygen [9].

Figure 2 compares the spectrum of the chemiluminescent emission of the luciferase catalyzed oxidation of luciferin and the absorption spectrum of hypericin in the red region of the visible spectrum. The overlap between these spectra indicates (based on Förster theory) that the "critical distance" for energy transfer between these two species is about 100 Å.

Luciferase can be generated using the luciferase gene. We expect to be able to regulate specifically the activation of the light source in order that hypericin is photoactivated only in cells where it is needed. Regulation of luciferase activation will be achieved by regulating the expression of luciferase, which will be under the control of a viral promoter. In this way, replication of the virus will transactivate the viral promoter, resulting in increased expression of luciferase and consequently leading to activation of luciferin and the photoactivation of hypericin. A crucial aspect of this approach is that these events are localized to virus-infected cells, thereby targeting the antiviral activity of hypericin.

ACKNOWLEDGEMENTS JWP is an Office of Naval Research Young Investigator

REFERENCES

1. Meruelo, D.; Lavie, G.; Lavie, D. *Proc. Natl. Acad. Sci. USA* 1988, **85**, 5230.
2. Carpenter, S.; Kraus, G. A. *Photochem. Photobiol.* 1991, **53**, 169.
3. Savikhin, S.; Tao, N.; Song, P.-S.; Struve, W. S. *J. Phys. Chem.* 1993, 97, 12379.
4. Racinet, H.; Jardon, P.; Gautron, R. *J. Chim. Phys.* 1988, **85**, 971.
5. Song, P.-S.; Walker, E. B.; Auerbach, R. A.; Robinson, G. W. *Biophys. J.* 1981, **35**, 551.
6. Gai, F.; Fehr, M. J.; Petrich, J. W. *J. Phys. Chem.* In press.
7. Borgis, D.; Hynes, J. T. *J. Chem. Phys.* 1991, **94**, 3619.
8. Falk, H. Personal communication.
9. Fehr, M. J.; Petrich, J. W.; Carpenter, S. *Bioorg. Med. Chem. Lett.* In press.

Distinguishing Between Energy- and Electron-Transfer Processes in Photosystem II Reaction Centres

D.R. Klug[1], *T. Rech*[1], *D.M. Joseph*[1,2], *J.R. Durrant*[1,2], *R.S. Knox*[1,2,3], *J. Barber*[2], *and G. Porter*[1]

[1]Molecular Dynamics, Centre for Photomolecular Sciences, Department of Biochemistry, Imperial College, London SW7 2AY, UK

[2]AFRC Photosynthesis Research Groups, Centre for Photomolecular Sciences, Department of Biochemistry, Imperial College, London SW7 2AY, UK

[3]Department of Physics and Astronomy, University of Rochester, Rochester, NY 14627-0171, USA

The photosystem two (PS II) reaction centre (RC) undertakes both electron and energy transfer processes as part of its function. Distinguishing between these two processes is an essential pre-requisite for meaningful studies of the primary photochemistry of this complex. We have previously reported the observation of a 100fs equilibration of excitation energy between P680, the primary electron donor, and some accessory chlorins in isolated PS II RC's [1]. In addition there have been several reports of much slower energy transfer processes in PS II RC's with lifetimes of 10 -100 ps. Uncertainty regarding how these slower energy transfer processes affect the observed lifetime of charge separation has complicated the current controversy over whether the charge separation occurs in 3 ps or 21 ps.

The results reported here employ photoselective excitation of either P680 or the accessory chlorins (694 nm or 665 nm excitation respectvely) in order to identify energy transfer processes between these pigments. In addition we employ a home built, highly sensitive, multi-channel diode array detector (MCD). Global analyses of data collected over a wide range of probe wavelengths have determined that, following photoselective excitation of P680, formation $P680^+Ph^-$ in PS II RC's proceeds primarily with a 21±1 ps time constant [2]. Following photoselective excitation of the accessory chlorins, the observed lifetime of radical pair formation increases to 27±3 ps. Figure 1 compares spectra of the amplitude of these 21 ps and 27 ps components. The difference between these spectra circa 670 nm indicates that the 27 ps component is accompanied by a much greater loss of population from the 670 nm absorbing chlorophylls than the 21 ps component. This observation, and the observed difference in the lifetime of radical pair formation, can only rationalised if, in addititon to the sub-picosecond energy transfer reported previously [1], there is at least one other chlorin which transfers excitation energy to P680 much more slowly.

A kinetic model which is consistent with these observations is shown in figure 2. The model includes four states: the excited singlet state of P680, the radical pair $P680^+Ph^-$ and singlet excited states of two pools of accessory chlorins $C670_I$ and $C670_{II}$. Excitation energy is rapidly transfered between P680 and the $C670_I$ chlorin pool. However, energy transfer from the $C670_{II}$ chlorin pool to P680 is relatively slow, which prevents equilibration of excitation energy between these pigments prior to RP formation, and results in a slower time constant of RP formation. Photoselective excitation of the C670 pigments will excite both the $C670_I$ and $C670_{II}$ chlorin pools, and will therefore result in a slower average time constant of RP formation than that which is observed when P680 is directly excited. This kinetic model is the simplest one which can take account of the observations reported here.

Springer Series in Chemical Physics, Vol. 60 **Ultrafast Phenomena IX**
Editors: P. F. Barbara · W. H. Knox · G. A. Mourou · A. H. Zewail

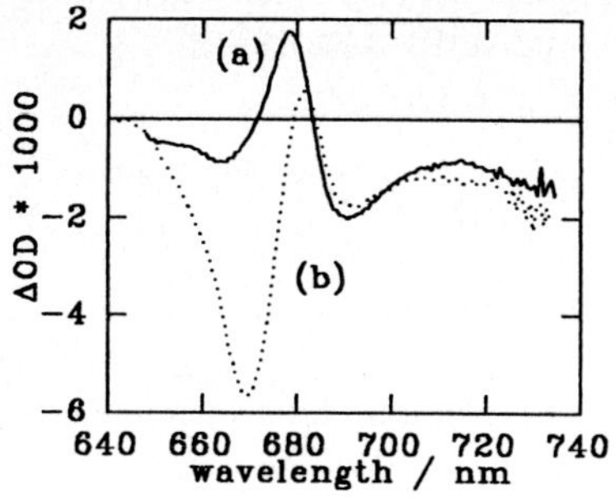

Figure 1. Spectra of the amplitudes of (a) a 21±1 ps component and (b) a 27±3 ps component observed using photoselective excitation of P680 and 670nm absorbing chlorins respectively. Isolated PS2 RC's were studied at 295K as previously [2].

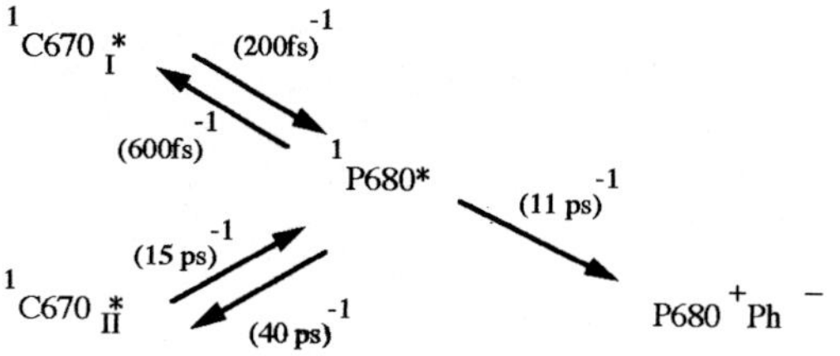

Figure 2. Kinetic model of isolated PS II RC's. The numbers refer to the rates (in ps^{-1}) of the respective energy & electron transfer processes.

It should however not be considered a complete kinetic model for the primary photochemistry of isolated PS II RC's, as we know that other processes occur in this complex with lifetimes of 1 -3 ps [2], which are not included in this model.

The results presented here are in agreement with those of other groups that some energy transfer processes are observed with lifetimes of 15 - 30 ps between some accessory pigments and P680 in PS II RC's. These slow energy transfers are effectively avoided by photoselective excitation of P680 as reverse energy transfer is slower than the forward electron transfer. However, our results crucially differ from those of other groups as we find that following photoselective excitation of P680, a 21 ps transient absorption component is still observed which is strongly associated with radical pair formation. Thus both slow energy transfer processes, and charge separation not limited by these slow energy transfer processes, can be observed in PS II RC with similar but distinguishable lifetimes. We conclude that if P680 is photoselected in PS II RC's, formation of $P680^{+}Ph^{-}$ is observed with a 21 ps lifetime, apparently not limited by any slow energy transfer steps.

References.

1) Durrant, J. R., Hastings, Joseph, D.M., Barber, J., Porter, G., and Klug, D. R. (1992) *Proc. Natl. Acad. Sci. USA 89*, 11632-11636.
2) Durrant, J. R., Hastings, G., Hong, Q., Barber, J., Porter, G., and Klug, D. R (1993) *Biochemistry 32,* 8259-8267.

Coupled-Oscillator Representation of Many-Body Effects in Ultrafast Optical Response

J.A. Leegwater[1] *and S. Mukamel*[2]

[1]Instituut Lorentz, Rijksuniversiteit, P.O. Box 9506, 2300 RA Leiden, The Netherlands

[2]Department of Chemistry, University of Rochester, Rochester, NY 14627, USA

A theory for the nonlinear optical response of an assembly of four level atoms representing an s-p transition with dipole-dipole interactions is developed. Simulation results for two pulse ultrafast four wave mixing measurements in neat atomic clusters and in liquids demonstrate that the electronic density of states can be obtained by combining linear absorption and four wave mixing measurements. We show that by reversing the pulse time ordering in a photon echo sequence we obtain a sensitive direct probe for interatomic interactions and correlations.

The interpretation of nonlinear optical measurements such as photon echo and pump-probe spectroscopv of a system of dilute chromophores in the condensed phase (e. g. doped crystals and dye solutions at low concentrations) is traditionally made using the optical Bloch equations. Homogeneous and inhomogeneous dephasing (T_2 processes) can be easily incorporated in these equations. The tremendous success of these equations in the interpretation of nonlinear optical measurements involving simple few level systems makes it tempting to apply them to more complex many body systems such as neat liquids and semiconductors. In these applications all many body effects are lumped into a frequency dependent dephasing rate. However, mapping a complex many body system into an effective two level system makes it very hard to develop physical intuition. Photon echoes, for example, are traditionally analyzed using the optical Bloch equations, but as the broadening of the linear absorption line for a neat cluster or a liquid is caused entirely by the intermolecular interactions among chromophores (rather than by the interactions with optically inactive bath molecules) this model does not apply. An alternative approach is based on a quasiparticle representation where the many-body system is mapped onto a set of coupled anharmonic (exciton) oscillators. This direct generalization of the Drude (harmonic oscillator) model for linear response [1,2] provides a completely different framework for analyzing such spectroscopies in terms of spectral densities. Nonlinear signals for ordered nanostructures of two level atoms were calculated using this approach [3,4]. Both frequency domain and time domain techniques have been investigated.

We consider a system of four-level atoms interacting via dipole-dipole interactions. The model can represent small clusters as well as liquids. Each atom has a ground state and a triply degenerate excited p state, separated by an optical transition frequency Ω. We chose this as a simple model for describing many-body effects in assemblies of polarizable atoms including exciton dynamics. Whenever needed we use parameters of Ar. We assume the following Hamiltonian [5]

$$H = \sum_{n\alpha} \Omega B^{\dagger}_{n\alpha} B_{n\alpha} + \sum_{n<m,\,\alpha\beta} J_{n\alpha,\,m\beta}(\mathbf{r}_{nm})\,(B^{\dagger}_{m\beta}B_{n\alpha} + B^{\dagger}_{n\alpha}B_{m\beta}) + H_{nucl} - \sum_{n\alpha} \mu_n E_{n\alpha} P_{n\alpha} \,. \quad (1)$$

Here $B^{\dagger}$ and B are creation and annihilation operators for the s-p system at atom m. The index n runs over the atoms and α runs over the internal degrees of freedom, $\alpha = x, y, z$. $B^{\dagger}_{n\alpha}$ changes the state of atom n from the ground s to the excited state p_α. J represents the interatomic dipole-dipole interaction

$$J_{n\alpha,\,m\beta}(r_{nm}) = -\mu^2 \left[\frac{3 r_{nm,\,\alpha} r_{nm,\,\beta}}{r^5_{nm}} - \delta_{\alpha,\,\beta} \frac{1}{r^3_{nm}} \right] (1 - \delta_{n,\,m}) \,,$$

where δ is the Kronecker delta, and μ is the magnitude of the transition dipole moment. The dipole coupling in Eq. (1) conserves the number of excitons (The Heitler London

Springer Series in Chemical Physics, Vol. 60 **Ultrafast Phenomena IX**
Editors: P. F. Barbara · W. H. Knox · G. A. Mourou · A. H. Zewail

approximation). H_{nucl} represents the nuclear motions

$$H_{nucl} = \sum_n \frac{p_n^2}{2m_n} + \sum_{n<m} V_{exc}(r_{nm}) \ .$$

Finally the last term in Eq. (1) represents the interaction of the electric field at r_n which is polarized in the α direction ($E_{n,\alpha}$) with the electronic polarization operator $P_{n\alpha}$,

$$P_{n\alpha} \equiv (B_{n\alpha}^\dagger + B_{n\alpha}) \ .$$

The commutation relations for the B operators can be found by studying the action of the B's on the basis $|s\rangle$, $|p_x\rangle$, $|p_y\rangle$, and $|p_z\rangle$. We immediately find that $B_x^\dagger |p_\alpha\rangle = 0$, and, using this, also that $B_x B_y^\dagger = 0$. We then have

$$\left[B_{n\alpha}, B_{m\beta}^\dagger\right] = \delta_{nm}(\delta_{\alpha\beta} - B_{n\beta}^\dagger B_{n\alpha} - \delta_{\alpha\beta} N_n) \ , \tag{2}$$

where $N_n \equiv \Sigma_\alpha B_{n\alpha}^\dagger B_{n\alpha}$ counts the number of excitations at atom n (0 or 1). For n≠m they commute and behave as bosons. The second and third terms on the right hand side represent deviations from a boson character. The second ($B^\dagger B$) term corresponds to a collection of three two level atoms on the same atoms. Eq. (2) shows that an s-p atom is not identical to a collection of three two level atoms on the same site, but contains an additional nonlinearity since each atom can carry only one excitation. This additional nonlinearity is given by the N_n term in Eq. (2) and contributes to the nonlinear response. It can be probed, e.g. through the dependence of the signal on the polarization of the incoming beams.

To calculate the third order nonlinear polarization

$$P_n^{(3)}(t) = \mu_n[\langle B_n(t)\rangle^{(3)} + \langle B_n^\dagger(t)\rangle]^{(3)} \ ,$$

we used the Heisenberg equation of motion for an operator A(t), $\hbar d\, A(t)/dt = i[A, H]$. The calculation requires the dynamics of single exciton states represented by B as well as of two exciton variables BB. Terms of normally ordered products of three and four B's are irrelevant for the third order response. We finally obtain the following closed set of equations

$$\frac{1}{i}\frac{d}{dt}\langle B_{n\alpha}\rangle^{(1)} = -(\Omega - i\gamma)\langle B_{n\alpha}\rangle^{(1)} - \sum_{m\beta} J_{n\alpha,m\beta}(t)\langle B_{m\beta}\rangle^{(1)} + \frac{\mu_n}{\hbar} E_{n\alpha}(t) \ . \tag{3a}$$

$$\begin{aligned}\frac{1}{i}\frac{d}{dt}\langle B_{n\alpha}B_{m\beta}\rangle^{(2)} &= -(\Omega_n + \Omega_m - 2i\gamma)\langle B_{n\alpha}B_{m\beta}\rangle^{(2)} \\ &-(1-\delta_{n,m})\sum_{i\gamma} J_{n\alpha,i\gamma}(t)\langle B_{i\gamma}B_{m\beta}\rangle^{(2)} - (1-\delta_{n,m})\sum_{i\gamma} J_{m\beta,i\gamma}(t)\langle B_{n\alpha}B_{i\gamma}\rangle^{(2)} \\ &+(1-\delta_{n,m})\frac{\mu_n}{\hbar}E_{n\alpha}(t)\langle B_{m\beta}\rangle^{(1)} + (1-\delta_{n,m})\frac{\mu_m}{\hbar}E_{m\beta}(t)\langle B_{n\alpha}\rangle^{(1)} \ ,\end{aligned} \tag{3b}$$

$$\begin{aligned}\frac{1}{i}\frac{d}{dt}\langle B_{n\alpha}\rangle^{(3)} &= -(\Omega_n - i\gamma)\langle B_{n\alpha}\rangle^{(3)} - \sum_{m\beta} J_{n\alpha,m\beta}(t)\langle B_{m\beta}\rangle^{(3)} \frac{\mu_n}{\hbar} E_{n\alpha}(t) \\ &- \frac{\mu_n}{\hbar}\sum_\beta [E_{n\beta}(t)\langle B_{n\beta}^\dagger\rangle^{(1)}\langle B_{n\alpha}\rangle^{(1)} + E_{n\alpha}(t)\ \langle B_{n\beta}^\dagger\rangle^{(1)}\langle B_{n\beta}\rangle^{(1)}] \\ &+ \sum_{m\beta\gamma} J_{n\gamma,m\beta}(t)\langle B_{n\gamma}^\dagger\rangle^{(1)}\langle B_{n\alpha}B_{m\beta}\rangle^{(2)} J_{n\alpha,m\beta}(t)\langle B_{n\alpha}^\dagger\rangle^{(1)}\langle B_{n\gamma}B_{m\beta}\rangle^{(2)} \ .\end{aligned} \tag{3c}$$

2γ is the exciton decay rate (inverse lifetime), which is introduced here phenomenologically and $\langle ... \rangle$ denotes the quantum average

We carried out simulations to explore the possible formation of an echo in clusters of s-p atoms. We used the Lennard-Jones form for the interaction potential $V(r) = 4\varepsilon[(r/\sigma)^{-12} - \varepsilon[(r/\sigma)^{-6}]]$. All fields are assumed to be linearly polarized along the same direction, and nuclear motions were not included in the following calculations. The natural timescale of our model is $\hbar/J$ with $J \equiv \mu^2/\sigma^3$. For Argon this timescale is ~5fsec. We studied an ensemble of clusters and a liquid at a temperature $T^* \equiv k_B T/\varepsilon = 0.2$, (corresponding to 16K for Ar). The ensemble of configurations were created using a molecular dynamics program where after a short thermalization period, a "box potential" was slowly turned off. The temperature was controlled by monitoring the average kinetic energies, and scaling the velocity if needed. Liquid simulations were done by a single run of a thermalized sample. The system interacts with two pulses (the first with wavevector $\mathbf{k}_1$ and the second with wavevector $\mathbf{k}_2$)

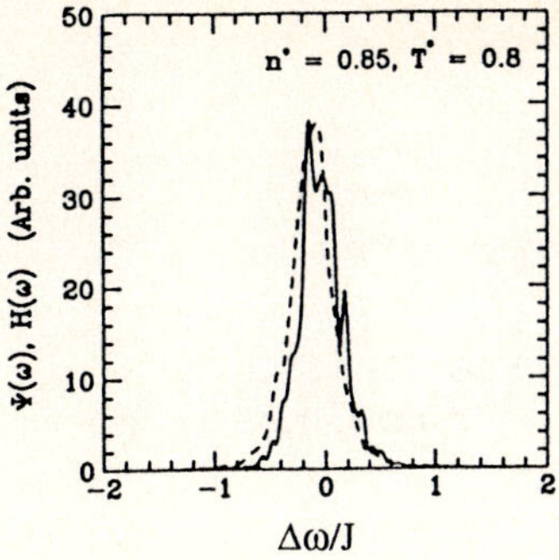

Figure 1
Solid: The Fourier transform of the echo amplitude $\psi(\omega)$ for a liquid at T^*=0.2 and one particular delay ($J\tau = 20$) compared to the "hyperabsorption spectrum" $H(\omega)$, (dashed).

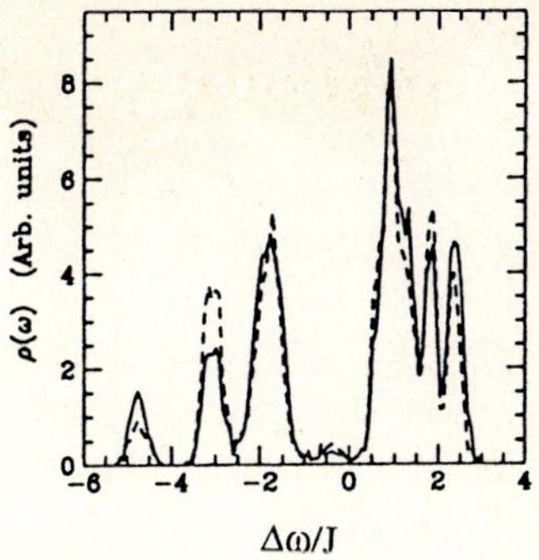

Figure 2
The density states for a cluster with 7 s-p atoms and Lennard-Jones interactions T^*=0.2 (solid) compared to spectroscopic observable $I^2(\omega)/H(\omega)$ (dashed).

separated by a delay τ. The photon echo signal is related to the third order polarization with wavevector $\mathbf{k}_s = 2\mathbf{k}_2 - \mathbf{k}_1$, and is proportional to $|P^{(3)}(\mathbf{k}_s, t)|^2$.

We have found the following physical interpretation for our simulations. The linear absorption spectrum is given by

$$I(\omega) = \sum_j \langle \delta(\omega_j - \omega) \, | \langle k=0 | \psi_j \rangle |^2 \rangle \ ,$$

where ω_j is the energy of mode j, $|\psi_j\rangle$ is its eigenfunction and $|k\rangle = \Sigma_i B_i^\dagger |0\rangle$, and $\langle k = 0 | \psi_j | 0\rangle$ represents the transition dipole of the j'th electronic state.

Let us consider the Fourier transform of the echo polarization for a given delay time τ

$$\psi(\omega) \equiv \int dt \, P^{(3)}(\mathbf{k}_s, t) \exp(i\omega t) \ .$$

Using standard expressions [2,4] we can always recast $\psi(\omega)$ in terms of sums over all global (electronic plus nuclear) states of the system. Such expressions are not practical for actual computations since the number of terms is very large and a complete calculation of all the states is a formidable task. Nevertheless they are still useful to gain some insight. Each term in the sum requires four interactions with the electric field. As an echo setup requires an exact rephasing, it is plausible that the levels can be considered independent for the dominant contribution. To test this idea we consider the following quantity

$$H(\omega) = \sum_j \langle \delta(\omega_j - \omega) \, | \langle k=0 | \psi_j \rangle |^4 \rangle \ .$$

This is the density of states scaled by the transition dipole moment to the fourth power. We argue that $H(\omega)$ should be at least qualitatively similar to $\psi(\omega)$. In Figure1 we compare $H(\omega)$ with $\psi(\omega)$ for a s-p liquid. The striking similarity confirms our conjecture. Is this a viable way to obtain the density of (electronic) states using purely spectroscopic experiments? To answer this question we performed simulations for $I^2(\omega)/H(\omega)$, the results for a small cluster with 7 atoms are presented in Figure 2 and for a liquid in Figure 3. If our conjecture holds, this quantity should be proportional to the density of states. As is obvious from the figures this is perfectly realized for a liquid, and is a very good approximation for a cluster.

When the order of the two pulses in the echo experiment is reversed (i.e. pulse 2 is first and we set $\tau < 0$) the nature of the signal completely changes [6]. When applied to a simple multilevel system (e.g. s-p atom) we find that it can no longer show an echo. For an isolated two level system the signal vanishes identically within the rotating wave approximation

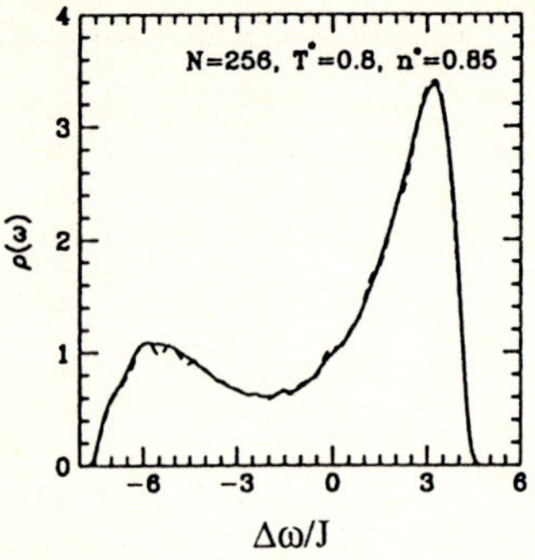

Figure 3
The density of states for a N = 256 liquid at T =0.8, n=0.85 (solid) compared to spectroscopic observable $I^2(\omega)/H(\omega)$ (dashed) of the same state point.

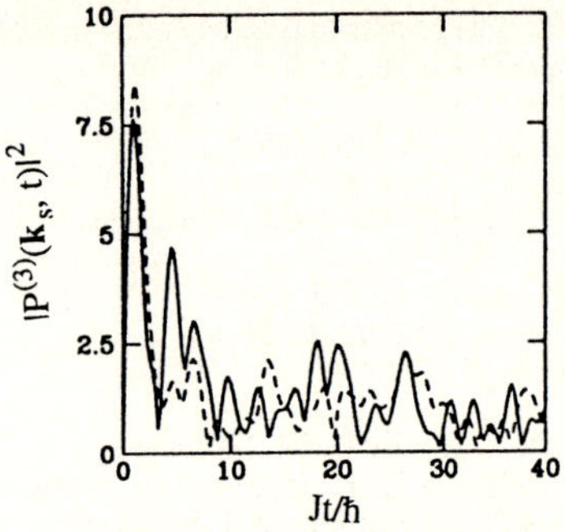

Figure 4
Two pulse four wave mixing signal for negative delays for two delays, $J\tau = -5$ (solid) and $J\tau = -20$ (dashed) at $\mathbf{k}_s = 2\mathbf{k}_2 - \mathbf{k}_1$ when pulse 2 comes first.

(RWA). For these reasons this particular pulse sequence is much less common and it does not even have a name! Simulation results for $\tau < 0$ are presented in Figure 4. What makes this technique particularly interesting is the fact that the entire signal is induced by intermolecular interactions. This could form an important new means for probing intermolecular motions in liquids. A detailed microscopic interpretation of the signal remains to be developed.

Nonlinear ultrafast spectra of molecular and semiconductor materials are remarkably similar. Yet, for historical reasons, their theoretical modeling is done using a completely different language and physical picture. Quantum chemists base their thinking for the most part on the global (many body) electronic eigenstates of molecular systems. It is then common to express the optical response using sums over states. Ultrafast spectroscopy of semiconductors is calculated using a quasiparticle electron hole picture. Since the methods for calculating eigenstates vary among disciplines it is very hard to compare systems with localized and delocalized electron states. It is possible to bridge the gap between these approaches by developing a systematic procedure that starts with the microscopic hamiltonian and maps the system onto a collection of coupled anharmonic oscillators. This way, the analysis and differences become more transparent and a direct comparison is possible. The present formulation can be generalized to various types of materials. (1) Molecular materials and assemblies can be represented as excitonic oscillators; This case was illustrated here (2) In off resonant spectroscopy in liquids, nuclear motions can be treated as oscillators [7,8]; (3) In semiconductors the oscillators describe electron hole pairs and Wannier excitons; and (4)Conjugated polyenes are intermediate between molecular and semiconductor materials and can be modeled as charge transfer exciton oscillators [9].

Acknowledgements

The Support of the National Science Foundation and the Air Force Office of Scientific Research is gratefully acknowledged. One of us (JAL) was in part supported by Stichting voor Fundamenteel Onderzoek der Materie (FOM), which is financially supported by the Nderlandse Organisatie voor Wetenschappelijk Onderzoek (NWO).

References

[1] H. A. Lorentz, *The Theory of Electrons* (Dover, New York, 1952); U. Fano, Rev. Mod. Phys. **45**, 333 (1974).

[2] N. Bloembergen, Nonlinear Optics (Benjamin, NY) (1985).

[3] J. A. Leegwater and S. Mukamel, Phys. Rev. A **46**, 452 (1992); N. Wang, V. Chernyak, and S. Mukamel, J. Chem. Phys., **100**, 2465 (1994).

[4] S. Mukamel, *Principles of Nonlinear Optical Spectroscopy.* (Oxford, New York, 1994).

[5] J. A. Leegwater and S. Mukamel, J. Chem. Phys. (in press).

[6] A. M. Weiner, S. DeSilvestri and E. P. Ippen, J. Opt. Soc. Am. B, **2**, 654 (1985); S. Schmitt-Rink S. Mukamel, K. Leo, J. Shah, and D. S. Chemla, Phys. Rev. A **44**, 2124 (1991).

[7] C. Kalzoupos, D. McMorrow and W. T. Lotshaw, J. Chem. Phys. **95**, 10395 (1991); K. A. Nelson E. P. Ippen, Adv. Chem. Phys. **75**, 1 (1989); N. F. Scherer, L. D. Ziegler and G. R. Fleming, J. Chem. Phys. **96**, 5544 (1992)

[8] Y. Tanimura and S. Mukamel, J. Chem. Phys., **99**, 9496 (1993); R. Inaba. K. Tominaga, M. Tasui, K. A. Nelson, and K. Yoshihara Chem. Phys. Lett (in press)

[9] A. Takahashi and S. Mukamel, J. Chem. Phys., **100**, 2366 (1994).

Femtosecond Photon-Echo Study of Solution Dynamics Using a Cavity-Dumped Ti:Sapphire Laser

W.P. de Boeij, M.S. Pshenichnikov, K. Duppen, and D.A. Wiersma

Ultrafast Laser and Spectroscopy Laboratory, Department of Chemistry, University of Groningen, Nijenborgh 4, 9747 AG Groningen, The Netherlands

Abstract. Using a 13 fs cavity-dumped Ti:sapphire laser femtosecond photon echoes were generated in solution. The results were modelled using the overdamped Brownian oscillator model.

Only recently the ultrafast dynamics of solute-solvent interaction has become accessible by direct time-domain nonlinear optical experiments [1]. However, the technology required to perform these experiments - based on compression of the amplified CPM laser - has been acquired only by a few laboratories, not the least due to the high cost of the necessary laser equipment. With our recent success in the construction of a cavity-dumped Ti:sapphire laser [2], which is capable to produce 13 fs pulses at a 200 kHz repetition rate with pulse energies exceeding 60 nJ, this situation is changing, as now a relatively cheap ultrafast laser can be made that is suitable for ultra-fast spectroscopy in the condensed phase.

The question may be raised why the direct 80 MHz output of a cw fs (non-cavity-dumped) Ti:sapphire laser cannot be used for the aforementioned experiments, as pulses of such a laser can be made as short as 11 fs. The main reason is that in *single-shot* four-wave mixing experiments like pump-probe or (stimulated) photon echo it is essential that in every excitation cycle a fresh sample is probed. Otherwise accumulation effects, as, for instance, a thermal grating may contribute to the observed signal. For solutions a fresh sample can be realized by using a solvent jet and an excitation source with the repetition rate of less than several hundreds kilohertz. Therefore, a cavity-dumped Ti:sapphire laser is an excellent choice for pump-probe or photon-echo experiments in the condensed phase. Nonlinear optical experiments will also greatly benefit from the fact that the pulse energy is increased by an order of magnitude compare with the non-cavity-dumped Ti:sapphire laser.

In this brief report we present some results of photon echo experiments on the dye HITCI dissolved in ethylene glycol. The experiments were performed in a similar manner as described previously [3]. The photon echo signals were detected simultaneously in conjugate phase-matched directions ($2\mathbf{k}_1$-$\mathbf{k}_2$ and $2\mathbf{k}_2$-$\mathbf{k}_1$) in order to determine precisely the zero delay point, which is important in the analysis.

The signals obtained using a cavity-dumped or a cw fs Ti:sapphire laser are depicted in fig.1a. In the former case the experiments were performed at a number of repetition rates and pulse energies to ensure that the observed response reflected the proper optical dynamics. The clear difference between the signals is, most likely, due to the effect of a thermal grating that builds up when the sample is not refreshed between pulse cycles. In addition, the shape of the signal using a cw fs Ti:sapphire

Springer Series in Chemical Physics, Vol. 60 **Ultrafast Phenomena IX**
Editors: P. F. Barbara · W. H. Knox · G. A. Mourou · A. H. Zewail

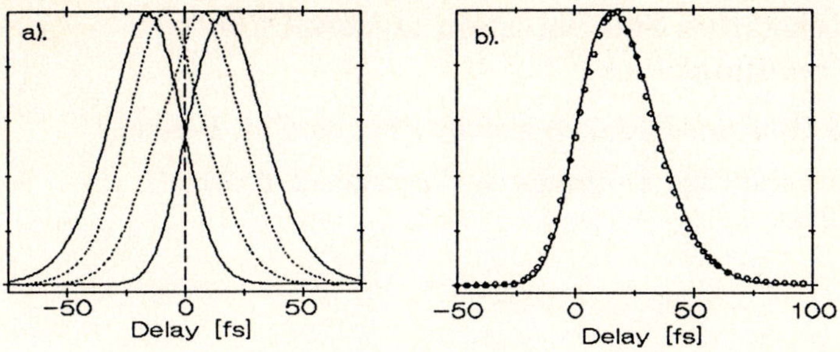

Fig1. Two pulse photon echo signal(s): a cavity-dumped (solid line) vs a cw fs (dotted line) laser (a) and a fit (solid line) to the experimental data (o) using a single overdamped Brownian oscillator (b).

laser was observed to change dramatically with varying the power of exciting beams. Similar discrepancies were observed in pump-probe spectroscopy. This proves that a limited repetition rate is essential for obtaining reliable information on the optical dynamics of a solute.

Fig.1b shows that the two-pulse photon echo signal can be simulated well by a single overdamped Brownian oscillator with parameters $\mathbf{\Delta}$=51 THz and $\mathbf{\Lambda}$=27 THz. However, unlike the case reported in [3], the calculated absorption and emission spectra turn out to be much narrower than the experimental ones, while the Stokes shift in emission is experimentally also much larger than predicted by this model. These discrepancies are caused by slow, diffusive-type solvent motions that are not accounted for in this simple one-Brownian oscillator or, equivalently, the stochastic optical dephasing model. A more extensive analysis of these and other nonlinear optical data on HITCI in ethylene glycol will be presented elsewhere [4]. One of the challenges in future work is to time-resolve the photon echo [5].

References

1. P.C.Becker, H.L.Fragnito, J.-Y.Bigot, C.H.Brito-Cruz, R.L.Fork and C.V. Shank, Phys.Rev.Lett. **63**, 505 (1989); J.-Y.Bigot, M.T.Portella, R.W. Schoenlein, C.J. Bardeen, A.Migus and C.V. Shank, Phys.Rev.Lett. **66**, 1138 (1991)
2. M.S.Pshenichnikov, W.P. de Boeij and D.A.Wiersma, Opt. Lett. **19**, 572 (1994)
3. E.T.J.Nibbering, D.A.Wiersma and K.Duppen, Phys.Rev.Lett. **66**, 2464 (1991); ibid. **68**, 514 (1992)
4. W.P.de Boeij, M.S.Pshenichnikov, K.Duppen and D.A.Wiersma, Chem.Phys. Lett., submitted for publication (1994).
5. K. Duppen and D.A. Wiersma. Opt. Lett. **13**, 318 (1988)

Mechanical Relaxation and Solvation Dynamics in Non-Polar Solutions

J.T. Fourkas, A. Benigno, J. Ma, D. Vanden Bout, and M. Berg

Department of Chemistry and Biochemistry, University of Texas, Austin, TX 78712, USA

Abstract. Transient hole burning experiments reveal a new mechanism for solvation. A viscoelastic model of the solvent explains both the phonon and structural relaxations observed experimentally.

We have used transient hole burning measurements to make the first measurements of solvation dynamics in a completely nonpolar system: dimethyl-*s*-tetrazine (DMST) in *n*-butylbenzene.[1,2] A new model for solvation based on the mechanical response of a viscoelastic solvent explains the main experimental features and predicts new features on the femtosecond timescale. Although the mechanical solvation mechanism is most prominent in nonpolar systems, it should play a role in all solvation [3,4] and electronic dephasing [5,6] processes.

Experimentally, the solvation dynamics are found to consist of two distinct components distinguished by their different temperature dependences. Although both components are subpicosecond at room temperature, as the temperature is lowered, one component remains subpicosecond, while the other slows dramatically.

The fast component is caused by the liquid-phase analog of solid-state phonons, as demonstrated in Fig. 1. The low-temperature side of the figure shows the amount of phonon-induced broadening of the electronic transition in the solid glass as measured by permanent hole burning. A fit to standard solid-state theories is extrapolated into the liquid phase at high temperatures. In the liquid phase, the broadening of the transient hole at 1.5 ps reflects the magnitude of the fast solvation component. The extrapolated solid-state results exactly matches the amount of subpicosecond solvation in the liquid.

Following the phonon-induced solvation, there is a second relaxation related to reorganization of the solvent structure. Through a combination of transient hole burning and time-resolved fluorescence, the time-dependent Stokes shift $S(t)$ resulting from this structural relaxation has been followed over 4 orders-of-magnitude change in relaxation time (Fig. 2). Because we have an absolute measurement of the Stokes shift, the

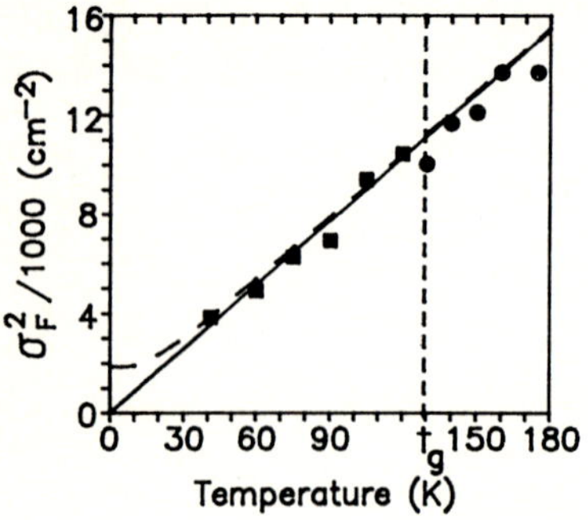

Fig. 1. Phonon-induced broadening of the DMST electronic transition in the solid glass (■). Fits to the full (dashed) or high-temperature (solid) theories of phonon-induced broadening are extrapolated above the glass transition (T_G) into the liquid phase. The transient hole burning widths at 1.5 ps (•) match the extrapolation of the solid-state phonon mechanism.

Springer Series in Chemical Physics, Vol. 60 **Ultrafast Phenomena IX**
Editors: P. F. Barbara · W. H. Knox · G. A. Mourou · A. H. Zewail

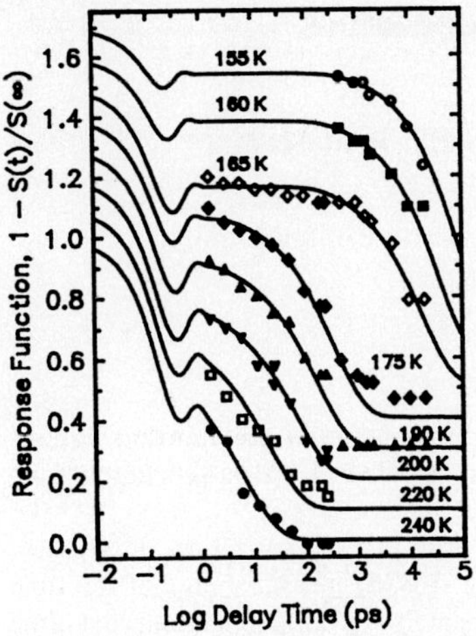

Fig. 2. The solvation response function at selected temperatures (points). The fits are generated by a viscoelastic theory of solvation. Note the logarithmic time scale. Each succesive temperature has been shift vertically 0.1 units.

earliest measurements accurately reflect the amount of solvation occurring within the experimental time resolution.

We have developed a new theory for solvation dynamics which accounts for these results. The solute occupies a spherical cavity inside a viscoelastic continuum. After excitation of the solute, the cavity expands or contracts as the solvent responds mechanically to changes in the solute-solvent interaction potential.

Fits of this theory to the experimental data are shown in Fig. 2. These fits are also consistent with the known mechanical properties of butylbenzene. The theory correctly predicts the occurrence of two distinct solvation timescales, the magnitude of the subpicosecond response and the relaxation time of the slower response as a function of temperature.

The theory also predicts interesting features in the subpicosecond relaxation. A rapid decay in ~100 fs is followed by a weak oscillation at ~300 fs. With current femtosecond techniques, these predictions can be tested.

References

1. J.T. Fourkas and M. Berg, J. Chem. Phys. **98**, 7773 (1993).
2. J.T. Fourkas, A. Benigno, and M. Berg, J. Chem. Phys. **99**, 8552 (1993).
3. P.F. Barbara and W. Jarzeba, Adv. Photo. Chem. **15**, 1 (1990).
4. M. Maroncelli, J. Mol. Liq., in press.
5. C.J. Bardeen and C.V. Shank, Chem. Phys. Lett. **203**, 535 (1993).
6. E.T.J. Nibbering, D.A. Wiersma and K. Duppen, Phys. Rev. Lett. **66**, 2464 (1991).

Determination of Two-Pulse Photon Echoes from Solvent Spectral Densities

D.C. Arnett, P. Vöhringer, R.A. Westervelt, M.J. Feldstein, and N.F. Scherer

Department of Chemistry, University of Pennsylvania, Philadelphia, PA 19104-6323, USA

Abstract: Photon echo measurement of a cyanine dye chromophore in acetonitrile solutions are reported. The measured spectral density is used to calculate the 2-pulse photon echo signal.

The nonlinear response function description of third-order polarization, $P^{(3)}(t_1,t_2,t_3)$, spectroscopy[1,2] shows that different four wave mixing (FWM) measurements of the time evolution of a solute-solvent system contain complementary information about the time correlation function of solvent fluctuations. In particular, chromophore optical dephasing is determined by the correlation function of the chromophore electronic frequency modulation and the coupling of the chromophore transition dipole to the bath. Solvent fluctuations occur on timescales corresponding to the Fourier transform of it's far-infrared or Raleigh-wing spectrum.[3] Experiments that are performed on such short timescales are capable of directly capturing the solvent effect on or response to chemical processes.[4] Photon echo methods are useful for studying optical dephasing of molecules in solution.[5-9]

We describe the results of experimental measurements and numerical calculation of the 2-pulse photon echo of the cyanine dye HITCI in acetonitrile. Analogous measurements have been performed in DMSO and alcohol solutions.[9] The experiments involve the simultaneous measurement of the echo signal in the $k_3=2k_1-k_2$ and $k_3'=2k_2-k_1$ phase matched directions for accurate determination of the temporal shift, an important fitting constraint. The short pulse source is a home-built[10] Kerr lens modelocked Ti:Sapphire laser [11] producing transform limited 18-22 fs duration gaussian pulses. The sample is a flowing jet (150μm thick) of the chromophore in solution. Solvent spectral densities,[12] obtained from impulsive optical Kerr effect (OKE) measurements, are used to generate the lineshape function used in the nonlinear response function formalism to simulate the echo signal.

Any FWM signal can be expressed as a convolution of the correct molecular response function, $R(t,\tau)$ with a sequence of excitation fields. $R(t,\tau)$ contains the temporal response of the system for pulses separated in time by τ and the optical lineshape function $g(t)$. We assume a well-separated pulse limit in the calculations. The real part of a lineshape function that allows incorporation of solvent dynamics through a spectral density function, $C(\omega)$, is given by [2]

$$g(t)=\int_0^\infty\left(\frac{1-\cos(\omega t)}{\omega^2}\right)C(\omega)\,d\omega \tag{1}$$

We assume the spectral density to be of the (ohmic) form

$$C(\omega)=N\omega^\alpha\exp\left(-\frac{\omega}{\omega_0}\right) \tag{2}$$

Figure 1 shows the spectral density of acetonitrile obtained as the imaginary part of the deconvoluted Fourier transform of our measured impulsive OKE response. The line through the data is the result of fitting to eqn. (2). The solvent spectral density is assumed to describe the spectrum of solvent fluctuations that contribute to the 2-pulse photon echo signals. The

Springer Series in Chemical Physics, Vol. 60 **Ultrafast Phenomena IX**
Editors: P. F. Barbara · W. H. Knox · G. A. Mourou · A. H. Zewail

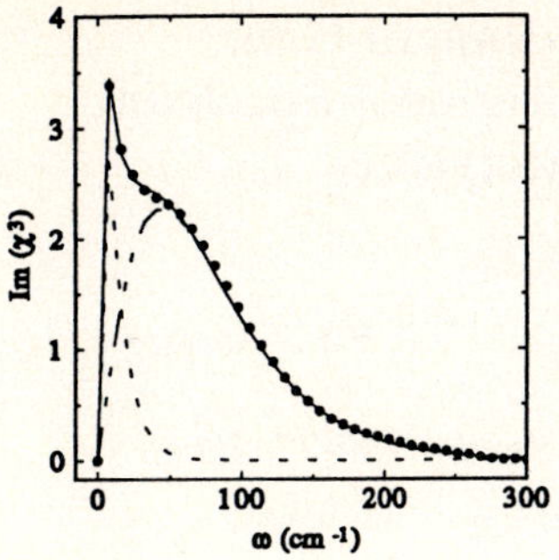

Figure 1

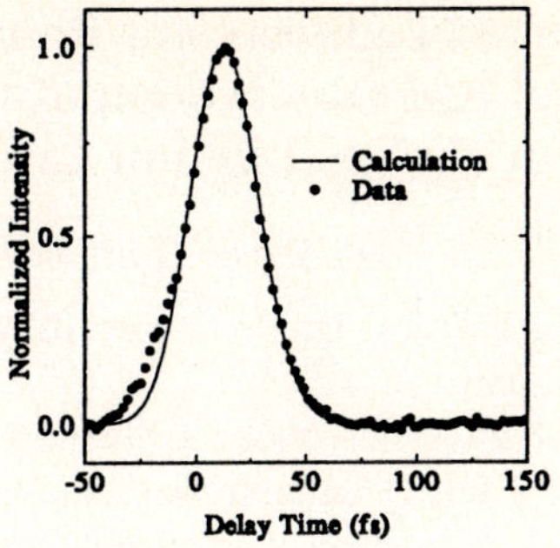

Figure 2

calculated echo signal, shown in Figure 2, is in very good agreement with the experimental signal. A thermal grating contribution to the signal causes a discrepancy at early times. The observed 2-pulse photon echo signal of this symmetric cationic cyanine dye is similar to the two-pulse echo signals observed in oxazine dyes.[11] The echo shape, the timescale for and nonexponential decay of the echo signal is in agreement with recent calculations [13] and simulations.[14]

The principal result of this study is that the 2-pulse PE signal can be "predicted" from the neat solvent spectral density. This conclusion implies that the solvent spectral density obtained from impulsive OKE measurements accounts for the solvent fluctuations that determine the evolution of optical coherence of HITCI in acetonitrile. Apparently, no frequency-dependent couplings are required. Solvent-dependent studies,[9] show that int the cas of HITCI the solute-solvent coupling scales with the solvent polarizability.

References

1. Y.J. Yan and S. Mukamel, Phys. Rev. A, **41**, 6485 (1990); J. Chem. Phys., **94**, 179 (1991).
2. W.B. Bosma, Y. Yan and S. Mukamel, Phys. Rev. A, 42, 6920 (1990).
3. P.A. Lund, O.F. Nielsen and E. Praestgaard, Chem. Phys., **28**, 167 (1978).
4. M.Cho, S.J.Rosenthal, N.F.Scherer, L.D.Ziegler and G.R.Fleming, J. Chem. Phys., **96**, 5033 (1992).
5. A.M. Weiner, S. Fe Silvestri and E.P. Ippen, J. Opt. Soc. Am **B2**, 654 (1985).
6. J.-Y. Bigot, M.T. Portella, R.W. Schoenlein, C.J. Bardeen, A. Migus and C.V. Shank, Phys. Rev. Lett. 66, 1138 (1991).
7. E.J. Nibbering, D.A. Wiersma and K. Duppen, Phys. Rev. Lett. 66, 2464 (1991).
8. T. Joo and A.C. Albrecht, Chem. Phys., 176, 233 (1993).
9. P. Vöhringer, D.C. Arnett, R.A. Westervelt, M.J. Feldstein and N. F. Scherer, J. Chem. Phys., submitted (1994).
10. P.Vöhringer, R.Westervelt, D.Arnett, M.Feldstein and N. F. Scherer, J. Phys. Chem., submitted 1994.
11. M.T.Asaki, C.P.Huang, D.Garvey, J.Zhou, H.C.Kapteyn and M.M.Murnane, Opt. Lett., **18**, 977 (1993); F.Krausz, Ch.Spielmann, T.Brabec, E.Wintner and A.J.Schmidt, Opt. Lett., **17**, 204 (1992).
12. D. McMorrow, W.T. Lotshaw, J. Phys. Chem., **95**, 10395 (1991).
13. M.˙Cho and G.R. Fleming, J. Chem. Phys., **98**, 2848 (1993).
14. N.E. Shemetulskis and R.F. Loring, J. Chem. Phys., **89**, 1217 (1992).

Development of Femtosecond Resonance-Raman-Gain Spectroscopy and Observation of the Vibrational Spectrum of Excitons with 1 ps Lifetime in Polydiacetylene

T. Kobayashi[1], *M. Yoshizawa*[2], *and Y. Hattori*[3]

[1]Department of Physics, University of Tokyo, Hongo 7-3-1, Bunkyo-en, Tokyo 113, Japan

[2]Department of Physics, Tohoku University, Aoba, Sendai 980-77, Japan

[3]New Chemistry R&D Laboratories, Sumitomo Electric Industries, Shimaya 1-1-3, Konohana, Osaka 554, Japan

Abstract. Geometrical relaxation of main chain configuration in polydiacetylene has been investigated. The Raman signal due to self-trapped exciton has been observed for the first time.

Conjugated polymers have localized excited states such as solitons, polarons and the self-trapped excitons (STE) after geometrical relaxation. We have investigated STE in polydiacetylene (PDA) using femtosecond spectroscopy [1–4] and resolved the formation of STE from free-exciton (FE) with a time constant ranging 50 to 200 fs. Time-resolved Raman spectroscopy has been appreciated to be a powerful method for studying structures of transient species and excited states. However, only a few time-resolved Raman experiments have been performed in conjugated polymers because of the difficulty due to very short lifetime of the excited states.

The femtosecond Raman gain spectroscopy was performed using three femtosecond pulses. The 1.97-eV pulse with 100-fs duration was generated by a colliding-pulse mode-locked dye laser and amplified by a four-stage dye amplifier [1]. The amplified pulse was split into three beams. The first beam (pump-1) generated excited states in PDA. A part of white continuum generated from the second one was amplified by a two-stage dye amplifier and used for the pump pulse of the Raman gain spectroscopy (pump-2, 1.78 eV). The probe pulse was white continuum generated from the last beam. The polarization of the three beams was parallel to oriented polymer chains of PDA-3BCMU deposited on a KCl crystal [4].

Raman gain spectra at several delay times after the 1.97-eV photoexcitation are shown in Fig. 1. At −0.5 ps, Raman signals due to the stretching modes of C=C and C≡C bonds in acetylene-like structure $(=CR-C\equiv C-CR=)_x$ are observed at 1440 and 2060 cm^{-1}, respectively. At 0.0 ps, broad Raman signal is observed below 1440 cm^{-1} down to 1000 cm^{-1}. At delay time longer than 0.2 ps the Raman gain signal has a clear peak at 1200 cm^{-1}.

Figure 2 shows transient Raman gain changes at 1200 and 1440 cm^{-1}. The negative change at 1440 cm^{-1} is explained by the depletion of the ground state. The time dependence is consistent with the decay kinetics of the STE. The solid curve is the best fitted curve using time constants of 150 fs and 1.5 ps. The time dependence of the Raman signal at 1200 cm^{-1} has two components. The long-life component decays within several picoseconds and is assigned to the STE. The short-life component has time constant shorter than the resolution time of 300 fs and is probably due to the nonthermal STE, because the 1.78-eV pulse can be resonant with the transition between the nonthermal STE and the ground state.

The theoretical calculation has predicted that Raman signals due to the localized excitations in *trans*-$(CH)_x$ are the reduction of the stretching vibration modes and new Raman lines at lower frequencies than the stretching modes [5]. The Raman signals observed in PDA are similar to

Springer Series in Chemical Physics, Vol. 60 **Ultrafast Phenomena IX**
Editors: P. F. Barbara · W. H. Knox · G. A. Mourou · A. H. Zewail

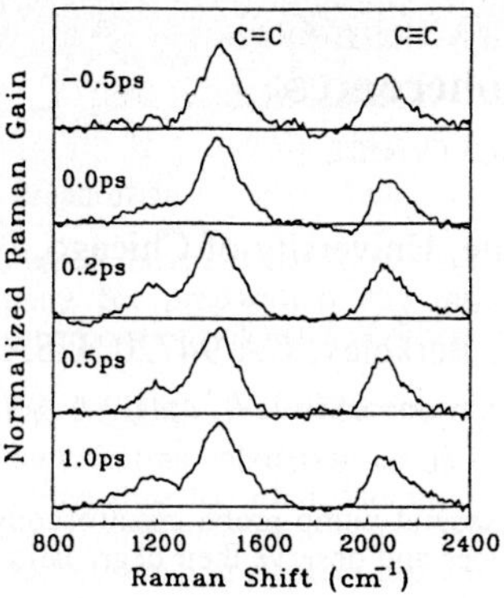

Fig. 1. Normalized resonance Raman gain spectra at several delay times after the 1.97–eV photoexcitation.

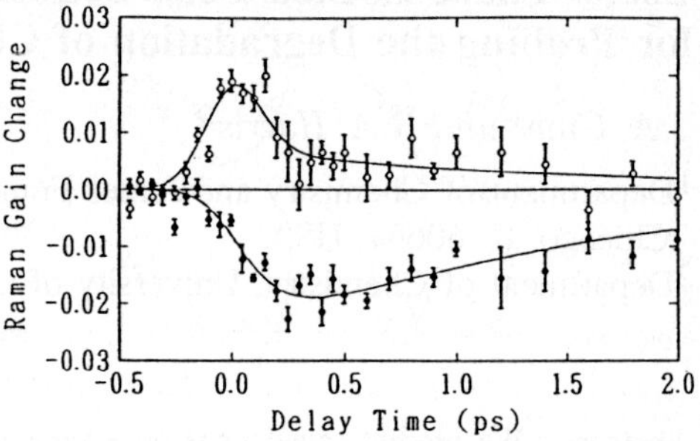

Fig. 2. Transient Raman gain changes at 1200 cm^{-1} (open circles) and 1440 cm^{-1} (closed circles) after the 1.97–eV photoexcitation. The solid curves are the best fitted curves with time constants of 150 fs and 1.5 ps.

this prediction. However, phonon modes of the STE in PDA have not been calculated. Here, the observed Raman frequency is compared with stretching modes in unsaturated hydrocarbons [6]. The C=C bond in *trans*-butene-2 (CH_3–CH=CH–CH_3) has a stretching mode with 1675 cm^{-1}, while the frequency of the C–C bond in *trans*–1,3-butadiene (CH_2=CH–CH=CH_2) is 1202 cm^{-1}. Therefore, the 1200 cm^{-1} Raman peak can be assigned to the C–C bond in the butatriene-like structure $(-CR=C=C=CR-)_x$. However, Raman signal due to the C=C bond in the butatriene-like structure cannot be observed. It can be explained by close frequencies of the stretching modes of the center C=C bond in butatriene (CH_2=C=C=CH_2) and the C≡C bond in dimetylacetylene (CH_3–C≡C–CH_3), 2079 and 2235 cm^{-1}, respectively.

In conclusion, we have developed a Raman gain spectroscopy with femtosecond resolution. The advantages of the method are as follows. (1) No disturbance by fluorescence and (2) No interference of signal with nonresonant background. This is extremely advantageous to the time-resolved CARS spectroscopy. The new Raman peak observed for the first time at 1200 cm^{-1} indicates the butatriene-like structure due to the formation of the STE after the geometrical relaxation from the acetylene-like structure in the FE state.

References

[1] M. Yoshizawa, M. Taiji, and T. Kobayashi, IEEE J. Quantum Electron. QE-25, 2532 (1989).
[2] T. Kobayashi, M. Yoshizawa, M. Hasegawa, and M. Taiji, J. Opt. Soc. Am. B7, 1558 (1990).
[3] M. Yoshizawa, K. Nishiyama, M. Fujihira, and T. Kobayashi, Chem. Phys. Lett. 270, 461 (1993).
[4] M. Yoshizawa, Y. Hattori, and T. Kobayashi, Phys. Rev. B47, 3882 (1993).
[5] A. Terai, Y. Ono, and Y. Wada, J. Phys. Soc. Jpn. 58, 3798 (1989).
[6] L. M. Sverdlov, M. A. Kovner, and E. P. Krainov, *Vibrational Spectra of Polyatomic Molecules*, (Wiley, New York, 1970), pp.282–323.

Deep-Ultraviolet Femtosecond Multiphoton-Ionization Studies of the Excited-State Dynamics of Small Gas-Phase Molecules

J.C. Owrutsky and A.P. Baronavski

Chemistry Division, Naval Research Laboratory, Code 6111,
Washington, DC 20375, USA

1. Introduction

Femtosecond studies using LIF and MPI [1] have lead to considerable progress in understanding small molecule gas phase photodissociation dynamics. In this letter we describe experiments that extend these studies to shorter wavelengths. Femtosecond deep UV (~200 nm) pulses have been used with (1+1) multiphoton ionization to directly measure the excited state lifetime for predissociative states of CS_2 (the S_3 state [2]) and of CH_3I (the $\tilde{B}$ state [3]). The isotopic (CD_3I) and vibronic dependences were measured in the latter case, confirming the multidimensional nature of the predissociation dynamics.

2. Experimental

A femtosecond laser system produces pump and probes pulses which are used for multiphoton ionization studies performed in a static gas cell containing parallel plates. Femtosecond pulses (80 fs, λ=618 nm) from a CPM laser are amplified to 150 μJ in a two stage amplifier pumped by a XeF excimer laser (351 nm, 300 Hz). The CS_2 studies were conducted using CPM frequency tripled (to ~206 nm) pump pulses and variable wavelength probe pulses. The latter were obtained by amplifying spectrally selected portions of continuum. In the CH_3I experiments, tunable deep UV excitation pulses (λ> 199 nm) were obtained by sum frequency mixing of the tunable amplified continuum and frequency doubled CPM pulses. The wavelength selection is controlled by the choice of amplifying dye and the BBO sum frequency phase-matching angle. The deep UV pulse duration was measured to be close to 300 fs by observing MPI signals for several molecules exhibiting fast photodissociation (e.g., aniline and acetone).

The pump and probe pulses are focused into a gas cell with fused silica windows containing a static pressure of the sample (0.4-0.6 torr of CS_2; 0.2-0.6 torr CH_3I). The photocurrent is collected on parallel plates that are separated by about 1 cm and biased to 200-400 V. The probe beam is synchronously chopped and the photocurrent is sent to a gated integrator and a lock-in amplifier.

3. Lifetime of the S_3 State of CS_2

The $S_3 \leftarrow S_0$ absorption band of CS_2 (220-190 nm) consists of ~16 sequences separated by ~410 cm^{-1}, which are due to excitation of ν_1 and ν_2 in the upper state and ν_2 hot bands. The dynamics of this predissociative band has been studied by various indirect methods. Hara and Phillips [4a] found that the fluorescence quantum yield ($\Phi = 10^{-4}$-10^{-3}) increases with wavelength (200-216 nm) indicating that lifetimes are shorter at

Springer Series in Chemical Physics, Vol. 60 **Ultrafast Phenomena IX**
Editors: P. F. Barbara · W. H. Knox · G. A. Mourou · A. H. Zewail

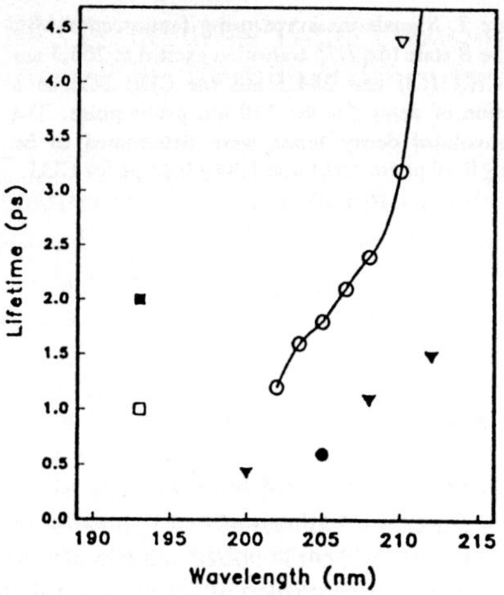

Figure 1. A plot of various lifetime measurements for the S_3 state of CS_2 as a function of excitation wavelength (190-215 nm): ● This work[2]; ○ Hara and Phillips[4a]; ▾ Li and Myers[5]; □ Yang *et al.*[4b]; ■ Waller and Hepburn[4c]; ▿ Liou *et al.*[4d]

higher excitation energies, i.e., with more vibrational excitation. They estimated excited state lifetimes based on the yields and the radiative lifetime. In their photodissociation product state anisotropy distribution studies at 193 nm, Yang *et al.* [4b] found an isotropic distribution of CS fragments in their room temperature study and estimated the lifetime to be 1 ps and Waller and Hepburn [4c] observed an anisotropic distribution of sulfur atoms in their low temperature study and estimated the lifetime to be 2 ps. The lifetimes measured by Li and Myers [5] using Raman depolarization ratios in several spectral regions of the S_3 state display a wavelength dependence consistent with that from the fluorescence quantum yield measurements. Liou *et al.* [4d] reported the lifetime at 210 nm obtained by simulating a rotationally unresolved jet-cooled LIF spectrum. We have directly measured the lifetime using time-resolved MPI for CS_2 with the pump pulses at 205.2 nm. The signal we observe can be well described as an instrument-limited rise followed by a single exponential decay with a lifetime of 600 ± 100 fs [2]. The same decay time was measured over the pump pulse tuning range (204.8-206.0 nm.) The plot in Figure 1 summarizes lifetimes measured in different spectral regions (193-215 nm) [2, 4, 5] of the $S_3 \leftarrow S_0$ spectrum of CS_2. The lifetime we measure is consistent with those reported for CS_2 by Li and Myers [5] using resonance Raman depolarization ratios for the neighboring spectral regions.[4a].

4. Vibronically Resolved Lifetimes of the $\tilde{B}$ State of CH_3I and CD_3I

Tunable deep UV ($\lambda > 199$ nm) femtosecond (1+1) MPI studies have also been carried out to directly measure the $\tilde{B}$ state lifetimes for the vibrationless [3a] and other vibronic levels [3b] of the $\tilde{B}$ state of gas phase CH_3I and CD_3I. Previous work on the dynamics of this state has been reported using indirect methods. Lifetimes on the order of a picosecond were reported by Wang and Ziegler [6] who used Resonance Raman depolarization ratios and by Syage [7] who found lower lifetime limits by measuring free jet spectral linewidths. Both approaches revealed predissociation rates that varied as a function of isotopomer and vibronic level, although some discrepancies exist. Using direct excitation and probing (as shown in Figure 2), we find that for the vibrationless state of CH_3I (CD_3I) the lifetime is

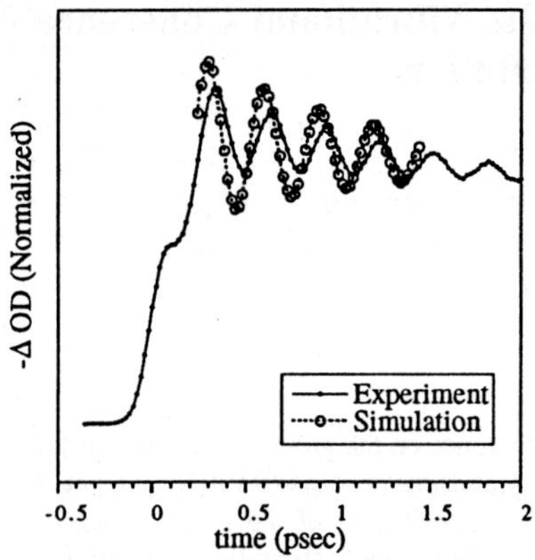

Figure 1: Transient absorption at 308 nm for triiodide in ethanol. The experimental and calculated signals for an initial thermal density are compared. The frequency, phase, and decay time are in good agreement.

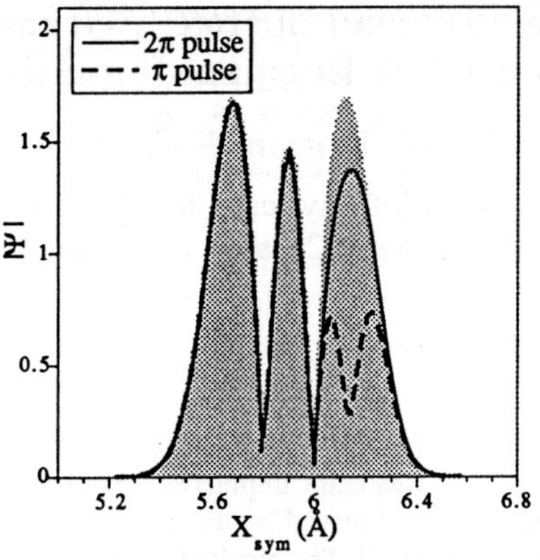

Figure 2: Predicted intensity cycling effects in the ground state of triiodide. The final wavefunctions, following excitation with π and 2π pulses are compared with the initial state. Following the excitation, dynamical evolution is observed in both cases.

agreement. An additional experimental observation is the amplitude to bleach ratio, which differs between the calculation and experiment.

The coherence measure calculated for this case, grows during the pump pulse interaction. The dynamics of the hole are visualized in phase space where a full cycle is completed in one vibrational period. Within several vibrational periods, vibrational dephasing diminishes the amplitude of both the positive and negative portions of the hole. This is accompanied by an exponential decay of the calculated coherence measure, with a 960 fsec time constant which agrees well with the decay time of the signal modulations demonstrating the quantitative significance of this parameter.

The numerical tools developed can be used for a comprehensive study of the effects of various pulse parameters and solvent induced processes on the signal. The intensity may be varied far beyond the weak pulse regime, into the domain of intense pump pulses where we predict substantial effects on the dynamics.

References

[1] Y. X. Yan, L. T. Cheng, and K. A. Nelson, in *Advances in Nonlinear Spectroscopy*, Eds. R. G. H. Clarke, R. E. Hester, Wiley, New York (1987); S. mukamel, *Annu. Rev. Phys. Chem.* **41**, 647 (1990); W. T. pollard and R. A. mathies, *Annu. Rev. Phys. Chem.* **43**, 497 (1992).

[2] For numerous examples: See *Ultrafast Phenomena VIII*, Eds. J.-L. Martin, A. Migus, G. A. Mourou, and A. H. Zewail, Springer Verlag (1993).

[3] U. Banin and S. Ruhman, *J. Chem. Phys.* **96**, 4391 (1993).

[4] U. Banin, A. Bartana, S. Ruhman, and R. Kosloff, *J. Chem. Phys.*.

[5] A. Bartana, R. Kosloff and D. Tannor, *J. Chem. Phys.* **99**, 196 (1993).

Femtosecond Spectroscopy of the Gain Band of a Styryl Dye: The Band Shape in Terms of Solvation Dynamics

B. Bingemann[1], *K. Kemeter*[2], *and N.P. Ernsting*[1]

[1]Institut für Physikalische und Theoretische Chemie, Humboldt-Universität zu Berlin, D-10117 Berlin, Germany
[2]Max-Planck-Institut für biolphysikalische Chemie, Postfach 2841, D-37018 Göttingen, Germany

Abstract. Transient absorption spectra of DASPI in methanol were recorded following 60 fs optical excitation. The evolution of the gain band is described in terms of its spectral moments. Oscillatory behaviour of all moments was observed at 50 °C. The spectrum of solvation modes is calculated from molecular dynamics simulations.

Optical excitation of styryl dyes (like DASPI, see Fig. 1) will be followed by solvent reorganization [1], intramolecular vibrational energy redistribution, vibrational energy transfer, as well as molecular isomerization. We measured the transient absorption and gain spectrum of DASPI in methanol where solvation is the dominant cause for the spectral shift. In analyzing the spectra we assume that ground-state bleaching and excited-state absorption have constant band shapes and that the gain band may be represented by a log-normal distribution. In this way the gain band is obtained as a function of time. Its band center can be described by an appropriate model for solvation dynamics. This model is also used to predict the evolution of the 2nd and 3rd central moments.

The dye solution is excited by a 60 fs pulse at 470 nm (20 μJ at 10 Hz). The probe consists of a supercontinuum generated at 470 nm in 2 mm of water. The pump-probe cross-correlation has a width of 100 fs. The measured spectra are distorted by residual group-velocity dispersion (GVD). A global fit to the measured data convolutes a spectro-temporal apparatus function, which was measured independently, with the spectral model mentioned above. In this way the effect of GVD is also corrected for.

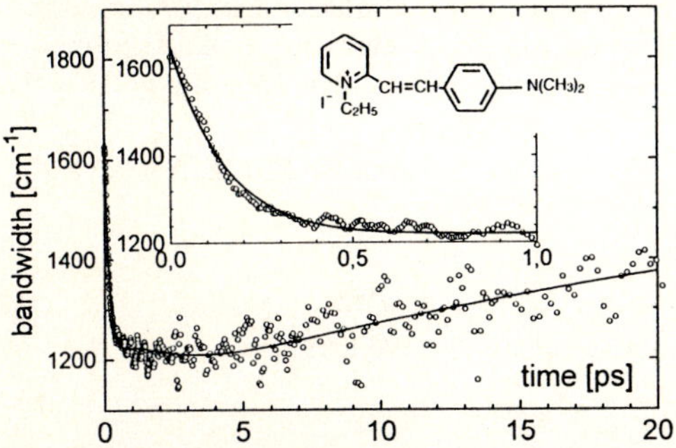

Figure 1: Evolution of the width of the DASPI gain band in methanol at -9 °C

Springer Series in Chemical Physics, Vol. 60 **Ultrafast Phenomena IX**
Editors: P. F. Barbara · W. H. Knox · G. A. Mourou · A. H. Zewail

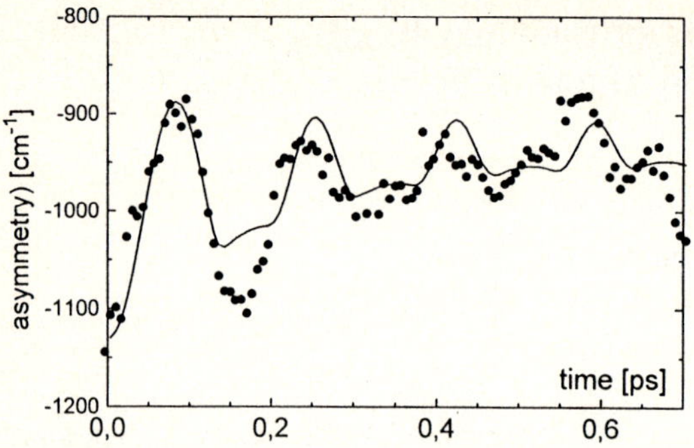

Figure 2: Evolution of the asymmetry of the gain band in methanol at 50 °C

A triexponential fit to the first moment appears adequate for methanol solutions below 40 °C, thus assuming a Debye continuum model of solvation dynamics. An extension of this model to the higher moments gives good agreement with the measured results. This is shown in fig. 1 for the bandwidth. Measurements in the temperature range 40 - 50 °C revealed oscillations in all three moments. This is shown in fig. 2 for the asymmetry. For a rough description, one Brownian oscillator [2] at 196 cm^{-1} has been added to the continuum model.

Non-equilibrium MD simulations of a model solute show triexponential behaviour of the spectral relaxation, the second time of which compares quantitatively with the experimental value. Higher moments can be reproduced qualitatively. We also simulate damped oscillations due to H-librations of methanol which give insight into further microscopic relaxation pathways.

References

[1] S.J. Rosenthal, X. Xie, M. Du, G.R. Fleming: J. Chem. Phys. **95**, 4715 (1991)

[2] M. Cho, S.J. Rosenthal, N.F. Scherer, L.D. Ziegler, G.R. Fleming: J. Chem. Phys. **96**, 5033 (1992)

Vibrational Relaxation of I_2^- in Polar and Non-Polar Solvents

P.K. Walhout, J.C. Alfano, K.A.M. Thakur, and P.F. Barbara

Department of Chemistry, University of Minnesota, Minneapolis, MN 55455, USA

Abstract New spectroscopic data of I_2^- vibrational relaxation following geminate recombination in polar and non-polar solvents indicates a breakdown of simple binary collision theory.

In this paper we present new results on the photodissociation, geminate recombination, and vibrational relaxation of I_2^- in various solvents. The photodissociation of I_2^- involves localization of the excess charge on a recoiling I· photofragment. This charge again delocalizes as the I· and I^- geminately recombine. The "charge flow" associated with recombination and subsequent large-amplitude vibrational motion leads to extremely fast, non-exponential vibrational relaxation of the nascent I_2^-. These results are in stark contrast to similar experiments performed on neutral I_2 which show nearly exponential vibrational relaxation rates that are an order of magnitude slower and are well-described by isolated binary collision (IBC) theory. Specifically, our results on I_2^- show:

- the effects of charge and charge flow lead to a fast, non-linear form of vibrational relaxation in both polar and non-polar solvents;
- support for the proposed claim [1] of trapped coherent motion in the excited state immediately following recombination;
- dephasing of ground state I_2^- vibrational coherence is complete in 300 fs.

Figure 1 shows pump-probe transient absorption spectra of I_2^- in dioxane and acetonitrile. The Ti:sapphire-based laser system used in these experiments has been described elsewhere [2]. I_2^- is synthesized by photodissociation of I_3^- with a 390 nm pulse. After a 4 ns delay during which the I_2^- is allowed to equilibrate, a variably delayed 780 nm pump pulse excites the I_2^- to a dissociative excited state. The probe pulse, either exciting the visible absorption band peak at 720-740 nm or the near-UV peak at 390 nm, interrogates the I_2^- as it recombines and relaxes.

The upper panel in Fig. 1 shows that the vibrational relaxation of I_2^- in dioxane is rapid despite the very non-polar environment. This data is complicated, however, by transient excited-state absorption of 720 nm photons [1]. A more accurate determination of the vibrational relaxation rate is obtained by probing the UV band. Competition for absorption of the 390 nm probe pulse is unlikely, hence any bleach recovery is due to ground-state vibrational relaxation. Bleach recovery occurs on a ~6 ps timescale in dioxane. For comparison, similar studies of neutral I_2 in non-polar solvents have yielded recovery times of 50 to several hundred picoseconds [3].

The fit to the two 390 nm probe transients of fig. 1 employ a modification to the linear spectral and kinetic model described in Kliner et al. which yielded an overall exponential decay of excess vibrational energy, consistent with IBC theory. A linear model fails for I_2^-, so we assumed the rate constant for vibrational relaxation, $k_{v \to v-1}$, increases with vibrational level v in an exponential rather than linear fashion. Hence $k_{v\to v-1}/k_{1\to 0} = v \cdot e^{v \cdot \alpha}$, where α is varied to provide the best

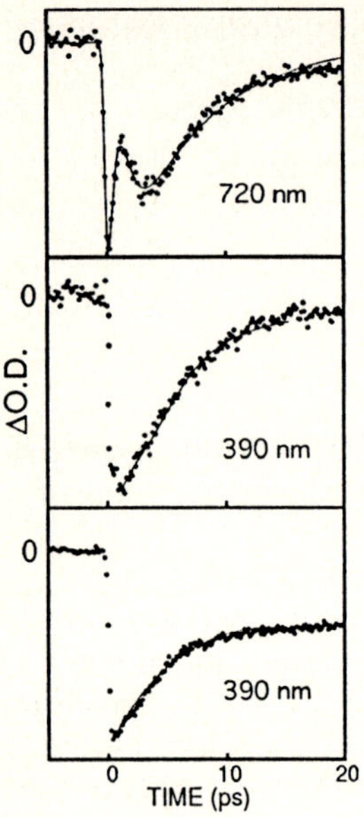

Figure 1. 780 nm magic angle pump/variable wavelength probe transient absorption spectra of I_2^- in dioxane (top two panels) and acetonitrile (bottom panel). Typical ΔOD values are 0.005. The top figure is fit with a 4-exponential function.

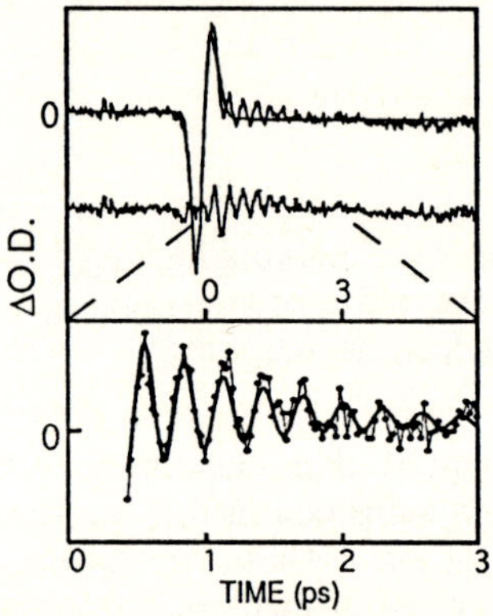

Figure 2. 780 nm pump/390 probe ΔO.D. signal of I_3^- in acetonitrile. The top panel shows the residuals to a multi-exponential fit. Oscillations are expanded in the bottom panel and fit to a damped sine wave of frequency 110 cm^{-1}.

fit. Dioxane fits best to α=0.15, which indicates a highly non-linear mechanism of excess energy loss, especially at the top of the ground-state well.

The 720 nm dioxane data of fig. 1 shows a "bump" similar to that seen in studies of I_2^- photodissociation and recombination in CO_2 clusters. The absence of a bump in the 390 nm probe dioxane data supports the analysis in [1] which reasoned that some trapped fraction of the recombined molecules coherently move along the lowest excited state to a point where they can absorb a 720 nm probe photon.

Figure 2 shows oscillations which are believed to be I_3^- vibrations excited by non-resonant impulsively-stimulated Raman scattering [5]. Our observation in these two-pulse experiments of I_3^- oscillations suggests our time resolution is fast enough to see 115 cm^{-1} I_2^- oscillations should they occur. No oscillations are observed in the three-pulse, 390 nm probe experiments done on I_2^-, which provides some evidence that following geminate recombination, hot I_2^- in the ground state has lost all vibrational coherence within the 300 fs resolution of our instrument response.

This work was supported by the NSF.

References

[1] J.M. Papanikolas, V. Vorsa, M.E. Nadal, P.J. Campagnola, H.K. Buchenau, W.C. Lineberger, *J.Chem.Phys.* **99**, 8733 (1993).

[2] D.A.V.Kliner, J.C.Alfano, P.F. Barbara, *J.Chem.Phys.* **98**, 5375 (1993); J.C. Alfano, Y.Kimura, P.K.Walhout, P.F. Barbara, *Chem.Phys.* **175**, 147 (1993).

[3] A.L. Harris, J.K. Brown, C.B. Harris, *Annu.Rev.Phys. Chem.* **39**, 341 (1988).

[4] U. Banin, A. Waldman, S. Ruhman, *J.Chem.Phys.* **96**, 2416 (1992).

Ultrafast Dynamics as Reflected in MPI-Photoelectron Spectra: Aniline, Azulene, Phenanthrene, and Carbon Disulfide

N. Thantu, B. Kim, and P.M. Weber

Department of Chemistry, Brown University, Providence, RI 02912, USA

Abstract: Two photon ionization photoelectron spectra of aniline, azulene, phenanthrene and carbon disulfide reveal characteristic signatures of the ultrafast dynamics occurring in the intermediate resonance employed in the ionization process.

Very short lived resonances can cause unobservably small yields of resonant multiphoton ionization (MPI), and severely alter the photofragmentation pattern of the resulting ions.[1] These effects, which are often observed when using high lying electronic states of medium size or large molecules, are usually interpreted as a consequence of fast intramolecular or photodissociation processes. In the present paper we investigate such dynamical phenomena by two-photon ionization photoelectron spectroscopy. To illustrate the range of possible effects we discuss the photoionization of aniline ($C_6H_5NH_2$), phenanthrene ($C_{14}H_{10}$), azulene ($C_{10}H_8$), and carbon disulfide (CS_2).

In the described experiments the molecules are seeded in helium carrier gas, expanded in a molecular beam, and ionized by two UV laser photons originating in a mode-locked, cavity dumped dye laser system. The photoelectron spectra are recorded using a time-of-flight scheme.[2]

Aniline, phenanthrene, and azulene are ionized via the vibrationless levels of their S_1, S_2 and S_3 states, respectively, by two photons of a single laser pulse. For aniline the bandwidth of the 7 ps laser pulse is sufficiently large to span the entire width of the resonant absorption feature, about 10^{-3} cm^{-1}. The laser pulse therefore prepares a molecular state that closely resembles the S_1 zero-order Born-Oppenheimer state. As expected, the photoelectron spectrum (figure 1.a) shows a vibrational structure that reflects ionization out of the pure S_1 electronic state. For phenanthrene the molecular absorption width (13 cm^{-1}) exceeds the bandwidth of the transform limited laser pulse, about 2.5 cm^{-1}. The laser therefore does not coherently couple all the eigenstates that make up the zero order S_2 state. The intermediate resonant state that is prepared by the laser is thus an unspecified molecular state composed of both S_2 and the lower S_1 electronic state. The photoelectron spectrum of figure 1.b reflects the relative contents of these two electronic states, weighted by their respective ionization cross sections: ionization out of S_2 leads to vibrationally resolved peaks near 8 eV, while ionization out of S_1 yields the broad peak at 8.65 eV. The S_3 state of azulene has a spectral absorption width of 140 cm^{-1}, indicating a lifetime of about 35 fs. This width greatly exceeds the bandwidth of the picosecond dye laser. Following the argument advanced above the laser should now prepare a state that contains very little character of the bright S3 electronic state, and a rather large fraction of the lower electronic state that couples to the S_3 state. The photoelectron spectrum of figure 1.c is consistent with this picture: the signal at 7.42 eV corresponding to ionization out of S_3 is

Springer Series in Chemical Physics, Vol. 60 **Ultrafast Phenomena IX**
Editors: P. F. Barbara · W. H. Knox · G. A. Mourou · A. H. Zewail

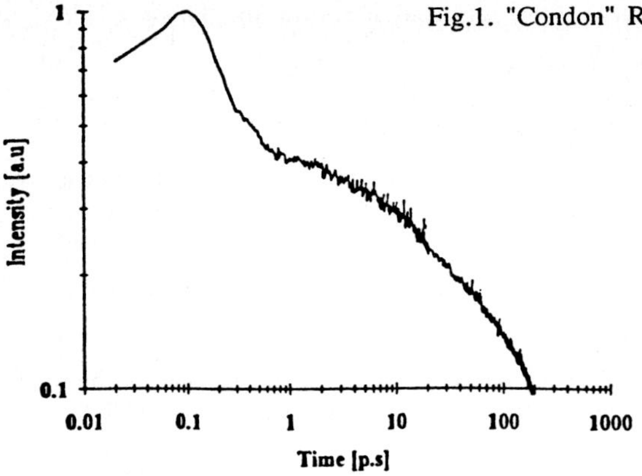

Fig.1. "Condon" RTPGS. Theoretical curves.

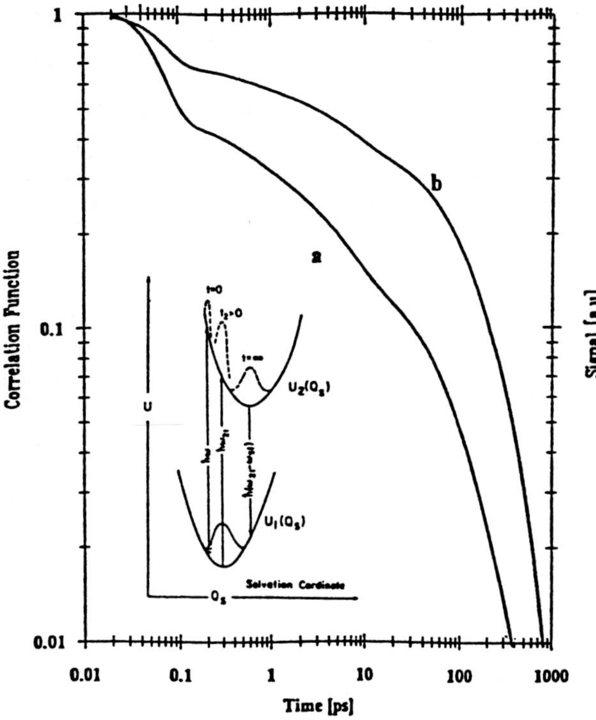

Fig.2. RTPGS of LDS 750 in 1,3-propanediol.

different electronic molecular states by vibrations) by transient polarization four-photon spectroscopy. The method is based on the following. The signal behavior in transient resonance four-photon spectroscopy does not depend on the polarization of the excitation pulses if one assumes that the electronic waves functions do not depend on the nuclear coordinates ("Condon" approximation). In contrast, when electronic wave functions and, consequently, the dipole moment of the electronic transition depend on the nuclear coordinates (due to the HT interactions), then in general, the signal depends on the polarization. Thus, we propose to conduct transient four-photon resonance experiments by two methods: (a) RTPGS with parallel polarizations of all pulses, and (b) induced resonance optical Kerr effect (OKE). The comparison of both signals provides pertinent information on the contribution of HT interactions. We have obtained expressions for the "non-Condon" signal and conducted OKE measurements of LDS 750 in monools and diols.

3.A new model for the description of solvation dynamics: the non-Markovian oscillator (NMO). The theory of RTPGS and the transient polarization spectroscopy of solvation dynamics connects a four-photon signal with the solvation correlation function S(t). Its calculation is an independent problem. We propose a new model (NMO) for the description of solvation dynamics. We used the NMO model for the description of a number of experimental and computer simulations of solvation correlation functions. Using a model comprised of two NMOs, we have described the experimental correlation function of Reference [1b] and computer simulations data given in references [3,4]. The NMO model allows to diminish essentially the number of fitting parameters in comparison with a model consisting of a multimode Brownian oscillator [3].

References

1. a) P.F. Barbara, and W. Jarzeba. Adv. Photochem., **15**, 1 (1990); b) S.J. Rosensthal, X. Xie, M. Du, and G.R. Fleming. J. Chem. Phys. **95**, 4715 (1991) .
2. B.D. Fainberg. Opt. Spectrosc. **68**, 305 (1990) ; Chem. Phys. **148**, 33 (1990) ; Phys. Rev. A 48,849 (1993) .
3. L.E. Fried, N. Bernstein, and S. Mukamel. Phys. Rev. Lett. **68**, 1842 (1992) .
4. E. Neria, A. Nitzan. J. Chem. Phys. **96**, 5433 (1992) .

Importance of Femtosecond Solvent Motions on Electron-Transfer Reactions: A Study of Electrolyte Solutions

H.P. Deuel, P. Cong, and J.D. Simon

Department of Chemistry, University of California at San Diego, 9500 Gilman Drive, La Jolla, CA 92093-0341, USA

Abstract Fast solvent motions in electrolyte solutions are investigated by optical Kerr effect spectroscopy. Comparison with electron transfer reaction rates obtained in these solutions suggest the dominance of the ultrafast solvent modes in determining the magnitude of the kinetic prefactors.

Electrolytes can influence chemical reaction dynamics in vastly different ways. In this report we focus on fast solvent motions in electrolyte solutions and their effect on the rate constants of electron transfer reactions.

As a particular example, we consider the back electron transfer that occurs upon the photoexcitation of the bimolecular donor-acceptor complex between stilbene and fumaronitrile. Excitation of the ground state complex creates a contact ion-pair. This intermediate undergoes competition between back electron transfer and diffusive separation to form free ions. Our group has reported a complete study on the effect of electrolytes on this chemical reaction [1]. In general, addition of alkali salts accelerates the rate of electron transfer. On the other hand, molecular salts (tetra-alkyl ammonium perchlorates) have no influence on the dynamics. For this particular reaction, the dynamics occur in the Marcus inverted region. In this region, a decrease in reaction exothermicity causes an increase in the reaction rate constant. Addition of alkali salts to acetonitrile increases the solution polarity. As a result, the ion-pair intermediate is stabilized to a greater extent than the ground state complex in the electrolyte solution. This causes a decrease in reaction exothermicity and an increased rate is expected, and is experimentally observed. However, it is not clear from this observation whether there is a corresponding change in the prefactor, which would reflect changes in solvent dynamic effects on the reaction rate. To address this question, a free energy parameterization of the reaction energetics in salt solutions was developed [2]. This enabled direct determination of the frequency factors associated with the rate constants, which within experimental error were found to be independent of salt concentration. Understanding this observation requires a detailed understanding of how added salt influences the molecular dynamics of the solution.

Recent evidence [3] supports the role of the fast solvent motions (librations, tens of fs) in determining reaction rates. In a study of the self-exchange reaction in a variety of metallocenes, Barbara and coworkers concluded that the fast solvent relaxation components revealed by C(t) measurements make important contributions to the magnitude of the prefactor [3a]. Even though C(t) measurements of electrolyte solutions with femtosecond resolution have not been reported in the literature, picosecond C(t) measurements do reveal some interesting features of the effects of electrolytes on solvation dynamics [4, 5]. In particular, Huppert and coworkers have determined C(t) for salt solutions in both acetonitrile and ethyl acetate [4]. For example, C(t) for a 0.09M $LiClO_4$ acetonitrile solution reveals a long time component (decay constant of 11.75 ns) that is not present in the neat solvent. Similar results were observed for ethyl acetate solutions. For concentrations as low as 10 mM, long time decay components appeared in the C(t) function. In THF solutions, the four salts $NaClO_4$, $LiClO_4$, $Mg(ClO_4)_2$ and Bu_4NClO_4 all show the same relaxation time for the concentration range 10 - 400 mM. These workers interpreted these data in terms of a salt-induced relaxation process that could be described by the Debye-Falkenhagen theory, which has been questioned by Chapman and Maroncelli recently [5]. Due to the limits of the time resolution of the photon counting technique, the dynamics on the time scale of a few picoseconds and faster was not resolved.

Comparing these picosecond C(t) measurements with our electron-transfer rate data, one is led to the conclusion that the slow relaxation of the ionic atmosphere does not contribute to the magnitude of the kinetic prefactor. However, the C(t) measurements do not provide any information of the effects of salts on the fast solvation dynamics. Such information is vital if

Springer Series in Chemical Physics, Vol. 60 **Ultrafast Phenomena IX**
Editors: P. F. Barbara · W. H. Knox · G. A. Mourou · A. H. Zewail

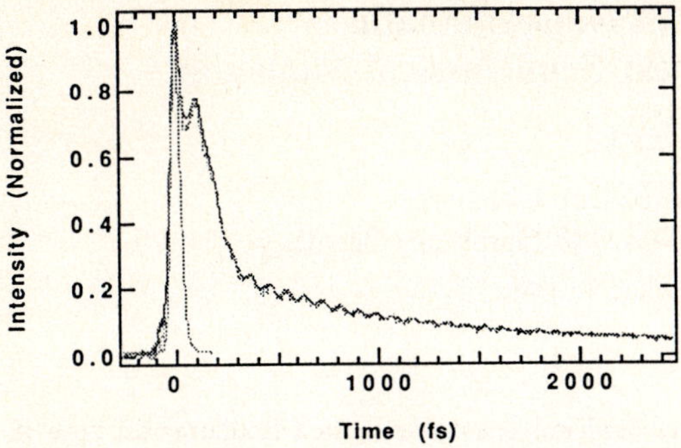

Figure 1. Comparison of the OHD-RIKES data for pure acetonitrile (———) with that of a 0.1 M lithium perchlorate solution (......). The autocorrelation of the laser pulse is also shown.

one is to assess whether or not these modes play a role in determining the kinetic frequency prefactor. OHD-RIKES can provide this information. In particular, the OHD-RIKES data contains information on rotational motions as well as translational and librational dynamics.

With the goal of exploring the electrolyte effects on fast solvent motions, OHD-RIKES transients were measured for acetonitrile and ethyl acetate salt solutions. Figure 1 shows the OHD-RIKES data for a 0.1 M $LiClO_4$ acetonitrile solution. Within our experimental resolution, addition of salt does not effect any of the femtosecond dynamic processes that contribute to the Kerr signal. Electrolyte solutions in ethyl acetate are similarly unaffected. At the concentrations studied, IR and NMR studies have shown that these salts undergo negligible ion-pairing [6].

The data shown in Figure 1 indicate that the fast relaxational processes in acetonitrile are unaffected upon addition of electrolytes. It is clear from the picosecond C(t) measurements that electrolytes strongly influence the picosecond and longer solvent motions [4]. If these modes were to dominate the kinetic prefactor in an electron transfer process, the magnitude of the prefactor should be highly sensitive to electrolyte concentration. However, the dynamical solvent modes responsible for the electron transfer barrier crossing need to be independent of electrolyte concentration. Only the ultrafast solvent modes show this independence. These results suggest that ultrafast solvent modes couple to electron transfer barrier crossing.

This work is supported by the National Science Foundation.

1. P. A. Thompson and J. D. Simon, *J. Am. Chem. Soc.* **115**, 5657 (1993).
2. P. A. Thompson and J. D. Simon, *J. Chem. Phys.* **97**, 4792 (1992).
3. (a) M. J. Weaver, G. E. McManis, W. Jarzeva and P. F. Barbara, *J. Phys. Chem.* **94**, 1715 (1990); (b) G. E. McManis, M. J. Weaver. *J. Chem. Phys.* **90**, 912 (1989).
4. (a) D. Huppert, V. Ittah and E. M. Kosower, *Chem. Phys. Lett.* **159**, 267 (1989); (b) V. Ittah and D. Huppert, *Chem. Phys. Lett.* **173**, 496 (1990).
5. C. F. Chapman and M. Maroncelli, *J. Phys. Chem.* **95**, 9095 (1991).
6. (a) M. K. Wong, W. J. McKinney, A. J. Popov, *J. Phys. Chem.* **75**, 56 (1971); (b) R. H. Erlich, E. Roach, A. I. Popov, *J. Am. Chem. Soc.* **92**, 4989 (1970); (c) M. S. Greenverg, R. L. Bodner, A. I. Popov, *J. Phys. Chem.* **72**, 2449 (1973).

Femtosecond Spectroscopy of Non-Adiabatic Photoisomerization Processes: Semiclassical Modeling

G. Stock, L. Seidner, and W. Domcke

Institut für Physikalische und Theoretische Chemie,
Technische Universität München, D-85748 Garching, Germany

The *cis-trans* isomerization in photo-excited polyenes represents a fundamental type of photoreaction which, despite of considerable experimental and theoretical efforts, is still not very well understood in many aspects [1]. Diffuse absorption bands and the absence of detectable fluorescence have hampered standard spectroscopic investigations, so that even for the smallest systems as ethylene and butadiene the relaxation pathways (i.e. the involved electronic states and nuclear degrees of freedom) and corresponding timescales are not yet clear.

As an illustration, Fig. 1a shows *diabatic* model potential-energy surfaces (PESs) of the ground and first excited electronic states as a function of the torsional coordinate φ (of periodicity π) and a harmonic nontotally symmetric coordinate Q_c. It is seen that the two electronic states interchange upon torsion, i.e. the upper electronic state for $\varphi = 0$ becomes the lower electronic state for $\varphi = \pi/2$. The nontotally symmetric mode Q_c couples the two electronic states. As a result of this coupling, the corresponding *adiabatic* PESs (Fig. 1b) exhibit two conical intersections, representing a simple example of a "photochemical funnel" [1]. Preparing initially a wave packet in the Franck-Condon region of the excited PES (e.g. by by an ultrashort laser pulse), it is generally believed that the system relaxes through these conical intersections into the adiabatic electronic ground state on a femto- or picosecond timescale [1].

Recently Seidner and Domcke introduced a quantum-mechanical approach to model the ultrafast electronic and vibrational relaxation dynamics of photo-induced *cis-trans* isomerization in a truly microscopic manner [2]. On the basis of a simplified model Hamiltonian (including two coupled electronic states, the torsional mode φ, the coupling mode Q_c, and up to two totally symmetric Condon-active modes Q_1 and Q_2), they were able to perform numerically exact wave-packet propagations on a picosecond timescale. In this work we want to set forth a semiclassical approach to the microscopic description of photoisomerization processes, which will allow us to extend the theoretical treatment

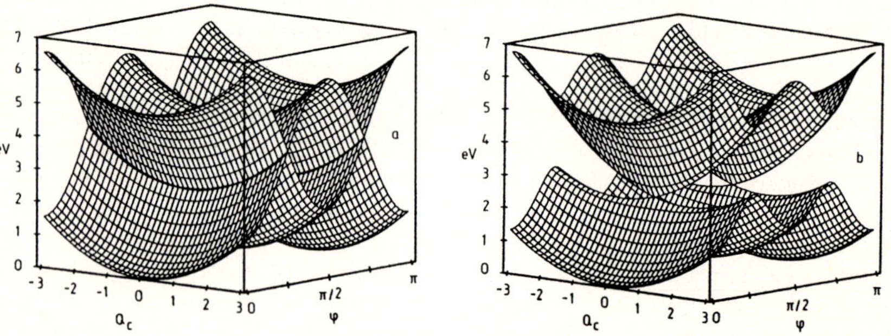

Fig. 1. Diabatic (a) and adiabatic (b) potential-energy surfaces of the model system.

Springer Series in Chemical Physics, Vol. 60 **Ultrafast Phenomena IX**
Editors: P. F. Barbara · W. H. Knox · G. A. Mourou · A. H. Zewail

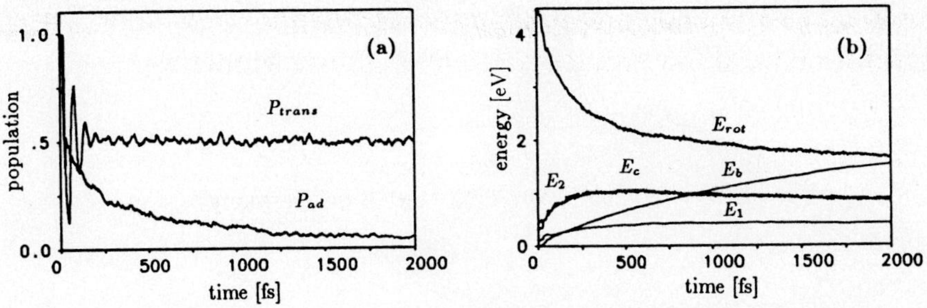

Fig. 2. (a) Time-dependent population propabilities of the *trans* confomer P_{trans} and the adiabatic electronic S_1 state P_{ad}; (b) time-dependent mean values of the vibrational energies E_{rot}, E_c, E_1, E_2.

of Ref. 2 towards larger molecular systems. We use a version of the classical electron analog model due to Meyer and Miller [3], which is a classical path method that treats both electronic and nuclear degrees of freedom on an equal dynamical footing, and has been successfully applied to a variety of model problems (see Refs. in [4]). In particular, the model has been shown to reproduce semiquantitatively the quantum calculations in Ref. 2. As a first demonstration of the capacity of the semiclassical approach, we add 16 weakly coupling harmonic "bath" modes to the four rather strongly coupling "system" modes φ, Q_c, Q_1, Q_2. The bath modes are meant to qualitatively account for additional intermolecular and intramolecular interactions.

Fig. 2 illustrates the electronic and vibrational relaxation dynamics of the resulting twenty-mode model during the first picosecond after impulsive electronic excitation by an ultrashort laser pulse. Panel (a) shows the *adiabatic* electronic population probability $P_{ad}(t)$ (i.e., the probability to be in the upper adiabatic electronic state, c.f. Fig. 1b), and the probability $P_{trans}(t)$ that the system is in the initially excited *trans* conformation. It is seen that the isomerization dynamics is almost completed after $\approx$ 200 fs, leaving $P_{trans}(t)$ fluctuating around its long-time limit of ≈ 0.5, while the adiabatic population $P_{ad}(t)$ continues to decay. To illustrate the corresponding vibrational relaxation dynamics, Fig. 2b shows the time-dependent energy content of the torsional mode (E_{rot}), the coupling mode (E_c), the two Condon-active modes (E_1, E_2), and the sum of the energies of the bath modes (E_b). Fig. 2b elucidates how the vibrational energy of the torsional mode is redistributed into the the other vibrational degrees of freedom. It is seen that the rapid initial decay of the adiabatic population $P_{ad}(t)$ is reflected in the vibrational relaxation of the system modes Q_c, Q_1, Q_2. While the energy in these strongly coupling modes stays approximately constant after $\approx$ 200 fs, the weakly coupling bath degrees of freedom need about two picoseconds to relax, accompanied by the decay of $P_{ad}(t)$ to its long-time limit of ≈ 0.05.

Femtosecond pump-probe spectrocopy of rhodopsin [5] and stilbene [6] has shown that the *cis-trans* isomerization dynamics is reflected in complex transient spectra, the microscopic interpretation of which is not yet clear. Recently, a general formulation of the spectroscopy of nonadiabatic excited-state dynamics has been developed in the framework of the classical electron analog formalism and applied to the description of femtosecond pump-probe experiments [4]. Assuming Gaussian pulses of 20 fs duration, Fig. 3 shows the time- and frequency-resolved pump-probe spectrum of the twenty-mode model. The pump-probe signal turns out to be rather complicated. It results from both emission from the upper diabatic electronic state and absorption from the lower diabatic electronic

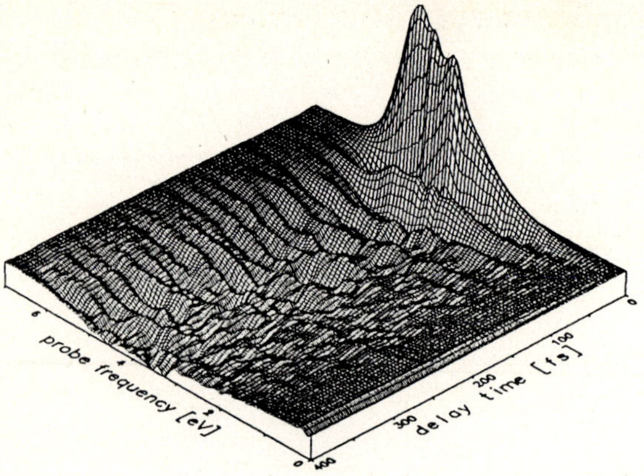

Fig. 3. Time- and frequency-resolved pump-probe spectrum of the 20-mode model assuming Gaussian pulse of 20 fs duration.

state as a consequence of the ultrafast interchange of the diabatic states upon torsion. An analysis of the pump-probe spectrum in terms of time-dependent wave-packet motion on the coupled PESs and its relation to intramolecular observables (as shown in Fig. 2) will be presented elsewhere. The simulations clearly demonstrate that femtosecond time- and frequency-resolved pump-probe spectrocopy of small polyenes has the potential to provide novel information on the ultrafast nonadiabatic isomerization processes in these systems.

References

1. J. Michl and V. Bonačić-Koutecký, *Electronic Aspects of Organic Photochemistry*, (Wiley,New York,1990).
2. L. Seidner and W. Domcke, Chem. Phys., in press.
3. H.-D. Meyer and W. H. Miller, J. Chem. Phys. **70**, 3214 (1979).
4. G. Stock and W. H. Miller, Chem. Phys. Lett. **197**, 396 (1992); J. Chem. Phys. **99**, 1545 (1993).
5. R. W. Schoenlein, R. A. Peteanu, R. A. Mathies, and C. V. Shank, Science **254**, 412 (1991).
6. S. Petersen, L. Bañares, and A. H. Zewail, J. Chem. Phys. **97**, 8801 (1992); R. J. Sension, S. T. Repinec, A. Z. Szarka, and R. M. Hochstrasser, J. Chem. Phys. **98**, 6291 (1993).

Femtosecond Spectroscopic Studies of Barrierless Excited-State Decay of Triphenylmethane Dye Molecules in Alcohol Solutions

M. Du, Y. Jia, and G.R. Fleming

Department of Chemistry and James Franck Institute,
The University of Chicago, Chicago, IL 60637, USA

Abstract. We observed solvent-dependent non-monotonic decays in the ground state recovery of triphenylmethane dyes. The nature of the decay is discussed.

Many chemical and biological reactions proceed on potential surfaces having zero or a low barrier [1,2] and are influenced by the surrounding medium and by vibrational relaxation during the course of the reaction. In this work, we chose triphenylmethane dye molecules, malachite green (MG) and crystal violet (CV), in alcohol solutions as model systems to investigate the time dependence of barrierless reactions.

Three experimental arrangements were used in this work: fluorescence up-conversion (FL), transient absorption (TA), and transient dichroism (TD) and birefringence (TB). FL and TA data were taken with 60fs pulses while TD and TB data were taken with ~30 fs pulses. The FL data could be well fit as sum of exponentials by a nonlinear least square fitting program. Attempts to fit TA, TD and TB data to a sum of exponentials were unsuccessful. A linear prediction singular value decomposition (LPSVD) program was used instead. The parameters for the exponentially damped cosinusoidal terms were determined for each data set.

TD and TB spectroscopy with heterodyne detection enhance and linearize the small amplitude signal and improve the signal-to-noise ratio. Providing there is no depolarization in the time window we study, a TD experiment gives the same information as a TA experiment. Both MG and CV are good saturable absorber dyes. The excited state absorption should be negligible in our experimental wavelength region. Thus the ground state recovery (GSR) is expected to be the major contribution to our TA and TD signals.

Fig.1 shows FL and GSR data for MG in water. The FL decay is highly non-exponential, while the GSR rises, then reaches a plateau before the signal decays. This lag in GSR is similar to two other examples of barrierless reaction: bacteriorhodopsin [3,4] and 1144-cyanine [5]. We note the difference in FL and GSR measurements: FL measures the population change within a narrow window along the potential surface while GSR should be more representation of total population change in the excited state. The TA data with the probe polarization parallel and perpendicular to pump polarization, respectively, yield no obvious difference. This result excludes the possibility that a transition dipole direction change is responsible for the "lag" observed.

Fig.2 displays the TD data of MG in alcohols. The "lag" is strongly dependent on solvent. One way to fit these peculiar curves is with a low frequency oscillatory component (in addition to the high frequency quantum beats, and exponential decays)

Springer Series in Chemical Physics, Vol. 60 **Ultrafast Phenomena IX**
Editors: P. F. Barbara · W. H. Knox · G. A. Mourou · A. H. Zewail

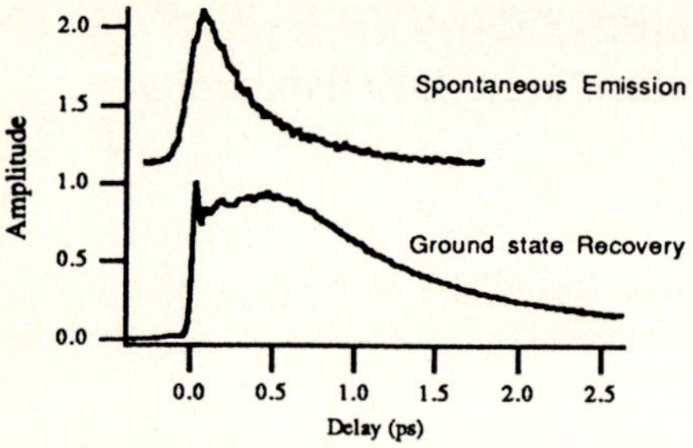

Fig.1 Comparison of fluorescence data and ground state recovery of MG in water

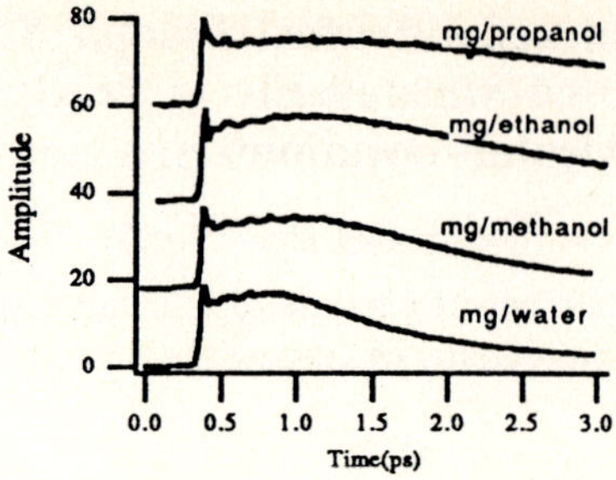

Fig.2 Transient dichroism data of MG in four different solvents

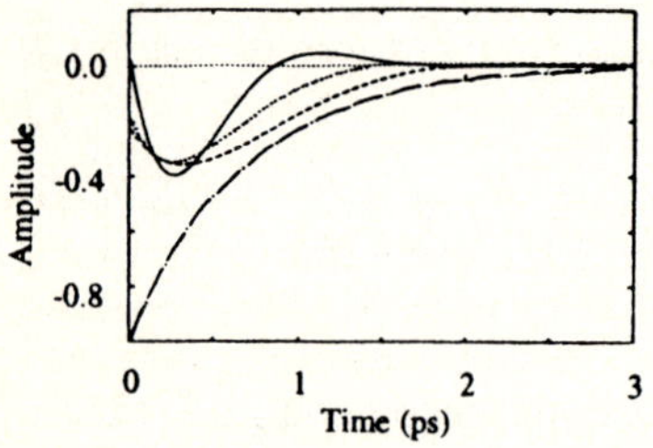

Fig. 3 Comparison of low frequency components of MG in four different solvents: solid line—water; dot line—methanol; dashed line—ethanol; broken line—propanol.

for MG in water, methanol and ethanol. No low frequency component is required to fit the data of MG in propanol (an exponential rise time fits the data well). Fig. 3 displays the normalized low frequency components of MG in different solvents from TD and the rise time of MG in propanol. The results here indicate two trends: one is that the "lag" is strongly dependent on solvent. The other is that the low frequency oscillation becomes lower in frequency as viscosity increases. This is interpreted as the frequency is determined by the effective moment of inertia when solvent molecules adiabatically follow the phenyl ring rotation. The TB data for MG show similar phenomena. CV in water and methanol gives a similar low frequency component but CV in ethanol does not require a low frequency component to fit the data.

References

[1] V. Sundstrom and U. Aberg, *J. Mol. liq.* (1993) in press

[2] B. Bagchi and G.R. Fleming, *J.Phys. Chem.* **94**, 9 (1990)

[3] S.L. Dexheimer, Q. Wang, L.A. Peteanu, W.T. Pollard, R.A. Mathies and C.V. Shank, *Chem. Phys. Lett.* **188**, 61 (1992)

[4] M. Du and G.R. Fleming, *Biophysical Chemistry* **48**, 101 (1993)

[5] U. Aberg and V. Sundstrom, *Chem. Phys. Lett.* **185**, 461 (1991)

Polarized Ultrafast Transient Spectroscopy of the Hydrated Electron: Quantum Non-Adiabatic Molecular-Dynamics Simulation

B.J. Schwartz and P.J. Rossky

Department of Chemistry and Biochemistry, University of Texas at Austin, Austin, TX 78712, USA

Abstract: Isotropic and anisotropic solvent fluctuations coupled to the hydrated electron are calculated to relax on different time scales, leading to a novel prediction for polarized ultrafast spectral transients which is directly testable by experiment.

The transient spectroscopy of the aqueous solvated electron has been the subject of intense experimental[1-3] and theoretical[4-8] interest. The coupling of solvent fluctuations to the electronic absorption spectrum makes the hydrated electron an outstanding probe of electronic solvation dynamics in the aqueous environment. The optical absorption spectrum of the hydrated electron is comprised of three s–>p like transitions which are broadened and split by coupling to solvent fluctuations.[6]

Experiments by Migus *et al.*[1] and Long *et al.*[2] first studied the equilibration of solvated electrons injected into liquid water by multiphoton ionization. These experiments observed a broad infrared absorption which decayed into the known equilibrium spectrum of the hydrated electron on a subpicosecond time scale. This result was subsequently reproduced by quantum non-adiabatic simulations,[4] which supported the assignment of the infrared absorption to an electronic excited state of the hydrated electron followed by radiationless relaxation to the ground state. More recent experiments by Alfano *et al.*[3] have photoexcited equilibrium hydrated electrons. In addition to providing insights into nature of the optical spectrum by bleaching dynamics (hole-burning), these experiments also contain spectral contributions due to transient absorption from the excited state as well as stimulated emission from the excited state back to the ground state. All of the above experiments used effectively unpolarized light (relative pump and probe pulse polarizations set at the magic angle) to probe the electronic dynamics.

In this paper, we present the results of quantum non-adiabatic molecular dynamics calculations of polarized transient spectroscopy of the aqueous solvated electron. The results indicate distinct time scales for the relaxation of anisotropic and isotropic solvent fluctuations, allowing polarized spectral holes to be burned in the hydrated electron absorption spectrum. When combined with contributions from transient absorption and stimulated emission, this separation of time scales results in a distinct spectral signature, transient absorption in one set of polarizations and transient bleaching in the other, which can be directly probed by experiment.

The details of the non-adiabatic dynamics method used are essentially identical to those employed in previous studies of electron localization after injection,[4,5] and are also described elsewhere.[8] Twenty trajectories launched from starting configurations in which the electronic absorption was resonant with the Ti:Sapphire laser (2.27±0.01 eV, corresponding for the model Hamiltonian to an experimental pump wavelength of 780 nm with a 200 cm^{-1} bandwidth[9]) were used to compute transient bleaching dynamics. Excited state trajectories initiated from these same

Springer Series in Chemical Physics, Vol. 60 **Ultrafast Phenomena IX**
Editors: P. F. Barbara · W. H. Knox · G. A. Mourou · A. H. Zewail

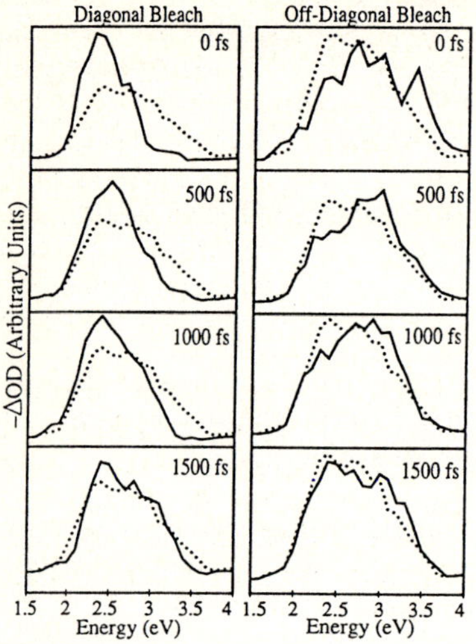

Fig. 1: Diagonal and Off-Diagonal Bleach Dynamics

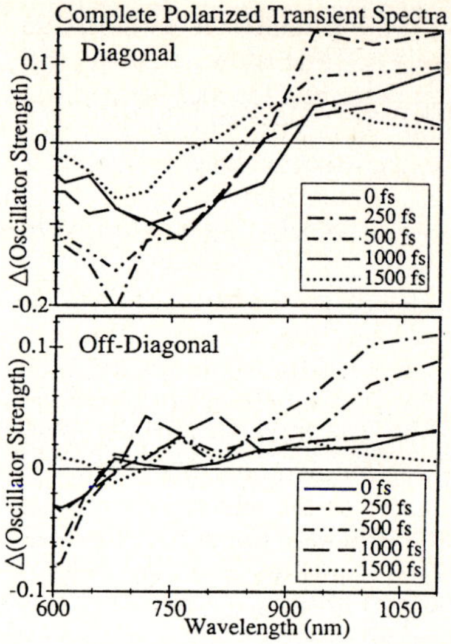

Fig. 2: Complete Diagonal and Off-Diagonal Spectra

starting points determined transient absorption and stimulated emission spectral contributions. Polarized spectra were computed from the appropriate transition dipole autocorrelation functions[8] and then combined[10] to produce "diagonal" (probing only to the initially prepared state) and "off-diagonal" (probing only the states with transition dipoles perpendicular to the initial state) spectral transients.

The diagonal and off-diagonal transient bleach spectra (no excited absorption or emission contributions) for the hydrated electron excited at Ti:Sapphire wavelengths are presented in Fig. 1 (solid curves). To emphasize the shape of the holes, the calculated equilibrium absorption spectrum, which would have been produced by uniform bleaching of the absorption band, is shown normalized to equal area at each time delay (dotted curves). The initial (t=0) diagonal hole is broadened by rapid solvent fluctuations, corresponding to essentially homogeneous bleaching of the lowest s–>p transition. This hole persists clearly past time delays of 0.5 ps before it begins to broaden and blue-shift. The off-diagonal spectra behave in a similar manner. The upper two s–>p transitions are bleached at early times, with absorption to the initial state growing in only at time delays past 500 fs. All of these spectra provide direct evidence that solvent fluctuations with different symmetries are relaxing on different time scales: The comparable widths of the diagonal and two off-diagonal holes indicate that the fast fluctuations causing the initial broadening are symmetric in nature, while the lack of uniform bleaching of the entire absorption band indicates that the anisotropic fluctuations which break the degeneracy relax much more slowly. In other words, the three excited p-like states do not interchange roles on a subpicosecond time scale.

Since the experimental spectral transients also probe contributions from transient absorption and stimulated emission, we have also computed the complete diagonal and off-diagonal transient spectra including these excited state components. These results are displayed on the experimentally accessible wavelength scale[9] in Fig. 2. The 300 fs instrumental resolution and non-adiabatic return to the ground state are reflected in the overall rise and decay of the spectral amplitudes. The slow, anisotropic bleaching dynamics result in the complex spectral shape and the lack of observed isosbestic point. In the 700-800 nm region, the diagonal spectra show an overall instrument-limited transient bleach while the off-diagonal spectra show a more slowly rising net transient absorption. This is an immediate consequence of the separation of time scales for the isotropic and anisotropic solvent fluctuations, and provides a direct experimental test of this phenomenon.

In summary, we have provided the first directly testable prediction of ultrafast polarized spectroscopic transients using quantum non-adiabatic molecular dynamics simulation. Anisotropic contributions to electronic solute-solvent coupling are found to take place on a slower time scale than isotropic fluctuations. This result is manifest as a diagonal transient bleach and off-diagonal transient absorption at wavelengths near and just blue of the excitation in a polarized pump-probe experiment on the aqueous solvated electron. This separation of fluctuation time scales by symmetry and its impact on polarized transient spectroscopy has important ramifications for theories of solvation dynamics as well as for the general understanding of the properties of the hydrated electron.

References

1. A. Migus, Y. Gauduel, J. L. Martin, and A. Antonetti, Phys. Rev. Lett. **58**, 1559 (1987).
2. F. H. Long, H. Lu, and K. B. Eisenthal, Phys. Rev. Lett. **64**, 1469 (1990).
3. J. C. Alfano, P. K. Walhout, Y. Kimura, P. F. Barbara, J. Chem. Phys. **98**, 5996 (1993); Y. Kimura, J. C. Alfano, P. K. Walhout and P. F. Barbara, J. Phys. Chem. **98**, 3450 (1994).
4. F. A. Webster, J. Schnitker, M. S. Friedrichs, R. A. Friesner, and P. J. Rossky, Phys. Rev. Lett. **66**, 3172 (1991); F. A. Webster, P. J. Rossky, and R. A. Friesner, Comp. Phys. Commun. **63**, 494 (1991).
5. T. H. Murphrey and P. J. Rossky, J. Chem. Phys. **99**, 515 (1993).
6. J. Schnitker, K. Motakabbir, P. J. Rossky, and R. A. Friesner, Phys. Rev. Lett. **60**, 456 (1988).
7. K. A. Motakabbir, J. Schnitker, and P. J. Rossky, J. Chem. Phys. **90**, 6916 (1989).
8. B. J. Schwartz and P. J. Rossky, Phys. Rev. Lett. **72**, 3282 (1994); B. J. Schwartz and P. J. Rossky, J. Phys. Chem. **98**, 4489 (1994).
9. Note that the calculated absorption spectrum is ~0.7 eV blue-shifted from experiment.[5,6] Figures in eV are the computed transition frequencies; nm figures were converted from eV with this blue-shift.
10. J. Yu and M. Berg, J. Chem. Phys. **97**, 1758 (1993).

Ultrafast Dynamics of the Excited Mesitylene·Br-Atom EDA Complex

W. Jarzeba, A. Hörmann, K.A. Thakur, and P.F. Barbara

Department of Chemistry, University of Minnesota, Minneapolis, MN 55455, USA

We recently made the first measurements of the electron transfer (**et**) kinetics of a bromide/aromatic cation *contact ion pair* ($Ar^{+}\bullet Br^{-}$) which is prepared by optical excitation of the charge transfer (CT) band of a Br atom/arene CT complex. [1]

$$Ar\bullet Br \underset{et}{\overset{h\nu}{\rightleftharpoons}} Ar^{+}\bullet Br^{-}$$

This class of reaction is a promising prototype for studying intermolecular **et** at the molecular level. It is relatively simple compared to other recently studied ultrafast intermolecular **et** systems, all of which involve donors and acceptors having more electronic, vibrational, and conformational degrees of freedom.

This paper describes femtosecond pump-probe absorption spectroscopy on **et** and relaxation of the Br atom/mesitylene CT complex. This complex is characterized by a strong CT absorption band centered at 525 nm and of 165-nm bandwidth.[2] In our experiments Br radicals were prepared by photolysis of α-bromoacetophenone. [3] After a 10-ns delay, sufficiently long to allow formation of mesitylene•Br complexes, the CT band was studied by fs-pump-probe laser spectroscopy. Various pump and probe wavelengths were employed, using 70-100-fs pulses.[4]

Fig. 1 shows transient absorption of the mesitylene•Br complex in 0.5-M mesitylene solution in methanol. After the excitation the ground state population is completely regenerated in a few hundred picoseconds and the observed bleach or absorption returns to the baseline. The recovery of the ground-state-CT absorption clearly shows three different relaxation times. The first two, 1.1 ps and 10 ps, are independent of the mesitylene concentration while the longest relaxation time increases as the donor concentration decreases. This is due to quenching of the ion pair by excess mesitylene molecules, similar to benzene•Br.[1] The transient at 395 nm corresponds to the absorption of the mesitylene cation (λ_{max} = 475 nm) formed after the CT excitation. The kinetics of the decay of the mesitylene cation is similar to the relaxation of the two long components in the bleach recovery observed at 550 nm. Therefore the two long components represent the kinetics of the back-electron transfer of the mesitylene$^{+}\bullet$Br^{-} ion pair. The fast 1.1-ps component observed at 550 nm is not present in the transient absorption of the monomer cation. A fast 1.1-ps absorption decay is however observed in the 700-1000-nm spectral region.

At least two possible interpretations of these results have to be considered. The first is that the fast kinetics at early time may be due to transient absorption of vibrationally "hot" ground-state molecules produced by a portion of the excited-state population returning to the ground state on a much faster time scale than the remaining population. The second possibility is that the fast dynamics are due to charge separation in the excited state, not ground-state return. In this model, the transient

Springer Series in Chemical Physics, Vol. 60 **Ultrafast Phenomena IX**
Editors: P. F. Barbara · W. H. Knox · G. A. Mourou · A. H. Zewail

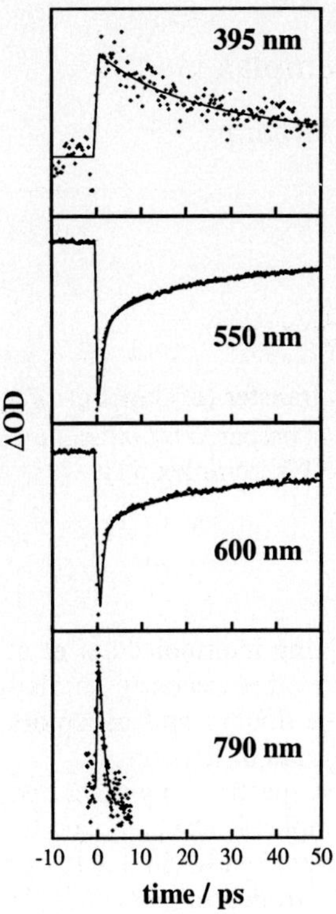

Fig. 1. Transient absorption at indicated wavelengths of the mesitylene•Br complex in 0.5 M solution of mesitylene in methanol. Pump wavelength 550 nm.

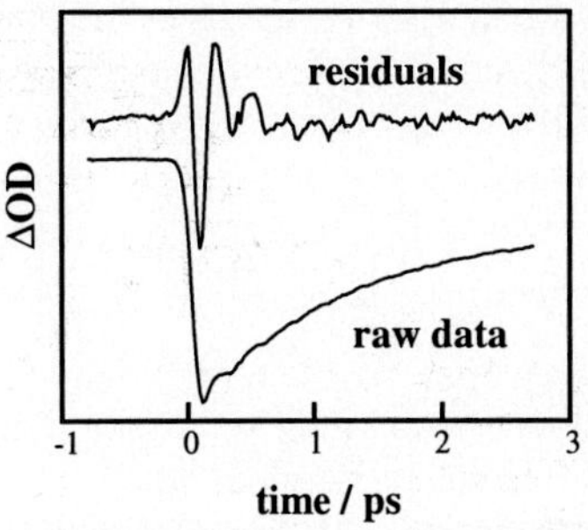

Fig. 2. Bleach recovery for the mesitylene•Br complex in 1 M mesitylene solution in methanol. The inset shows the data with the multi-exponential part of the fit subtracted. 550-nm pump and probe wavelengths

absorption in the near infrared and the apparent bleach recovery at 550 nm would be due to excited-state absorption of the complex.

Fig. 2 presents pump-probe transient absorption experiments at 550 nm for the complex in 1 M mesitylene solution in methanol. Non-exponential bleach recovery is observed with an oscillation superimposed on this signal. The period of the oscillation is about 260 fs corresponding to 120 cm^{-1} and it decays in ≈700 fs. A similar oscillation with the same phase shift was observed at 620-nm-probe wavelength. The amplitude at $t = 0$ corresponds to a bleach of the ground-state absorption. We attribute the oscillations to coherent excitation of the ground-state D-A stretching mode, likely due to impulsive stimulated Raman scattering.

This research was supported by the Basic Energy Sciences program of the Department of Energy. AH thanks the Swiss National Science Foundation for a grant.

References

[1] W. Jarzeba, A. Hörmann, P. F. Barbara, in preparation.
[2] K. D. Raner, J. Lusztyk, K. U. Ingold *J. Phys. Chem.* **93**, 564 (1989).
[3] W. G. McGimpsey, J. C. Scaiano *Can. J. Chem.* **66**, 1474 (1988).
[4] A. E. Johnson, N. E. Levinger, W. Jarzeba, R. E. Schlief, D. A. V. Kliner, P. F. Barbara *Chem. Phys.* 176, 555 (1993).

Index of Contributors

Printing: Mercedesdruck, Berlin
Binding: Buchbinderei Lüderitz & Bauer, Berlin

Springer Series in Chemical Physics

Editors: Vitalii I. Goldanskii Fritz P. Schäfer J. Peter Toennies

1 **Atomic Spectra and Radiative Transitions** By I. I. Sobelman
2 **Surface Crystallography by LEED** Theory, Computation and Structural Results. By M. A. Van Hove, S. Y. Tong
3 **Advances in Laser Chemistry** Editor: A. H. Zewail
4 **Picosecond Phenomena** Editors: C. V. Shank, E. P. Ippen, S. L. Shapiro
5 **Laser Spectroscopy** Basic Concepts and Instrumentation By W. Demtröder 3rd Printing
6 **Laser-Induced Processes in Molecules** Physics and Chemistry Editors: K. L. Kompa, S. D. Smith
7 **Excitation of Atoms and Broadening of Spectral Lines** By I. I. Sobelman, L. A. Vainshtein, E. A. Yukov
8 **Spin Exchange** Principles and Applications in Chemistry and Biology By Yu. N. Molin, K. M. Salikhov, K. I. Zamaraev
9 **Secondary Ion Mass Spectrometry SIMS II** Editors: A. Benninghoven, C. A. Evans, Jr., R. A. Powell, R. Shimizu, H. A. Storms
10 **Lasers and Chemical Change** By A. Ben-Shaul, Y. Haas, K. L. Kompa, R. D. Levine
11 **Liquid Crystals of One- and Two-Dimensional Order** Editors: W. Helfrich, G. Heppke
12 **Gasdynamic Laser** By S. A. Losev
13 **Atomic Many-Body Theory** By I. Lindgren, J. Morrison
14 **Picosecond Phenomena II** Editors: R. M. Hochstrasser, W. Kaiser, C. V. Shank
15 **Vibrational Spectroscopy of Adsorbates** Editor: R. F. Willis
16 **Spectroscopy of Molecular Excitions** By V. L. Broude, E. I. Rashba, E. F. Sheka
17 **Inelastic Particle-Surface Collisions** Editors: E. Taglauer, W. Heiland
18 **Modelling of Chemical Reaction Systems** Editors: K. H. Ebert, P. Deuflhard, W. Jäger
19 **Secondary Ion Mass Spectrometry SIMS III** Editors: A. Benninghoven, J. Giber, J. László, M. Riedel, H. W. Werner
20 **Chemistry and Physics of Solid Surfaces IV** Editors: R. Vanselow, R. Howe
21 **Dynamics of Gas-Surface Interaction** Editors: G. Benedek, U. Valbusa
22 **Nonlinear Laser Chemistry** Multiple-Photon Excitation By V. S. Letokhov
23 **Picosecond Phenomena III** Editors: K. B. Eisenthal, R. M. Hochstrasser, W. Kaiser, A. Laubereau
24 **Desorption Induced by Electronic Transitions DIET I** Editors: N. H. Tolk, M. M. Traum, J. C. Tully, T. E. Madey
25 **Ion Formation from Organic Solids** Editor: A. Benninghoven
26 **Semiclassical Theories of Molecular Scattering** By B. C. Eu
27 **EXAFS and Near Edge Structures** Editors: A. Bianconi, L. Incoccia, S. Stipcich
28 **Atoms in Strong Light Fields** By N. B. Delone, V. P. Krainov
29 **Gas Flow in Nozzles** By U. G. Pirumov, G. S. Roslyakov
30 **Theory of Slow Atomic Collisions** By E. E. Nikitin, S. Ya. Umanskii
31 **Reference Data on Atoms, Molecules, and Ions** By A. A. Radzig, B. M. Smirnov
32 **Adsorption Processes on Semiconductor and Dielectric Surfaces I** By V. F. Kiselev, O. V. Krylov
33 **Surface Studies with Lasers** Editors: F.R. Aussenegg, A. Leitner, M.E. Lippitsch
34 **Inert Gases** Potentials, Dynamics, and Energy Transfer in Doped Crystals. Editor: M. L. Klein
35 **Chemistry and Physics of Solid Surfaces V** Editors: R. Vanselow, R. Howe
36 **Secondary Ion Mass Spectrometry, SIMS IV** Editors: A. Benninghoven, J. Okano, R. Shimizu, H. W. Werner
37 **X-Ray Spectra and Chemical Binding** By A. Meisel, G. Leonhardt, R. Szargan
38 **Ultrafast Phenomena IV** By D. H. Auston, K. B. Eisenthal
39 **Laser Processing and Diagnostics** Editor: D. Bäuerle